EARTH SCIENCES IN THE 21ST CENTURY

PACIFIC OCEAN MEGA ECOTONE OF NORTHERN EURASIA

AN EVOLUTIONARY MODEL OF A CONTINENTAL BIOSPHERE

Earth Sciences in the 21st Century

Additional books and e-books in this series can be found on Nova's website under the Series tab.

EARTH SCIENCES IN THE 21ST CENTURY

PACIFIC OCEAN MEGA ECOTONE OF NORTHERN EURASIA

AN EVOLUTIONARY MODEL OF A CONTINENTAL BIOSPHERE

ERLAND G. KOLOMYTS

nova science publishers
New York

Additional color graphics may be available in the e-book version of this book.

Library of Congress Cataloging-in-Publication Data

ISBN: 978-1-53616-493-0

Published by Nova Science Publishers, Inc. † New York

CONTENTS

FOREWORD

The monograph presented by the author is one of the results of multiyear study of geographical ecotones. What is meant here is a transitional zone at the global scale: the Pacific Ocean oro-climatic mega ecotone of North Eurasia. This transitional zone includes the eastern tectonically active margin of the continent and insular land. It is characterized by the submeridian extension of morphostructural zones of different ages and the distinct latitudinal and longitudinal-sectoral differentiation of climate. The mega ecotone is part of the global belt of convergence of matter-and-energy flows on land surface. The processes of interaction and mutual transformation of the two major types of geographical environment (oceanic and continental) in this mobile belt of the planet is extremely dynamic. Hence, insular, peninsular and coastal-continental territories are a very good natural laboratory for studying different geological stages of the development of continental biosphere. The modern stage is most marked precisely in the Pacific Ocean Mobile Belt.

The book offers a detailed representation of the author's empirical concept of the ecology of boreal forests in the island and marginal-continental landscapes of North-Western Pacific.

The multidimensional empirical-statistical modeling of landscape-ecological connections, the structure and function of floristic and phytocoenotic formations in the boreal zone at the eastern margin of the Asian continent and the adjacent volcanic and continental islands has been performed for the first time. The known methods of discrete mathematics (information theory and "fuzzy" set theory) were used to establish the main features of spatial organization and evolution of island and marginal-continental boreal biogeocoenoses.

The study was carried out at the local (topological) level of organization of natural complexes. The scientific and methodological significance of this method is due to the fact that it allow researchers to use the known techniques for modeling inter-component and inter-complex relationships and thereby to introduce a particular measure in the analysis of structural and functional organization of geo(eco)systems, as is rather widely represented in this monograph. In landscape geography, quantitative estimates have been used so far only with regard to the morphology of natural complexes, without concerning the causal mechanisms of this morphology, i.e., landscape connections per se.

The solid factual material of large-scale landscape-ecological surveys performed by the author in three experimental areas (Kunashir Island of the South-Kuril Ridge, the Komsomolsk Nature Reserve in the Lower Priamurye, and Putyatin Island in Peter the Great Bay) shows that the landscape-ecological analysis of the data of traditional field landscape studies and the discrete empirical-statistical models constructed on their basis make it possible to rather strictly determine the sources and to describe the regularities of phytocoenological and landscape-forming, including evolutionary, processes. It is evidence of not only scientific but also methodological significance of this monograph.

In the analysis, the author makes extensive use of the literature on tectonic-volcanic, geomorphological, soil and biotic processes, as well as paleogeography, of Priamurye, Primorye and Kuril Islands, also trouncing upon the issues of island landscape science in general. An informative interpretation of the results of landscape-ecological modeling is based on the fundamental studies of the 30–60-s of the last century performed by the pleiad of outstanding Russian geobotanists and paleogeographers: V.L. Komarov, A.N. Krištofovič, V.V. Vasilyev, B.A. Ivaskkevich, V.B. Sochava, B.P. Kolesnikov, etc. First of all, it concerns the problems of origin, florogenesis and phytocoenology of Manchurian mixed forest as the major formation in the south of the Russian Far East.

On the other hand, the empirical-statistical models obtained by the author imparted much greater strictness to the revealed regularities of formation, evolution and spatial distribution of forest ecosystems in the Northwestern Pacific, as well as to gain more knowledge in this field. The models made it possible to present more comprehensively the regularities and causal mechanisms of formation of buffer forest communities, which are so characteristic of ecotone systems. The methods of discrete empirical-statistical modeling allowed the author to more strictly substantiate and essentially deepen the classical concept of the pattern of evolution of forest formation in this ecoregion, which is based on the floristic richness of forest phytocoenoses and includes the diversity of their successions.

The strategic trend of this study is undoubtedly of fundamental significance for the theory and methods of geographical ecology. In the island-arc sector of the continental-oceanic mega ecotone, one can trace the successive stages of primary subaerial biogenesis, with a particular range of floristic and phytocoenological structures typical of the ocean-to-continent transition zone. Here there is the origin of natural-territorial organization of continents. The mechanisms of this organization have yet been little studied both in this country and abroad. In essence, this novel trend in biogeography and geographical ecology must be aimed at investigation of the "supporting mechanisms" of formation, development and disruption of landscape-ecological connections on land, the search of driving forces and determination of the tempos of evolutionary processes, the elucidation of taxonomic and age-related correlations between the structure and function of biogeosystems, their ontogeny and phylogenetic successions.

The solution of these problems is of general scientific significance. It will promote the development of scientific and methodological bases of controlling natural processes and the development of new ecological balances in nature.

The results of empirical-statistical modeling presented in this book altogether made it possible to deepen and impart specific content to the conceptual ideas put forward at different times by the author of this book, as well as by B.A. Yurtsev and V.D. Panfilov, about the Pacific Ocean mega ecotone of Northern Eurasia as a focus of evolutionary phytocoenotic and landscape-forming processes and phenomena in the continental biosphere. E.G. Kolomyts has developed an evolutionary ecological-geographical model by the example of one of the regions of the planet's mobile belt. This systemic model is somewhat akin to the biosphere evolution theory developed in the fundamental works by M.I. Budyko.

The publication is a certain stage in development of the theory of biogeographical and landscape ecotones as the foci of evolutionary trends in the biosphere and, undoubtedly, will attract great attention of many experts in the field of geographical ecology, as well as landscape-ecological modeling and prediction.

Academician V. M. Kotlyakov
Prof. G. S. Rozenderg, corresponding member of the Russian Academy of Sciences

PREFACE

A comprehensive evolutionary landscape-evolutionary concept of organization of terrestrial geo(eco-)systems in the Pacific Ocean Mobile Belt – the tectonically and climatically active ocean-continent interface as a focus of evolutionary ecological and geographical processes and phenomena at the modern stage of development of the biosphere has been set forth. The empirical statistical modeling of forest natural complexes has been performed using the materials of large-scale landscape-ecological surveys performed at experimental sites on the Pacific Ocean mega ecotone of Northern Eurasia. The theses put forward on the basis of evolutionary landscape studies have been substantiated. The three spatial-temporal sections with a common trajectory ofsubaerial landscape-genesis have been described: (1) neo-pacific progressive – the origin and evolution of island-arc geo(eco)systems of the Neogene/Quaternary period in the oceanic environment; (2) sub-pacific marginal-continental – the subsequent ascending development of the continent as a young (Mesozoic) mountain-valley morphostructure; and (3) sub-pacific regressive – the final study of damped evolution on the continental islands with "senile" erosion-denudation relief. The climatogenic mechanisms of evolutionary processes in different sectors of Pacific Ocean mega ecotone are described.

ACKNOWLEDGMENTS

This book was supported by the Russian Foundation for Basic Research (project no. 14-05-00032-a).

The author is also grateful to anonymous reviewers and to the editor for constructive comments that have substantially improved the quality of this manuscript.

INTRODUCTION: THE CONTEMPORARY STATE OF THE PROBLEM

The monograph presents the results of research into the spatial-temporal organization of terrestrial ecotone geo(eco)systems in the Pacific Mobile Belt as a focus of evolutionary ecological and geographical processes and phenomena at the modern stage of development of the biosphere. This trend of complex physical geography has been put forward previously by the author as a conceptual idea (Kolomytc 1987, 1988). The problem of studying the organization of biogeocoenotic and landscape systems of marginal-continental and insular land in the tectonically and climatically active zone of contact between the ocean and the continent is set forth by the example of Pacific Mega Ecotone of North Eurasia. The proposed statements are factually substantiated on the basis of the results of large-scale landscape-ecological surveys performed at experimental sites in the marginal-continental and island-arc sectors of the mega ecotone, as well as the known bioclimatic and geological-geomorphological data.

In the book we have considered the two spatial-temporal sections of geological history of the formation of continental biosphere in the boreal Eurasian region of the Pacific Ocean mega ecotone: (1) the origin and evolution of island-arc geo(eco)systems of the Neogene/Quaternary period in the oceanic environment, and (2) the subsequent ascending development of the continent as a young (Mesozoic) mountain-valley morphostructure.

The idea of *evolutionary geoecology* did not appear out of nowhere. It had already been mentioned in various biogeographical surveys of insular landscapes over the past years. For example, the work of (Ignatyev 1979) they put forward the idea of landscape development on young volcanic islands as a dynamic system of related factors: soil-vegetable, geomorphological, and geochemical. At the same time, the above factors form a system with a positive feedback and, therefore, their development is possible only insofar as landscape development occurs in general. It has been noted that "...there is a possibility of considering natural-territorial complexes of different ages and, after comparing them, draw some conclusions about the peculiarities of nature development in general" (Ignatyev, 1979, p. 98).

The experience of paleogeographical and biogeographical exploration of Beringia and Hultenia encompassing the closest (more than once joining in the past) part of Northeastern Asia and Northwestern America has made a substantial contribution to interpretation of the Pacific Ocean mega ecotone of Eurasia as an evolutionary model of continental biosphere (Hulten 1968; Hopkins 1972; Yurtsev 1974; Brigham-Grette 2000; Krestov 2004). It was established that marginal-continental and insular territories of the extratropical part of the

Pacific Ocean Ring were the "generators" and "keepers" of continental and oceanic elements of biota against the background of pulsed regimen of the abiotic environment. The resultant empirical landscape-ecological models of the Priamurye sub-Pacific have confirmed that the modern orographic and climatic contrasts by the oceanicity–continentality gradient, which are highly pronounced at the mega ecotone, maintain the function of this "florogenetic node" and provide the stable existence of mixed (buffer) natural ecosystems, with preservation of numerous relict species of phytobiota and entire communities.

Being part of the global belt of convergence of matter and energy fluxes on the Earth surface and being connected with the tectonically labile Earth crust, the Pacific Ocean mega ecotone of Eurasia, since the middle of the Mesozoic Era, has been *one of the sources of origin of continental biosphere*. Here there have been favorable conditions for the development of focal biocoenotic processes (Yurtsev 1974) and the formation of plant and animal taxa of a high systematic rank (Panfilov 2005), as well as for the emergence of new stable formations in phytocoenological structures (Krištofovič 1946).

The present research proceeds from the most general conception of ecology "...as a science of structure and function of nature" (Odum 1975, 10). This conception assumes broad integration of the ecological and landscape-geographical methods of research, with interpretation of natural environment from the viewpoint of general systems theory, which allows establishing the mechanisms of ecological relationships in a particular geospace (Sochava 1970).

We also proceed from the extended interpretation of the concept of evolution (Мазинг 1970), which encompasses the super-population (biocoenotic and landscape) levels of life organization, where many "external" abiotic factors become an "internal" environment and are included in the ecosystem of this level. In the course of evolution, new ecological relationships arise and develop at the scale of not only separate populations but also entire communities (Pianka 1978). The theory of species series as part of the concept of "natural-historical species" is of great significance for investigation of this phenomenon. This concept was developed by V.L. Komarov (1940, 1947) by the example of Southeastern and Central Asia and opened the way to solution of the complicated problems of floro- and phytocoenogenesis of this specific and contrasting plant kingdom (Kozhevnikov 1997). In the Komarov's concept, a species is interpreted not morphologically but phylogenetically and has an ecological-geographical content closely related to space and time; therefore, "...the series method gives an idea of the course of evolution ... and makes it possible ... to restore the natural process of fragmentation of organisms through divergence of character, adaptation to environment, etc., which underlies the process of species formation" (Komarov, 1940, 62). In light of this concept, one of the main objects of our evolutionary-ecological study was the florogenetic complexes of the Priamurye Sub-Pacific.

The basic criteria of estimation of the evolutionary progress of landscape-ecological systems are as follows: (i) the complexity and ordering of their mono- and polysystem structure (Solntsev 1981, 1997); (ii) the primary productivity and intensity of the biological cycle (Bazilevich et al. 1986; Bazilevich 1993); (iii) the degree of autonomy (relative independence) and stability of ecosystems (Puzachenko 1983; Solntsev 1997); (iv) the efficiency of utilization of matter and energy resources (Shwarts 1973; Gerasimov 1985); and (v) the "power" of environment-forming influence of their coenotic "core" (Kolesnikov 1956; Isakov et al. 1986). These criteria were used in the evolutionary-ecological analysis of boreal forests of the Priamurye Sub-Pacific. The basic feature of evolutionary forest-forming process thereat, according to the concepts of functional ecology (Kerzhentsev 2010), was considered to be the appearance of a novel, previously lacking taxon of classification.

The landscape–ecological approach to modeling the functional parameters of local ecosystems, realized inthegiven investigation, is based on Timofeeff-Ressovsky's biochorological concept (Timofeeff-Ressovsky and Tyuryukanov 1966) concerning the spatial organization of the matter and energy cycles by the living matter in biogeocenoses as discrete, elementary structural units of the biosphere. According to theoretical developments in this field, the biosphere is regarded as a statistical ensemble of biogeocenoses as biochorological units that weakly interact with one another but have a highly ordered internal organization (due to stabilizing selection).

Each ecosystem at the level of facies or their groups is assigned specific functional characteristics related to the structure of local matter and energy fluxes. This approach permits the description of the behavioral patterns of forest communities under different geomorphological and edaphic conditions on the basis of comprehensive empirical data collected in the course of field studies, while retaining the statistical methods of their analysis.

The scientific and methodological significance of the topological landscape-ecological approach is that it allows for a wide application of the known modeling techniques for inter-component and inter-complex relationships (D.L. Armand 1975; Andreev 1980; Puzachenko, Skulkin 1981; Kolomyts 1984, 2008; etc.) and, thereby, for introduction of a definitive measure into the characteristics of structural and functional organization of geo(eco)systems. In landscape geography, quantitative estimations have been used hitherto only in the context of morphology of natural complexes (Ivashutina, Nikolaev 1972; Roosaare 1982; Viktorov 2006; Ganzey 2010), without any concern for the causal mechanisms of emergence of this morphology, i.e., landscape connections per se.

Our evolutionary landscape-ecological analysis is based on the construction of *discrete empirical-statistical models of natural ecosystems* (Rozenberg 1984). These models may adequately reflect the processes of natural selection as the main driving force of evolutionary process. During self-organization, the system is replaced by a qualitatively new one; therefore, self-organization becomes a mechanism of evolution of natural ecosystems (Armand 1988; Shestakov, Drozdov 1992). Though system functions are not directly represented in self-organized models, these processes can be assessed by discrete input and output parameters of the biological cycle, as we have implemented.

The accumulated extensive and comprehensive empirical material characterizes the regularities and conditions of the formation of plant cover, soils and landscapes of the Kuril Islands, as well as the neighboring insular and marginal-continental territories of the Pacific Ocean mega ecotone of Northern Eurasia. Along with numerous descriptions, there are the most general analytical estimates aimed at understanding the deep processes of floro-, phyto- and landscape genesis in the Northwestern Pacific. The numerous geobotanic studies of the past years (Hulten 1933; Tatewaki 1933, 1957; Vorobyev 1963; Popov 1963; Bobrov 1972, 1980; Urusov 1988, 1993; etc.) and recent decades (Pietsch et al. 2003; Jacoby et al. 2004; Nakamura, Krestov 2005; Urusov, Chipizubova 2008; etc.), including the Atlas of the Kuril Islands (2009), provide the highly versatile characteristics of plant cover structure and dynamicsof thePleistocene and Holocene. The analogous geobotanic descriptions can be found for other ecoregions of the Northern Pacific, in particular, Hokkaido (Kato 1952; Tatewaki 1958; Kojima 1981, 1991; Igarashi 1994), the Commander and Aleutian Islands (Hulten 1960; Krestov 2004), as well as the coastal-insular regions of British Columbia

(Hopkins 1972; Pojar et al. 1987; Stinson et al. 2011). At the same time, the known quantitative indicators of organization of plant communities are often used. For example, the work of (Kojima 1991) shows the climatic characteristics of the area of Sakhalin spruce in Hokkaido, the floristic structural similarity of basic forest types and their edaphotopic ranking, and a scheme of interaction between forest phytocoenoses and soils obtained on the basis of correlation coefficient matrices.

The already elaborated aspects of the theory of soil formation in the continent-to-ocean transition zone are of great significance for developing the concept of evolutionary landscape genesis (Lashkov 1948; Sokolov 1973; Ivanov 1976; Targulyan 1982; Ivlev et al. 1987 a, b; etc.). These aspects have been substantiated by the study of soil morphology and different soil-geochemical models.

The most significant results in the construction of local and regional numerical models were obtained in domestic geobotanic studies of sub-Arctic sub-Pacific and Primorye. The discrete polysystem modeling of tundra communities in one of the regions of Central Chukotka revealed the dynamics of primary successions, as well as the features of progressive and regressive development of microgroups (Komarova, Syomkin 1976). The analogous set-theoretic analysis of florogenetic complexes of the South Chukotka as a refugium for plants of the cryo-humid climate in the glacial epochs of Pleistocene made it possible to describe fluctuations of the boundaries between the Arctic and Boreal floristic areas and to reveal the causes of genetic diversity of the local populations of cryophytes (Yurtsev 1978). One should also note the experience of set-theoretic modeling of the structure of subboreal forest vegetation in the south of Primorsky Krai for different conditions of atmospheric humidification, with construction of the respective electronic maps (Puzachenko, Skulkin 1981) of unquestionable predictive significance.

Henceforth, the large regional geobotanic generalizations were obtained as a result of quantitative analysis of the plant cover of Northern Asian-American Pacific (Krestov 2005, 2006). The set-theoretic and cluster analysis of the lists of vascular plant species, as well as the estimates of heat and moisture supply of different floristic complexes, were a basis for bioclimatic zoning of this territory and determining the hierarchy of phytochorological units within the ranks of areas, subareas, and provinces. However, these results are related only to the zonal types of vegetation at the subplanetary scale. The low spatial resolution of cartographic materials prevented from revealing not only the local but even regional diversity of plant cover in separate territories of the insular and marginal-continental land, including some islands of the Kuril Ridge.

Our research was aimed, in particular, at eliminating such blind spots. The survey performed at the Far-East boreal ecotone (Kolomyts 2015, a) resulted in the construction of empirical-statistical models of spatial organization of forest geo(eco)systems in the marginal-continental sector of the Pacific Ocean mega ecotone of Northern Eurasia. The multidimensional mono- and polysystem modeling of the structure and function of boreal forests of the Lower Priamurye performed by the example of the Komsomolsky Nature Reserve confirmed the known (Krištofovič 1946; Vasilyev 1946; Sochava 1946; Kolesnikov 1956; Bobrov 1980; etc.) and revealed the new topological and regional regularities of their evolutionary ecology (Kolomyts 2015 a). The large-scale landscape-ecological survey has been carried out here in 2011, 2012 and 2014.

The territorial object of this research is the Far-East branch of the sublatitudinal boreal bioclimatic ecotone of Eurasia, according to the published data (Bazilevich et al. 1986;

Kolomyts 1995), which is part of the zonal-belt structure of the sub-meridian Pacific Ocean mega ecotone. Here, the boreal ecotone encompasses the mountain and valley systems of the Lower Priamurye, which are included by V.B. Sochava (1980) in *Sub-Pacific,* i.e., the marginal continental geographical space of the Mesozoic geological age, transitional from insular, peninsular and coastal-oceanic Cenozoic volcanogenic *Neo-Pacific* to the much more ancient (Paleozoic) *Paleo-Pacific* being part of the Siberian platform.

The regularities of evolutionary forest-forming processes resulting in the known florogenetic richness and phytocoenotic diversity of forest cover of the ecoregion were described. The mechanisms of appearance of buffer (mixed) forest communities, including spruce/broad-leaved and nemoral fir/spruce forests of the Manchurian-Okhotsk phratrie were revealed. It has been shown that the forests of this phenomenal buffer flora differ from the forests of basic floras in the higher indices of structural and functional development and have reached the state most similar to the evolutionary climax.

The situation with the modeling of landscape systems of the insular, peninsular and marginal-continental Pacific is much worse. In the studies of natural-territorial complexes (NTC) of the Lower Priamurye and volcanic regions of the Kamchatka Peninsula and the Kuril Islands, both in the past (Korsunskaya 1958; Zhuchkova et al. 1973; Nikonov 1975; Zonov 1977; Grabkov, Khudyakov 1993) and at present (Ivanov 2005, 2014; Kotlyakov et al. 2009; Gasnzey 2009, 2010, 2014; Razzhigaeva et al. 2009; etc.), the conventional descriptive methods of landscape science have been and still are predominant. These methods were used for the component-wise characterization of natural conditions of the Kuril Island Arc, the typification of its regional NTCs, the construction of a series of medium- and large-scale geomorphological, geobotanic and landscape maps of separate islands, and for making the schemes of landscape zonation. Finally, the paleoclimatic stages of plant cover dynamics in the insular territories under consideration were described on the basis of paleobotanic analysis of Pleistocene and Holocene spore-pollen spectra.

In many geobotanic and landscape studies, the factors critical for the insular phytobiota (isolation from the continent, cold sea currents, active volcanic activity, low temperatures of the vegetative period, severe wind regime, unfavorable young volcanogenic edaphotops, etc.) are mentioned and the general contrasts and instability of natural environment resulting in a distinct mosaicity of biogeocoenotic and landscape systems are emphasized. The measures of the impact of these critical factors have not yet been revealed, since the landscape studies on the islands and continental margins have so far been almost entirely descriptive, without the thorough quantitative analysis of intercomponent and intercomplex connections. This problem cannot be solved by applying different measures of landscape diversity of the islands: complexity, atomism, and other parameters of landscape pattern (Ganzey, 2010). Such mathematical-cartographical approach undoubtedly deepens our knowledge of the spatial differentiation of island landscape systems but does not touch at all upon the driving forces of their formation nor discloses the multidimensional hierarchic system of landscape-ecological relationships.

All in all, summarizing the results of the works on the problem of interest, we have to state that these works mark only the first, inventory stage of landscape and landscape-ecological studies in the Russian Far East. It is obvious that now we should pass over to the analytical stage, i.e., understanding of the causal mechanisms of natural interactions via the numerical empirical modeling.

The experience of implementation of this stage is described in the materials of this book. The Mendeleev Volcano situated in the southern part of the Kunashir Island, the South Kuril Island Arc, was taken for the analysis. Here, a large-scale landscape-ecological survey was carried out in the first half of August, 1985, on the experimental test ground of the Mendeleev landscape using the method described in (Kolomyts, 1995, 1998, 2008), when more than 50 parameters of the structural-functional state of forest biogeocoenoses were collected from 32 sample plots. The list of structural and functional characteristics of geo(eco)systems is given in the Supplement.

The experimental test ground of the Mendeleev volcano is notable for the fact that the floristically and phytocoenologically diverse groups of plants are concentrated here within a relatively small area, which include many species close to the northern boundaries of their distribution (Vasilyev, Rozenberg 1977). This fact allows us to trace the emergence of buffer forest communities so typical of the boreal belt of the Pacific Ocean mega ecotone (Kolomyts 2015 a) already at the initial, island-arc stage of subaerial biogenesis.

The research methodology fits into the overall concept of landscape ecology (Hef, 1968), the main content of which is "...elucidation of the landscape budget and the principles of ecological function of geosystems," which makes it possible to "...establish the mechanism of interrelationship between many components of the geosystem" (Sochava 1970, 7–8). The new, deeper approaches to the study of landscape-ecological connections in island-arc volcanic landscapes have been developed using the methods of discrete empirical statistical modeling in the scope of small natural complexes (microlandscapes). The fundamentally novel data have been obtained from the information and set-theoretical models of ecological relationships between forest plants/soil and geomorphological/climatic factors. The analysis on the basis of these models and the synthesis of phytocoenotic, soil and microlandscape structures under particular geological/geomorphological and macroclimatic conditions made it possible to find out the level of structural and functional unity or, on the contrary, disconformity of the natural components and complexes formed on the morphostructures of different age in the contemporary endogenous regime of passive volcanism, as well as in the climate with different degrees of oceanicity.

As opposed to the authors putting forward the basic postulates of island biogeography and island landscape science (Voronov 1972; Ignatyev 1979; Dyakonov 1990; Dyakonov, Puzachenko 2005; Ivanov 2005, 2014), we consider the plant cover and landscapes of geosynclinal volcanic islands not in contrast to the analogous formations on the continents but in their dynamic unity – as *certain elements of a single evolutionary physical-geographical process* (Grigoryev 1966), which forms new territories of the continental biosphere in the mobile belt of the planet.

The modeling results provide a much stricter and more ordered concept of the "fuzziness–discontinuity" of altitudinal-zonal boundaries and their "leveling" by local factors (Ivanov 2004, 2014; Ganzey 2009) and reveal the specific role of these factors in the formation of regional natural-territorial structures.

The Neo-Pacific – Sub-Pacific – Paleo-Pacific triad distinguished by V.B. Sochava (see above) is not only the subdivisions of geographical space but also the categories of *structural-functional states* of these parts of the continental biosphere, which form an ordered genetic series of its geological evolution. Since each member of this series covers quite a considerable period of time and is characterized by the great diversity of natural processes and events, the distinguished categories will be referred to as *mega states* of the continental biosphere.

Thus, our concept of evolutionary landscape science (Kolomyts 1987, 2015 b) is based on the comparative analysis and synthesis of the states of geographical environment by these three drastically different stages of continental evolution. Each state is described by the system of intercomponent and intercomplex landscape connections on the basis of empirical-statistical models constructed by the materials of large-scale landscape-ecological surveys on representative experimental test grounds. Such models provide us with a spatiotemporal profile of the states of the given category of geographical environment within its own mega state. Mainly the first two mega states of geographical environment are considered, where the dynamic tendencies of the evolutionary and regressive series of morpholithogenesis and biogeocoenogenesis are most marked. The third state is mentioned for comparison only.

The island-arc model of formation of a sequential space of the continental biosphere described in this book characterizes only one of the stages of this long-term geological process. However, it is exactly the stage when the primary landscape genesis is implemented not only by the morphotectonics and the hydrological regime but also by the intensive "work" of soil-biotic complexes. At this stage, the *active volcanism* represented by intensive eruptive processes is completed and proceeds to *passive volcanism*. The determining endogenous factor (against the background of continuing periodic ash falls) now is the high-temperature energy-intensive deep *fluid systems* (Letnikov 2000), with their output channels (fumaroles) forming solfatare fields so typical of most not only active but also passive volcanoes of the Kuril Island Arc. However, even for the *fluid stage*, our model describes a rather stable biogeocoenological and landscape structure of this ecoregion.

The causal mechanisms of lowering of the boundaries of altitudinal bioclimatic belts and the southward shift of natural zones on extratropical neo-Pacific islands compared to the neighboring continent were revealed. The evolutionary significance of island-arc bioclimatic system was established.

The results of *numerical modeling of the forest-forming influence of volcanic geothermal wells*, i.e., their effects on the hydrothermal regime of soils, as well as on phytocoenotic and soil processes determining the structure of forest communities, productivity, and biological cycle, are of particular novelty. Quantitative assessment of the effects of magmatic energy of active volcano on the thermal regimes of soil and near-surface air layer, as well as on the water balance and forest formation in an island volcanic landscape, has been made for the first time. The abnormally high productivity of boreal forests of the Kunashir Island has been shown, which brings them closer to the subtropical forests of Northwestern Pacific and provides, together with volcanic geotherms, the survival of these "climatically unjustified" communities under the conditions of "cold" oceanicity. The concept of geothermal energy of active volcanoes as a factor of accelerated evolution of island phytobiota has been substantiated empirically.

The examined landscape-ecological situation with forest formation under the conditions of passive volcanism represents the initial, island-arc trajectory of emergence and development of another piece of the continental biosphere. All of the newly formed subaerial natural complexes in oceanic mobile belts pass through this period; therefore, the Mendeleev volcanic landscape is not something unique but a typical model of one of the stages of the primary terrestrial (subaerial) landscape genesis.

The comprehensive structural and functional analysis of island volcanic landscape has been performed. Within this landscape, the rudiments of the regional properties of marginal-continental and intracontinental geographical spaces are formed at the local level. The island-arc geosystems are the sources of formation of the known floristic and phytocoenotic diversity of the Far East phytobiota, as well as the foci of origin of buffer biogeocoenological systems: the phenomenal event typical of the continental/oceanic mega ecotone.

The issues of the theory of evolutionary landscape study are considered as a novel trend of complex physical geography; its subject must be the processes and events of landscape formation in the tectonically and climatically active ocean-continent interface.

PART I: PACIFIC OCEAN MEGA ECOTONE AS THE OBJECT OF LANDSCAPE RESEARCH

Chapter 1

LANDSCAPE-ECOLOGICAL ORGANIZATION OF CONTINENT-OCEAN MEGA ECOTONE

ABSTRACT

The problem of studying the structural and functional organization of geo(eco)systems of islands and marginal continental lands in different morphotectonic and macroclimatic sectors of the Pacific Ocean mega ecotone of Northern Eurasia has been expounded. This mega-ecotone is considered as a global model of modern development of the continental biosphere. Some ecological-phytocoenotic effects of oro-climatic interactions on the mega-ecotone have been described. The theoretical and methodological aspects of evolutionary landscape science as a novel trend of complex physical geography have been considered. The research object here must be the landscape-forming processes and events in the tectonically and climatically active contact zone between the ocean and the continent.

1.1. INTRODUCTION

Ecotonic biogeosystems are the zones of transition between contrasting natural environments. These are the most dynamic structural/functional subdivisions of the biosphere, with enhanced "tension" of matter-energy interactions and with the most pronounced manifestation of new, evolutionary tendencies in the environment (Odum 1971; Sochava 1979; Kolomyts 1987; Zaletaev 1997). Landscape contents are a critically important element of the wide range of modern notions of ecotones (Solovyeva, Rozenberg 2006). As a result, a concept of landscape-ecotone, or geoecotone, has been developed (Preobrazhensky 1986).

The lateral fluxes and the respective inter-complex connections along geoecotones create the spatiotemporal order of natural complexes at all hierarchical levels: from continental or oceanic sector to landscape facies (Milykov 1981). Longitudinal climatic sectors or morphotectonic belts, on the one hand, and soil-geochemical catenas with the respective phytocoenotic spectra, on the other hand, comprise the two extreme levels of hierarchy of geographical ecotones.

The overwhelming majority of studies, both in Russia and abroad, are concerned with local phytoecotones (Odum 1971; Wittaker 1975; Armand 1992; Solovyeva, Rozenberg

2006) due to the problems of discreteness and continuum of vegetation cover, biodiversity, as well as the distribution of species and communities along the gradients of the abiotic factors and at the interface between contrasting natural environments. Various ecotone classification schemes developed in this context were considered in the work (Solovyeva, Rozenberg 2006). It is also known about the studies of ecotones of the higher taxonomic rank, e.g., zonal geoecotones and oroecotones (Ricklefs 1976; Milykov 1981; Kolomyts 1987, 2005; Zaletaev 1997; Bailey 1998; etc.).

The challenge issued in this presentation is the study of a more complex ecotone object: the hierarchic system of geoecotones of the regional and topological levels under different morphostructural and macroclimatic conditions (by the example of the Pacific margin of Eurasia), which is a new stage in development of the theory and methods of ecotone research. Here, the age row (at the geologic time scale) of geoecotone objects is represented by certain nodal stages of the evolutionary trajectory of exogenous landscape genesis in the continental biosphere. The reported strategy of scientific research is aimed at establishment of zonal-regional and local regularities of landscape organization of insular and marginal-continental land in different morphotectonic and macroclimatic sectors of the Pacific Ocean megaecotone of Eurasia, which is considered as a natural laboratory for studying the modern stage of development of the geographical envelope.

1.2. Oro-Climatic Interactions along the Mega Ecotone and Their Ecological Consequences

The Pacific Ocean oro-climatic mega-ecotone of Northern Eurasia covers the eastern, tectonically active margin of the continent and the insular land within the temperate and sub-Arctic geographic belts. This transition zone is characterized by the submeridional extension of morphostructural belts of different ages and the marked latitudinal and longitudinal-sectoral differentiation of the climate. It is part of the global belt of convergence of matter-energy fluxes on the Earth surface: the so-called mobile belt of the planet (Rice 1978; Gerasimov 1986) The interaction between and transmutation of the two main types of geographical environment (oceanic and continental) in this "continental-oceanic suture zone" is extremely dynamic (Khudyakov 1977; Meshcheryakov 1981; Gerasimov 1986); therefore, insular, peninsular and littoral-continental territories (as well as the coupled shelf areas) are a natural laboratory for studying the modern geological stage of biosphere development, since this stage is most marked exactly in the Pacific mobile belt. The known biogeographical and landscape studies in oceanic mobile belts (Zhuchkova et al. 1973; Ignatyev, 1979; Markov 1980; Ivanov 2014) give grounds for this scientific research.

The ecotones of lower taxonomic levels (macro- and meso-ecotones), which are clearly distinguished within the Asian-Pacific Ocean mega-ecotone, have been formed by latitudinal and longitudinal-sectoral contrasting differentiation between radiation-thermal and circulatory fields in the atmosphere, as well as by submeridional ordering of morphostructural belts (differing in age and modern tectonic activity) and the zones of exogenous relief formation, with the respective paragenesis of soils and vegetation.

In accordance with the main provisions of geotectonics (Rice 1978; Gerasimov 1986) and global climatology (Blüthgen, 1966), all continental–oceanic ecotones are of the two types:

The Atlantic regime is characterized by the discordant bedding of the coast relative to the pre-Mesozoic morphostructure of the platforms; the wide continental shelf as a continuation of the platform; the absence of volcanism; and the maintenance of platform regime against the background of land fragmentation and subsidence.

The Pacific transition zone of Eurasia, in contrast to the Atlantic one (Rice 1978), is characterized by the following features.

1. The mega-ecotone is confined to destructive plate boundaries (the regions of subduction in the Zavaritsky–Benioff zones), i.e., to the zones of convergence of the ascending global lithodynamic flux and lithospheric contraction. It determines: (a) the long-term tectonic instability of the given continental margin, with intense influence of the modern volcanism on the biotic medium (Markhinin 1980); (b) the high geomorphological contrast (paragenesis of island-arc ranges of mountains and deep oceanic trenches); and, finally, (c) the distinct meridional age-related sectorality of land (Gerasimov 1986). What is meant here is the belt of not only the ancient and modern formation of continental crust, but also the age-related growth (increase) of the continent per se – a reciprocal process that continues throughout the entire Phanerozoic. Spatial diversification and contrast of landscape-forming conditions are created here primarily by morphotectonics.

The regime of the Pacific continental margins is characterized by the considerable permeability of earth crust associated with abyssal fractures; the development of magmatism; volcanism as andesitic and andesitic-basaltic lava; the chains of volcanic islands; and the maximum contrast range of block-wave motion resulting in the neighborhood adjacency of mountain ranges and deep ocean trenches, mountainous islands and marginal seas.

In the zone of transition from the northwestern part of the Pacific Ocean to the continent of Asia, the following regional sectors are distinguished (Erlikh et al. 1974): (1) island-arc, of the present-day geosynclines (the Kuril Island Arc); (2) of the early continental orogenesis (Eastern Kamchatka and Hokkaido); (3) of the mature orogenesis (Sakhalin); and (4) of the platform development of morphostructures (the continental margins of Asia). Geosynclinal belts are characterized by direct reflection of the overwhelming part of morphostructures in the relief (Melekestsev 1974).

2. This transition zone is an arena of acute rhythmic hydrothermal monsoon-type in teraction between the continent and the ocean. Drastic seasonal changes in the land-sea interactions result in the high diversity of landscape-forming environments on relatively small territories and the frequent overlapping of ecological niches for soil-biotic components. The abiotic environment "proposes" quite a number of possible regimes and structures for the biota; hence, there is the enhanced competitiveness of species and entire communities, the rich blend of flora and fauna, certain abundance and spatial diversity of life forms. The western boundary of the Pacific Ocean monsoon region passes along the upper reaches of the Amur River and the Great Hingan Range (Alisov, 1956). Within this area, there are three alternating ecological-phytocoenological spaces (from east to west): Neopacifica → Subpacifica → Paleopacifica, differing in age and degree of the Pacific Ocean influence on vegetation cover (Sochava, 1980). Three western concentric borders take shape here for different extents of expansion of taiga-forest formations with a predominance of the Ajan spruce (the core of its areal is in the lower Amur basin and on Sakhalin (Sochava 2005).

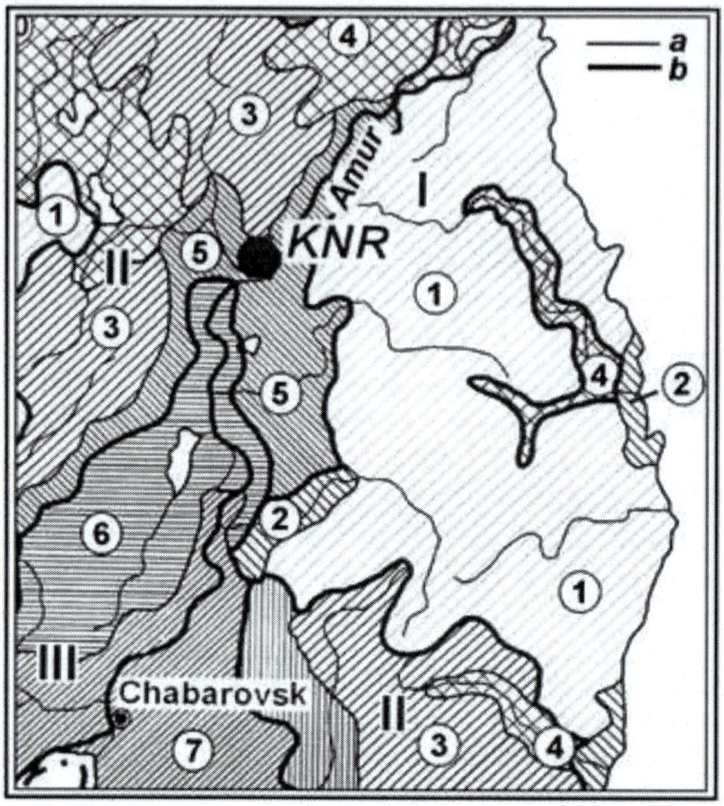

Figure 1.1. The fragment of the ecological-phytochoenotic map for southern part of the Pacific Ocean mega-ecotone of Northern Eurasia (Sochava, Bayborodin 1977), with author's additions. Symbols: I–III, the zonal types of plant formations; 1–7, the groups of plant formations. I. Middle-taiga forests. Mountain-taiga forests of limited development: 1, middle-mountain larch ledum and rhododendron derivative forests; 2, larch ledum/green-moss forests, fir/spruce forests, sphagnum mari and grass bogs. II. South-taiga and subtaiga forests: 3, middle-mountain fir/spruce green-moss forests and their restorative series; 4, south-taiga larch forests, grass/dwarf-bush and larchmari; 5, subtaiga nemoral grass larch forests, cedar/broad-leaved forests, maris and meadow bogs. III. Subnemoral forests, plain and mountain: 6, birch and aspen grass forests with the fragments of primary ones, with reedgrass meadows, yerniks, bogs; 7, broad-leaved and coniferous/broad-leaved tall grass forests and their restorative series. Additional legends: *a*, the boundaries of formation groups; *b*, the boundaries of zonal types of formations; *KNR*, Komsomolsky Nature Reserve.

3. There is an interesting experience of paleogeographical and biogeographical studies of Beringia, which covers the closest parts (joining more than once in the past) of Northeast Asia and Northwest America (Hopkins 1972; Yurtsev 1974; Brigham-Grette 2000). This experience showed that the continental-marginal and insular territories of extra-tropical mega-ecotone of the Ring of Fireof the Pacific Ocean were the "generators" and "keepers" of continental and oceanic elements of the biota against the background of pulsating abiotic environment. Periodic alternation of natural situation initiated the new form-building processes, accelerated the directed (adaptive) and random transformations of plant and animal gene pools, and opened up the possibilities for hybridogenesis. The modern orographic and climatic contrasts along the oceanity–continentality gradient, which are well-marked on the mega-ecotone, support the work of this "florogenetic node" and ensure the sustainable

existence of mixed (buffer) natural ecosystems, with conservation of numerous relic species of phytobiota and the plant communities comprising them.

4. Thus, Pacific mega-ecotone of Northern Eurasia shows up *one of sources of continental biosphere forming* beginning in middle of Mesozoic. Next processes created necessary conditions for that: (a) the frequently repeated joining and isolation of insular and marginal-continental land, which caused periodic isolation of biocoenotic complexes and interchange between them; it accelerated the evolutionary development of plant and animal populations and contributed to the diversity of their modern adaptations; and (b) intensive orogenesis and volcanism, which created the primary altitudinal differentiation of biota on the mountainous islands of the Pacific Ocean. Ecological niches remained permanently vacant, with the migration of modifications of biotic communities from the neighboring altitudinal zones. The primary altitudinal-zonal groupings of organisms and ecosystems were a basis of formation of the zonal types of geographic environment on the plains of all continents. These are some of the basic points of the concept of evolutionary biogeography put forward by D.V. Panfilov (2005).

5. The mega-ecotone per se, being mainly a mountainous territory, is a three-dimensional structure. It was created by three types of ecotonization processes in the biosphere: latitudinal-zonal (macro-climatic), longitudinal-sectoral (macro-orographic), and altitudinal-zonal (meso-oroclimatic), which is clearly reflected in the formation structure of vegetation cover (Figure 1.1). The marked regional and local contrasts of natural-territorial structures appear at the geoecotone under the conditions of dominant climate-forming influence of the continent, which is expressed in the very concept of "monsoon-continental climate." It results in the high mosaicity of natural complexes, as well as frequent zonal and altitudinal-zonal inversions favoring the formation of biotic communities of the ecotone. The Amur River basin is a sphere of contact between the six phratries of plant formations: Beringian, Angaridian, Ural-Siberian, Manchurian, Mongolian-Chinese, and mountain-tundra (the buffer formations are widespread here) (Sochava 1980). All subtaiga dark coniferous forest zonesin middle mountains areconsidered transitional from south taiga to subnemoral and nemoral forests (Kolesnikov 1969; Sochava 1977) by their hydrothermal conditions.

6. The formation of the Pacific Ocean geoecotone is closely related to the history of geographic environment in the Meso-Cainozoic era, when not only the fundamental orogenic processes occurred on the eastern margin of the Asian continent but also the essential modern features of monsoon-continental climate were obtained. The near-Pacific continental margin in the Cainozoic era underwent serious neotectonic transformations associated with crust warping, increasing crustal thickness, rearrangement of Paleozoic and Mesozoic structures, and creation of newly formed orogenic depression morphostructures. The recent appearance of Amur basin has been formed already in beginning of Neogene (Nikolyskaya 1969). Development of the zones of enhanced magma permeability and jointing resulted in basaltic volcanismandformation of taphrogenic and riftogenic hollows: Zeysko-Bureinskaya, Sredneamurskaya, Amuro-Tugurskaya, Anadyrsko-Penzhinskaya, and the Tatar Strait. The shift of crust granitization from intracontinental regions to continental margin resulted in the formation of elongate orogenic-arch uplifts of asymmetric structure. In the Late Cainozoic era, it favored the development of asymmetric landscapes on the macroslopes of large

highlands (Sikhote-Alin, West Priokhotye, Koryakskoye highland) (Khudyakov 1972). Latitudinal zonality was complicated by longitudinal sectorality formed due to meridionally elongated orographic barriers.

The bioclimatic systems of the major meridian sectors of the mega-ecotone are of different age. The modern zonal and altitudinal-zonal structure on the continental margin, with the mixed flora of representatives of sub-Arctic, boreal and subtropic species, was formed in early Palaeogene: 60–65 million years ago (Arkhipov, Nikolaev 1972; Ushakov and Yasamanov 1984). In the southern part of the Kuril Arc, the bioclimatic zonality similar to the modern one was established only in the late Pleiocene, i.e., no more than 2 million years ago (Luchitsky, 1974).

Latitudinal zonation of landscapes along the Pacific Ocean megaecotone, which is typical of internal regions of Eurasia, is shaded by their longitudinal sectorality, many of the physico-geographical boundaries coincide with the geological borders, and the monsoon-continental climate maintains historically blended northern and southern species within each zonal type of plant formations, favoring the development of their sufficiently broad ecological plasticity. Owing to this fact, for example, the spruce/fir forests of the Middle and South Sikhote-Alin, the stone-birch communities of Kamchatka and the high Chosenia floodplain forests along the banks of the Kolyma, Anadyr and Penzhina rivers, being the transformed relics of tertiary vegetation of the vast areas of Eurasia (Kolesnikov 1969; Reutt 1970), are the stable elements of modern Far East landscapes.

Ecological plasticity is a characteristic feature of the main forest-forming species of boreal and subnemoral forests. In the dark coniferous taiga, the Ajan spruce is rather cold-resistant but highly "oceanic," because it is adapted to high relative humidity (more than 55–60%) in spring and summer. Cedar/broad-leaved forests represented primarily by Korean cedar pine and Mongolian oak have a highly varied phytocoenotic structure depending on the type of location, as well as litho- and hydroedaphic conditions, which made it possible to distinguish an entire series of "geomorphological types" of this ancient, historically established subnemoral formation in the south of the Far East (Markov 1980). The most xerophytic cedar-oak communities have no competitors on rocky sun-heated mountain slopes; at the same time, Mongolian oak tolerates frosts down to –60° without serious consequences. Finally, it is known that the peculiar biological feature of cedar prostrate shrubs is to "lodge" and be submerged in snow after the first frosts (see Figure on the front cover of book), which protects it from low temperatures and winter winds (Kolesnikov 1969).

1.3. Bioclimatic Structure of the Mega Ecotone

The immediate vicinity of the winter Asian baric maximum, with the cold pole of the Northern hemisphere, and prevalence of cold sea currents near the northeast coast of Eurasia (Kuril–Kamchatsky, Primorsky) are the cause of two large-scale physical-geographical phenomena that determine the regional bioclimatic structure of the megaecotone. First, due to low thermal energy resources, there is a general substantial shift of natural zones to the south with their simultaneous meridional extension (Ivanov 2014). North-taiga stonebirch and larch forests and thin forests, often waterlogged, with a very low productivity, are widespread over the northern and middle Kuril Islands, as well as on the most part of northern continental territories up to the lower Priamurye (see Figure 1.1). The middle-taiga forests of the plains (GF 4–6) in Kunashir and Sakhalin and along the Sea-of-Japan coast of the continent descend to the 52th–50th parallel. For comparison, let us note that mixed forests and forest steppe are predominant at the same latitudes on the Russian Plain. The forest types of flat landscapes are predominant over the entire southern part of the Far East: namely, at the latitudes corresponding to the steppe and desert zones in the interior of the continent, e.g., in West Siberia (Sochava 1977).

The predominance of light coniferous and bog/light-forest landscapes, which are gradually replaced by moss/small-shrub tundras in the north and turn into the landscapes with polydominant dark-coniferous/broad-leaved forests with substantial admixture of subtropical flora in the south, is a characteristic feature of the monsoon-continental spectrum, or type, of latitudinal zonality, which can be retraced both on the plains and along the mountain ranges. Altitudinal zonation of southern Sikhote-Alin and the southern extremity of Western Sakhalin and in Kunashir mountains reproduces this regularity in miniature and, with advance to the north, the spectrum of altitudinal zonation is sequentially "truncated" from below due to the general decrease of upper boundaries of the belts.At the same time, the pinching-out of altitudinal zones on islands proceeds much more rapidly than on the margin of the continent. For example, cedar elfin wood tangles in the extreme south descend from the height of the subgoletz zone (1200–1500 m of absolute height) to the sea level near the latitude 47° in the strip of the Kuril IslandArc and around 58–59° N onthe coast of the Sea of Okhotsk.

Second, the cold and humid "respiration" of the ocean and its marginal seas in summer and dramatic cooling of the continent in winter are related to the frequent phenomenon of latitudinal and altitudinal-zonal inversion of landscapes when, for example, sub-Arctic plant groups develop on the coasts and in the intermountain troughs of the taiga zone (Ivanov, 2014). In Chukotka, large river valleys in the tundra zone have, on the contrary, the warming effect on soils and vegetation and, therefore, are occupied by flood plain forests and shrubs, which sharply contrast the surrounding moss/small-shrub tundras.

As a rule, the landscapes of more southern appearance maximally advance northwards along the slopes facing the continent. The opposite mountain slopes face cold coastal currents; therefore, orographic barriers create the most favorable conditions for the biota just on the "continental" but not "oceanic" slopes of mountain ridges and close to their feet (see Figure 1.1). Here, the spectrum of altitudinal belts is more diverse, the upper forest line is higher, soil formation processes are more intensive when the cryogenic factors weaken, and the species compositions of flora and fauna are substantially richer.

Thus, the geographic concepts of oceanicity and continentality are rather complicated and ambiguous when applied to estimation of the level of structural development and functioning of landscapes. The extreme manifestation of oceanicity with predominant windy and cloudy weathers, cold fogs and drizzling rains is physiological unfavorable for tree vegetation growth. Hence, e.g., in the narrow coastland near the foot of the eastern slope of the Middle Sikhote-Alin there is a clearly marked lower natural forest line (shrub/herbaceous oak forests, see below), similar to the forest-tundra boundary at its upper limit in the mountains and to the northern boundary on the plains (Maksimova 1982). At their lower line, arboreal plant communities are replaced by shrub and then herbaceous communities (see Figure 1.1). Shrub-grass formations are known to be widespread on oceanic islands, including the Kuril and Commander Islands (Figure 1.2). The extreme oceanicity essentially simplifies the vertical component structure of landscapes and reduces their bioproductivity. The traces of oceanicity on the Middle Amur plains are observed up to the western boundaries of distribution of sphagnum larch forest mari (Kolesnikov 1969).

Figure 1.2. South Kurile Islands, the Shikotan Island. Sparse stunted trees with flag- and umbrella-shaped crowns and crooked trunks against the background of the dominant herbaceous vegetation.

The suboceanic climate is much more favorable in this respect. It combines moderately cold winters with high snowiness and weakened winds, on the one hand (which preserves the vegetative organs of plants from frost-killing and drying up), and summer cloudy weathers with the sufficient sums of biologically active temperatures (up to 1800–2000°) and a moderate range of humidity factors (0.7–1.5), on the other hand. These climatic conditions maintain stable photosynthesis throughout the entire vegetation period. The processes of organic matter decomposition, humification, soil weathering, and fersiallitization of the soil profile are also more intensive (Ivlev et al. 1986). Thus, the suboceanic climate corresponds to a certain *zone of the optimum of exogenous continental-oceanic interaction*. This zone is well observed along the "continental" slopes of the largest mountain ranges of the continental margin: Sikhote-Alin and its extension along the left bank of the Amur River (see Figure 1.1), the Dzhugdzhur Range and the ranges of Kolyma Uplands adjacent to the Sea-of-Okhotsk coast and, finally, the Vostochny (Eastern) Range in Kamchatka. The significance of barrier effect generally decreases northwards, which corresponds to "fuzzification" of the monsoon climate features at high latitudes.

There is also an analogous *marked zone of the optimum on the continent-facing slopes of volcanic mountains of insular land* located in the properly oceanic longitudinal sector of the geoecotone. Here, the warm and cold sea currents have a greater effect than the orographic factor (Markov 1980; Ivanov 2014). One of the branches of the warm Kuroshio Current reaches the west coasts of the South Kuril Islands, while their east coasts are washed by the cold Kuril–Kamchatka current. In line with the above, four landscape zones (meadow, stone birch forest, alder/alder–cedar prostrate shrubs and mountain–tundra) are observed on the western (the Sea of Okhotsk) slope of volcanoes of Simushir (the Kuril Ridge), while only two zones (alder–cedar prostrate shrubs and fragmentary mountain–tundra among the primary volcanic deserts) are observed on the eastern (the Pacific Ocean) slope. The upper belt on the eastern slope is 200 m lower compared to the western one, where the life forms of plants are better developed. The suboceanic climate features are associated with broad distribution of the landscapes of forests of prostrate cedar elfin wood in the mountains of the northern Priokhotye, Kamchatka and Koryakskoye Upland and dark-coniferous/broad-leaved forests with the elements of subtropic flora (Amur cork tree, magnolia, Aralia) on South Sakhalin and the South Kuril Islands.

1.4. Forest-Forming Processes in the Mega Ecotone

When describing in detail these processes in the south of the Far East, B.P. Kolesnikov characterized the forests of this vast ecoregion "... as the forests of mountain–valley landscape and monsoon climate, ... most spatially variable and extremely dynamic, i.e., with exclusively multiform and changeable habitus" (1956, 136). Just *multiformity* and *dynamicity* are the characteristic features of forest formations of the boreal and nemoral Eurasian sub-Pacific region from their intracontinental plain and mountain analogs on the Russian Plain and in West Siberia, on the Ural and Altai, in the Sayans and Transbaikalia, in the Carpathians and Alps. These analogs are characterized by the relatively lesser number of ecologically homogenous phytocoenotic components and rather structurally uniform types of forest.

In contrast to Western and Eastern Europe, as well as West Siberia, there has been no solid inland ice sheet in the south of the Far East since the early Quaternary Period; therefore, the development of plant cover was not interrupted here. According to the spore-pollen data from coal-bearing sediments, mixed coniferous/broad-leaved forests grew in the Middle-Bikinsk depression (Khudyakov 1972) as early as in the Miocene (about 25 million years ago), with the involvement of both subtropical species and representatives of moderately thermophilic (subboreal) and even moderately psychrophilic (boreal) forms. Thus, there had been an intermixture of very contrasting floras in Primorye and Priamurye (the Amur River basin) long before the Quaternary glaciations, undoubtedly as a result of influence of the already existing monsoon climate.

The plant cover of the Primorye–Priamurye Sub-Pacific region continuously developed as it stayed under these conditions for a long period of time. The transition of the flora and vegetation from the Tertiary to Quaternary period was relatively smooth (Krištofovič 1946; Sochava 1946), with the migration of separate elements of the Tertiary flora from north to south and back north and with the penetration of cold climate species (Kolesnikov 1969). The continental margin of Northern Eurasia came to be at the crossroads of migration pathways (Yurtsev 1974). Most of the present species constituting the broad-leaved/Korean cedar communities had ancestor taxa in the Tertiary palaeofloras of the Late Miocene and Pliocene. During the Pleistocene Ice Age the rich mesophytic flora in this area lost most of its mesothermic species. The vegetation of the maximum stage during the last glacial period consisted of drought-tolerant meadows and larch woodlands on the plains and dark-conifer taiga in the lower mountain belts. During the following warming, broadleaved temperate species invaded this territory from warmer and wetter regions in the south (Nakamura, Krestov 2005).

Such specific and long-term forest-formation processes resulted in the gradual conversion of tertiary thermophilic suboceanic formations into more moderate formations adapted to drastic seasonal hydrothermal contrasts. Even representatives of the modern cryogenic-taiga flora, "… larch forests in the process of evolution combined the tertiary species with different genetic relationships and centers of origin and their descendants" (Kurentsova, 1973, 173). According to the figurative expression of A.N. Krištofovič, the abundance of flora in the south of the Far East "… is only the result of its reaching the saturation point of climatic possibilities" (cited by: Kolesnikov, 1969, 218).

At the same time, climatic fluctuations manifested themselves mainly in the temperature changes against the permanently high moisture content (Akhmetova 1977; Ganzey 2008), which favored the transgression of representatives of the humid Okhtskaya flora, first of all the Ajan spruce, into relic cedar/broad-leaved forests. Such was, e.g., the phase of cedar/spruce/broad-leaved forests at the turn of the Atlantic and subboreal periods of Holocene (4.7–4.1 thousand years ago), when there was cooling with the yet high moisture content (Korotky et al. 1999). This forest-forming process as a result of Pacific Ocean influence still goes on in the Priamurye Subpacific Region at present. The transgression of representatives of the Angaridian continental flora (in particular, Dahurian and Kayander larch) proved to be a more localized phenomenon associated with the climatic inversions under the conditions of mountain-valley relief. Nevertheless, the areal of larch extended during the century-long changes, with its penetration into mixed forests (Kurentsova 1973).

The result of the forest-forming processes described above was the well-known floristic abundance and coenotic diversity of cedar/broad-leaved and spruce/broad-leaved forests in the south of the Far East (Figures 1.3 and 1.4). Primarily mixed cedar/broad-leaved and, to a lesser extent, coniferous forests are characterized here by the multispecies composition and ecological-biological heterogeneity of phytocoenotic components at all layers, which leads to the complexity and entanglement of interspecies relationships and the diversity of forest-forming processes. This diversity is composed by age-related, secular and restorative changes of forest communities. The results of these relationships are the vertical closeness (solidity) of vegetation layers, the grouping arrangement of individuals, the presence of differently aged stands in a forest community, the periodicity of tree growth in general and of any generation and, finally, the aggravated interspecific competition between individuals and, as a consequence, the general decrease in the site quality (bonitet)and productivity of forests, which does not correspond to the potentialities of habitat environment.

The dominants of broad-leaved/conifer forest ecosystems differ in growth forms and life strategies. Up to 12 tree species growing together can be found in the upper sublayer of one forest stand (Nakamura, Krestov 2005). The main species forming the dominant layer are *Abies holophylla*, *Betula costata*, *Fraxinus mandshurica*, *F. rhynchophylla*, *Juglans mandshurica*, *Kalopanax septemlobus*, *Phellodendron amurense*, *Pinus koraiensis*, *Quercus mongolica*, *Tilia amurensis*, *T. mandshurica* and *Ulmus japonica*. Their usual height is 25–35 m, but on rich sites *Abies holophylla* and *Pinus koraiensis* may reach a height of 45 m, exceeding the height of other tree species and forming a sparse layer above the canopy. The century phytocoenotic replacements eventually resulting in substitution of one forest formation for another consist of a number of stages and, therefore, can be presented as a certain *genetic series of forest groupings* (Kolesnikov 1956). Uneven age structure is a characteristic feature of old-growth broad-leaved/Korean pine forest (Nakamura, Krestov 2005). The trees in such stands grow in cohorts reflecting the gap patternof stand dynamics (Figures 1.3 and 1.4).

Figure 1.3. Low Priamurye, Komsomolsk nature reserve, cedar-broadleaf Forb forest with multispecies composition and presence of relict flora.

On the other hand, due to multispecies composition, age replacements of forest stands occur in different habitats according to a single spiral scheme, with no less than 2- to 2.5-century cyclicity (Kolesnikov 1956). The spirality (closure failure) of these cycles manifests itself in the long-restorative and even stably irreversible successions and evokes a general tendency to displacement of cedar from stand structure by deciduous trees in Primorye and by coniferous trees (mainly the Ajan spruce) in the Middle and Lower Priamurye (Figure 1.5). The emerging long-restorative and even stable naturally irreversible forest communities consist of representatives of other forest floras. For the Priamurye, it means the displacement of Manchurian flora by either Okhotian or Angaridian flora. During the cyclic dynamics of forest stands the shade-tolerant nemoral species, as well as shade-intolerant species, are able to complete their life cycles. Because normally in old-growth forests the cycles do not coincide, diversity within a forest community with stable edapho-climatic conditions is maintained (Nakamura, Krestov 2005).

Figure 1.4. Low Priamurye, Komsomolsk nature reserve, young spruce-broadleaf forest with dense undergrowth and high Nemoral Forb.

The multispecies composition and structural diversity result from the superposition (mutual overlapping) of age, secular and restorative (demutational) replacements of forest communities. At the same time, age replacements cause gradual quantitative transformations of habitat conditions from generation to generation: first of all, the accumulation of organic matter and the changes in hydrothermal parameters of edaphotopes and soil trophicity. Such environment-forming activity of phytobiotic components (Isakov et al. 1986) creates prerequisites for irreversible century-long replacements of the main forest-forming species. This process has become especially widespread at the Far-East boreal ecotone, with its floristic richness of mixed forests and with their complex different-aged and multilayered structure; the striking example is the above-mentioned replacement of cedar by the Korean and Ajan spruce, with the development of cedar/spruce and spruce/broad-leaved communities.

In other cases, especially after fires, the age and century-long replacements of cedar/broad-leaved forests are accompanied by displacement of cedar with small-leaved species (yellow and Amur birch and aspen), with broad-leaved species (oak, ash, lime) pushed to the sidelines and mesophilic Manchurian plant species replaced by psychromesophilic taiga species (Kolesnikov 1956).

Figure 1.5. Low Priamurye, Komsomolsk nature reserve, silver/fir-spruce Fern-Nemoral/Forb forest with cedar.

1.5. On Development of the Theory of Evolutionary Landscape Study

The Pacific Ocean megaecotone of Eurasia, being an element of the global belt ofconvergence of matter and energy flows on the Earth surface, is one of the *sources of formation of the continental biosphere*. It concentrates the conditions for the development of focal biocoentoic processes (Hansen and di Castri 1992) and spontaneous hybridization, for the appearance of stable new formations in the natural and territorial structures (Khudyakov 1972). According to D.V. Panfilov's concept of evolutionary biogeography (Panfilov 2005) such conditions in the geological history were as follows: (1) the frequently repeated joining and isolation of insular and marginal-continental land, which caused periodic isolation of biocoenotic complexes and interchange between them; it accelerated the evolutionary development of plant and animal populations and contributed to the diversity of their modern adaptations; and (2) intensive orogenesis and volcanism, which created the primary altitudinal differentiation of biota on the mountainous islands of the Pacific Ocean. Ecological niches remained permanently vacant, with the migration of modifications of biotic communities from the neighboring altitudinal zones. The primary altitudinal-zonal groupings of organisms and ecosystems were a basis of formation of the zonal types of geographic environment on the plains of all continents.

The study of modern landscape connections in morphostructures and morphosculptures of different ages is the way to understanding the phylogenetic regularities of the long-term geological trajectory of terrestrial landscape formation. This trajectory begins with the stages of geosynclinal island arcs (the brilliant example is the Kuril–Kamchatka island arc), with volcanic relief and formation of the primary subaerial landscapes on volcanic rocks. They are followed by the stages of development of epigeosynclinal orogens (Khudyakov 1977). These are the mountain ranges of Sakhalin and the Sikhote-Alin with its extension along the Amur River left bank, which are characterized by young erosion-denudation and accumulative marginal-continental morpho- and biogeoecosystems.

This is the first stage of formation of continental landscapes of the *evolutionary series, or progressive development*. At this stage, one should expect a general increase in the level of biogeosystem organization: the complication and ordering of their structure, the increase in spatial differentiation of landscape connections, and the enhancement of the system-organizing role of soil-biotic components (Kolomyts 1987). As a result, the efficiency of exploitation of environmental sources by biogeocoenoses increases, predetermining the biosphere evolution (Shwarts 1973). Hence forth, there is a much longer change of the stages of origination and reformation of residual/denudation "senile" landforms, when the territory passes into the state of intracontinental ancient platforms (Siberian, East-European).

In parallel with the above, the reverse processes of landscape–geomorphological disintegration on the continental margin can also be traced. They are manifested in the replacement of roof plastic deformations by faults and differentiated block movements, in the fragmentation of separate marginal parts of the continent and their foundering below sea level (Khudyakov 1977), with the formation of *continental islands of the regression series* (the islands of Peter the Great Bay, Shantarskiye and Yamskiye). Analogous effect of "attenuating" landscape evolution is observed when land stays too long at the island-arc stage and sedimentary rocks are formed (the islands of the Lesser Kuril Ridge (Luchitsky 1974). The trajectory of regressive natural complexes is characterized by depleted composition, simplification of landscape-ecological connections, and enhanced role of the abiotic factors. In this case, we can obtain the natural analog of anthropogenic changes in continental geosystems: the processes of disintegration of landscape connections have different characteristic times but are analogous in essence.

The presented concept of the Pacific Ocean mega-ecotone of Northern Eurasia as a focus of evolutionary processes in the continental biosphere was published by the author in his book (Markhinin 1980). Similar ideas had been put forward somewhat earlier by B.A. Yurtsev with regard to florogenesis of the northern wing of the mega-ecotone, Beringia, which he denoted as "…a peculiar florogenetic laboratory of the northern areas of the Earth" (1974, 144).

The modern landscapes of the Pacific Ocean mega-ecotone of Eurasia have been formed in a very broad age spectrum of morphotectonic structures (Khudyakov 1977) with two notable geological–geomorphological borders separating the sectors of the Cainozoic, Mesosoic, and still earlier foldings (Sochava 2005). In the direction from oceanic islands to the marginal and then interior regions of the continent, the landscape connections of land are formed on a more and more ancient lithogenic base, under the conditions of weakening of the neotectonic movements and successive burial of the main source of mineral nutrition for phytobiota (crystalline bedrocks) under the increasing sheath of loose sediments, which are much more depleted in this respect, that have passed through numerous cycles of siallitization (clay formation), denudation ablation and accumulation.

On the other hand, the role of radiation climate-forming factors enhances and the influence of circulation (advective) factors decreases in the westward direction. Therefore, continentality of the climate generally increases. As a consequence of interference between both groups of factors in the direction from the margins to the interior of the continent, the total intensity of biogeochemical cycles of matter in the soil-plant system weakens (under certain zonal conditions), the processes of weathering and soil formation slow down and, finally, the biological productivity of landscapes decreases. All the above occurs against the background of weakening of the direct landscape-forming role of morphotectonic factors and increasing significance of exogenous geomorphological processes. Landscape appearances show less and less traces of endogenous forces (tectonic benches, volcanic cones, lava beds, etc.), while residual-denudation and accumulative landforms become more and more prevalent.

All intracontinental territories of Eurasia passed, in one way or other, through the above described sequence of changes of landscape situations in their geological past. In the past, the continental-oceanic seam zones were the Alpine–Caucasus–Himalayan mountain belt, as well as even more ancient folded Hercynides of the Urals and Kazakhstan (Gerasimov 1986). The latest orogenic-geosynclinal region of the Pacific Ocean mega-ecotone is characterized by the initial stages of development of the next continental part of the geographical envelope of the Earth during age-related buildup of area of the given continent. In essence, the matter in question here is the *primary landscape genesis on land*: a rather poorly studied branch of complex physical geography. In the meantime, the primary landscape genesis is directly related to the fundamental problems of continental biosphere formation and development, understanding the initial mechanisms of surface physical-geographical process, according to the interpretation of A.A. Grigoryev (1966), which created in the past and continuously forms in the present the landscape appearances of continents.

The comparative analysis of modern landscapes of the Far East (island-arc volcanic, continental-insular and marginal-continental) will probably reveal the regularities of formation and development of landscape connections on land, as well as their degradation and disappearance. Undoubtedly, V.B. Sochava (1980) meant precisely the evolutionary direction of sublatitudinal variation of landscape connections on the Pacific Ocean geoecotone, distinguishing three sectorial geospaces: Neo-Pacific, Sub-Pacific, and Paleo-Pacific. It can be illustrated by the following example. In the subtaiga coastal strip of the Middle Sikhote-Alin, in the direction from the Japanese Sea shore to low mountains, one can trace a certain topological series of associations, with the incumbation series of indigenous life forms of Mongolian oak (Maksimova 1982): herb-cereal meadow → bush-clover/hazel cereal/herb association → herb/hazel low (up to 1 m) bushy oak forest → herb/bushy crooked oak forest (3–6 m in height) with wind-shaped crowns → straight-stemmed oak forest (up to 15–17 m in height). Considering the first and second associations to be oceanic (neo-Pacific representatives) and the third and forth associations to be suboceanic and continental (sub-Pacific and paleo-Pacific representatives), respectively, we obtain the local model of evolution of the given zonal type of vegetation cover on the mega-ecotone in the direction from the youngest to more ancient formations.

Thus, the Pacific Ocean mega-ecotone of Eurasia is a highly favorable object for addressing the problems of *evolutionary landscape science*. This novel field of complex physical geography must be aimed at the study of "supporting mechanisms" of the formation, development and degradation of landscape connections on land, the search of motive forces determining the tempos of evolutionary process (both ascending and descending), the revealing of taxonomic and age-related correlations between the structure and function of geo(eco)systems, their ontogeny and phylogenetic shifts. We believe that the solution of these problems will contribute to the development of scientific and methodological bases of controlling natural processes and creating new ecological equilibriums in nature.

Our initial working propositions of the future concept of evolutionary landscape science provide for the study of the structure and function of island geosystems in their spatial-genetic diversity and as components of the continent-to-ocean transition zone. At the same time, the study of spatiotemporal organization of island landscapes (both evolutionary development and regression series), together with the assessment of potential natural resources of islands and recommendations on the maintenance of their ecological balance, must be aimed at eventual creation of complex and sectoral physico-geographical models simulating the structural and functional changes in continental landscapes at the initial stages of their formation.

In particular, the coastal–continental islands of the regressive series, with their general descending trajectory of the system of landscape connections, may serve as a natural model of anticipated events for many intracontinental regions of the Far East in the foreseeable future, with the estimated intensification of development of their natural resources and expansion of technogenic zones. These islands, being at different stages of anthropogenic transformation and territorially closed, are a favorable object for addressing the scientific and methodological problems of resistance of natural landscapes to external impacts.

The study of island and marginal-continental geo(eco)systems of the progressive series (the Kuril Islands, Kamchatka, etc.) will be aimed at understanding the regularities of formation and development of landscape connections on land during its gradual transition from insular to continental regime. Here, the nature can demonstrate the ways and stages of formation of landscape systems of different levels of development, which will be directly related to the problem of artificial creation of new ecological balances in nature. As is known (Ignatyev 1979; Ivanov 2014), the process of species formation and phytocoenotic successions is more rapid on oceanic islands compared to the continents.

In the landscape-ecological analysis of the Pacific Ocean mega-ecotone, it would be advisable to compare the data on the intracontinental East-Eurasian system of transition zones: e.g., on the bioclimatic boreal ecotone of the Volga River basin (Kolomyts 2005), which is tectonically stable and is characterized by "senile" farewell-rock erosion-accumulative landforms. Both megaecotones are components of the global Eurasian "ocean–continent" ecotone system (Polynov 1952; Milykov 1981), with the spatial ordering of geotextures of different ages (Meshcheryakov 1981) and modern macroclimatic fields, as well as with the respective change of zonal/regional and local vegetation and soil spectra. Such transition from young island geosystems of the evolutionary series to marginal-continental natural complexes of epigeosynclinal orogens and then to intracontinental landscapes, already having a rather long history of subaerial development, represents the basic geological stages of formation of the natural-territorial structure of the continents (Sochava 1980; Kolomyts 1987).

The most important indicator of geo(eco)system state is known to be its primary bioproductivity. The coastal and island landscapes of the Pacific Ocean mega-ecotone differ from their intracontinental analogs in the higher phytomass reserves and productivity. For example, these indices for boreal forests are 500–800 tons/ha and 10–18 tons/ha·yr in the former case and more than 300–350 tons/ha and 5–9 tons/ha·yr in the latter case (Bazilevich 1981; Kolomyts 2005). At the same time, the portion of green phytosynthesizing mass on islands is much higher compared to the mass of the above-ground skeletal part of plants and the root mass. They are also characterized by much greater (compared to the continent) involvement of the lower, especially near-ground, biogeohorizons in organic matter production.

The very structure of phytobiota also undergoes the respective changes. The phenomenon of island ecosystems is known (Vorobyev 1963; Stepanova et al. 1981) to be high- and tall-grass communities with the elements of giantism. These are, for example, the high-grass dropwort coenoses on the Kamchatka and the Kuril bamboo underbrush in Sakhalin and South Kurils. Herbaceous formations, as opposed to forest ones, are characterized by much higher solar energy utilization and water use efficiency. They also accumulate much greater amounts of nitrogen and ash substances per unit of products (Bazilevich 1981). All the above makes island landscapes with herbaceous phytocoenoses more economic energetically and contributes to their broad distribution at the initial stages of landscape genesis.

1.6. General Scheme of Primary Landscape Genesis

It is a rather complicate and multistage process, because various natural components and factors do not simultaneously get involved in the work on evolutionary transformation of the young land and the ratio of their roles in this physico-geographical process changes continuously. Volcanic islands develop already as a certain dynamic system of interrelated soil-plant, geomorphological and geochemical factors (Ignatyev 1979). At the same time, the phytobiota with microbocoenoses and soil fauna are the pioneers of transformation of the surface of the primary geological substrate (Figure 1.6, *A*). The processes of initial biogenic invasion of land and its preparation for further hypergene transformation are substantially accelerated here due to the mineral composition of the initial lithological material. Lava and ash-scoria-gas products of volcanic eruptions contain chemical elements most important for organic synthesis: hydrogen, organic carbon, nitrogen, phosphorus and sulfur, as well as trace elements forming prebiological structures. All of them are often in the form of salts readily soluble in water, which are actually prepared to be assimilated by living organisms (Manyko 1974; Markhinin 1980). Ashes also contain huge amounts of microbial cells, which contribute to the development of blue-green and diatomic algae, the pioneers of soil cover formation on the young land (Markov 1980). Ashes are the major portion of volcanic material, usually spread by air masses far beyond the region of eruption.

Further exogenous transformation of insular land is associated with macrophytic plants, which form the developing, mainly monodominant phytocoenoses: from the pioneer open communities to the "mature" communities reaching, in one way of other, their climax state (Ivanov 2014) and acquiring the respective zonal features (Manyko, 1974) and autonomy from the lithogenic basis of landscape (Figure 1.6, *B*). The rates of biota development may be substantially different under different zonal and local conditions of insular land. However, the age-related phytogeographical differences between small natural complexes, e.g., with forest vegetation, can be traced back deep into the distant past only to the Upper Holocene boundary, below which they actually do not manifest themselves (Ignatyev 1979). The modern soils that had been developing during the entire Holocene or even since Upper Pleistocene become diagnostically significant before Upper Holocene. At the same time, the normal (autochtonous) trend of paedogenesis was predominant at the initial stages of soil formation (Ivlev et al. 1987), because any significant processes of exogenous denudation and accumulation had not yet been developed.

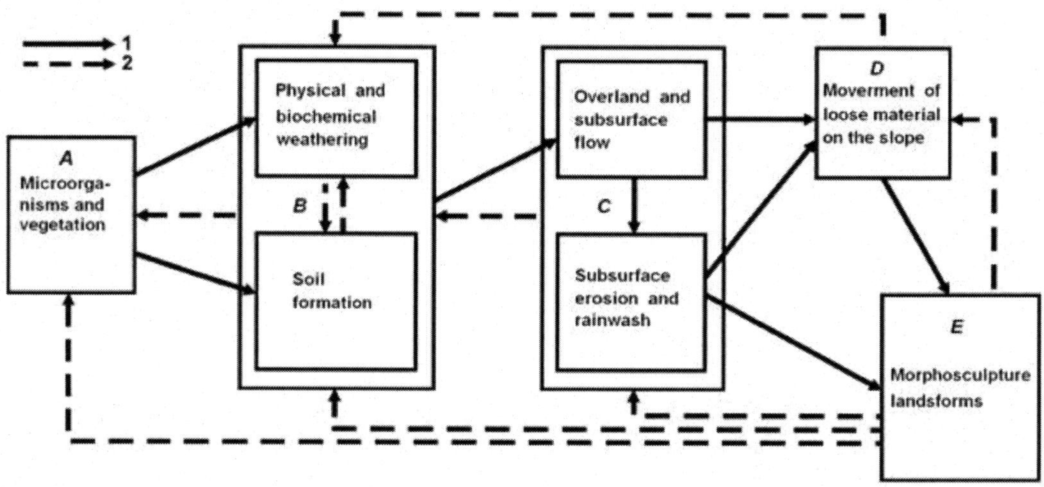

Figure 1.6. The scheme of cause-and-effect relations between natural components and factors in the primary landscape genesis (Kolomyts, 1987). Relations: 1, direst; 2, reverse. A–E, the stages of the primary landscape genesis (explained in the text).

The next age category of terrestrial geo(eco)systems consists of erosion landforms (see Figure 1.6, *C*). Surface runoff and linear erosion develop rather slowly and manifest themselves during the formation of the weathering crust of heavy mechanical composition. The primary volcanogenic substrate (lava and pyroclastic) does not favor the concentration of groundwater in the negative landforms due to its jointing and good water permeability; therefore, the developing phytocoenoses are under the conditions of variable hydrological regime of "dry rivers" and are fed mainly with atmospheric moisture (Manyko 1974).

The development of morphosculpture is accompanied by the appearance of denudation and accumulation trends of integrated paedo-lithogenesis, which makes a qualitative leap in complication of the entire landscape structure (Figure 1.6, *D* and *E*). The secondary differentiation of natural complexes begins, while the formation of more complex phytocoenoses results in the narrowing of ecological niches of the dominants and the appearance of polydominant communities (Ignatyev 1979).

Biota, in particular, vegetation is known to play a stabilizing role in the structural-functional organization of continental landscapes by creating, due to its edificator properties, the natural-territorial structures that are rather stable in space and time (Sochava 2005). It seems not to concern the involvement of biotic components in the primary landscape genesis. Here, as it follows from the above, microflora and vegetation act as the initial and, moreover, accelerating factors of landscape development. They determine the intensity of biochemical weathering and soil formation processes determining the tempos of soil development, surface and ground runoff and, accordingly, the rate of the linear erosion and slope processes giving rise to the formation of exogenous morphosculpture.

In turn, morphosculpture during its development has an adverse effect on hydrological, deluvial-proluvial, soil-geochemical processes and, via them, on the biotic unit of the geosystem. It is known, for example, that soil characteristics and distribution depend on the maturity of relief, the physical properties of bed rocks, and the new cycles of erosion (Gerrard 1981). The maturity of erosion-denudation landscape is indicated by the two characters (Borzov 1954; Milykov 1981): (1) asymmetry of river valleys and (2) two- or multilayered landscape structure within a single natural zone or a single altitudinal belt. Landscape layering manifests itself in the directed changes from watersheds to the lower degree of relief compartmentalization and ground water table, as well as in the washout of the earlier deposits and outcrop of more ancient bedrocks, etc. The feedback system may be both positive and negative, which will cause the respective rearrangement of the structure of natural complexes.

Let us give an example of young tectonic-volcanogenic geosystems of the Late Pleistocene/Holocene age, which we have examined on the slopes and nearby the foot of the Mendeleev volcano located in the south of the Kunashir Island (the southern outpost of the Greater Kuril Ridge). Here, at an extremely high intensity of manifestation of volcanic processes, with instant (on the geologic time scale) creation of new landforms, there are the most complete coincidences between the age of relief and the age of deposits, on the one hand, and between landforms and geological structure, on the other hand (Luchitsky 1974).

Golets (mountain-tundra) and subgolets (cedar elfin wood) ridgetop and hilltop landscapes of the Mendeleev volcano and the fragments of its soma (the Mechnikov Mountain) situated above the absolute elevation of 350–400 m, as well as peripheral geosystems of the Golovnin caldera, have been formed on Middle Pleistocene basaltic andesites subjected to tectonic and erosive destruction. The underlying "floors" are formed by a strip of stone-birch light wood meadow with cedar elfin wood and a broad belt of coniferous/broad-leaved forests with tall grass meadows (Kuril bamboo underbrush) on steep slopes made by pyroclastic rocks and then on volcanogenic-proluvial sloping plains (up to ~200-m elevations). Riparian coniferous/broad-leaved forests occupy subhorizontal 80- to 100-m marine terraces, with pyroclastic rocks washed by the sea.

All transit and accumulative landscapes of the region are of approximately the same age: 35–40 thousand years (Luchitsky 1974). This time proved to be enough for the development of dynamically mature soils on piedmont plains and marine terraces. Considerable humus accumulation, deep penetration of illuvial processes and lessivage into soil formation rock and distinct differentiation of the entire soil profile have been noted. As is known (Ivlev et al. 1987), under the humid, moderately cold and snowy conditions of highlands, the vertical soil profile under forest-meadow vegetation reaches the late stages of its formation soon after the first millennium. In addition, soil formation processes on loose substrates are very intensive, because the dispersed medium turns out to a certain extent prepared for the formation of the soils.

In the book we have considered the three spatial-temporal sections of geological history of the formation of continental biosphere in the boreal Eurasian region of the Pacific Ocean mega ecotone: (1) the origin and evolution of insular-arc geo(eco)systems of the Neogene/Quaternary period in the oceanic environment, (2) the subsequent ascending development of the continent as a young (Mesozoic) mountain-valley morphostructure, and (3) the final regressive branch of the total trajectory of subaerial landscape genesis – the fragmentation of separate marginal parts of the continent and their sinking below sea level, with the formation of continental islands with a "senile" erosion-denudation relief. Thus, the three main steps of the known Davis–Penck–King geographical cycle have been encompassed. The proposed statements are factually substantiated on the basis of the known bioclimatic and geological-geomorphological data, as well as the results of large-scale landscape-ecological surveys performed at experimental sites in the marginal-continental and insular-arc sectors of the mega ecotone. Hence, it was possible to saturate the concept with particular contents and to develop its fundamental theoretical and methodological aspects on the basis of empirical-statistical modeling and prediction of the structural-functional organization of biogeosystems and landscapes at different stages of geological history of continental biosphere formation in the Pacific Mobile Belt.

Chapter 2

METHODS OF EMPIRICAL-STATISTICAL MODELING

ABSTRACT

The landscape-ecological processes and phenomena presented in this monograph are based on the construction of discrete empirical–statistical models of natural ecosystems. Geoecological analysis was performed mainly with two types of empirical models: (a) informational models describing geocomponent interrelations, i.e., the monosystem organization of natural complexes, serving as a basis for the regional bank of their ecological niches that characterizes their parametric space; and (b) fuzzy set theory-based models describing the polysystem units of landscape organization by operations with ecological niches as descriptive vectors. The integral models of polysystemic organization were created too. The matrices of inclusion relations were used to calculate the reserve of "homeostaticity" as a measure of potential stability of geo(eco-)system to a given factor. The structural schemes of temporal states of regional and local geo(eco-)systems and the schemes of their climatogenic transformations were developed on the basis of this measure. The proposed empirical simulation modeling of global–regional changes in the local geo(eco-)systems based on the schemes of binary ordinations is aimed at a certain advance toward elucidation of the local mechanisms of these changes.

2.1. SYSTEMS OF LOCAL AND REGIONAL LANDSCAPE COUPLINGS

Empirical-statistical analysis of phytocoenological and landscape organization of mega ecotone was based on the materials of large-scale landscape-ecological surveys carried out by the author and his colleagues at three experimental test sites: (1) in vicinity of the active Mendeleev volcano in Kunashir Island (South Kuril Ridge); (2) in the territory of Komsomolsk Nature Reserve in the Lower by-Amur (see Figure 1-1); and (3) Putyatin Island in Peter the Great Gulf (South Primorye).

The local (topological) geoecological space of researched ecoregion is characterized by a certain set of forest phytocoenoses, their distribution by micro-relief forms, and the respective hydrothermal regimens of soil. *The landscape facie* (Berg 1958), or *biogeocoenosis* (Sukachev 1960, 1972), is the main local (topological) operational unit in our research. This elementary unit of geographical ecology accepted in Russia corresponds to the "Site," "Ecoelement," and "Landtype Phase" categories in the Australian–Britain, Canadian and USA classifications, respectively (Klijn and Udode Haes 1994).

Fiveto seven *floristic*, *phytocoenotic* and *microlandscape* (*facies*) groups covering the major diversity of forest types and *types of locate sites (geotopes)*were formed on the basis of 35–55 sample areas in each of the test sites. The given groups settled down in the vector system of local landscape couplings (micro- and meso-catenas) in the direction of edaphic moistening increase (Figure 2.1, *A*): from eluvial (E) and transeluvial (TE) types of local sites through slope transit (T) and transaccumulative (TA) to riverain accumulative (A) and bed superaqual (Saq) types of local sites, according to the classification (Polynov 1956; Glazovskaya 1964). The eluvial-accumulative sites on interfluve padingsare alsocharacterized by excessive edaphic moistening. We combinethese sites together with superaqual ones.

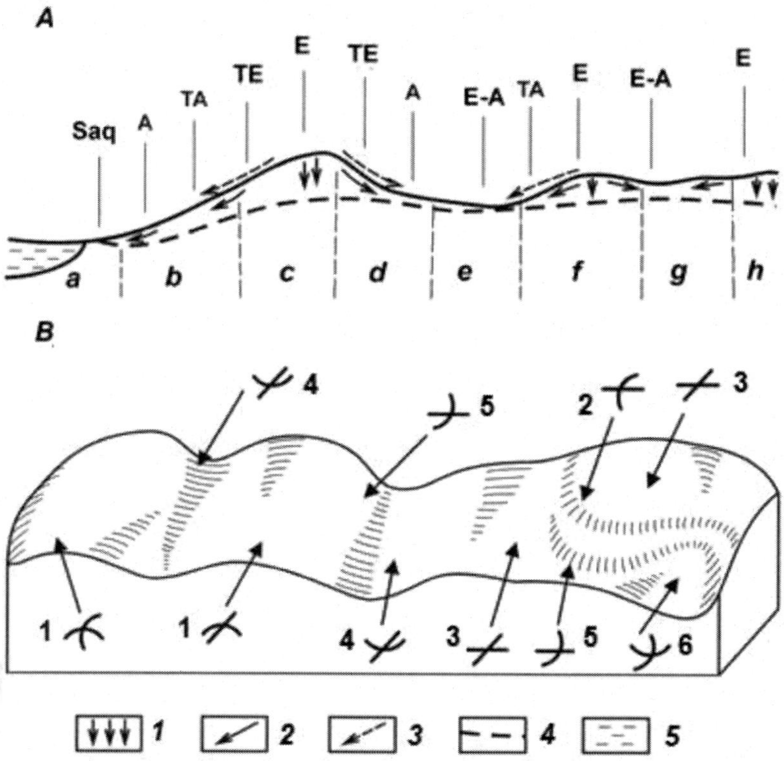

Figure 2.1. Distribution of the types oflocal sites (geotopes) on catenas under the conditions of plain relief. Moisture and substance transfer in soil and friable deposits: *1* – vertical; *2* – along a slope; *3* – litho-dynamic and air flows. *4* – level of ground water table. *5* – water body. The local sites typesare as follows (Polynov, 1956; Glazovskaya, 1964): E – eluvial; TE – transeluvial; T – transit; TA – transaccumulative; A – accumulative; E-A – eluvial-accumulative; and Saq – superaqual. *a – h* – stows (landtype associations), or *urotshistshes*, in Russian transcription (Isachenko, 1965), are meso-topological units; each of them may spread to several types of sites and consist of certain facial series. Other legends, see in the text.

The group of *flat-interfluve (plakor) phytocoenoses* and *facies* was distinguished among them accorging to Vysotsky's conception (1909, 1960). They usually occupy eluvial sites and reflect the zonal-regional background of a given territory. The facial groups of other geotopes have been considered as *exrazonal* categories (Milkov 1981) as representatives of other zonal types of geographic environment, not only neighboring but also rather distant.

The examination of catenary organization geo(eco-)systems permits going deeper into the natural laws of their connections with relief, climate and historical factors (Tishkov 2005). The meso-site types of the regional level have the same system of landscape couplings (Polynov 1956). The methodical construction worked out in *Empirical Simulation of Regional Bioclimatic Trend by Local Ecosystems* (Kolomyts 2008) is based on the demonstrated landscape couplings of the topological level.

At the same time, certain *forms of microrelief* (minor surface landform) were distinguished – by the extent of their dissipating or, on the contrary, accumulating influence on water and ground masses. This catenary system includes six *geotopes* (Figure 2.1, *B*) – forms of micro-relief: (1)  – crests with even and bulging longitudinal profile; (2) 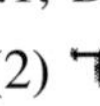– terrace ledges with even diametrical profile; (3)  – flat parts of slopes; (4)  – bottoms of hollow and river valleys; (5)  – foot of terrace and crests; and (6) – half-close arm-chair-shape accumulative forms. The first three geotopes correspond to eluvial and transeluvial types of sites; other correspond to types TA, A and Saq.

2.2. Parameters of Plant Matter Cycles

During the modeling, ahigh emphasis was placed on the phytocoenotic functional block – the parameters of annual and perennial cycles of plant matter turnover, which are especially important for prognostic ecological investigation. Such turnovers are the most important indicators of ecosystem resistance to external disturbance (Odum 1971; Gerasimov 1985). The multidimensional ecological analysis of various phytomass categories (live and dead, foliage and skeletal, on separate biogeohorizons, etc.) was performed and the integral indices of mass exchange were determined. Let us enumerate the measured and calculated discrete parameters of the biological cycle, with indication of their designations.

A. Phytomass (tons/ha):

(1) wood mass, *BW;*
(2) skeletal tree–shrub phytomass, *BS;*
(3) total tree–shrub phytomass, *BB;*
(4) total live above-ground phytomass, *BL;*
(5) total green mass, *BV;*
(6) total live phytomass (above- and underground), *BC;*

B. Productivity (tons/ha per year):

(7) total annual production of forest phytocoenosis, *PC;*
(8) annual growth of skeletal phytomass, *PS;*
(9) annual production of green mass, *PV*;
(10) annual production of grass stand (*PG*).

C. Complex indices of biological cycle:

(11) productivity to phytomass ratio, $CE = PC/BC$;
(12) allometric coefficient, $CA = log(PC)/log(BC)$;

(13) coefficient of annual phytomass turnover (rotation) – the ratio of green mass production to the total above ground live phytomass, *CR = PV/BL*;

(14) coefficient of annual destruction (known as the litter-fall index as well) – the ratio of a forest litter mass (*ML*) to a mass of green substance produced annually, *CY = ML / PV*, years;

(15) coefficient of utilization of dead above-ground phytomass – the ratio of the difference between the two total above-ground phytomasses, live and dead, to the total live above-ground mass, *CU = (BL – BD)/BL*.

The first four complex indices have already been known, and we have proposed the rest of them.

Factual material was collected at each point (test site), processed, and subsequent calculations were made in accordance with the known methodical regulations.

It is necessary to elucidate the complex indices of short-term biological cycle. The *CE* parameter is a measure of efficiency of plant matter in sustainable operation of the entire ecosystem. Therefore, we call it an *index of ecological efficiency (IEE) of autotrophic biogenesis*, according to Eu. Odum (1971). The coefficient *CR* indicates the portion of live above-ground phytomass involved in its annual turnover. The coefficient *CY*is also known as the litter–fall index (Bazilevich and Rodin 1971), which is expressed as a period (in years) necessary for accumulation of the given mass of forest litter (dead plants) at the existing annual rates of plant mortality and organic matter decomposition. The *PC*parameter characterizes the balance between the ratio of creative and reduced branches of the plant matter perennial cycle during the succession of several generations of stands.

2.3. Landscape-Ecological Modeling

The problems of predictive dynamics of forest ecosystems under changing climate conditions have not yet been sufficiently developed. The known simulation models of forest responses to climatic impacts (Bonan et al. 1992; Holten et al. 1993; Tarko 2005; etc.) meet the preset narrow framework of habitat conditions and, therefore, the results of such modeling are insufficient for prediction of the state of forest community as a whole and do not cover the spatial diversity of succession changes under the same background influence. The developed local landscape-ecological predictionis based on the construction of *discrete empirical-statistical models* of natural ecosystems (Rozenberg 1984; Kolomyts 1995). These models make it possible to operate with a relatively small number of the most informative characteristics and to obtain results probably not as accurateas those of simulation modeling but with a higher (geographical) level of territorial generalizations.

The landscape-ecological analysis was performed using a series of empirical models: (a) informational, serving as a basis for the regional and local banks of ecological niches of geo(eco)systems, which characterizes their parametric space, and (b) fuzzy set theory, describing the polysystem unit of natural–territorial organization by operations with ecological niches as descriptive vectors. The complex interconnections were revealed and the territorial ecological space of regional and local geo(eco)systems was described on the basis of information-statistical and fuzzysettheory-based models.

2.4. Monosystem Models

2.4.1. Informational Measures of Component Interconnection

The information theory method is used on the preliminary (prognosis-preceding) stage of predictive analysis. The methods of theory of informationare known (Brillouin 1956; Khilmi 1966; Puzachenko, Moshkin 1969) to be the most adequate for the principles of organization of geo(eco-)systems and have proved their efficiency, especially in the study of intercomponent landscape connections (Preobrazhensky 1969). The information-statistical analysis of empirical material was performed by the methods described (Ashby 1956; Kustler 1957; A.D. Armand 1975; Puzachenko, Skulkin 1981). These methods were employed to uncover thelandscape component interconnection over a territory considered as an indivisible system.

The main parameters of these connections are as follows: (1) the *normalized coefficient of interrelationC(A;B)* of two attributes taken in pairs: the object under considereation – phenomenon *A* (*X*), and the influencing factor *B*(*Y*); and (2) the *partial coefficient of connection* $C(a_i / b_j) \equiv C(x_i / y_j)$ as the means of "information code disclosure" determining the degree of spatial connection between separate gradations (states) of a phenomenon (x_i) and a factor (y_i). These gradations are in the form of elements of the following sets: $A = \{a_1, a_2, \ldots, a_i, \ldots, a_n\}$ and $B = \{b_1, b_2, \ldots, b_j, \ldots, b_m\}$. Obviously, each phenomenon and each factor has a certain set of states.

The first parameter indicates the degree of general closeness of spatial interrelations between the attributes of the states of forest ecosystems. The second parameter allowed us to reveal partial connections in the factor–phenomenon binary system and thereby to define particular kinds of ecological niches for each value (gradation) of the phenomenon *A* in the space of values of factor *B*. In the initial working matrix, the vector of states (gradations) of the factor *B* forms lines and the vector of states ofthe phenomenon *A* forms columns. The matrix cells contain the probabilities, or frequencies, $p_{ij}$of co-occurrence (probability) of the given states of *A* and *B*. The sums by the columns and lines yield a priori (independent) probabilities $p(a_i)$ and $p(b_j)$, respectively; they are used to calculate the values of conditional probabilities $p(a_i /b_j)$. At the same time, a priori probabilities $p(a_i)$ characterize the phenomenon *A* under the assumption of its absolute independence of the factor *B*, when $p(a_i / b_j) = p(a_i)$. The measure of difference between the conditional and a priori distribution of *A* is the function $I(A/b_j)$ equal to:

$$I(A/b_j) = H(A) - H(A/b_j), \tag{2.1}$$

where

$$\mathrm{H(A)} = -\sum_{i=1}^{N} p(a_i) \log_2 p(a_i)\,, \tag{2.2}$$

$$\mathrm{H(A/b_j)} = -\sum_{i=1}^{N} p(a_i/b_j) \log_2 p(a_i/b_j) \tag{2.3}$$

Parameter $I(A/b_j)$ is the measure of force of the general effect of this factor on the phenomenon. The higher is the $I(A/b_j)$ value, the more rigid is their connection. Parameters $H(A/b_j)$ and $I(A/b_j)$ are calculated for each line of the working matrix. The total amount of information, which is transferred from factor to phenomenon and characterizes the closeness of their connection, is equal to:

$$T(AB) = \sum_{j=1}^{m} p(A/b_i)\, I(A/b_i) \qquad (2.4)$$

The amount of information is estimated in binary units (bits); however, it is more convenient to use the relative measures of connection expressed in the quotas of 1. In the present study, *the normalized coefficient of interrelation* (alsoreferred to as the coefficient of maximum likelihood) has been used:

$$C(A;B) = \frac{2^{T(AB)} - 1}{2^{H(min A,\, B)} - 1} \qquad (2.5)$$

Here, $2^{T(AB)}$ is the number of general states of A and B, while $H(A)$ and $H(B)$ are the common measure of diversity of characters A and B, respectively. Thecoefficient ofthe receipt of information$C(A/B)$ has itself certain meaning (A.D. Armand 1975):

$$C(A/B) = \frac{T(AB)}{H(A)} \qquad (2.6)$$

The $C(A;B)$ parameter was a basis for constructing the generalized information-statistical models of landscape-ecological connections in regional and local ecosystems (see below, Figures 6.2, 6.3, etc.).

The reliability of analytic results significantly depends on the proper use of mathematical apparatus in the calculations. In accordance with the requirements of information analysis, the statistical excerpt of a limited number of variables (in this case, less than 100 points) can be divided to no more than 5–6 classes (gradations). Our verification by criterion χ^2 showed that parameter $C(A;B) < 0.07$ often seems to be inexact (interrelation of attributes below the 90% signification level) even at such a comparatively coarse quantization. Therefore, all connections with the coefficients below the specified threshold were excluded from the analysis.

2.4.2. Modeled Distribution of Ecological Niches

The second parameter (the partial coefficient of connection) made it possible to reveal the informational structure of relations in the binary "factor–phenomenon" system and thus to find some kind of*ecological niches* (inthis case, *hydrothermal niches*) of each value (gradation) a_i of the phenomenon A in the space of values$b_j$of the factor B. In ecology, it is quite permissible to apply two of the three known (Odum 1971; Begon et al. 1986) aspects of the concept of ecological niche: (1) spatial, close to the concept of siteinhabitancy, and (2) resource determined by position of an object in relation to the gradients of a particular substance/energy factor (temperature, moisture content, physicochemical properties of soil, etc.). Taken together, both these aspects characterize the niche as a certain space of vital environmental factors. We may accept the concept of ecological niche as "... a certain space of vitally important environmental factors, where the population of a certain species can exist" (Svirezhev 1982, 37).

Nevertheless, the term niche is applicable not only to individual plants or animals but also to entire biotic communities, ecosystems, and their selected structure/function attributes (Ashby 1956). According to R.H. Whittaker (1975), we may generalize the notion and space of a niche, but the principles of the concept remain the same. The essence of such generalization in case of plant formations considered below is that attention is focused on the populations of coenosis-forming edificator species (after which the formations are named) rather than on the entire variety of constituent plant species.

Each niche is presented in the form of a row (descriptive) vector – a "fuzzy" set of the states of phenomenon *A* caused by the given factor *B*. The row vector consists of standardized partial association coefficients:

$$C(a_i / b_j) = \frac{p(a_i / b_j)}{p(a_i)} \tag{2.7}$$

It is agreed that the association between$a_i$and b_i is significant at $C(a_i / b_j) > 1$. The higher the probability, the greater the coefficient; therefore, the components of row vector $C(a_i / b_j)$are considered as "weighting" coefficients.

In the matrix of significant (>1) partial coefficients of connection, the phenomenon and the factor are graded by columns and lines, respectively. Each column vector describes an ecological niche of certain gradation of the phenomenon in the space of values of the given factor. Factor gradations with the maximum values of coupling coefficient form a certain region of phenomenon dominance (hereinafter, *ecological dominant*), while other gradations belong to the "fuzzy" part of the niche. According to the statistical conception of ecosystem stability (Puzachenko 1998), the most probable value of ecological niche of the given gradation is considered to be most stable, corresponding to the equilibrium process of variation of the state of phenomenon by the gradient of factor states. However, it is feasible only under condition that distribution of the states of phenomenon within a niche does not differ from normal distribution, which in fact occurs rather rarely (see below). In most cases, ecological dominants cannot be considered as the most stable states of objects.

If define the ecological domination in each vector-column of a niche by symbol "•" and "fuzzy" links of the niche by symbol "×" (or "+"), then we will obtain a formalized matrix of climatic niche. In this form, the matrix clearly demonstrates the character and closeness of connections (see below Tables 6-3, 6-5, etc.). In this case, the connection is close to the linear one, with the maximum intersection of the ecological niches of landscapes in the range of relatively low summer temperatures typical of the middle and southern taiga. According to Armand and Vedyushkin (1989), it means that the trigger properties and competitive relations of zonal types of vegetation must be maximally manifested in their association with thermal resources of the vegetation period at the northern wing of the boreal ecotone.

The series of ecological niches by each phenomenon describe *a phase space* of the factor–phenomenon system (Ashby 1956). It is accepted as well that each value $C(x_i/y_j)$ is a certain equivalent of probability of the landscape-zonal kind or biogeocoenosis group *i* by the given graduation *j* of a geophysical parameter.

Thevector graphs of *A* and *B* interrelations were constructed on the basis of the matrixes of ecological niches (see below Figure 5.8). It is a graphical expression of a binary ordination system. The matrixes of ecological niches were also used to obtain the *taxonomic base norms* of the above hydrothermal parameters: the values average weighted by territory. The central values of all gradations for the given geophysical parameter and the matrixes of normalized (by columns) partial association coefficient were used for calculating taxonomic norms.

The main pattern of binary ordination in the phase space is given by a curve plotted through the ecological dominants of a phenomenon (see below Figures 5.8–5.10, etc). Additional (dotted) lines connect the boundaries of ecological niches, distinctly representing the volume (width) of niches (as a number of realized positions of column vector) in the space of changes in factor states. Obviously, the broader niches indicate that the given state of phenomenon is more tolerant to the variation of factor states and, in this sense, more stable. On the contrary, the narrower niches (in the ultimate case consisting of the dominant only, without "fuzzy" parts) indicate strict determination of the given state of phenomenon by a particular factor, i.e., its high sensitivity to the variations of this factor.

The new approaches to substantial interpretation of the results of information-statistical modeling of natural ecosystems were developed. The respective graphs were plotted by the matrix of significant (>1) partial coupling coefficients, and gradations of the phenomenon were arranged in the order of increase or decrease in their ecological dominants by gradations of the given factor. Then the envelope curve was drawn through the dominants, giving a clear insight into the most significant tendency of the relation under study.

Many characteristics of forest ecosystems have not only the fundamental ecological niche for some factor or another (with the respective ecological dominant and the "fuzzy" part of the niche) but also a kind of *enclave* separated from the niche on the graph of partial coefficients by more than one gradation of the factor (see below Figure 5.7). The appearance of breaksin phytocoenotic areals shows the migration of species and entire communities carrying out the evolution of plant cover in general (Vasilyev 1946). Consequently, enclaves are the evidence of continuous migration processes on the given areas of plant cover and the dynamics of the given forest formations. Each of these enclaves can be considered as transgressive localization of the phenomenon under consideration at different values of the given factor but under the effective influence of other factors creating similar conditions for existence of the phenomenon.

Assuming that the dominant part of ecological niche corresponds to the optimal conditions of its formation (Pusachenko 1986), we supposed that the location of each of such optima is a "focus" of spatial distribution (transgression) of the given phenomenon towards the existing enclave. The direction of enclave deviation from the niche was considered as a vector of localized transgression of the phenomenon. The attention was also paid to the cells of binary ordination of the phenomenon with the factor, where $C(x_i/y_j) \leq 1$. These were the regions of *sporadic distribution* of the given phenomenon by the gradient of changes in the factor under consideration, frequently with the emergence of an enclave.

The system of ecological niches was used to calculate the *taxonomic norms* of functional characteristics of the given object. Thereto, the central values of all gradations of the factor were used, as well as the matrix of the normalized partial coupling coefficients of the object introduced into the calculations as "weight" coefficients. The normal value or, in other words, the weighted average value of the functional feature (Ramensky 1971), was determined as a sum of products of central values of all its gradations by the "weight" coefficient corresponding to the given gradation of the phenomenon (object).

The phase space of ecological niches reflects the dynamics of their volume under changing factor states. While moving along the lines of the matrix or along the line of ecological determinants (which is the same), one can follow the variation of "homeostasis reserve" of the phenomenon and, accordingly, its sensitivity to the changes in factor states. Abrupt narrowing or broadening of the niche demonstrates that with each subsequent state the phenomenon changes the character of its response to the variation of factor states, becoming more or, on the contrary, less sensitive to this factor. It appears that irreversible changes occur in the mechanisms of interaction between the components of phenomenon system and its behavior becomes nonlinear (Zeydis et al. 2001). The nonlinear character of transition of the states of phytocoenotic, soil, and landscape complexes as a whole by the arid gradient is typical of their binary relations with the most of abiotic factors under consideration. It adds complicating uncertainty to the landscape-ecological prediction based on the principle of actuality (Puzachenko 1998).

2.5. Polysystem Models

2.5.1. Methods of Calculation

The second aspect of natural-territorial organization describes polysystem models (Preobrazhensky 1969; Syomkin 1973; Andreev 1980; Kolomyts 1983, 2008). Most generally, these models are based on the insight into the similarities and differences between elements of the final set of attributes adequate to some or other state of ecosystem. The models of system states expressed in the measures of ordering and similarity are anyhow discrete. Their advantage is that they can represent the diversity of qualitative properties of initial structural elements of geo(eco-)systems.

Description of natural complex as a composite system provides researcher with a set of attributes usually forming indistinct, "fuzzy" sets. Here, the similarity and difference can be revealed statistically using the "weight" of each attribute. We have used two measures of pairwise ordering of the objects: the measure of similarity (or difference) and the measure of inclusion of one object into another. The binary relations of topoecosystems (*A*, *B*, *C*, etc.) were estimated by the set of their functioning parameters acting as their elements. If there are two ecosystems *A* (A_1) and *B* (A_2) consisting of the same number *n* of elements a_i and b_i, respectively, which have certain numerical values, then the measure of similarity $N(a_i,b_i)$ of these elements taken by pairs, as well as the inclusion measures $J(a_i;b_i)$ and $J(b_i;\ a_i)$ of element a_i into b_i and vice versa, can be found from the ratios:

$$N(a_i,b_i)= \frac{2(a_i \cap b_i)}{a_i + b_i};\ J(a_i;b_i)= \frac{a_i \cap b_i}{a_i};\ J(b_i;\ a_i) = \frac{a_i \cap b_i}{b_i}. \tag{2.8}$$

Here, $a_i \cap b_i$ is intersection of the values of elements, determined as a minimal value of two elements a_i and b_i. The measures of difference, or distance, between the elements is $D = 1 - N$. In the general case, $J(a_i; b_i) \neq J(b_i; a_i)$. If both of these measures exceed a certain threshold (critical) value J_{crit} of the inclusion measure (a certain "banality" threshold Z_{crit}), then it is possible to speak about the tolerance of elements (objects), i.e., about their relative similarity at a given J_{crit} (Andreev, 1980); we believe that $Z_{crit} \equiv J_{crit}$. For each pair *A* and *B* of the vectors of ecological niches or any other descriptive sets, each containing *n* elements (quantitative attributes), the above measures of binary relations are calculated by the formulas (Syomkin, Komarova 1977):

$$N(a_1, a_2) = \frac{2\sum_{j=1}^{N}\sum_{k=1}^{S_j} \min\left\{P_j^k(a_1), P_j^k(a_2)\right\}}{\sum_{j=1}^{N}\sum_{k=1}^{S_j} P_j^k(a_1) + \sum_{j=1}^{N}\sum_{k=1}^{S_j} P_j^k(a_2)}; \quad (2.9)$$

$$J_1(a_1; a_2) = \frac{\sum_{j=1}^{N}\sum_{k=1}^{S_j} \min\left\{P_j^k(a_1), P_j^k(a_2)\right\}}{\sum_{j=1}^{N}\sum_{k=1}^{S_j} P_j^k(a_1)}; \quad (2.10)$$

$$J_2(a_2; a_1) = \frac{\sum_{j=1}^{N}\sum_{k=1}^{S_j} \min\left\{P_j^k(a_1), P_j^k(a_2)\right\}}{\sum_{j=1}^{N}\sum_{k=1}^{S_j} P_j^k(a_2)} \quad (2.11)$$

Here, $P_{k,j}(a_j)$ is the "weight"*j*, the *k*-th attribute at a_i element; *r*is the common number of attributes, each of then having S_j values.

The respective graphs of similarity measures and orientation graphs (or-graphs) of inclusion measures for the objects under consideration are plotted by matrices *N(A,B)* and *J(A;B)*. As concerns the matrices of inclusion, the line and column indices of landscapes denote elements a_i and b_j, respectively, in parameters *J(A;B)* or, on the contrary, elements b_j and a_i in parameters *J(B;A)*. In both cases, the arrow under the matrix shows the direction of inclusion (Table 2.1).

2.5.2. Matrices and Graphs of Inclusion Relations between the Objects

Set-theoretical modeling of relatively homogenous natural-territorial complexes based on the known methodical guides (Syomkin 1973; Andreev 1980) will be demonstrated by the example of 7 groups of abstract forest phytocoenoses (Table 2.1). Let us will believe that the differences between inclusion measures (like between similarity measures) below the level of

0.89 are insignificant. However, stricter analysis requires introduction of a certain threshold of indistinguishability. All values of inclusion measures above and below the preset threshold will be denoted by 1 and 0, respectively. At the same time, 1 and 0 will denote the presence and absence of an arrow in the or-graph of inclusion (or a line in the graph of similarity), respectively.

Table 2.1. The measures of inclusion between 7 groups of forest phytocoenoses

	1	2	3	4	5	6	7
1	–	0.90	0.90	0.90	0.88	0.85	0.69
2	0.91	–	0.91	0.87	0.88	0.85	0.77
3	0.85	0.85	–	0.84	0.83	0.80	0.72
4	0.91	0.87	0.90	–	0.86	0.84	0.72
5	0.94	0.93	0.93	0.90	–	0.94	0.78
6	0.92	0.91	0.92	0.90	0.96	–	0.82
7	0.79	0.88	0.89	0.83	0.85	0.87	–

——► Direction of the inclusion.

The above mentioned critical threshold of indistinguishability (Z_{crit}), when the or-graph (or the graph for similarity measures) is still connected, i.e., each point (node of the graph) has at least one connection (edge of the graph) with some other point, is obviously of the greatest interest. In this case, $Z_{crit} = 0.89$.

The degree of exoticism–banality of landscape kinds, i.e., the levels of their structural-functional development by the given set of factors, is revealed by the following operations on the matrix of inclusion relations. Using the found value of Z_{crit}, we will convert this matrix (see Table 2.1) to a new contiguity matrix *M,* i.e., into the matrix of direct connections, where 1 and 0 stand for the initial measures:

Matrix M

	1	2	3	4	5	6	7
1	–	1	1	1	0	0	0
2	1	–	1	0	0	0	0
3	0	0	–	0	0	0	0
4	1	0	1	–	0	0	0
5	1	1	1	1	–	1	0
6	1	1	1	1	1	–	0
7	0	0	1	0	0	0	–

Then we will transpose the matrix *M* into a matrix M_t (Sumner 1978). Then, after element-wise addition of the matrices *M* and M_t we will obtain the matrix $M + M_t$:

M MM^T M+M^{T,}

	1	2	3	4	5	6	7
1	–	1	1	1	0	0	0
2	1	–	1	0	0	0	0
3	0	0	–	0	0	0	0
4	1	0	1	–	0	0	0
5	1	1	1	1	–	1	0
6	1	1	1	1	1	–	0
7	0	0	1	0	0	0	–

\+

1	2	3	4	5	6	7
–	1	0	1	1	1	0
1	–	0	0	1	1	0
1	1	–	1	1	1	1
1	0	0	–	1	1	0
0	0	0	0	–	1	0
0	0	0	0	1	–	0
0	0	0	0	0	0	–

=

1	2	3	4	5	6	7
–	2	1	2	1	1	0
2	–	1	0	1	1	0
1	1	–	1	1	1	1
2	0	1	–	0	0	0
1	1	1	1	–	2	0
1	1	1	1	2	–	0
0	0	1	0	0	0	–

The ($M + M_t$) matrix shows strongly connected groups; such connections are indicated by twos. In this case, there is one group of strongly connected matrix elements: $U_1 = \{1,2,4\}$ and $U_2 = \{5,6\}$. They have tolerance connections (mutual inclusion) and form the following row of inclusion relations: 5–6, 1–2 and 1–4. Thus, these phytocoenological groups show the maximum similarity.

Then, on removing *U* elements from the initial matrix *M* we will obtain a reduced matrix M'. After analogous operations of matrix transposition and addition, we will obtain:

M'

	3	7
3	–	0
7	1	–

\+ M^T'

3	7
–	1
0	–

= M'+M^T'

3	7
–	1
1	–

The ($M' + M_t'$) matrix elements equal to 1 denote weakly connected objects. In this case, there is also a single group of such elements: $m = \{3, 7\}$. They are maximally different and make the following row of inclusion relations (with the signs of tolerance).

Finally, we return to the initial matrix *M* to distinguish in the latter the groups of strongly (*U*) and weakly (*m*) connected elements, arranging them in a certain manner:

M

	1	2	4	5	6	3	7
1	–	1	1	0	0	1	0
2	1	–	0	0	0	1	0
4	1	0	–	0	0	1	0
5	1	1	1	–	1	1	0
6	1	1	1	1	–	1	0
3	0	0	0	0	0	–	0
7	0	0	0	0	0	1	–

Our matrix proved to have 16 such connections in the following pairs of landscape kinds: 1–3, 2–3, 4–3, 5–1, 5–2, 5–3, 5–4, 6–1, 6–2, 6–3 and 6–4. As we can see, such connections are predominant and demonstrate clearly marked differentiation of ecological niches.

The complete or-graph of inclusion relations for phytocoenological groups under study was plotted on the basis of operations with the initial matrix (Figure 2-2). It isa*structure scheme of domination* of given objects in their territorial interrelations. The oriented edges of the graph (the arrows) show the direction of inclusion. As one can see, the inclusion relations are generally intransitive: e.g., 4 → 3 and 7 → 3; however, 4 −/→ 7.

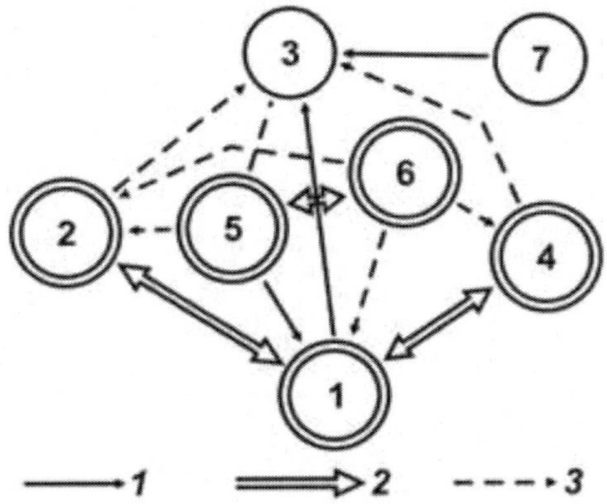

Figure 2.2. Structure scheme of domination of geo(eco-)system groups. Ecological groups on the or-graph of inclusion relations. Groups: 1 – strongly connected; 2 – ordinary; 3 – weakly connected. Connections: 4 – tolerance (mutual inclusion); 5 – critical; 6 – weak.

The matrices of inclusion measures of landscape kinds and the respective or-graphs are considered as initial set-theoretical models of ecological spaces of floristic phratries, groups of phytocoenoses and groups of microlandscapes by some or other set of input factors. The matrices and graphs characterize certain *partial structures of these topological geospaces* relative to particular landscape-forming factors. These partial structures will be objects of further landscape-ecological analysis.

2.5.3. Distinguishing of Ecological Groups of Geo(Eco-)Systems

The *N(A,B)* measure of similarity between two objects characterizes essentially *the degree of their isomorphism* by a given factor. At the same time, their inclusion measures *J(A;B)* allow the ordering of geo(eco)systems into definite rows, or some *ecotypes* (Odum 1971; Bykov 1973). The ecotypes of geo-ecological units in the system of their factor ordination characterize the level of structural-functional development and the level of efficiency of using the resources of the abiotic environment. At *J(A;B)* > *J(B;A),* one may state that geo(eco)system *A* is more widely distributed in the space of a given factor and more developed to its changes compared to system *B*; at *J(B;A)>J(A;B)*, on the contrary, it proves to be more pessimal and, accordingly, less stable. Thus, the case in point is the phenomenon of homomorphism in the binary relations between the objects. Object *A* is homomorphically represented by object *B* in the former case and vice versa in the latter case.

The ecotypes, or *ecological groups,* of landscape faces are revealed by means of simple operations on the matrices of inclusion measures (Andreev 1980; Kolomyts 1998) with obtaining re-sorted single matrices of adjacency of objects. These matrices are used for creation of structural schemes of domination of geo(eco)systems at some or other thresholds of banality Z and for determination of critical values Z_{crit} of this threshold (see below).

The system of inclusion relations comprises strongly and weakly coupled objects and critical couplings, disappearance of which leads to disintegration of the entire or-graph of inclusion relations. The whole set of objects was divided into four ecological groups:

1) the dominant objects with the maximum number of ingoing arrows on the or-graphs, i.e., having the maximally stable ecological niches most similar to the niches of all other objects; according to Odum (1971), this ecological group can be referred to as *eurytopic*;
2) the subdominant objects with the average number of ingoing arrows (often in the presence of outgoing arrows) showing enhanced stability, i.e., the *subeurytopic* group;
3) the temporal(ordinary) objects fixed with a proportional number of ingoing and outgoing arrows on the or-graph;
4) and (5) the subreduced (*substenotopic*) and reduced, or "banal" (*stenotopic*), objects with the maximum number of outgoing arrows and single or almost no ingoing arrows; they are at the lowest level of the hierarchy of inclusion relations and characterized by narrow (reduced) ecological niches, which gives them the minimum stability. This row characterizes essentially different *levels of ecosystem functioning*.

One should mention the meaningof the critical threshold of banality Z_{crit}. This threshold denotes the lower limit of the measures of inclusion of system elements, when the structural and functional integrity of the system still persists. Lower Z_{crit} values must denote the relatively high degree of inter-complex interactions in a given ecological space. It provides the territorial integrity of the latter even at relatively low values of inclusion measures. Here, one may expect the greatest diversity of spatial couplings with predominance of critical couplings and presence of *tolerance couplings* forming the typological "cores" of ecological space.

At a relatively high banality threshold, one should speak about the weaker total degree of inter-complex interactions, in spite of significant mutual overlapping of the ecological niches of system elements. The system preserves its structural-functional integrity only under rather substantial crossing of these niches, which is manifested by the high threshold of Z_{crit} values. Weak connections are predominant in the structural scheme, in the absence of strong and insignificant number of critical couplings. Ecological space seems to be disintegrated into separate structural elements, indicating the total low level of system organization.

By distinguishing the ecological groups of landscapes, biogeocoenoses, or separate geocomponents based on the measure of their significance in the given ecological space, we simultaneously determine the fields of their *homeostasis:* the state of dynamic equilibrium, potential stability under some or other variations of parameters of this space. The value of significance of some or other geo(eco)system characterizes, in essence, the "reserve of homeostaticity," within which a natural complex, in accordance with the Le Chatelier's principle, may vary in its phytocoenotic structure while preserving the qualitative distinctness and the level of the minor biological cycle. Stability expressed by reserve ofhomeostaticity of geo(eco)system will be furtherreferred to as *potential stability* of geo(eco)system. The total reserve of homeostaticity as a measure of potential stability of geo(eco)system must enhance under the increase in volume of its ecological niche.

2.6. Integral Models of Polysystemic Organization

The combination of information and set-theoretical methods of analysis makes it possible to obtain complex discrete models of the state of geographical space. Let us present the scheme of further calculation procedure by the example of elemental landforms (geotopes). More than 20 structural and functional indicators of the state of the respective forest biogeocoenoses were taken for each of them. They included: (a) the characteristics of "structural metabolism," according to Odum (1971), i.e., soil morphological and phytocoenotic structural units, as well as phytomass reserves; (b) the parameters of functioning of plant communities (green mass productivity); and (c) the hydrothermal indicators of the state of edaphotopes. The following three complex parameters were determined for each geotope based on the matrices of similarity and inclusion relations between its biogeocoenosis and the coenoses of other geotopes: (1) the measures of biogeocoenosis isomorphism (invariance) I_k; (2) the metabolic index of biogeocoenosis $S^m{}_k$ characterizing the general level of its structural and functional development; and (3) the significance index of biogeocoenosis in the given altitudinal-zonal geospace G_k representing the reserve of its homeostaticity (elastic resistance). The ideology and methods for calculating the above parameters are given below.

Ecological objects belong to an extensive group of natural systems with relatively independent behavior of their components; therefore, geo(eco)systems are rather weakly connected natural formations (Armand 1988), which is one of the basic conditions for manifestation of polystructural properties, or polymorphism in a broader sense (Kolomyts et al. 2012).

Polymorphism is expressed, in particular, by the relative mutual independence (autonomy) of structural and functional characteristics of natural ecosystems. Under varying external conditions, geo(eco)system first of all changes its structure in order to maintain the previous level of functioning, including the level of bioproduction as its most important invariant property (Timofeeff-Ressovsky–Tyuryukanov 1966; Sochava 1978; Svirezhev 1982; Kerzhentsev 1999).

In the cases when two ecosystems under consideration are similar in their structural and/or functional characteristics, we can say that they are *isomorphic* to each other. The concept of isomorphism of two systems suggests interchangeability of their elements, providing the similarity of their structure and function (Paffengolts 1973). If an ecosystem shows isomorphism with respect to quite a number of other systems, it may be considered to possess certain *invariance* of its properties in a given geoecological space.

This property of polymorphism allows the information theory method to be used for calculations of landscape component interconnection on the territory, which is considered as any indivisible system (Andreev 1980). The main parameter of these connections is the normalized coefficient of interrelation $C(A;B)$ between the two attributes taken in pairs. One of them is considered as phenomenon A and the other is considered as factor B. Obviously, each phenomenon and each factor has a certain multitude of states, and parameter $C(A;B)$ is a measure of diversity for these space couplings. Parameter $C(A;B)$ expresses also the diversity of microstates of object A (a_1, a_2, a_3 …) or object B (b_1, b_2, b_3, …) themselves.

We have used the matrices of similarity measures to calculate the measure of structural-functional isomorphism (invariance) of biogeocoenoses I_k. For each k geotope with the respective biogeocoenosis, the Shannon information measure I_k was defined as:

$$I_k = H_k / H_{max}, \tag{2.12}$$

where

$$H_k = -\sum_{i=1}^{n} N(k,i)\cdot \log_2 N(k,i);\ H_{max} = \log_2 N. \tag{2.13}$$

Here, $N(k,i)$ is the measure of similarity between the i and k geotopes; n and N are the number of connections of the k geotope and the number of geotopes, respectively. In our case, $N = 6$ and $n = 5$. Geotope may be maximally invariant if its elemental ecosystem is maximally similar to other systems of the same rank and, in this sense, maximally isomorphic to them.

The I_k parameter is used for comparative assessment of the degree of isomorphism between the systems with different number of differentiated gradations. It characterizes the *variety of similarity* between biogeocoenoses of the given geotope and all biogeocoenoses of other geotopes by a different set of their structural and functional characters. The more diverse is such a set, the higher is the degree of complexity of similarity relations, i.e., this geotope can be to a greater extent considered as a structural and functional invariant of the given altitudinal zone. Invariant geotope is not necessarily a geotope with the most developed biogeocoenosis. Invariance of a geotope characterizes the degree of structural-functional variety of similarity between its biogeocoenosis and the coenoses of all other geotopes in the ecoregion of interest.

The high I_k values indicate the most equal levels of similarity between this object and all other objects, which are close to a certain optimal value of similarity ($N(k,i) \rightarrow N(k)opt \approx$ 0.38, when the product $N(k)\cdot \log_2 N(k)$ is found to be maximal (0.5307 bit). If this object has a very high similarity (0.90–0.95 and more) with some objects and, on the contrary, low similarity ($\leq$ 0.65-0.70) with other objects, then the final measure of variety of similarity will be minimal ($\sum [N(k,i)\cdot \log_2 N(k,i)] \rightarrow$ min).

In addition to geotopes, the analogous measures of a variety of similarity were calculated also for phytocoenotic groups per se, irrespective of their location. In this case, the degree of invariance of each k meadow community I_k by the set of its structural-functional characteristics was considered.

Two other complex parameters of the state of biogeocoenoses were determined using the initial matrices and/or orgraphs of inclusion relations. Here, we will also give necessary explanations.

Both parameters are calculated on the basis of the structural scheme of geospace dominance. For each altitudinal-zonal geo(eco)system, it is necessary first of all to determine the degree of structural-functional development of its components: the groups of biogeocoenoses or phytocoenotic groups. Thereby, it is possible to assess the general level of organization of an altitudinal-zonal system, as well as the field of its spatial biogeo- or phytocoenotic contrasts. As it follows from the factorial-based dynamic analysis (Kolomyts 2015, b) , the degree of development of a mountain forest biogeocoenosis is greater in case of the higher soil temperature in the vegetative season (at a rather high humidity of edaphotope), the larger herbage projective cover and average height (as well as floristic and phytocoenotic diversity), the thicker organo-mineral soil layer with a lower content of lithomass and, finally, the larger total and fractional reserves of living phytomass and the higher productivity of green mass. Eventually, the general structural and functional state of each biogeocoenosis is expressed in a certain value of its metabolic index S_k. On the structural schemes of altitudinal-zonal geospaces, the resultant pattern of the level of biogeocoenosis organization is fixed by the edges of orgraphs of inclusion relations. The connection arrows are directed from less developed local ecosystems towards more developed ones.

The number of input (n_i) and output (n_j) connections was determined for each geotope (or each phytocoenotic group) on the structural schemes of dominance. The total number of connections in the system (N) was also calculated. Based on the above, the metabolic index S_k of the biogeocoenotic group of the *k* geotope, or the *k* group of phytocoenosis, was found from a simple ratio:

$$S^m{}_k = P_i - P_j + 1, \tag{2.14}$$

where $P_i = n_i/N$, $P_j = n_j/N$. Obviously, $0 \leq S^m{}_k \leq 2$.

The value of parameter $S^m{}_k$ may be considered as a generalized structural-functional index of the above *ecological groups* of phytocoenoses on some geotopes or other. Hence, it is possible to differentiate certain *metabolic ecotypes of biogeocoenoses*. It is known (Bykov 1973; Odum 1983) that the concept of ecotypes is used to single out the ecosystems adapted to local soil-climatic conditions. At $S^m{}_k \gg 1$, the comparative level of structural-functional development of a biogeocoenotic group at the considered geotope is sufficiently high. Such group falls into the category of *subdominant* and can be named as a *submacrobolic ecotype*. If $n_j = 0$, it becomes a *dominant,* i.e., *a macrobolite*. As the metabolic index decreases, the group of biogeocoenoses first becomes a *rank-and-file member* of the altitudinal-zonal set of such groups, i.e., a *mesobolic ecotype*. Finally, at $S^m{}_k \ll 1$, meadow ecosystems are characterized by the least developed structure and the low intensity of functioning; hence, they are classified into the categories of *reduced ecotypes: submicrobolites* and *microbolites*. In the latter case, the respective geotope, or a phytocoenotic group, has no input connections on the structural scheme of dominance ($n_i = 0$).

Finally, the third complex parameter G_k of polysystemic organization of natural complexes characterizes the *significance of each geotope in representation of the complete diversity of the states of input ecological factorsby its biogeocoenosis* (i.e., absolute height, slope exposure and steepness, hydrothermal conditions). The value of significance was used for the ranking (a kind of classification) of topoecosystems, i.e., determination of their position in the structure of a given altitudinal-zonal geoecological space.

The G_k parameter represents one of the aspects of geo(eco)system resistance to external perturbations. According to the most common definition (Ricklefs 1976), system resistance is a ratio between variability in the environment and variability in the system per se. Here, the matter in question is the ability of the system to withstand environmental perturbations or to rather quickly neutralize the consequences of disturbances in its structure and function caused by this perturbation. The so-called *resistant* and *elastic stability* is demonstrated in the former and latter case, respectively (Odum 1983; Thompson et al. 2009). When we consider the response of the geo(eco)system to the external hydrothermal signals that have disturbed it from the previous dynamic equilibrium, we are concerned first of all with resistant stability of this system.

When determining the significance of one or another biogeocoenosis in a given altitudinal-zonal geospace, we estimate the "reserve of its homeostaticity" (Armand 1989), within which an elemental natural complex can vary in its phytocoenotic structure while preserving its qualitative definiteness and the level of the small biological cycle in accordance with the Le Chatelier principle (Gorshkov 1994).

The Shannon function of diversity was used to assess the measure of significance G_k. The measure of significance of the *k* topoecosystem in a given altitudinal-zonal space was calculated on the basis of probabilities of input and output connections on its matrices $J_1(A;B)$ and $J_2(B;A)$. At the same time, it was accepted that the input and output connections were mutually independent. It allowed us to use the property of additivity of individual information (Ashby 1956). The general significance of connections in a given biogeocoenosis, which characterizes the diversity of altitudinal-zonal space properties represented in this biogeocoenosis, comprises the negentropy (Brillouin 1956) of in-degrees (incoming arrows on the orgraph) and the entropy of out-degrees (outgoing arrows). The relative reserve of homeostaticity G_k for the group of biogeocoenoses at the *k* geotope was calculated by the formula:

$$G_k = [-P_i \cdot \log_2 P_i + P_j \cdot \log_2 P_j + 1] / [\log_2 N + 1], \tag{2.15}$$

where P_i and Pj are the probabilities (in unit fractions) of input and output connections, respectively, and N is the total number of connections. The G_k parameter is used for direct assessment of the degree of susceptibility of a biogeocoenosis developed at a given geotope to the changing hydrothermal conditions.

The S_k and G_k parameters are closely interrelated. By differentiating the metabolic ecotypes of elemental geo(eco)systems on the basis of the generalized measure of their structural-functional development, we simultaneously determine their homeostatic potential contributing to their stability under certain variations of climatic and geomorphological conditions of the given geospace. The higher level of phyto- or biogeocoenosis organization corresponds to the more intensive biological cycle, which serves as a mechanism of elastic stability neutralizing external perturbations and returning the ecosystem back into the previous state after "elimination" of the influence. In general, the elastic stability of biogeocoenoses under variations ofthe abiotic environment must decrease from their macrobolic ecotypes to microbolic ones.

2.7. Methods of the Cartographic Modeling

The microlandscapes, total productivity and soil temperature parameters were mapped by L.S. Sharaya using the novel methods of geomorphometry (Shary et al. 2002; Sharaya 2007, 2009; Zalihanov et al., 2010), i.e., the science of terrain modeling, which has been most actively developing in the past 30 years. The versatile properties of relief are expressed in more than 20 characteristics calculated from DEM of the Earth surface.

An extended set of geomorphometric variables is presented in the monograph that has not been used in other papers on ecological modeling. This set has permitted to detect statistically significant interrelationships between functional properties of forest ecosystems and environmental factors, including topography. These interrelationships have allowed to reveal new rules of spatial ecosystem organization. Based on these interrelationships, detailed maps of functional and structural ecosystem parameters are calculated for the base and prognostic periods. Computation of the matrices for mapping is based on statistical connections between the parameters of ecosystem state and the topographic attributes. The resulting multiple regression equations contain a number of morphometric variables, which are the leading predictors of spatial differentiation of certain ecosystem parameters.

The large-scale maps of productivity, soil temperature and temperature abnormalities were calculated and constructed using NASA satellite data on the relief: the elevation matrices with a 90-m resolution obtained as a result of a USA–Italy–Germany SRTM project (Shuttle Radar Topography Mission). The *matrix maps* describing in detail the distribution of different characteristics of the Mendeleev landscape were calculated. The image of a matrix map was formed in the GIS program by the appropriate legend, with the procedures of *interpolation and extrapolation of the measured data by the matrices of factors* with the closest statistical relationships revealed by regression analysis.

PART II: ECOLOGY OF ISLAND-ARC VOLCANIC LANDSCAPE

Chapter 3

REGIONAL BIOCLIMATIC SYSTEM OF THE KURIL ISLANDS AND ITS EVOLUTIONARY SIGNIFICANCE

ABSTRACT

It was shown that on the island-arc stage of continental biosphere development in the North-West Pacific local geomorphological conditions created the centers of origin of diverse phytocoenological structures of the higher (zonal-regional) level at the initial stages of continental biosphere development. The causal mechanisms of the general descent of the boundaries of altitudinal climatic zones and the southward shift of natural zones on insular extratropical Neo-Pacific compared to the neighboring continents are discussed. The phenomenal feature of bioclimatic system at the stage of insular landscape formation is the direct subordination of phytobiota to vertical hydrothermal gradients. It forms a system of future altitudinal zonality of the continent already within a framework of low-mountain island-arc landscape.

3.1. CLIMATE-FORMING FACTORS

The regional bioclimatic system is understood as an ordered set of spatial relationships between plant formations, as well as their groups and classes (Sochava 1979), and landscape-geophysical factors. The elements of the system are the climatic niches of phytocoenological units mentioned above, which are distributed along the hydrothermal gradients and representthe spatial (Odum 1975) and resource (Svirezhev 1982) aspects of ecological niches. The regional bioclimatic system imposes the boundary conditions of the formation of phytocoenotic units of the topological level (associations, their groups and classes) under the refracting influence of local geomorphological and edaphic factors (Kolomyts and Surova 2018 b).

The regional bioclimatic system of the Southern Kuril Islands is formed, first, by the warm (the Sea of Okhotsk, Soya) and cold (the Pacific Ocean, Kuril, or Oyashio) sea currents washing them on the northwest and southeast, respectively, and, second, by the monsoon circulation of air masses, with a drastically marked wind rose: northwestern continental in winter and southeastern oceanic in summer. Rather contrasting climates of the Sea-of-

Okhotsk and Pacific coasts, as well as the adjacent slopes of volcanic ridges and separate volcanoes, are formed due to the interference of the above factors.

The geographical positions of the islands become highly important in this context. The Southern Kuril Islands, together with the Hokkaido Island, are situated in the focus of frontal systems of the northwestern part of the Pacific Ocean – in the narrow western part of the northern moderate belt in the Pacific Ocean, where the meridian countermovement of warm and cold waters drastically increases the energy activity of the ocean with enhanced moisture and heat transfer to the atmosphere (Lappo, Lebedev 2005). The main trajectories of tropical cyclones bringing heavy rainfall and increasing the atmospheric humidity factor up to 3.5–4.5 and more pass exactly through this zone (see Figure 3.3 *c* hereinafter).

The regional diversity of climates is determined first of all by air mass circulation. The wind roses of the Kunashir Island (Figure 3.1), as well as most islands of the Kuril Ridge, are characterized by drastic seasonal contrasts, with the dominance of northwestern wind currents in winter and the winds of S–SE–E–NE courses in summer. It was considered as a manifestation of monsoon climate pattern (Lashkov 1948; Korsunskaya 1958). It was believed that the Sea-of-Okhotsk coast of South Kuril Islands differs from the Pacific-Ocean coast in the colder and drier winter (the winter continental monsoon) and warmer summer. In summer, the Pacific coast is more humid and cloudy, with frequent fogs, due to the cold and moist winds of oceanic monsoon. Later on it was shown that the climate on the islands is generally oceanic (with mild winters and cool summers) and the monsoon pattern of changes in the direction of prevailing winds is caused by the seasonal shifts (East ↔ West) of cyclone trajectories (Vitvitsky, 1969). Nevertheless, in the vegetation period, the Sea-of-Okhotsk sector remains warmer and drier than the Pacific one. The "cold" oceanicity, which is unfavorable for the growth of forest vegetation, is still more marked on the islands of the Lesser Kuril Ridge washed by the cold Kuril Current (see above Figure 1.2).

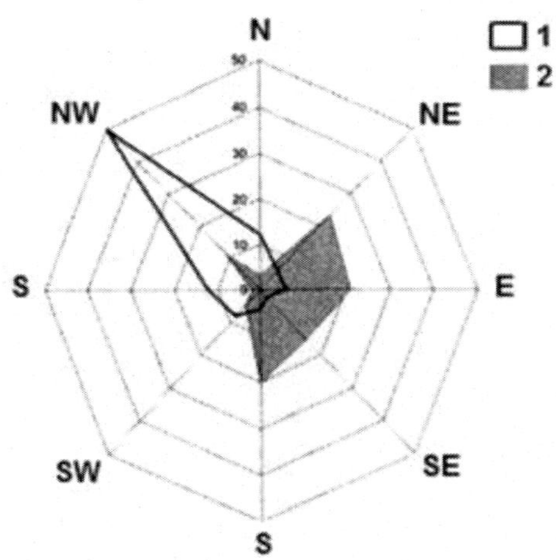

Figure 3.1. The wind rose by the Yuzhno-Kurilsk Station (Kunashir). 1 – January; 2 – July.

The latest cartographic materials (Tunegolovets et al. 2008; Kotlyakov et al. 2009) have made certain adjustments to the climatic situation described. It has been established that, first, the trajectories of the major cyclones in this region pass in the general direction from southwest to northeast and must bring nearly equal amounts of rainfall to both island sectors (Figure 3-2). Second, the southern half of Kunashir is washed by the branch of the warm Soya sea current on both sides. This branch approaches first the northwestern coast of the island and then, after passing along the Kunashir Strait, the southeastern coast. Though, upon the turn to the northeast, the current fades away; therefore, its thermal effect on the Pacific sector of the island must decrease, which is confirmed by the contrasts in the structure and phenological regime of phytobiota between the marine and oceanic sectors. Nevertheless, the effects of marine currents are not so strong (Ivanov 2005, 2014) as to substantially distort the altitudinal-belt (zonal) bioclimatic regularities on volcanoes and volcanic ridges of Southern Kuril Islands.

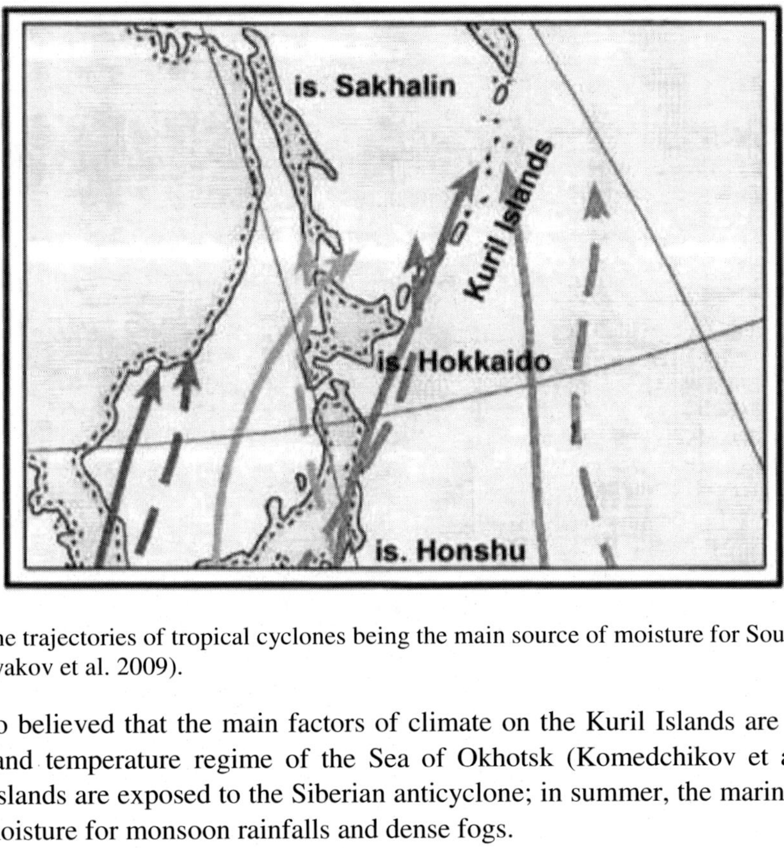

Figure 3.2. The trajectories of tropical cyclones being the main source of moisture for Southern Kuril Islands (Kotlyakov et al. 2009).

It is also believed that the main factors of climate on the Kuril Islands are the air mass circulation and temperature regime of the Sea of Okhotsk (Komedchikov et al. 2012). In winter, the islands are exposed to the Siberian anticyclone; in summer, the marine tropical air brings the moisture for monsoon rainfalls and dense fogs.

3.2. Calculations of Hydrothermal Parameters

The regional hydrometeorological data for South Kuril Islands, as well as the data of our landscape ecological survey, were used in bioclimatic analysis. The distribution of monthly average temperatures of the warmest month (August) t_{VIII} and the annual precipitation r_{year} by absolute height for the Sea-of-Okhotsk and Pacific-Ocean sectors of the Mendeleev Volcano (Figure 3-3) was calculated by L.S. Sharaya on the basis of WorldClim data (Hijmans et al., 2005) using the Analytical GIS Eco Program (Wood et al. 2009). The WorldClim planetary matrices were formed through interpolation of meteorological data. The algorithm of thin-plate smoothing spline was used as implemented in the ANUSPLIN software (Hutchinson 2004). The procedure of interpolation for land involved latitude, longitude and height as independent variables. For the South Kuril region, the WorldClim matrices were transformed into the Gauss–Kruger projection (zone 25) with a resolution of 600 m, which fully satisfies the particularity of WorldClim data for the latitudes of Kunashir. Then, the multiyear normal values of the mean August air temperatures were reduced to those of 1985, when our thermometric survey of the soil was conducted.

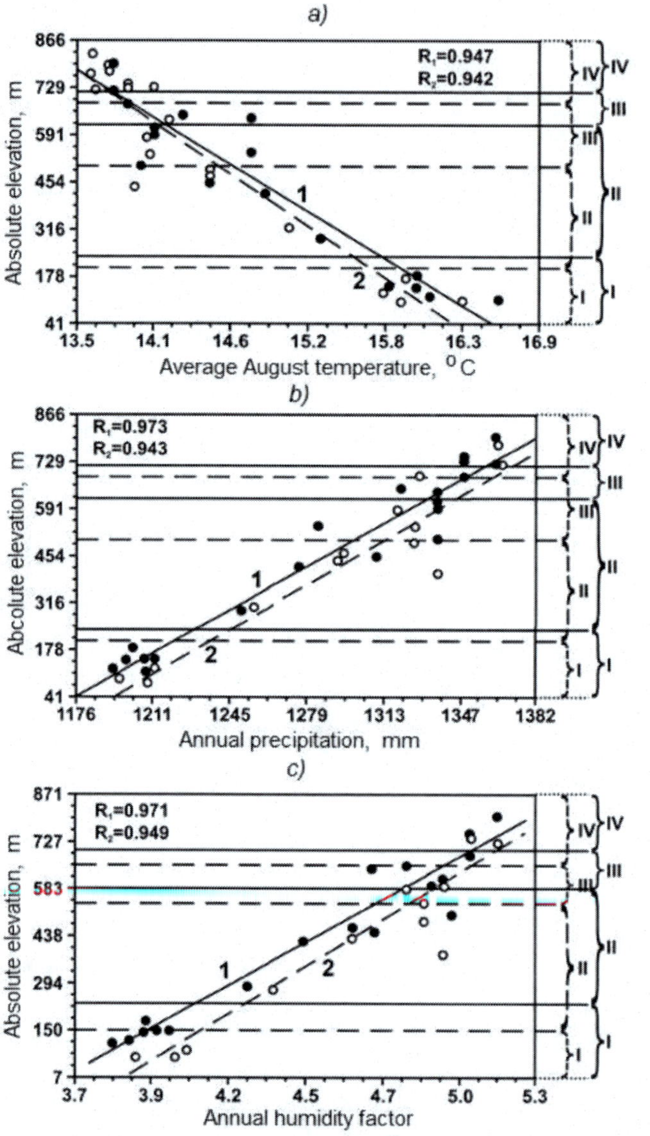

Figure 3.3. The Mendeleev Volcano landscape. Distribution of the mean August air temperature (tVIII), annual precipitation (ryr) and Vysotsky-Ivanov humidity factor (Fhum) byabsolute height(Habs). The averaged linear regression equations: (a) tVIII = 0.162 – 0.00033·Habs; R = 0.925; (b) ryr = 1187 + 0.2396·Habs; R = 0.946; (c) Fhum = 3.77 + 0.0019·Habs; R = 0.945. Altitudinal belts: I, coniferous/broad-leaved forests; II, dark-coniferous forests; III, stone birch crooked forest; IV, cedar elfin wood. The boundaries between the altitudinal belts in the sectors of: 1, the Sea of Okhotsk; 2, the Pacific Ocean.

The annual potential evaporation $E_O(yr)$ is calculated as follows. First, the monthly and yearly evaporation capacities were calculated for Yuzhno-Kurilsk station by N.I., Ivanov's formula (1958):

$$E_O(mon) = 0.0018 \cdot (25 + t)^2 \cdot (100 - f), \quad (3.1)$$

where t and f are the monthly average temperature and relative humidity of the air, respectively. The annual amplitude of monthly evaporation rates is rather large and determined by the fluctuations of relative air humidity by 16%. The maximum values $E_O(mon)$ = 40–52 mm are typical of the second half of summer and autumn (September–November), when f = 73–86%, i.e., is close to the annual minimum. In the first half of summer (June–July), relative air humidity increases to 94–95% and evaporation rate drastically decreases (making 12–13 mm per month), in spite of high air temperatures. The warmest month, August, takes an intermediate position. In this month, like in all other seasons of the year, the $E_{O(mon)}$ parameter is by 60–65% determined by temperature.

It has also been established that the annual monthly evaporation rate (25 mm) at Yuzhno-Kurilsk station is close to the evaporation rate of August (21 mm), allowing us to take the relationship between $E_O(yr)$ and t_{Aug} as directly proportional. On the basis of t_{Aug} values, the annual evaporation rate was determined for each of the 32 test grounds of experimental area by the following formula (with high correlation coefficient R):

$$E_O(yr) = 19.215 \quad t_{Aug} - 0.057; \quad R = 0.999. \quad (3.2)$$

The annual evaporation rate was also calculated by the data of 8 weather stations in the Kuril Islands:

$$E_O(yr) = 20.74 \quad t_{Aug} - 34.19; \quad R = 0.971. \quad (3.3)$$

The results of calculations by both formulas differ insignificantly.

Then, Vysotsky–Ivanov formula was used to calculate the atmospheric humidity factor F_{hum}:

$$F_{hum} = r_{yr} / E_0(yr), \quad (3.4)$$

where r_{yr} is the annual precipitation.

The F_{hum} values at the experimental testing ground were in the range of 3.58–5.15, i.e., much higher than the known humidification values for oceanic plain subtaiga landscapes of the southern Far East (Isachenko 1985). An analogous level of atmospheric humidification is typical of "taiga highlands" (Sochava 1963): subcontinental middle-mountain south taiga regions of East Siberia, e.g., the Khamar-Daban Range (absolute height1442 m, F_{hum} = 4.23). Thus, the boreal forest communities of low-mountain island-arc Neopacific were formed under the altitudinal-zonal hydrothermal conditions analogous to those of intracontinental "taiga highlands." This fact confirms the hypothesis (Urusov 1988) that the primary floristic complex on the South Kuril Islands was a high-mountain complex subsequently transformed by the intrusion of nemoral and ultranemoral species during oceanic transgressions and climatic changes.

The expenditure part of water balance the territory is composed of evaporation and the surface and groundwater flows. According to the theory of evaporation and transpiration (Budyko 1971), in case of sufficient and excessive atmospheric humidity ($F_{hum} > 1$), the total evaporation from the surface of closed canopy is represented almost solely by transpiration. Then, under the conditions of excessive atmospheric moisture and at a summer soil moisture content exceeding certain critical values W_{crit}, the evaporation rate is equal to total evaporation E_C(Zubenok 1976). For a 1-m soil layer, $W_{crit} = 150$ mm, i.e., much lower than the values of summer soil moisture content for all plots of the experimental testing ground (see below). Thus, we accept that: $E_C(\text{yr}) \equiv E_O(\text{yr})$.

The annual surface flow S_{yr} was calculated by the following multiple regression equation with a high correlation coefficient (R):

$$S_{yr} = 0.1028\ r_{yr} - 37.724\ t_{Aug} + 794;\ R = 0.90. \quad (3.5)$$

The summer productive soil moisture reserves (mm) for the layers of 0–20 cm (*W-20*), 0–50 cm (*W-50*) and 0–100 cm (*W-100*) were calculated by the following three hydrothermal parameters (Kolomyts 2003):

$$(W\text{-}20) = 0.0257\ r_{yr} + 0.0676\ S_{yr} - 0.0136\ E_O + 0.8;\ R = 0.88; \quad (3.6)$$

$$(W\text{-}50) = 0.0631\ r_{yr} + 0.1746\ S_{yr} - 0.0231\ E_O + 7.6;\ R = 0.89; \quad (3.7)$$

$$(W\text{-}100) = 0.0964\ r_{yr} + 0.3216\ S_{yr} - 0.0802\ E_O + 33.5;\ R = 0.86. \quad (3.8)$$

Finally, the values of gross water content ($W_{gross} = r_{yr} - S_{yr}$) and groundwater flow $U_{yr} = W_{gross} - E_c$) were determined.

3.3. THE PROBLEM OF OCEANICITY AND CONTINENTALITY

This problem has not yet been sufficiently developed for the eastern margin of the Asian continent, including the insular and peninsular territories going far into the water area of the Pacific Ocean and its marginal seas. It is associated not only with the latitude and longitude system of sea and ocean currents and with the respective air mass circulation but also with the complicating influence of the mountain-valley terrain. The latter terrain forms intermontane-kettle continental and mountain suboceanic climates on the continental margin and the adjacent insular land, as well as the climates of "warm" and "cold" oceanicity on the coasts and in the foothills. Let us add here the dynamic events of the "barrier foot" and "barrier shade" created by the effects of orography on monsoon air mass circulation; as a result, the above oceanic types of climate become still more humid or, on the contrary, more arid.

The bioclimatic contrasts have a distinct localization even on the Kuril Ridge islands. As it has already been mentioned, the Sea-of-Okhotsk coasts of Kunashir and Iturup in the warm half of the year are much warmer than the Pacific Ocean coasts. These sublatitudinal hydrothermal contrasts are quite clearly manifested also on the Mendeleev Volcano (Figure 3.4). Accordingly, the vegetation period in the former case is longer than in the latter case, while the homonymous phases of development of the same plant species (e.g., maple, mountain ash, honeysuckle, bird cherry tree, Shiksha (crowberry), Actinidia, lianas, etc.) come 8–12 days earlier (Vorobyev 1963).

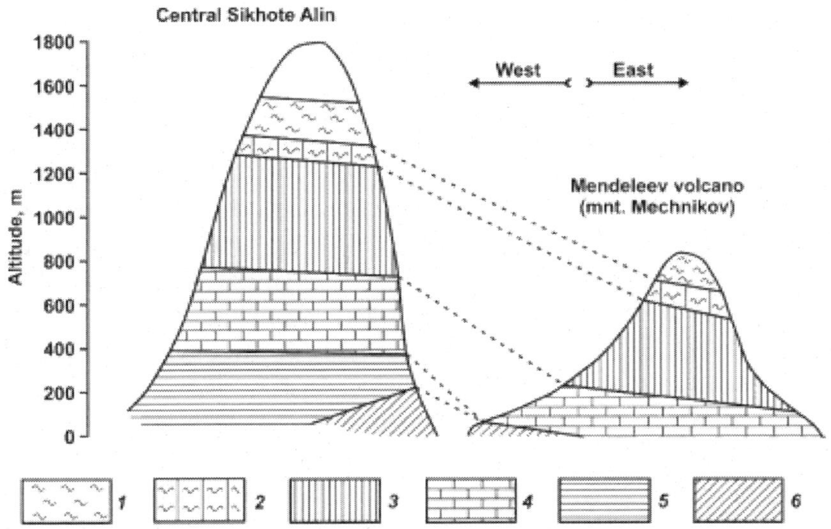

Figure 3.4. The schemes of altitudinal zonality in the Middle Sikhote-Alin (Kolesnikov 1961), and the Mendeleev Volcano in Kunashir. Altitudinal bioclimatic belts: 1 – subgoltsy creeping forests (cedar elfin wood); 2 – stone birch crooked forest (often with cedar elfin wood in undergrowth); 3 – dark-coniferous fir/spruce forests (on the Middle Sikhote-Alin, with an admixture of larch); 4 – spruce/broad-leaved forests; 5 – cedar/broad-leaved forests (absent on Kunashir); 6 – broad-leaved forests, mainly derivative.

In accordance with the above, we consider the altitudinal-zonal structure of plant cover of the Mendeleevo testing ground for the two opposite sectors: the northwestern Sea-of-Okhotsk sector, which is warmer and drier in the vegetation period, and the southeastern Pacific sector, which is colder and more humid. It should be noted, however, that the temperature contrasts are most marked on the plains and in the foothills and gradually smoothed out with rising up the mountains, coming to naught in the subgoltsy belt (see Figure 3.3 *a*). The pluviometric sector contrasts can be traced over the entire height of volcanic low mountains. Accordingly, sector contrasts are maintained in the humidity factor, though slightly decreasing with height (see Figure 3.3 *c*).

The advection of maritime air masses causes frequent and durable fogs in the spring-summer season, giving a significant increase in precipitation (not taken into account by weather stations) in the lower mountain belt and in the foothills (on the coasts). In the regions of "foggy shade" (e.g., under the shelter of the Otdelnaya–Mendeleev–Mechnikov mountain system), in summer there are much more sunny and, accordingly, warm weathers compared to the open coast, which must contribute to the higher structural and functional development of phytobiota.

The substantial import of precipitation by fogs seems also to smooth out the vertical pluviometric gradients. It is represented, e.g., by the large altitudinal-zonal amplitude of distribution of dark-coniferous species inKunashir: from the stone birch crooked forests on volcanic slopes to piedmont mixed forests coming nearly to the coast line.

Hence, the correlation between dark-coniferous (fir/spruce) and oak forests as indicators of climate oceanicity and continentality, respectively, seems to be an important problem. In sub-Pacific, a considerable part of the Sea-of-Japan slope of the Southern and Middle Sikhote-Alin is occupied by oak forests, especially the coast (with islands) and the foreranges.In Kunashir Island (the more oceanic Neo-Pacific), thewidespread mixed (broad-leaved/dark-coniferous) forests descendalong the middle and high marine terraces (80–100 m in height) directly to the coastline. Oak is predominant only in the forests of the Pacific Ocean coast of the island, with a simultaneous decrease in the height of forest stands (to shrub-thicket oak forest on the coast, with ugly stems and "flag" crowns). Low oak forests grow on the coastal ledges of terraces, at their foot along the coastline (by the internal slopes of beach ridges) and along the beds of short river valleys cutting terrace edges. In miniature, it resembles the floro- and phytocoenotic change on the eastern macroslope of the Middle Sikhote-Alin (Maksimova 1982). Such oak forests occur much less frequently on the coast of the Sea of Okhotsk. According to the data of Vorobyev (1963), oak associations with maple grow in internal regions of Iturup Island, with the weaker oceanicity and the presence of flat and hilly sites. Closer to the sea, they are replaced by stone birch forests.

It is beyond doubt that the major cause of such mosaic distribution of forest vegetation in the moderate belt of the Pacific Ocean mega ecotone should be sought in the refracting influence of submeridionally stretching mountain ridges on the monsoon, generally sublatitudinal, air mass circulation, with predominant influence of continental katabatic winds in the cold half of the year. Oak is a xerophilic and frost-resistant species and ipso facto is an indicator of enhanced continentality of the climate. It is a matter of regional peculiarities of heat and moisture transfer by air masses. The Sikhote-Alin eastern slopes are an area of descending air flows not only in winter but also in mid-seasons, with adiabatic drying of vegetation. The drying is stronger closer to the coast.

The continental Putyatin Island (Peter the Great Bay) is a brilliant example of the "oro-wind" distribution of oak forests and spatial differentiation of vital forms of these forest stands. The dark-coniferous and mixed forests of Sikhote-Alin occur in the regions with the upward movement of westerly air masses (the Eldorado region, the upper reaches of the Ussuri River) or in the intermontane depressions sheltered from the prevailing katabatic winds (the area of the Khrustal'naya River basin).

Consequently, the coastal oak forests of Sikhote-Alin emerged as a result of winter continental monsoon and periodic breakthroughs of strong withering continental winds in mid-seasons (which we observed in Putyatin Island in the second half of September, 1984). The hydrothermal conditions of eastern slopes of the ridge are unfavorable for dark-coniferous taiga even in summer due to the circulation of summer oceanic monsoon in the south of the Far East. Here, not only in winter but even in summer (consequently, all year round), air masses move generally eastwards; therefore, the Sea-of-Japan slopes of Sikhote-Alin and, to a lesser extent, the Pacific slopes of Southern Kuril ridges and volcanoes are in the "rain" shade, which is an additional factor of bioclimatic aridization together with the cold Primorskoye and Kurilskoye ocean currents.

This oro-climatic phenomenon is typical not only of the Neo-Pacific boreal bioclimatic belt but also of the sub-Arctic belt, as is exemplified by the patterns of vegetation distribution over volcanic slopes on the Simushir Island (Grabkov et al. 1984). There are 4 formations on the western (Sea-of-Okhotsk) slope of the volcano: the meadow (the absolute height of 20–50 m), the stone birch forest (80–150 m), the alder/cedar elfin wood (150–1130 m), and the mountain tundras (1130–1360 m). The eastern (Pacific) slopes have only 2 formations: the cedar elfin wood/alder (50–930 m) and the rare spots of mountain tundras among naked volcanic surfaces.

Thus, the mountain orography of marginal-continental and insular land, namely, the sub-meridian stretching of ridges parallel to the coastline, causes the longitudinal-sectoral inversion of elements of the "continentality/oceanicity" bioclimatic system. The water area of the Sea of Japan exerts a direct effect on land only within a narrow coastal strip, coming to naught already at a small distance from the coast. Obviously, "oceanicity" should be calculated in the direction of the main moisture and heat transfer but not by a formal distance from the coastline. Here, the water balance of differently oriented river basins may be a good indicator of the degree of influence of the ocean (Gartsman 1971).

3.4. The Structure of Bioclimatic System of Neo-Pacific Island Landscapes

It is known that the smaller the island and the more remote it is from the mainland, the more its climate is like that of the ocean; therefore, the basic bioclimatic, hydrological and morphosculptural features of islands are determined by their areas (Ivanov, 2005, 2014). With increasing island area, the degree of moist oceanic air mass transformation by the latter and the level of moisture condensation are enhanced, the altitudinal bioclimatic differentiation of landscapes begins from higher altitudes (Dyakonov, Puzachenko 2005) and, finally, the altitudinal zones are extended. Simultaneously with the increasing areas of water catchment basins, the runoff also increases and becomes more stable and, accordingly, the erosion-accumulation processes in the river network become more intensive, creating a more developed fluvial morphosculpture. Analogous patterns can be also observed on greater spatial and temporal scales: during the transition of subaerial landscapes from island-arc (Neo-Pacific) to marginal-continental (sub-Pacific) stage of development. A striking example is the Kunashir (Mendeleev Volcano)/Middle Sikhote-Alin binary operation system considered below.

In addition, there is a definite relationship between the structure of altitudinal zonality and the height of low-mountain construction per se. On the Dokuchayev Ridge of Kunashir, the increase in the upper boundary of dark-coniferous forests from 400 m (the Fregat Mountain) to 630 m (the Gerasimov Mountain) is almost directly proportional to the increase in mountain height from 501 m to 906 m (Grabkov et al. 1986).

The researchers have already mentioned the very clear and fairly complete manifestation of altitudinal-zonal bioclimatic zonality on the Mendeleev Volcano, in spite of its relatively low height (Manyko, Rozenberg 1970). At the same time, the territory of the volcano is in the general series of zonal boundaries becoming naturally lower as the general oceanicity of the climate increases and the height of mountain systems on the Kuril Islands decreases (Vasilyev, Rozenberg 1977).

Thewinter panorama of the Mendeleev Volcano is shown in Figure 3.5. There are three distinct altitudinal belts: the mountain dark-coniferous forest, the stone birch crooked forest and the subgoltsy with cedar elfin wood wintering under snow. Let us consider the structure of altitudinal zonality on the volcano (Table 3.1, Figure 3.4). In general, vegetation belts in the Sea of Okhotsk sector are at higher altitudes compared to the Pacific Ocean sector. According to the basic concepts of mountain climatology (Alisov and Poltoraus 1974), it may be supposed that the climate of the first sector is less oceanic, with more marked inter-belt boundaries. On the opposite slope of the volcano, under the conditions of higher oceanicity, the inter-belt geobotanic boundaries are not only lower (the spruce/fir forest belt reaching the coast in some places) but also much "fuzzier." At the same time, the vast *phytoecotones (the rudiments of buffer forest communities)* are formed precisely under these conditions. Three such ecotones were found in the sample areas on the Mechnikov Mountain slopes: one binary ecotone separating the belts of cedar elfin wood and stone birch crooked forest (the absolute height of about 640 m) and two ternary ecotones, with equal participation of spruce, stone birch and cedar elfin wood, at the upper boundary of the dark-coniferous belt (absolute heights of 580–590 m). Finally, in close proximity to the Pacific Ocean coast, on the ancient marine terraceoverlapped with pyroclastic material, there was a buffer broad-leaved/spruce/fir community at the altitudes of about 70 m in the mixed forest belt, with dominant broad-leaved species. The latter example demonstrates that even at the island-arc stage of subaerial landscape development there was a massive invasion of representatives of the Manchurian nemoral-forest flora into the Okhotian dark-coniferous flora, with the formation of buffer dark-coniferous/broad-leaved forests.

Table 3.1. The absolute heights (m) of the boundaries between altitudinal vegetationbelts of the Mendeleev Volcano and the surrounding territories

Altitudinal belt	Sector	
	Sea of Okhotsk	Pacific Ocean
Cedar elfin wood	700–720	670–680
Stone birch crooked forest		
	610–620 (in some areas, decreases to 540)	520–580 (in some areas, decreases to 480–500)
Dark-coniferous forests		
	230–250	90–150 (in some areas, decreases to 70)
Coniferous/broad-leaved forests		

The altitudinal-zonal boundaries presented in Table 3.1 are highly generalized and abstracted from the effects of other factors, which are extensively reported in the literature on island biogeography. Of course, this altitudinal-zonal system does not imply the absence of certain associations outside its main boundaries. The stone birch and cedar elfin wood communities are especially eurytopic in this respect; in Kunashir and Iturup Islands, they intrude into the dark-coniferous taiga belt here and there and even go down to the very sea coast (Figure 3.6), being indicative of the phenomena of substantial local refraction of altitudinal zonality by geomorphological, microclimatic and edaphic factors. The breakthrough of shrubby and creeping vegetation into taiga forest vegetation seems to have long historical roots and to be determined by the biological resistance of the latter vegetations, their local adaptation in the post-glacial period during warming and the movement of the taiga belt upper boundary up the slopes (Urusov 1988).

Figure 3.5. Common appearance of Mendeleev volcano in winter season. Altitudinal geobotanic zonation is expressed enough distinctly. The piedmont zone of fir-silver forests is on the foreground. Higher on low part of slopes this is the zone of stone-birch elfin woodland with cedar prostrate shrubs. Yet higher, right up to apex, this is zone of cedar prostrate shrubs which spend the winter in snow.

Figure 3.6. The Kunashir Island. Andesitic-basaltic lava outcrops on the marine ter-race. Below: Kuril bamboo thickets; above: the low-growing cedar elfin wood cover descending to the sea.

The altitudinal belts in Kunashir are generally lower compared to the Middle Sikhote-Alin (see Figure 3.4) due to a significant difference between these regions in the degree of oceanicity–continentality, which fits into the general pattern of these changes for highlands. At the same time, the level of atmospheric moisture content of this territory is of crucialimportance. The comparison of altitudinal-zonal bioclimatic conditions of, e.g., the Alps and Greater Caucasus, as well as Western and Central Caucasus, gives a visual representation for the European sector of Eurasia (Gerasimov et al. 1980). In the external (moist) regions of the Alps, up to the high-mountain stratum, annual precipitation is greater by 500–1000 mm than on the northern macroslope ofthe Western Caucasus; therefore, the upper boundary of the forest belt corresponding to the annual precipitation of about 2000 mm is at the absolute heights of about 1700 m and 2200–2300 m, respectively. At the same time, the upper boundaries of subalpine crooked forests are at the altitudes of 2200 and 2800 m, respectively. The simplest calculations made on the basis of these data have shown that the lowering of altitudinal belts in the Alps by 300–500 m compared to Greater Caucasus corresponds to the increase in annual precipitation by 200–400 mm, i.e., the pluviometric stage of the lowering of altitudinal belt boundaries is 75 m / 100 mm on average.

Tables 3.2 and 3.3 present the temperature parameters and annual precipitation, as well as heat to moisture ratios, recorded at weather stations in the south of the Far East, which characterize different zonal/subzonal landscapes of the continental margin, Sakhalin and Kuril Islands. The stations are situated in descending order of the sums of biologically active temperatures, which made it possible to rank them among the respective types/subtypes of landscapes (Isachenko 1988).

With regard to the sums of active temperatures and annual precipitation, the above weather stations were placed on a simplified diagram of climaareals of natural ecosystems in the territory of the Former USSR (Bazilevich et al. 1986), which allowed us to establish thepositions of marginal–continental, continental–insular and island-arc landscapes in the hydrothermal field of the boreal belt of the Pacific Ocean mega ecotone (Figure 3.7).

The diagram of climaareals clearly shows the two "beams" of an inversely proportional relationship between heat supply and moisture content. Beam (*a*) describes the hydrothermal trend in the direction from lowland plains of the Lower Priamurye to middle mountains of the Middle Sikhote-Alin continental macroslope and to the Sakhalin Island. This trend fits into the gradient hydrothermal field of transitions from East European oak forests to subtaiga and then to south- and middle-taiga geo(eco)systems. During this transition, the sums of active temperatures decrease and annual precipitation increases, with the respective increase in K_{hum} from 0.70–0.90 to 1.15–1.30. According to the climatic zoning by B.R. Alisov (1956), this beam characterizes the landscape-geophysical conditions of subcontinental sector of the Russian Federation. Such conditions are consequently typical of the boreal sub-Pacific of the Pacific Ocean mega ecotone: its marginal-continental sector with the adjacent continental islands.

Table 3.2. The hydrothermal characteristics of forest ecoregions of the boreal belt on the Pacific Ocean mega ecotone of Northern Eurasia

Weather stations (absolute altitude, m)	Hydro-thermal characteristics *)				
	t VII-VIII	$\sum t \geq 10°$	E_0, mm	r_{year}, mm	F_{hum}
Subboreal landscapes					
Khabarovsk (87)	21.1	2672	755	657	0.87
Komsomolysk-in-Amur (21)	20.0	2413	650	577	0.89
Журавлевка (215)	19.8	2366	633	879	1.39
Boreal south-taiga landscapes					
Terney (27)	17.4	1802	463	939	2.03
Yuzhno-Sahalinsk (23)	17.3	1779	457	870	1.90
Yuzhn0-Kurilysk (23)	15.8	1426	390	1255	3.22
Boreal middle-taiga landscapes					
Ulunga (763)	16.3	1544	409	759	1.85
Poronaysk (3)	15.8	1426	390	1027	2.63
Kurilysk (38)	15.3	1309	373	1108	2.97
Subarctic landscapes					
Urup (73)	11.1	322	360	1230	3.42
Simushir (26)	10.4	157	379	1706	4.50

*) $t_{VII-VIII}$ – average temperature of most war month; $\sum t \geq 10^0$ – sum of the biological active temperatures, ^{0}C; E_O, mm – annual potential evaporation, mm; r_{year} – annual precipitation, mm; F_{hum} – Vysotsky– Ivanov's annual humidity factor.

Table 3.3. The comparison of the main climatic characteristics of altitudinal-zonal plant communities of Mendeleev volcanic landscape and analogous formations of the marginal-continental sector of the Pacific Ocean mega ecotone of Northern Eurasia

Regions	Altitudinal zonal plant communities and formations (in brackets – altitude, m)		Complex parameters		
			t VII–VIII r_{year}, mm	$\sum t \geq$ 10°	F_{hum}
Kunashir Island, Mendeleev volcanic landscape. Plant communities	Lowland and piedmont dark/coniferous-broad-leaf forests with subtropical elements (20–250)		16.0 1207	1473	3.43
	Lowland and low-mountain dark-coniferous (spruce/fir) forests with nemoral elements (120–600)		14.6 1299	1144	4.64
	Subalpine stone birch elfin woodland with admixture of spruce and with cedar elfin in undergrowth (540–700)		14.4 1320	1097	4.70
	Subalpine cedar elfin wood (650–800)		14.1 1347	1027	5.02
Low Priamurye and Middle Sikhote-Alin. Plant formations	Lowland and piedmont broadleaf and coniferous-broadleaf tall grass forests	st. Neran (221)	19.6 865	2319	1.40
		st. Zhuravlevka (215)	19.8 875	2366	1.38
	Middle-mountain fir/spruce dwarf-bush/grass forests with nemoral elements	st. Ulunga (547)	16.3 749	1544	1.83
		st. Irumka (516)	16.7 761	1638	1.78
	Northern-taiga middle- and high-mountain larch light forests, cedar elfin woods	ст. Sophiysky Priisk (903)	15.1 757	1262	2.06

Footnote: t VII–VIII – average temperature most warm month, °C; $\sum t \geq 10°$ и – sum of biological active temperatures, °C (sum of the average daily temperature equal or higher 10°); r_{ear} – annual precipitation; F_{hum} – Vysotsky-Ivanov's annual atmospheric humidity factor. Denominations of continental plant formations are given by (Sochava 1977).

The second beam (*b*) begins, like the first one, with sub-continental territories; however, then it goes through the climaareals of forest ecosystems of much more humid (sub-oceanic and oceanic) regions. It covers the broad-leaved forests in the foothills of southwest Ukraine and the mountain spruce forests of the Carpathian Mountains and successively "strings" the middle mountains of the western macroslope of the Middle Sikhote-Alin and the Sea-of-Japan coast of the continent, then the Southern Kuril Islands and, finally (with a large spread), the Middle Kuril Islands.

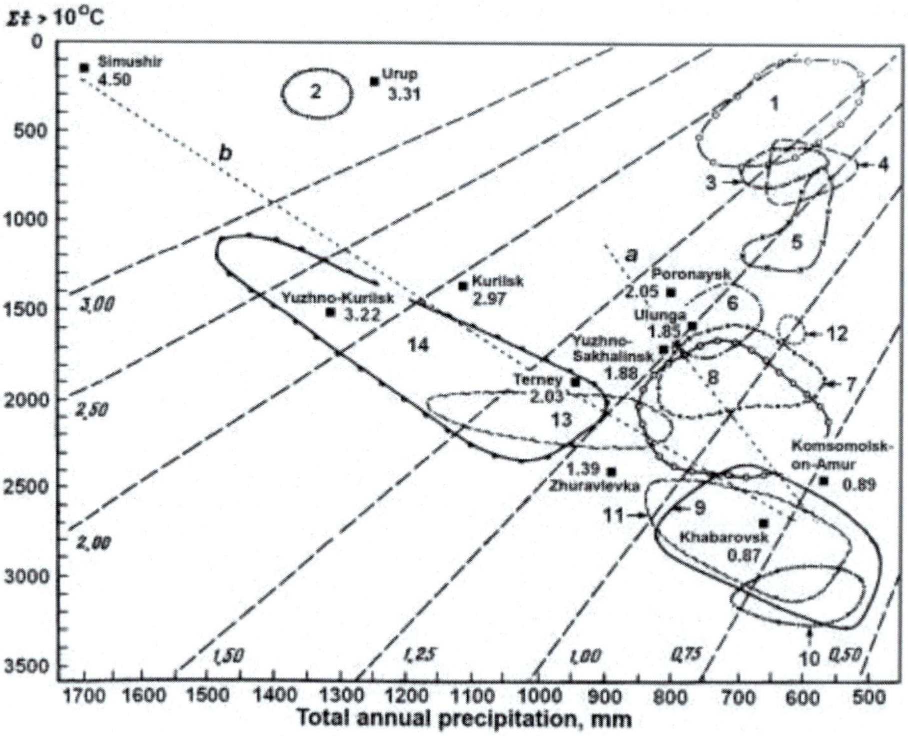

Figure 3.7. The positions of weather stations in the southern part of the Far East in the system of climaareals of natural ecosystems in the territory of the former USSR, according to Bazilevich et al. (1986), with additions. Natural ecosystems: *1* and *2* – plain and mountain tundras; *3* and *4* – birch and spruce forest tundra; *5* and *6* – north-taiga spruce and dark-coniferous forests; *7* - south-taiga forests; *8* – sub-taiga coniferous/broad-leaved forests; *9* and *10* - east-European and western oak forests; *11* – hornbeam/oak mesophilic forests; *12* - pre-Ural coniferous/broad-leaved forests; *13* – beech forests of the southwest Ukraine; *14* – mountain spruce forests of the Carpathian Mountains. *3.0*, *2.5*, etc. – the lines of humidity factor for climaareals. 4.40, 3.31, etc. – humidity factors calculated for the Far-East weather stations. The "beams" of correlation relationships between heat supply and atmospheric moisture content: *a* – for sub-continental territories; *b* – for sub-oceanic and oceanic territories.

This beam, within the framework of Neo-Pacific, describes the latitudinal-zonal change of bioclimatic conditions that would be interesting to compare with the altitudinal-zonal hydrothermal field at the experimental test ground of the Mendeleev Volcano (see Table 3.2). The Yuzhno-Kurilsk station gives humidity factor values (3.22) rather close to the values for the surrounding piedmont-plain coniferous/broad-leaved forests of Mendeleev landscape (3.43 on the average). However, the more northern Urup and Simushir stations situated in the zones of birch crooked forests and cedar elfin woods, respectively, give the lower F_{hum} values (3.31 and 4.50) compared to their altitudinal-zonal analogs in Mendeleev volcanic landscape (4.70 and 5.02). In the island-arc region, as one can see, the altitudinal-zonal hydrothermal field is more contrasting than the latitudinal-zonal field, which must cause more marked phytocoenological differences on the slopes of volcanic mountains and ridges compared to the transitions in the bioclimatic system of altitudinal zonality.

3.5. Altitudinal Bioclimatic Stages

The data from Tables 3.2 and 3.3 and Figure 3.7 were used to calculate the pluviometric stage by comparing weather stations on the Southern Kuril Islands with those on plains and in low mountains of the Lower Priamurye and the Middle Sikhote-Alin, respectively. The upper boundaries of the coniferous/broad-leaved and dark-coniferous/taiga belts on the Mendeleev Volcano were shown to be lower by 550 and 620 m, respectively, compared to the analogous marginal-continental region mentioned above (see Figure 3.4), which was adequate to the increased annual precipitation in the foothills by 400–650 mm. The average values of the pluviometric stage obtained for the upper boundaries of coniferous/broad-leaved forests and the fir/spruce forest belt (Table 3.4) slightly exceed the stage calculated for the sub-oceanic mountain territories of southern Europe (see above), which is quite typical of the monsoon-oceanic region, being evidence of the reality of the altitudinal zonality/atmospheric water content ratios obtained for southern Kunashir.

Table 3.4. The altitudinal pluviometric (*a*), thermometric (*b*) and hydro-thermometric (*c*) stages of the lowering of zonal-belt boundaries in the south of Kunashir compared to the Middle Sikhote-Alin

Upper altitudinal zonal boundaries	Altitudinal stages (n / 100 m)		
	a	*b*	*c*
Dark-coniferous/broadleaf forests	10 mm	0.76^0	0.405
Dark-coniferous (spruce/fir) forests	89 mm	0.68^0	0.360

The thermometric stage was calculated by analogy, based on the mean temperature of the warmest month: for Yuzhno-Kurilsk station (15.8°) and for the Ussuri and the Lower Amur plains (around 20.0° on average for eight stations). Finally, the stages of the complex hydrothermal index (the Vysotsky-Ivanov humidity factor) were calculated.

It would be interesting to note that the thermometric stage in Kunashir is more than twice higher than the altitudinal temperature gradient, which is here about 0.35°/100 m, i.e., generally close to the low temperature gradient typical of insular territories (Ivanov 2005). Thereat, the entire system of altitudinal-zonal boundaries shifts downwards and the altitudinal zonality range becomes contracted. It is precisely the great spread of the thermometric stage compared to the altitudinal temperature gradient that reflects the decrease in condensation level on the islands, as is postulated by Dyakonov and Puzachenko (2005). The smaller is the island area, the more efficient is this process. Thus, the stone birch crooked forest belt in the south of Kunashir begins at the altitudes of 520–620 m (see Table 3.3) but descends to the altitudes of 30–40 m on the neighboring smaller Shikotan Island, in much the same way as in the larger, more northern, Iturup Island (Vorobyev 1963).

The obtained values of landscape-geophysical stages for the regional shifts of the two most important altitudinal-zonal boundaries can be accepted as tentative for predictive calculations of statistically significant shifts in altitudinal belts on the Mendeleev Volcano according to particular scenarios of global climate changes. As we can see, the upper boundary of the fir/spruce forest belt must be more sensitive to global climatic signals than its lower boundary; hence, e.g., one should expect the narrowing of this belt in case of warming and/or increase in atmospheric moisture content or broadening in case of cooling and/or aridization. Accordingly, the areals of the lower coniferous/broad-leaved forests, as well as the higher subgoltsy stone birch crooked forest with cedar (and alder) elfin wood, must change too.

It would be interesting to note that the group of weather stations on the continental coast of the Sea of Japan, the southern and middle parts of Sakhalin, and Southern Kuril Islands is within a comparatively narrow range of the sums of biologically active temperatures but considerably different in atmospheric moisture content, which is the major climatic factor of drastic biogeographical differences between the island-arc Neo-Pacific and the marginal-continental and continental-insular Sub-Pacific.

The comparison of the presented diagram of climaareals (see Figure 3.6) with the ecological-phytocoenotic complexes in the south of the Far East (Sochava 1977) raises a number of fundamental problems that require special consideration. So, the lowlands of Middle and Lower Priamurye (Khabarovsk and Komsomolsk-on-Amur stations), on the one hand, and littoral plains in the southern part of Kunashir (Yuzhno-Kurilsk station), on the other hand, are ranked within the same ecological-phytocoenotic complex of nemoral and subnemoral coniferous/broad-leaved forests. At the same time, the climaareals of these complexes differ very dramatically, especially in annual precipitation and, accordingly, in humidity factor.

For comparison, let us give the boundaries of altitudinal belts in the two southern territories of the Far East being at the same latitude as Kunashir: Southern Sakhalin and the southern part of the Middle Sikhote-Alin (Kolesnikov 1961; Vitvitsky 1961). The mountain ranges of the former and latter regions were formed at the turn of the Mesozoic and Cenozoic eras and in the epoch of Mesozoic folding, respectively. In Sakhalin, the belt of spruce/fir forests rises to a level of about 700 m and the belt of stone birch forests and crooked forests rises to 900–1000 m, with the belt of cedar elfin wood above them (up to 1100–1200 m). In the Middle Sikhote-Alin, the above vegetation belts go even higher. Even on the ridge slopes facing the Sea of Japan, with bioclimatic borders being lower by 150–200 m compared to the continental (western) slopes, the mountain belt of spruce/broad-leaved forests occupies the range of heights from 350–400 m to 650–700 m, while the upper boundary of the dark-coniferous forests located above has an altitude of 1200–1250 m. After a comparatively narrow strip of stone birch crooked forests, beginning from the altitude of 1300–1350 m, there is the cedar elfin wood belt with the upper boundary of 1550–1600 m.

Thus, under the climatic conditions of insular boreal Neo-Pacific, the basic range of altitudinal zonality (from nemoral forests to subalpine crooked forests and elfin woods) is reduced and compressed simultaneously, if compared with the analogous ranges of zonality of large continental islands and the more so the continental margin. On volcanogenic islands, this range keeps within the limits of the low-mountain layer of macrorelief. This altitudinal-zonal phytocoenotic compression is considered as a phenomenon of extra zonality of island biotic system formed by pendulum-like changes in the formations and even types of plant cover during the alternation of Quaternary glacial and interglacial epochs (Urusov and Chipizubova 2008). It should be noted that the transition from coniferous/broad-leaved forests via taiga to subgoltsy elfin wood, which is extremely limited in space, occurs notwithstanding the decreasing vertical temperature gradient above the islands.

The phenomenon of compression and general lowering of the system of altitudinal bioclimatic belts has been traced not only on the super-regional scale of meridian sectors of the Pacific Ocean mega ecotone but also at the level of separate landscape districts of the Kuril Ridge, e.g., within Kunashir Island. The comparative analysis of altitudinal zonality of separate mountain massifs of the island along the Dokuchaev Ridge has shown that the rise of the lower boundary of birch crooked forests from 400 m to 630 m is proportional to the increase in mountain height from 500 m to 915 m. It is associated with the soil-ground moisture content, which is determined by slope length (Grabkov et al. 1986). Suchan explanation is rather questionable. According to the laws of climatology (Alisov and Poltoraus 1974), the continentality of a mountain climate increases and the upper boundaries of altitudinal belts shift upwards with the increasing massiveness of the mountain.

3.6. Evolutionary Significance of Island-Arc Bioclimatic System

Thus, the main spectrum of altitudinal bioclimatic zonality originates at a rather early stage of subaerial landscape development, when the tectonic-volcanic massifs emerging from the ocean have not yet gone beyond the limits of lowland relief. In this case, it is neogenic age of the morphostructural basis of the Kunashir Island, with quaternary volcanism planted on it. The climatically determined compression of altitudinal zonality on the Mendeleev Volcano, as well as along the entire Kurile Ridge, *shifts the regional bioclimatic contrasts* (which are usually widespread in mountains inside the continents and even on their margins) *to the local (topological) level.* It can be assumed that the topological bioclimatic system of island-arc Neo-Pacific is a kind of focus ("core") of the origin of continental (zonal, sectoral and altitudinal-zonal) biogeographical and landscape structures. Later on, with transition to the marginal-continental regime, the hierarchic level of these structures increases from topological to regional level, as is manifested in the general rise of bioclimatic borders and simultaneous broadening of altitudinal belts.

The special geobotanic studies of the Klyuchevskaya Group of volcanoes in Central Kamchatka (close to the northern boundary of the boreal belt) have shown (Vyatkina et al. 2014) that the time necessary for the formation of the conditionally indigenous forest communities at the initial stages of subaerial biogenesis is: (1) no less than 150–200 years for stone birch crooked forests; (2) 250–300 years for cedar elfin wood with single forest stands; and (3) 2000–2500 years for upright coniferous forests. At the same time, forest communities develop much more rapidly on a loose pyroclastic substrate than on lava platforms of different age. In the southern ecoregion of Kunashir and close to the southern boundary of the boreal belt, the above time intervals should be significantly reduced.

Each new part of continental biosphere in the Northwest Pacific emerges under environmental conditions highly unfavorable for the zonal types of forest phytobiota: unstable morphotectonic regime, critical hydrothermal and edaphic conditions. However, precisely these regimes and conditions are the strong stimulating factors of evolutionary processes in young island-arc geo(eco)systems yet having no rich florogenetic and phytocoenological diversity.

All of the evolutionary changes in biota are based on its close interaction with the environment. In the history of organic world on the Earth, as is known, the new forms of life (species, populations and communities) emerged in the geological and climatic epochs critical for the biota: orogenesis, intensification of volcanic activity and global cooling (Budyko 1984). In these epochs, the stabilizing selection predominating in antecedent warm climatic periods changed for disruptive selection creating a certain diversity of species and communities (allogenesis) with higher specialization, i.e., adapted to the tighter conditions of existence (Yablokov and Yusufov 1976; Pianka 1981). At the same time, the rate of selection increased with the decrease in the quantity of populations (Shmalgauzen 1968). It was due to the factor of pressure of population waves (Timofeeff-Ressovsky et al. 1969). This process can be considered as a "... homeostatic evolutionary response of a population" (Ricklefs, 1979 213).

The novel types of ecological systems in the temperate latitudes emerged in the last glacial period (Budyko 1977). Moist taiga forests with ancient (Eopleistocene) well-developed soils have existed in Kuril Islands since Early Pleistocene (Urusov 1988), which allows us to consider this period as the beginning of evolutionary processes in the plant cover of these islands.

Chapter 4

VOLCANOGENIC BASIS OF ISLAND-ARC LANDSCAPES

ABSTRACT

The empirical statistical models of the island-arc stage of continental biosphere development in the North-West Pacific have been created by the example of experimental testing area near the active Mendeleev Volcano in Kunashir (South Kuril Ridge). This region characterizes one of the initial stages of subaerial landscape genesis. At this stage, active volcanism with its intensive eruptive processes is completed and changes into passive volcanism. The determining endogenous factor is the energy-consuming deep fluid systems, with their conduits (fumaroles) forming solfatare fields. The fumaroles activity of the volcano is determined by the enhanced fluid conductivity of the vocano–tectonic zone at the final stage of its development, when the fluid mass transfer of gaseous and liquid solutions becomes a seld-organizing system. The volcanic islands of the Greater Kuril Ridge were formed by the linear-nested arrangement of volcanoes typical on the basis of at least nine volcanic cones of different age (from Early Pleistocene to present-day) and, accordingly, different levels of activity. The present-day relief of Kunashir is a chain of large and small volcanoes of different age (from middle-Pleistocene to present-day) connected with abrasion-accumulative lowland-plain vents.

4.1. MORPHOTECTONICS OF THE KURIL ISLAND ARC

The Kuril Islands are included into the system of island arcs of the Pacific Ocean Mobile Belt. They are the above-water part of epigeosynclinal mountain ranges separating the kettle of the marginal Sea of Okhotsk from the Kuril-Kamchatka Trench in the Pacific Ocean (Figure 4-1). Kuril volcanic physico-geographical region (Grabkov and Khudyakov 1993) is included in the higher regional natural complexes of the Pacific Ocean mega ecotone of Northern Eurasia. It is a contemporary geosynclinal system with a typical monsoon and marine climate, in contrast to the Kamchatka and Koryakskoye Upland regions also having continental climates. According to the legends to the Landscape Map of the former USSR (Isachenko 1985), the Kuril country is occupied by boreal type landscapes (subtaiga, i.e., transitional to the subboreal type) and their Pacific subtype.

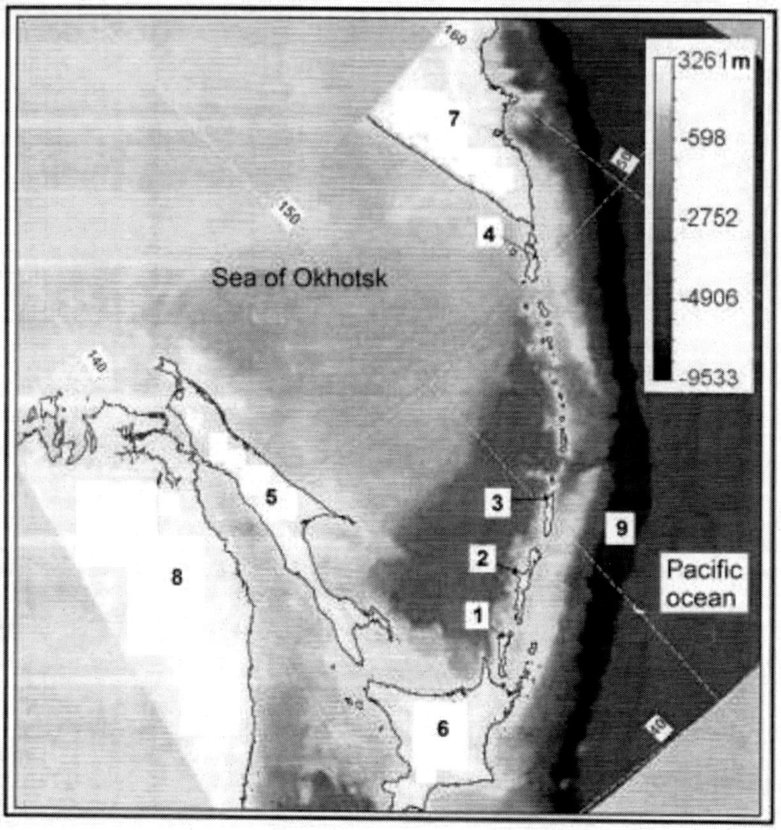

Figure 4.1. The map of lithosphere relief in the region of the Kuril island arc. The islands of the Kuril Ridge: 1 – Kunashir; 2 – Iturup; 3 – Urup; 4 – Paramushir. Other land territories: 5 – Sakhalin; 6 – Hokkaido; 7 – South Kamchatka; 8 – Lower Priamurye. 9 – Kuril-Kamchatka deep-water trench.

Geosynclinal islands are formed at the second, inversion (Belousov 1978) stage of the geosyncline regime of the earth crust and the surface of the lithosphere, as well as under the conditions of mio- and parageosyncline regimes (intermediate between properly geosyncline and platform regimes). Islands are developed as a result of partial inversion of the geotectonic regime of subsidence accompanied by magmatism and regional metamorphism, crust fractures, orogenesis and lateral thrusts. This process is accompanied also by flysch deposition, i.e., terrigeneous sediments, at small depths as a result of erosion of insular land.

The epigeosyncline orogenic regime is the final stage of the endogenous cycle of development of the surface of the lithosphere. It succeeds the second stage of the geosyncline regime (after partial inversion), being its direct continuation. The permeability of earth crust decreases and changes from dispersed to linear type (narrow deep fractures). Acid and intermediate effusions are predominant, with basalts playing a subordinate role, as exemplified by the central and eastern Kamchatka.

The regime of young platforms is characterized by the absence of magmatic rocks in the sedimentary cover and by the absence of metamorphism. Subgeosynclines and subgeoanticlines sometimes have an inheritance relationship with downfolds and elevations of the previous geosynclinal development; at the same time, the downfolds and elevations are often limited by deep fractures. Sakhalin is an illustrative example of the above.

As is known, there are two types of regimes of continental margins: Atlantic and Pacific. The Atlantic regime is characterized by the discordant bedding of the coast relative to the pre-Mesozoic morphostructure of the platforms; the wide continental shelf as a continuation of the platform; the absence of volcanism; and the maintenance of platform regime against the background of land fragmentation and subsidence.

The regime of the Pacific continental margins is characterized by the considerable permeability of earth crust associated with abyssal fractures (deep faults); the development of magmatism; volcanism as andesitic and andesitic-basaltic lava; the chains of volcanic islands; and the maximum contrast range of block-wave motion resulting in the neighborhood of mountain ranges and deep ocean trenches, mountainous islands and marginal seas.

In the zone of transition from the northwestern part of the Pacific Ocean to the continent of Asia, the following regional sectors are distinguished (Erlikh et al. 1974): (1) island-arc, of the present-day geosynclines (Kuril Island Arc); (2) of the early continental orogenesis (Eastern Kamchatka and Hokkaido); (3) of the mature orogenesis (Sakhalin); and (4) of the platform development of morphostructures (the continental margins of Asia). Geosynclinal belts are characterized by direct reflection of the overwhelming part of morphostructures in the relief (Melekestsev 1974).

The island-arc systems are the zones of formation of the major geochemical reservoirs of our planet; therefore, they play the key role in plate tectonics (Martynov et al. 2009), which creates the shape of continents. This very fact predetermines the representativeness of island-arc volcanic landscapes as objects for studying the initial stages of formation of the continental biosphere and disclosure of the initial mechanisms of subaerial landscape genesis.

Island arcs have the overthrust-shear fault nature of tectonic deformations. At the same time, the South Kuril Islands and the Aleutian Arc are similar in principle in the dynamics of development of their tectonic systems (Erlikh et al. 1974). The Kuril and Commander Islands have basically inherited the structural plan of the Paleogene and Neogene, and the epigeosynclinal development still continues there.

There are 68 volcanoes on the Kuril Islands, 36 of them being active and potentially active. The volcanic activity has not weakened over time. In the past 300 years, there have been 28 strong and catastrophic eruptions; for the most dangerous explosive eruptions, the radius of the affected area may reach up to 25–30 km (Komedchikov et al. 2012).

According to the modern concepts (Avdeyko et al. 2001), the Kuril Island Arc is a system of volcanic belts of different age, which are situated above the zones of subduction – the V-shaped underthrusting of oceanic crust beneath continental crust. The Kuril Islands are a typical example of the double-arc tectonic system (see Figure 4.1). The Kunashir, Iturup and other islands belong to the western (inner) arc, which currently undergoes total relative downwarping, with the predominance of strike-slips faults. It includes all modern volcanoes creating the overcompensated, relative to downwarping, accumulation of volcanic sediments. Volcanism is generally concentrated in graben-synclines (Erlikh et al. 1974). The Eastern (external) arc (the Little Kuril Chain – the Vityaz Ridge) has the predominant uplifts with reversed strike-slip faults in the absence of present-day volcanism. The tectonic-volcanic relief (and the respective landscapes) and the denudation-tectonic relief are predominant in the former and latter case, respectively. The Greater Kuril Islands are the uplifted blocks of the Tertiary basal complex with the linear rows of Quaternary volcanoes.

Thus, volcanic landscapes per se develop under the conditions of "inverted downfold" against the background of the total uplift of the morphostructure, i.e., under the conditions of compensated downwarping of the territory. There is an analog of tectonic uplift of the territory – due to the accumulation of dense (magmatic) and loose (pyroclastic) rock masses of volcanogenic material on top. As a result, the "accumulative ranges" are formed (Erlikh et al. 1974). The localization and contrast of manifestations of different types of volcanism create the diversity of volcanic landscapes. The frequent burial of soil under new layers of effusives and explosions is a cause of youth and instability of the soil-lithogenic basis of natural complexes.

On the contrary, volcanism is almost absent in the regions of tectonic uplifts (horst anticlinoriums) with erosive-tectonic relief and in the deepwater trenches with noncompensated downwarping. In the areas of development of the denudation-tectonic relief (Shikotan and some parts of Kunashir, where volcanism occurred in the Pliocene), the landscapes are formed by shearing off the bedrock layers, deepening of the weathering crust and development of erosion systems. Here, the general tendency to smooth the local contrasts is predominant. Though both types of territory development give similar effects with regard to hypsometric characteristics (Erlikh et al. 1974), the final landscape structures may be quite different.

Volcanism is closely associated with tectonics. Basaltic eruptions are predominant in the periods of tectonic stabilization, while acidic volcanism with stratovolcanoes and pyroclastic rocks are predominant in the periods of dramatic uplifts of mountain systems. Consequently, coastal basaltic cliffs along the Kunashir shore (see above Figure 3.5) demonstrate that the coastal territories of the island were formed in the period of a stable tectonic regime – in Late Pliocene and Early Pleistocene; the Mendeleev Volcano itself is a heritage of the subsequent new potent uplift in Middle Pleistocene and Holocene. Hence, there is a distinct total zonality in the distribution of different forms of volcanism. The major centers of acidic volcanism are concentrated in the frontal volcanic zone, e.g., in the somas of the Mendeleev Volcano in the south of Kunashir. The periphery of volcanic landscape was formed on the rocks of basaltic volcanism.

As is known (Gorshkov 1967), the power supplies of volcanoes in island arcs and the causes of diversity of volcanic formations are beyond the limits of the Earth crust, namely, in the upper mantle; therefore, volcanism is a "through-crust" process. "The origin of diversity of lavas is magma per se and the processes occurring in the magma body" (ibid. 269). The peculiarity of morphological structure and the high dynamicity of volcano formation, its significant influence on the entire island ecosystem are accounted for by high energy consumption of the lithogenic component and the waters. It is associated with the position of the volcano as a top of the vertical dynamic system – a thermogen that drains the depths of the Earth (Grabkov and Khudyakov 1993).

Volcanoesare formed on large magmatic columns, or diapirs, which rise in the asthenosphere in regular rows by means of gravitational convection (buoying). In the volcanic belt of South Kamchatka, this row is especially striking: the average distance between large volcanoes is 29 km (Fedotov 1991). There are 104 volcanoes on the Kuril Islands, apart from submarine volcanoes; 39 of them are active (Aprodov 1982). The effect of volcanism on the natural-territorial complexes of the island is directly proportional to the area and volume of volcanic bodies. Volcanic cones and adjacent lava and pyroclastic plains make up to 90% of the area and volume of all Kuril volcanic landscapes (Grabkov and Khudyakov 1993). The percentage of volcanic landscapes of the island area is as follows: Iturup – 35; Urup – 66; Simushir – 44; Shikotan – 79; Onekotan – 90; Paramushir – 28.

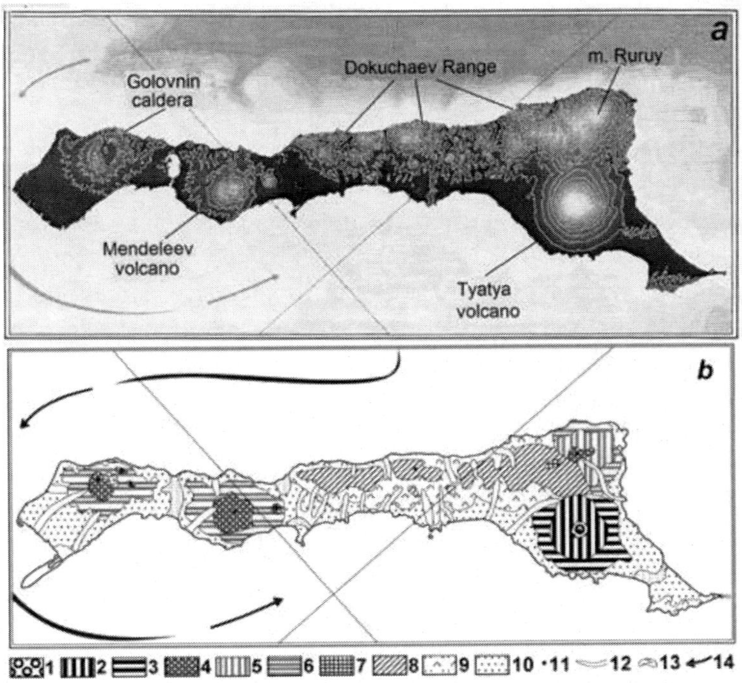

Figure 4.2. Geomorphometry (*a*) and geomorphological maps (*b*) of Kunashir Island.
The legends to the geomorphological map (Kotlyakov et al. 2009): Mountain landscapes. Volcanic relief. Denudation-volcanic (weakly destroyed): 1 – the craters and steep upper parts of the cones; 2 – middle mountains, low mountains; 3 – lava-pyroclastic plains, bajadas (foothill trails). Denudation-volcanic (strongly destroyed): 4 – the craters and upper parts of the cones; 5 – middle mountains, low mountains; 6 – lava-pyroclastic plains, bajadas (foothill trails). Erosion-denudation relief (denudation-tectonic): 7 – middle mountains; 8 – low mountains, mid-altitude hill country; 9 – low-altitude hill country. Plain landscapes: 10 – Pleistocene abrasion-accumulative plains. Other designations: 11 – goltsy; 12 – river valleys; 13 – lakes; 14 – warm sea currents.

The present-day relief of Kunashir Island is a chain of large and small volcanoes of different age (from middle-Pleistocene to present-day) connected with abrasion-accumulative lowland-plain vents. These vents emerged in Late Holocene: 2200–2300 years ago (Razzhigaeva et al. 2011). Geomorphometric processing of the satellite image of Kunashir showed (Figure 4.2 *a*) that this island was formed by the linear-nested arrangement of volcanoes typical of the Greater Kuril Ridge (Gorshkov 1967) on the basis ofat least nine volcanic cones of different age (from Early Pleistocene to present-day) and, accordingly, different levels of activity. The most ancient volcanic cones compose the Dokuchayev Range, which is unreasonably indicated on the geomorphological map (Figure 4.2 *b*) as an erosion-denudation (denudation-tectonic) relief, without any allusion to its originally volcanogenic genesis. Low-lying isthmuses combine the volcanic cones into a single island system of Kunashir.

Thus, in the mobile belt of the planet, *volcanoes are the foci of accretion of the continental biosphere*. The lithogenic basis of young island-arc landscapes is directly related to the deep endogenous processes in the "living planet" Earth, which determines the direct landscape-forming influence of the morphotectonic factor (Grabkov and Khudyakov 1993). The extremely high frequency and intensity of eruptions and tempos of relief formation (on the geological time scale) cause an exceptionally complete correspondence between the forms of the earth's surface and the geological structure (Braytseva et al. 1974). Periodic recurrence of volcanic activity may bring to naught the results of denudation processes closely associated with the landscape formation.

The poorly developed denudation processes in the regions of present-day volcanism, as has already been said, are due to mainly tectonic downwarping (or delayed uplift) of these areas compared to the neighboring territories. Under such morphotectonic conditions, the destruction and ablation of bedrock layers proceed slowly, while the development of denudation-accumulative exogenous is considerably suppressed – largely due to the poorly developed river network of the territory. Hence, the predetermining role of volcanogenic-accumulative morphosculpture in landscape structure is much more marked.

One should also add the considerable resistance of lava rock masses to wash-out, as well as the high water permeability of loose pyroclastic material, which prevents the formation of water-bearing strata in the upper layers of mother rock and, consequently, a stable channel runoff. As a result of rapid infiltration of precipitations, the surface water runoff ceases even on steep volcanic slopes, and the development of an accumulative-humus soil layer on ashfalls slows down the erosion and the formation of fluvial relief (Mukha 2007). The slopes of the Mendeleev Volcano are almost devoid of watercourses, while the great diversity of genetic types of lakes in Kunashir (Korsunskaya 1956) are known (Borzov 1954) to indicate the disconnection of catchment basins, the very poorly developed hydrological network of the territory. At the same time, Shikotan has a dense river network having developed under the conditions of denudation-tectonogenic relief on the dislocated rocks of Upper Cretaceous–Paleogenic age (Melekestsev et al. 1974).

The volcanogenic-accumulative relief is characterized by thick pyroclastic sediments with accelerated soil formation processes. These conditions are characterized by the maximum coincidence between the age of relief and the age of sediments. The denudation-volcanic forms of relief are characterized by noncoincidence between the age of relief and the age of sediments, which is more marked if volcanic forms are more destroyed (Melekestsev et al. 1974). Consequently, as the relief of young geosynclinal islands change from the volcanogenic-accumulative types of relief to the denudation-tectonic relief of continental regions, the disproportion between the age of relief and the age of soil-forming bedrocks increases.

4.2. On the History of Volcanic Forms of Relief

The history of the Kuril Island Arc in the recent epoch has had three stages (Melekestsev et al. 1974; Erlikh et al. 1974):

1. Late Pliocene – the high stand of land close to the present-day level; intensive differential motion with strong explosive volcanism (around Kunashir, Iturup and Onekotan); fractionation of the ridge into a series of horsts and rift valley (graben) straits; the transtension and strong submarine volcanism at the western boundary of the Greater Kuril Ridge;
2. Late Pliocene/Early Pleistocene– heavy downwarping of the ridge; separate parts of land remained in Paramushir, Onekotan and, probably, Kunashir; and

3. Middle Pleistocene/Holocene– new heavy elevation by 500–750 m (at a rate of about 2 mm per year); the maximum absolute heights for the entire Quaternary Period were obtained.

In Kamchatka, the Early Quaternary (original) volcanism was mainly basaltic and andesite-basaltic, creating a widespread "plateau–effusive" complex. Explosiveness and acidic volcanism are untypical of this period (Brytseva et al. 1974). Many mountain ranges of the peninsula are formed by basaltic outpourings. However, such volcanism has not been revealed in Kuril Islands. Here, volcanic plateaus (in Paramushir) are younger; they overlap a 200–220-m Late Pleistocene marine terrace.

In the Middle and Late Quaternary period there was a strong outburst of acidic explosive volcanism over the entire Kuril-Kamchatka area. Extensive pumice covers and the largest collapse calderas were formed in South and East Kamchatka. Glacial deposits overlay the ignimbrite covers of the beginning of Middle and Late Pleistocene. The youngest ignimbrite covers are about 17 thousand years old. In Kuril Islands (Kunashir, Iturup, Urup, Paramushir, etc.), the most ancient lava covers belong to Middle Pleistocene. They are younger than the marine terraces of 200–220 m and 300–320 m in height, aged $Q^1{}_3$.

The absolute age of plant debris buried by pyroclastics during the formation of the Mendeleev and Golovnin calderas in the south of Kunashir is about 40 thousand years. The contemporary (Holocene) stage of volcanism is a continuation of the acidic one, as evidenced by the extrusive domes of stratovolcanoes (the post-caldera phase). Basaltic (areal- and shield-type) volcanism is also renewed.

The volcanic relief of the Kuril-Kamchatka Arc is characterized by the directed-cyclic change of forms over time (Braytseva et al. 1974). There are two large cycles with a single general trend: from shield to pyroclastic volcanoes via lava covers. It is accompanied by the reduction of circular volcano-tectonic depressions, while the areas of pyroclastic and pumice covers decrease and increase, respectively. The degree of sintering of pyroclastic material is reduced. The first cycle comprised Early Pleistocene and the first half of Middle Pleistocene. The second (uncompleted) cycle started 50–60 thousand years ago, again with basaltic outpouring.

The cycles differ in volcanic landscapes themselves. In the first cycle, the predominant landscapes are positive volcanic forms and vast lava plains. In the second cycle, there is a partial inversion of the relief, i.e., volcanic massifs are replaced by the negative forms – volcano-tectonic depressions (Braytseva et al. 1974). These depressions go with the surrounding pyroclastic plains, which bury and vary the previous relief. Later on, the relief was complicated with new basaltic outpouring. Both cycles are well-marked in Kamchatka; only one cycle (mainly the second one) is marked in Kuril Islands and, on the contrary, the first cycle is marked in Commander Islands.

According to I.V. Melekestsev's paleogeographical maps (1974), there was the lowest stand of the ocean level in the Kuril-Kamchatka region in the second half of Late Quaternary glaciation (after 40 thousand years ago), and Kunashir, together with the entire Lesser Kuril Ridge, was connected to Hokkaido. Consequently, the insular regimes of Kunashir and Shikotan were disturbed not long ago. Paramushir was connected to Kamchatka. Other islands of the Greater Kuril Ridge remained isolated since the very beginning of Pleistocene.

In the epochs of quaternary cold snaps in South Kuril Islands, including Kunashir, glaciation spread only to the upper layer of the mountains, as evidenced by the traces of glacial forms of relief remaining on the Mendeleev Volcano (Nikolyskaya 1969). The piedmont plains, as well as the lower and middle layers of low mountains, remained extraglacial areas, which ensured the existence of paleobotanic refugiums and the possibility of extensive migration of plant communities from the Asian continental margin to Sakhalin and Kuril Islands connected to the continent in the epochs of glaciations and sea regressions. Thus, the low-mountain plant cover of the South Kuril Islands has developed continuously since the very beginning of their emergence and has never been isolated from the vegetation of the entire Far East South. This fact allows us to consider Mendeleev experimental testing ground in the general context of evolutionary landscape genesis on the Pacific Ocean mega ecotone of Northern Eurasia.

4.3. The Mendeleev Volcano

The Mendeleev Volcano (absolute height, 890 m) coincides with the longitudinal fault along the crests of the inner ridge of the Kuril geosynclinal system (Bezrukov et al. 1958). It is among 27% of all active single volcanoes on the Kuril Islands and has a structure of the Somma–Vesuvius type, which is most representative of an island arc (Gorshkov 1958), i.e., belongs to extrusive volcanoes (see on the front cover of given book) or, more precisely, to the type of active "clustered" (multi-vent) isolated stratovolcanoes (Melekestsev 1980; Zlobin et al. 1997). A rather large cone (up to 12 km in diameter) appeared on the tertiary foundation (Aprodov 1982). At the beginning of formation of the volcano, andesites and andesitobasalts were poured out but, gradually, the rocks became more and more acidic, and the final stage was characterized by explosions and ejections of acidic pumices (Gorshkov 1967).

The acidic volcanism, which developed in the middle of Late Pleistocene interstadial, brought a new pumice cover on the island. The washed-up pyroclastic material of this stage became part of the deposits of an 80–85-m marine terrace (Brytseva et al. 1974). On the coast, this terrace forms steep rocky cliffs with hexagonal columns of quartz andesite (see above Figure 3-5).

The total terrain of the ecoregion has a distinct NW ↔ SE asymmetry. The Pacific pedestal of the volcano is narrower and steeper, with narrow marine terraces and short deeply cut valleys of small streams. Here and there, the lava tongues of late eruptions are laid over the surface of marine terraces (Korsunskaya 1958). The Sea-of-Okhotsk pedestal, on the contrary, is more gentle and broader; in the foreland, it is deeply cut with a well-developed river network, with turf-covered creek valleys and sea cliffs. Thus, the Sea-of-Okhotsk submontane sector of the southern part of Kunashir has a much more developed river network compared to the Pacific Ocean sector. It may be supposed that the landscapes of the Sea-of-Okhotsk sector of the southern part of Kunashir have a more ancient foundation and the longer development pathway compared to the landscapes of the Pacific sector.

The Mendeleev stratovolcano is set on a system of faults, normal faults and fracture zones of the Earth's crust determining the high magma permeability of the latter. The three successively formed volcanic cones were later destroyed as a result of explosive eruptions with the formation of caldera-crater depressions of 6×9, 3×3.5 and 1×1 km in size (Kozlov and Zharkov 2012). The age of the most ancient caldera is about 40000 years (Braytseva et al., 1974). The formation of the last caldera-crater depression about 2500 years ago was followed by the growth of an andesitic-dacitic extrusive dome (Abdurakhmanov et al. 2004).

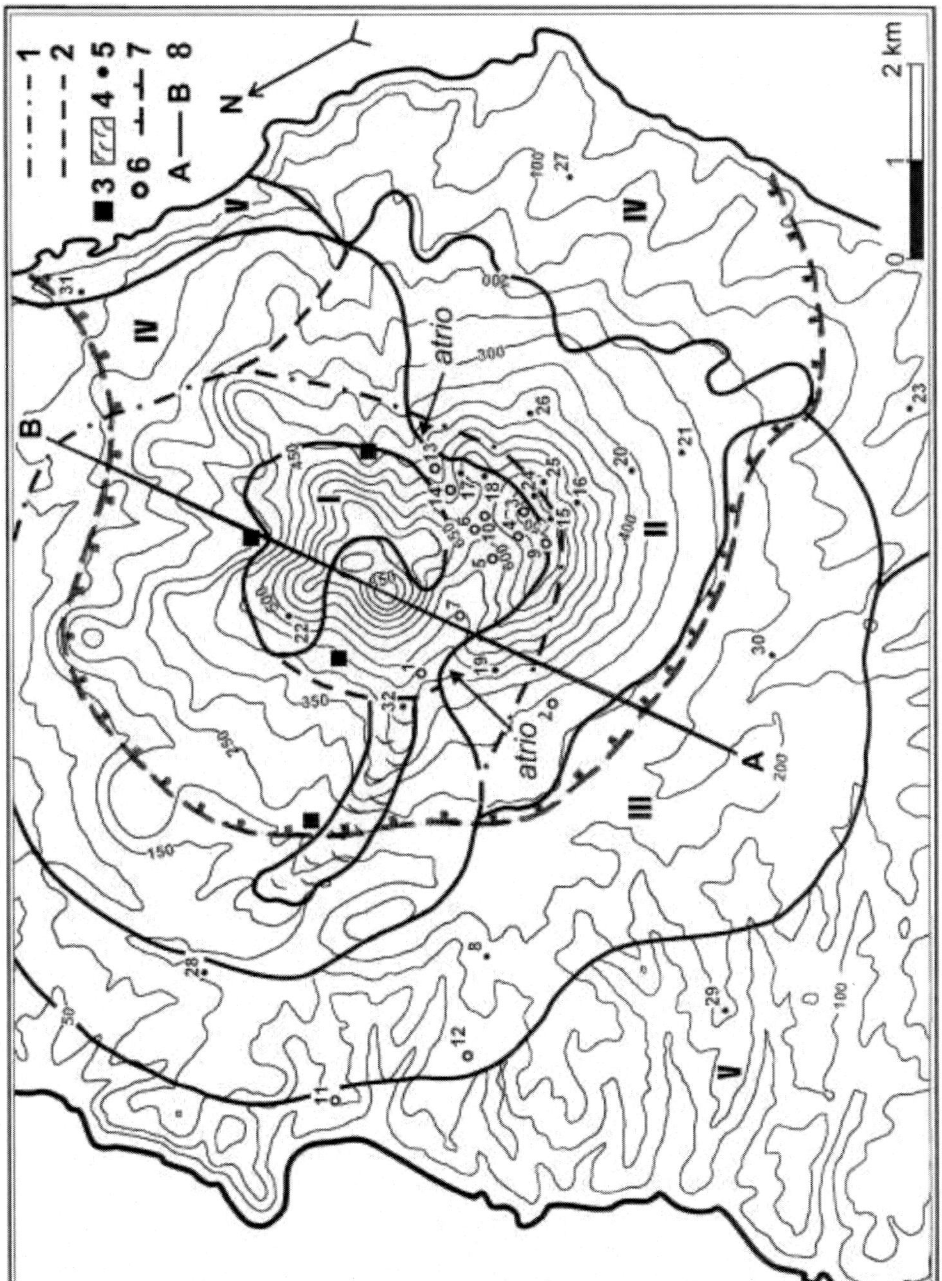

Figure 4.3. The hypsometric map of the Mendeleev Volcano and its surroundings (by L.S. Sharaya), with indication of the boundaries of the main geological structures and solfatare fields, as well as the test grounds of landscape ecological survey. Geological structures (Gorshkov 1967): I – central cone; II – second somma; III – first somma; IV – nonsegmented mass (partly tertiary); V – foundation (tertiary terrigeneous–pyroclastic rocks). Other designations: 1 – the outlines of ancient caldera; 2 – the outlines of young caldera; 3 – solfatare fields; 4 – pyroclastic flow; 5 – test grounds; 6 – test grounds with temperature anomalies in soil; 7 – the boundaries of geothermal fields of the volcanic cone (Safonov et al. 1998); 8 – the line of seismic profiling (see below Figure 4. 4). *23*, *30*, *21*, … – the numbers of sample plots (points).

The volcano is composed of two somas of different age (the annular crests with outer slopes) and a central cone, which is complicated by strong destruction of its separate parts and the presence of collateral formations (Gorshkov 1958). The most ancient volcanic formation (of the preglacial period) is a large caldera (a vast depression on the volcano top) of 6–7 km in diameter, with its strongly destroyed ridge remaining only in the northwest. The formation of caldera was accompanied by the formation of the first somma cloak, i.e., loose pyroclastic strata of the Golovnin Formation, according to E.K. Markhinin (1956). This somma currently encircles the lower parts of the SW–W–NW slopes of the Mendeleev Volcano and partially covers the piedmont plains of its Pacific sector (Figure 4.3).

In the epoch of Late Pleistocene (Würm) interstadial, i.e., about 39000–40000 years ago (Gorshkov 1967; Braytseva et al. 1974), a new cone of no less than 6–8 km in size appeared at the southern edge of the first caldera, which in its northern part was destroyed by directed blast. The second caldera (3–3.5 km in diameter) was formed, and the synchronous second somma on the S–SE–E slopes of the volcano (from the tops to theabsolute heightsof 150–200 m) overlaid the first somma. Here, the crest of the second caldera remained as a small semicircular ridge, (see Figure 1.6), with a plateau-like type of the *Mechnikov Mountain* (the highest point is 798 m of absolute height). The most of our test grounds were situated on the slopes of this somma (down to its foothills, with access to the coastal marine terraces), as well as in the atrio – a semicircular depression separating the somma from the central dome (see Figure 4-3). The northern half of the second somma partially filled the first caldera, while the position of the second caldera is fixed by the four major solfatares – fumarole fields.The presence of caldera indicates the existence of a large hollow magmatic focus under the volcano (Fedotov 1991), while fumaroles are indicators of the presence of productive geothermal reservoirs in the depths of the volcanic dome (Droznin et al. 2004). The Northeastern and Northwestern fumarole fields (solfatares) and most active; there, magmatic energy is discharged not only by releasing volcanic gases and thermal solutions but also as periodical microquakes (with a focus no less than 20 km deep).

Volcano is considered active if the quiescent stage in its eruptive history is no longer than 3200–3500 years (Melekestsev et al. 2001). Consequently, the Mendeleev Volcano is one of the currently active volcanoes. The last stage in volcanic development was the formation of explosion craters along the ring faults in the periphery of the extrusive dome, resulting in the formation of active fumarole fields with gas and hot water (90–110°C) vents on the N, NW and NS slopes of the volcano. There are four well-marked explosive craters set on the ellipsoidal trajectory of the fault, which outlines the entire eruptive system of the Mendeleev Volcano (Markhinin 1956). The main present-day fumarole activity of the volcano shows itself in these four blasting cones, as well as in the crater of the central cone (Figures 4.4 and 4.5).

The latest (pyroclastic) eruption of the Mendeleev Volcano occurred in 1880, and its steam and gas activity was observed in 1901, 1946 and 1977. The current state of the volcano figuratively called "the Sleeping Beauty" by people, with reference to volcanic landscape itself, will be denoted as a *passively active state* and its fumarole activity will be denoted as a manifestation of *passive magmatic energy*. The intensive gas-hydrothermal activity of the volcano began to develop exactly at the final Late Quaternary (andesitic) stage of its history (Chudaev and Chudaeva 2009) and, thus, characterizes its passive active state.

Figure 4.4. The solfatare field on the northeastern slope of the Mendeleev Volcano. The belt of stone-birch/spruce forests changing higher into the belt of cedar elfin wood is in the background.

Figure 4.5. Fumaroles in the mouth of Mendeleev Volcano.

The fumarole activity is caused by the high fluid conductance of volcanic-tectonic zone at the final stage of its progressive development, when fluid mass transfer becomes a self-organizing system (Letnikov 2000). Due to the intrusion of fluid from the subducted oceanic platform into the mantle wedge, the typical island-arc magmas are melted in the high temperature zone. The most of hydrous minerals are dehydrated under the frontal zone, and the aqueous fluid migrates upward, directly into the zone of magma formation in the mantle wedge (Avdeyko et al. 2001). It is just the beginning of formation of volcanic hydrotherms.

The steam and gas vents are often situated linearly, thereby accentuating their confinement to fracture systems (Markhinin 1956). The IR thermal imaging of the northeastern fumarole field of the Mendeleev Volcano in 2010 showed the outlines of the most heated regions over the entire field (Kozlov and Zharkov 2012). Altogether, 180 thermograms were obtained for the volcano, with detailed representation of the temperatures at the sites of the northeastern slope of this extrusive dome.

The internal structure of the volcano is a system of near-surface and deep magmatic foci (Figures 4.6 and 4.7). The major peripheral focus of ~2 km in diameter is situated directly under the volcano at a depth of 4.5 km; the deep magmatic focus is located at a height of 30–60 km. The upper lava complex of the ancient somma (I, see Figure 4.3) is formed by andesitic flows with subordinate pyroclastics (Markhinin 1956). The basic composition of lavas is quartz dacite with olivine. The underlying part of the volcanic cone with predominant area is represented by the upper effusive andesitic complex (II). The volcano bottom is outlined with an intermediate pyroclastic complex (III) consisting of tuffs, tuff conglomerates, and coarse pyroclastics. When reaching the coastal plain areas, these formations intermingle with the sea-washed terrigenous deposits and create a terraced erosion-accumulative complex.

Since the entire volcanic massif is sheared closer to the Pacific Ocean coast, the Sea-of-Okhotsk sector of the submontane morphostructural belt is more gentle and extended, with a much more developed river network (Korsunskaya 1958). Finally, along the Pacific Ocean coast there is a narrow band of the lower effusive complex (IV) formed by andesitic-basaltic lava flows up to 8 m thick, which underlie the Middle Pleistocene marine deposits and are exposed in sea cliffs (see Figure 3.5).

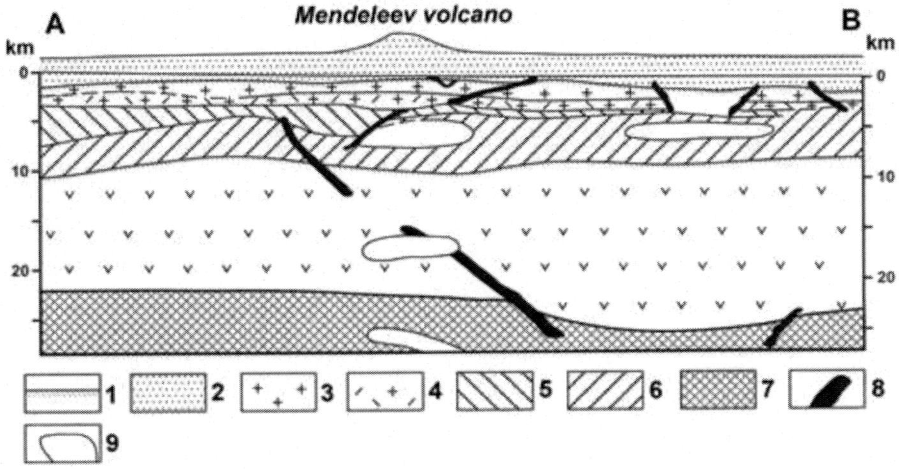

Figure 4.6. The depth profile of the lithosphere across the Mendeleev Volcano by the AB profile (see Figure 4.3) (Zlobin et al., 1997). 1 – the roof of the basaltic layer; 2 – the volcanogenic sediment layers; 3 and 4 – the upper and lower parts of the granite-gneiss layer; 5 and 6 – the upper and lower parts of the granulite-gneiss layer; 8 – the basaltic layer; 9 – the areas of melting of the lithosphere.

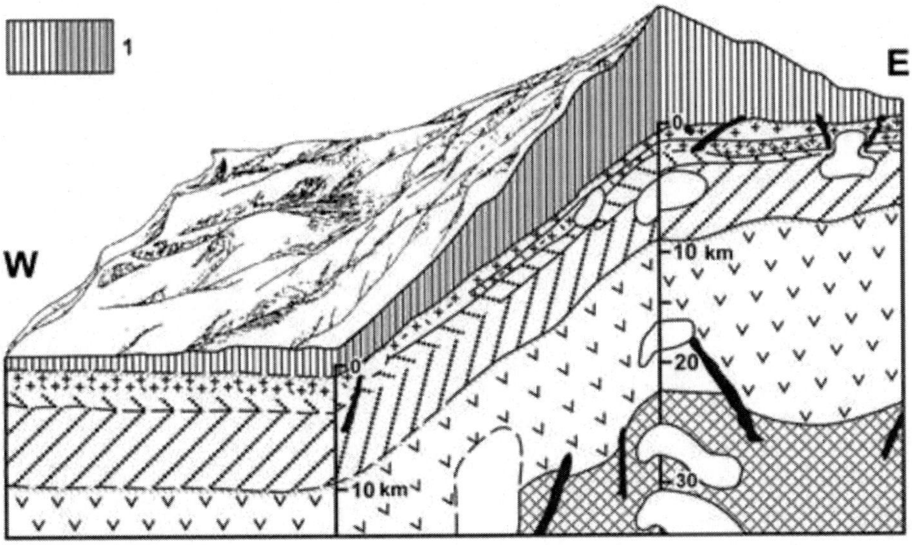

Figure 4.7. The 3D model of the lithospheric structure under the Mendeleev Volcano (Zlobin et al., 1997). 1 – the volcanogenic sediment layers. Other symbols are as in Figure 4.6.

The fumarole fields are known (Droznin et al. 2004) to be the objects of thermal discharge of the steam-and-gas and magmatic activities of the volcano and are connected with productive geothermal reservoirs. Let us note this important circumstance exerting, as we will see below, a substantial effect on soil temperatures and causing the respective structural and functional changes in the soil-plant cover of young volcanic landscape.

At present, the active volcano on the Kunashir Island is only the known Tyatya Volcano. Its last large-scale eruption was in 1973. However, the solfatares of even currently inactive Kuril volcanoes, like the Mendeleev Volcano, carry up approximately the same amounts of juvenile water and other steam-and-gas magma components per year as those that are released in explosions during eruptions (Markhinin 1980). The hot springs at the bottom of the Mendeleev Volcano with a temperature of 30–60°and more are connected with the circulation of deep thermal waters. Jarosites and geyserites are formed along their flow paths, each covering an area of about 3000 m^2.

Chapter 5

Soil-Vegetation Cover and Microlandscapes in the Island-Arc Volcanic Ecoregion

Abstract

The empirical statistical models of the island-arc stage of continental biosphere development in the North-West Pacific have been created by the example of experimental testing area near the active Mendeleev Volcano in Kunashir Island (the South Kuril Ridge). The vegetation and soil covers and the microlandscapes have been described. It was shown that the starting (island-arc) stages of subaerial biogeo- and landscape genesis were charcaterized by high floristic and botanical-geographical diversity with the formation of buffer plant communities.

5.1. Forest Vegetation

The South Kuril Islands (Kunashir, Shikotan, Iturup and Urup) previously belonged to the Kuril-Sakhalin district of the Japanese–Korean Oceanic Province of the Far-East coniferous/broad-leaved forest region (Vorobyev 1963). According to this scheme, the South Kuril Islands, together with Hokkaido, form a single botanic-geographic region, which generally corresponds to the paleogeographic unity of these islands since the second half of the Late Pleistocene and the Early Holocene (Melekestsev 1974). However, according to the latest botanical-geographical demarcation (Kotlyakov et al. 2009), Kunashir has been included into the South Kuril-Hokkaido district and the South Kuril region, with the predominance of dark-coniferous/broad-leaved forests enriched in southern east-Asian elements (*Magnolia hypoleuca, Quercus crispula, Q. dentate, Acer tschonoskii, A. japonicum, Betula maximowicziana, Daphniphyllum humile*). Thus, the region of our experimental test site on the Mendeleev Volcano, with reference to plant cover, is most similar to the neighboring Hokkaido: about 78% species in common with the latter (54% with Sakhalin and 44% with Priamurye). In Kunashir Island, along the low-lying isthmus connecting the Pervukhin and South Kuril Bays, there is also an important botanical-geographical boundary between the South Nemuro–Kunashir and Central Kunashir–Iturup subregions (Vorobyev 1963). This boundary coincides with the first-order bioclimatic border in the Northern Pacific, which separates Southern Kunashir from other parts of the island and, accordingly, temperate macro biocilmates from boreal macro bioclimates (Figure 5.1).

The South Kuril Islands, especially Kunashir, are an example of extremely high diversity and contrasts between the *non-relict types of vegetation* of the temperate belt in the Northwestern Pacific (Popov 1963; Pietsch et al. 2003), from mountain tundras to coniferous/broad-leaved and broad-leaved forests with abundant evergreen subtropical and relict species (see Figure 5.1). Among the latter, there are the species both present and absent on the continent: holly (*Ilex crenata* Thunb), skimmia (*Skimmia repens* Nakai), yew (*Taxus cuspidata* S), magnolia (*Magnolia abovata* Thunb), onoclea (*Onoclea sensibilitis* L.), etc. However, the number of endemics is insignificant because of heavy volcanic activity of the Kuril Islands. The morphological similarity of endemics to species of the neighboring territories (Sakhalin, Hokkaido, and Kamchatka) is a result of recent isolation of the islands (Vorobyev 1963). Not only broad-leaved but also coniferous forests of South Kuril Islands are saturated with relict species (Barkalov 2009). It is accounted for by the fact that they have retained the appearance and species composition of the mountain preglacial forests of East Asia due to the actual absence of Quaternary glaciations on the islands and the mild oceanic climate (Popov 1963; Nakamura, Krestov 2005).

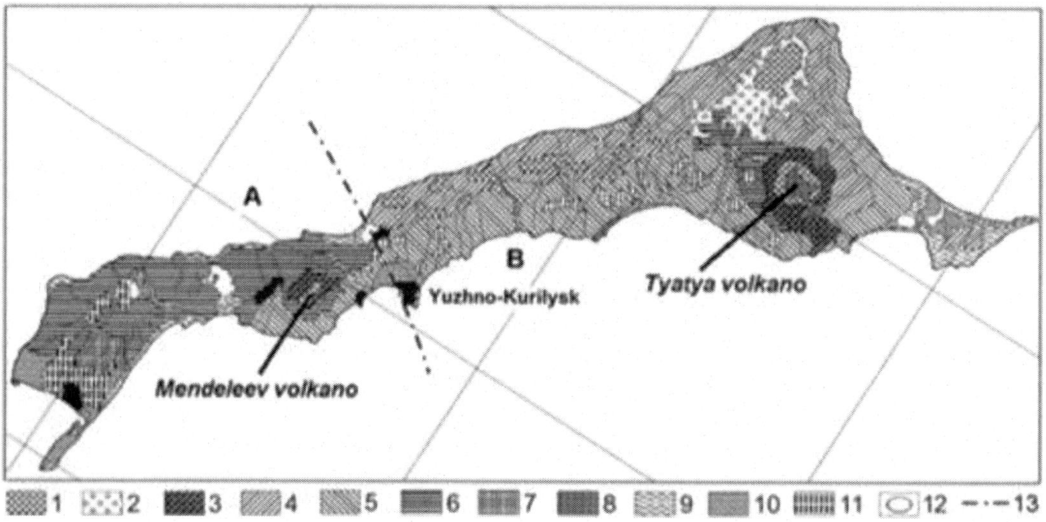

Figure 5.1. The flora of the Kunashir Island, according to (Kotlyakov et al. 2009), with simplifications. Mountain tundras, meadows and elfin woods: 1 – cedar and alder elfin woods, sparse, dead-cover, at the upper limit, often in complex with heaths and shrubs; 2 – elfin wood/bamboo thickets, here and there with stone birch light and crooked forests. Forests, light and crooked forests: 3 – stone birch forests, here and there crooked forests, with herbs, elfin woods and bamboo; 4 – fir-spruce and spruce-fir forests, mountain and piedmont, with stone birch and bamboo; 5 – fir-spruce and spruce-fir forests of lowlands, with low fir woods on the coast and with spruce forests on waterlogged and volcanogenic sites; 6 – fir-spruce and spruce-fir forests with ferns and mosses, in some areas with lianas, and with the fragments of dark-coniferous/broad-leaved forests (oak, dimorphant, magnolia, yew) at the foot of the slopes and in river valleys; 7 – oak-birch-maple forests enriched in broad-leaved species. Vegetation on volcanic rocks: 8 – vegetation at different stages of successions of old and ancient (hundreds and thousands of years) overgrowing effusives. Vegetation of lowlands and sea coasts: 9 – shrubby-forb, reedgrass and herb meadows; 10 – herb, hummock, sphagnum bogs and shrubby/sedge moist meadows. Anthropogenic vegetation: 11 – agricultural lands. Other symbols: 12 – lakes; 13 – the boundary between the temperate (A) and boreal (B) macro bioclimatic regions of Northwestern Pacific, according to (Krestov 2006).

The belt of moist taiga forests with tertiary hardwoods and lianas was formed over the entire coast of the Sea of Okhotsk as early as in the pre-Quaternary period and had remained almost unchanged to this day, closing on Sakhalin, South Kuril Islands and Hokkaido with Japanese taiga per se. This fact suggests the existence of a center of formation of subalpine and boreal east-Asian species on this coast, to the south from 47°, in the Cretaceous–Paleogene. Such center formed a highland floristic complex later enriched in nemoral and ultra nemoral species during tectonic motions and glacio-eustatic fluctuations of the ocean level (Urusov 1988). In the epochs of Pleistocene cold snaps, the forest boundary became lower but the ensuing subalpine coenoses remained closed and preserved the numerous boreal and nemoral elements. It resulted in the formation of well-developed sod-humus soils on high marine terraces, with considerable accumulation of humus (Ivlev 1977).

Figure 5.2. Satellite imagery of plant cover on Mendeleev volcano and its encirclement. 1–6 – numbers of groups of forest types (see in the text); *F* – fumarols.

Thus, the floristic richness and phytocoenotic diversity of forest ecosystems, as well as adequate soil development, so typical of the marginal-continental sector of the Pacific Ocean mega ecotone, emerged already at the initial island-arc stage of subaerial biogeogenesis, and the loci of formation of buffer communities were the phytocoenoses of not only dark-coniferous but also subalpine belt. Most of the groups of forest types and floristic phratries had a mixed composition, i.e., buffer forest communities developed already at the island-arc Pacific stage of aerial biogeocoenogenesis and were the basis of young land forest cover.

In general, the experimental test site on the Mendeleev Volcano characterizes such a spatio-temporal section since the starting (island-arc) stages of subaerial biogeo- and landscape genesis, which is a basis of quite considerable species diversity of the forming zonal and extrazonal types of plant cover (Kolomyts and Surova 2018 a,b). The plant groups concentrated in the vicinity of the volcano are characterized by high floristic and botanical-geographical diversity (Manyko and Rozenberg 1970); therefore, the floristic and phytocoenotic spectra of the test site are rather broad.

Distribution of these groups of forest types over the Mendellev volcano landscape is shown in Figure 5.2.

Here, we have distinguished six groups of forest types (the symbols and taxonomic norms of absolute heights are given in parentheses): (1) mixed dark-coniferous/broad-leaved forests (– CB; 184 m; Figures 5.3–5.., see the insert); (2) dark-coniferous (spruce/fir) forests (– DC; 379 m; Figure 5.6); (3) buffer stone birch/spruce forests (– SbS; 575 m); (4) stone birch elfin woodland with cedar elfin wood (– SbCe; 685 m; Figure 5.7); (5) cedar elfin wood (– CeW; 718 m; Figure 5.8); (6) buffer cedar-elfin-wood/birch/spruce communities (– CeSbS; 575 m).

These types belong to the respective floristic phratries: (1) – Manchurian–Okhotian (MO); (2) and (3) – Okhotian (Okh); (4) – Beringian–Kamchatian (BK); (5) – Beringian (Ber); and (6) – Beringian–Kamchatian–Okhotian (BKO).

Figure 5.3. The mixed dark-coniferous/broadleaf forest in the piedmont zone of the Mendeleev volcano landscape.

Figure 5.4. Japanese yew pointed (*Taxus caspidata*) – relict subtropical tree species in zone of broadleaf and mixed forests on seaside and piedmont planes of South Kunashir.

Figure 5.5. Lianas in coniferous-broadleaf forest on sea by side plains of South Kunashir.

Figure 5.6. The mountain spruce/fir forest in the lower layer of the Mendeleev volcano. There is a well marked alteration of generations of the stand.

Figure 5.7. Stone birch elfin woodland with Kuril bamboo on sea by side planes of Kunashir island.

Figure 5.8. The subgoltsy zone with cedar elfin wood and goltsy in the upper zonal layer of the Mendeleev volcano.

As has been noted earlier, the structure of altitudinal zonality at the experimental test site of the Mendeleev Volcano (see Table 3.1 and Figure 3.4) is typical of South Kuril Islands (Manyko and Rozenberg 1970). For example, in the south of Iturup Island situated in the same SouthKuril geobotanical region as Kunashir, the upper boundary of coniferous/broad-leaved forests is at the height of 200–250 m, while the belt of dark-coniferous taiga goes up to 500–600 m (Urusov 1988). Let us consider the main floristic and phytocoenotic features of the distinguished altitudinal belts.

The buffer Manchurian–Okhotian phratrie is represented by coniferous/broad-leaved forests being fragments of the lowest bioclimatic belt of the Mendeleev Volcano and its surroundings. Prior toLate Holocene (2–3 thousand years ago), this ecoregion (as well as the entire southwest of Kunashir) had been occupied by purely nemoral forests (Razzhigaeva et al. 2011) of the Manchurian flora with subtropical relicts. The transgression of fir/spruce forests and the emergence of buffer Manchurian–Okhotian communities was related to the cold snap in the first half of Late Holocene.

In these forests, the predominant broad-leaved species are *Quercus dentata* and *Q. crispula* and maple (*Acerpictum*) in the second layer. There are also elms (*Ulmus laciniata, U. propinqua*), cherries (*Cerasus sachalinensis*), Sakhalin cork tree (*Phellodendron sachalinense*), bird cherry trees (*Padus ssiori*), etc. The nemoral forest stand has an admixture of dark-coniferous species: Sakhalin fir (*Abies sachalinensis*) and two spruce species (*Picea microsperma, P. glehnii*). There are many lianas (see Figure 5-5): *Actinidia kolomikta, A. arguta,* and *Hydrangea petioralis*. It is also characterized by the presence of exotic subtropical species: magnolia (*Magnolia abovata*) and dimorphant (*Kalopanax septemlobus*), as well as the most ancient Far East relict (see Figure 5.4) – Japanese yew (*Taxus cuspidata*). The predominant species of the lower layer are ferns (*Dryopteris austriaca*) and nemoral forbs (*Cacalia kamtschatica, Veronica sibirica, Aconitum maximum, Majanthenum dilatatum, Oxalis acetosella, Trillium kamtschaticum,* etc.), the thickets of aralia (*Aralia cordata*) and impenetrable Kuril bamboo (*Sasa kurilensis*) sometimes up to 1.5–2 m in height (see below Figure 7.4).

The dark-coniferous forests of the Okhotian phratrie make up the main forest resources of South Kuril Islands (Popov 1963). They occupy the largest part of Kunashir, including the test site on the Mendeleev Volcano, and form an altitudinal belt following coniferous/broad-leaved forests. The northern limit of their distribution is in the south of Iturup (Vorobyev 1963). The forest-forming species arethe Ajan, small-seeded and Glehn's spruces (*Picea ajanensis* Fisch., *P. glehni* Mast.), Sakhalin and Mayra fir (*Abies sachlinensis* Mast., *A. mayriana* Miyabe et Kudo). Hydrangea (liana) and fern fir forests (Vorobyev 1963) with the above-mentioned subtropical relicts (dimorphant and yew) are widespread in Kunashir. The lower layer of fir/spruce forests can be dead-cover/bamboo and dwarf shrub/forb. In the latter case, it includes red bilberry, Canadian dogwood, Kamchatka bead ruby, Kamchatka bilberry, hare's-eared wood sorrel, spiral-pine, etc.

These forests may be both poly- and monodominant, with certain cycles of changes in their species composition (Popov 1963). Under the most favorable soil and climate conditions and at more waterlogged sites, the predominant species is fir, which also dominates at the early age stages of a forest community. Spruce dominates in the age of 120–140 years, and fir becomes dominant again after the decline of the 180–200-year-old generation. In polydominant communities, spruce usually takes the first layer and fir takes the second layer. Other species occur here too: dwarf elm (*Ulmus laciniata*) and oak (*Quercus crispula*), large stems of stone birch (*Betula ermanii*), alder Maksimovič and maple (*Acer pictum*). Under the extreme conditions of the upper boundary of dark-coniferous forests (on windswept and cooled sites), spruce is most resistant and forms single-composition tree stands. Glehn's spruce occupies boggy areas or salinated (close to fumaroles) sites, which are the least favorable for other coniferous species.

The mountain dark-coniferous forests were often exposed to clean felling and fires in the past. In the middle of the 1940s, the coniferous forest stands on the eastern slope of the Mendeleev Volcano were no more than 75 years old, mainly 40–50 years old (Korsunskaya 1958).

On the piedmont plains and in the lower mountain layer there occur rather exotic fir-spruce forests with the admixtures of subtropical relicts: magnolia, dimorphant, yew, skimmia, hollies, Araliaceae, as well as with liana abundance (Vorobyev 1963). In general, dark-coniferous forests are enriched in thermophilic species (Manyko, Rozenberg 1970). Such communities are essentially transitional forms from dark-coniferous taiga to spruce/broad-leaved forests as buffer phytocoenotic communities – the evolutionary phenomenon of the boreal belt on the Pacific Ocean mega ecotone (Kolomyts 2015 b). This fact indicates that the evolutionary forest-forming processes, which are based on species migration and interpenetration of communities (Krištofovič1946; Vasilyev 1946), are quite clearly manifested not only in the marginal-continental sub-Pacific but also at the earlier stage of subaerial biogenesis. The basic evolutionary processes of forest formation creating the respective floro- and phytocoenotic structures are already established in the island-arc neo-Pacific, followed by their more diversified development on the continental margin.

The Kamchatianflora is represented by crooked and, less frequently, straight stone birch forests widespread in Kuril Islands (Figure 5.7). These communities consist of Erman's birch (*Betula ermanii* s.l.) with the admixtures of alder Maksimovič and Sakhalin willow, with lianas (hydrangea), often with the undergrowth of cedar elfin wood. At some sites they are supplemented by the shrubby forms of Japanese yew (*Taxus cuspidata* S. et Z.), as well as yellow maple (*Acer ukurunduense* Trautv. et Mey), wahoo (*Euonymus*), Kuril cherry (*Cerasus kurilensis*), and dwarf Kuril bamboo (Korsuinskaya 1958). In Kunashir, stone birch is above the dark-coniferous forest belt; however, here and there it descends right to the seashore (Vorobyev 1963). The stone birch forest belt is more clearly marked on the western and northwestern (Sea of Okhotsk) slopes of the volcano (Manyko, Rozenberg 1970).

Finally, cedar elfin wood (*Pinus pumila*) – the typical representative of sub-Arctic Beringian flora – forms independent subgoletz communities (see Figure 5.8, see the insert). It is of Angaridian origin and had appeared in KurilIslands even before they separated from Japan, Sakhalin and the continent of Asia, i.e., is a rather ancient formation. In the modern epoch, cedar elfin wood slightly expands its areal (Tikhomirov 1946). On the slopes of the Mendeleev Volcano, its thickets often pass through the belt of stone birch crooked forests and intrude into dark-coniferous and even coniferous/broad-leaved forests. It is due to the broad ecological amplitude of cedar elfin wood, which is determined by: (a) its cup-shaped crownand the ability to lodge after the first frosts and be covered by snow, which makes it resistant to wind exposure and snow corrosion and (b) the development of adventitious roots as a tool for peculiar movement of individuals towards more favorable conditions of existence (Tikhomirov 1946).

However, cedar elfin wood significantly changes its life forms while descending from the peaks to the bottom of the volcano. In the subgoletz belt, its height varies from 2.5–3.0 m to 4.0–4.5 m, with actually impassable thickets of tree stands and with a dense tangle of branches (when working at such a test site, one sometimes had to walk on tops of thick tree stands without touching the earth surface); at the foothills and on coastal plains, it is no more than 0.5–1.0 m (see Figure 5.8). In the latter case, such intrazonal associations of cedar elfin wood were not included in the list of test sites and, therefore, were not represented as independent communities in our landscape-ecological survey.

InKunashir, stone birch forests usually occur more often as separate plots; therefore, dark-coniferous communities often turn directly into the thickets of cedar elfin wood (Korsunskaya 1958). On the southern and southwestern slopes of the Mendeleev Volcano and the Mechnikov Mountain, in the subgoletz belt, the bottoms of narrow gullies (barrancos) are occupied by the narrow bands of stone birch crooked forest, while the gently convex spaces between the narrow gullies are covered with cedar elfin wood.

On the other hand, the forests of the mountain birch *Betula ermanii*, with the undergrowth of cedar elfin wood at the upper limit of their distribution, demonstrate particular transitions to independent associations of *P. pumila* (Tikhomirov 1946). In both cases, the buffer communities are formed: binary (KbE, KbKs) and ternary (KsBE). It is indicative of genetic affinity between the Okhotian, Kamchatian and Beringian phytobiotas and demonstrates that migration processes in the plant cover of the upper belts of island mountains, as well as in their lower belt and on the piedmont plains, are equally developed. The compositions of some groups at the lower boundary of stone birch crooked forest is typical of coniferous forests (with the involvement of yew, maples, holly, and a number of herbaceous plants), which assumes historical lowering of the upper boundary of the dark-coniferous forest zone in South Kuril Islands (Vasilyev and Rozenberg 1977).

Distribution of groups of forest types (*GFT*) forMendeleev volcano landscape is presented in Figure 5.9. This map wasplotted using the equationof multivariate logistic regression (Kolomyts et al. 2016):

$$\ln(GFT) = 0.002617 \cdot Z_{+14.13} - 0.8907 \cdot \cos(A_{45})/GA_{-2.79} +$$
$$+ 0.01125 \cdot [\mathrm{GrW}(10\text{–}20)]_{+2.44} - 0.03613 \cdot t\,(40)_{-1.59} + 0.1289.$$
$$R^2 = 0.919;\ Degr = 5.0\%);P<10^{-6}, \qquad (5.1)$$

Here, ln(*GFT*) is thelogarithmic series of groups of forest types distributed by altitude; Zis theabsolute height; $\cos(A_{45})/GA$ is the northwestern component of exposure to gentle sloping; GrW(10–20) is themoisture vertical gradient in the soil layerof 10–20 cm; and*t* (40) is the soil temperature at a depth of40 cm. The subscripts of each predictor are *t*-statistics; their module determines the significance of contribution of predictors to the spatial variability of temperature. R^2 is the coefficient of determination and *P* is the statistic characterizing the probability of error in the establishment of statistical dependence.

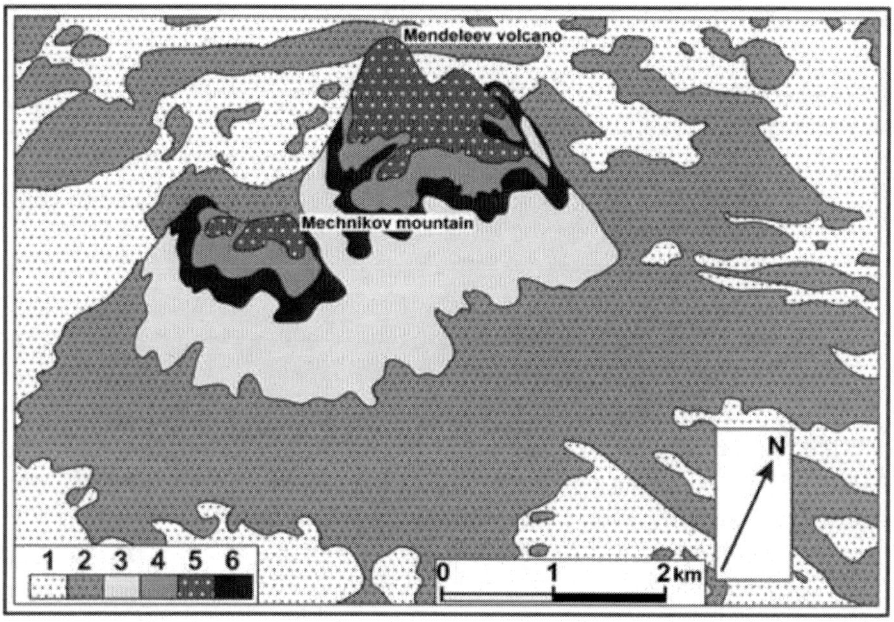

Figure 5.9. The map of the groups of forest types in the Mendeleev volcanic landscape according to Zalihanov et al. (2010). For symbols and explanations, see the text.

The model (5.1) describes 92% of *P* variability in space by means of comprising factors. The model was successfully verified (tested) by Allen's cross-validation (Allen 1974): the empirical criterion (*Degr*< 50%) shows that the predictive strength of the model is sufficient for predictive mapping or calculating the characteristic at new points.

5.2. Soils and Soil-Forming Processes

The formulated statements of the theory of soil formation in the zone of continent-ocean transition are of great significance for development of the concept of evolutionary landscape genesis (Lashkov 1948; Sokolov 1973; Ivanov 1976; Targulyan 1982; Ivlev et al. 1987; etc).

These statements have been substantiated by the studies of soil morphology and different soil-geochemical characteristics.

First of all, it is necessary to characterize the soil-forming rocks of the experimental test site. The Mechnikov Mountain (somma II of the Mendeleev Volcano) and the periphery of the Golovnin caldera are based on andesites and basaltic andesites. These are the remains of small stratovolcanoes of different types and ages, which appeared in the first half of Middle Pleistocene (Melekestsev et al. 1974). Consequently, all altitudinal belts of the Mendeleev Volcano up to the height of about 400 m were formed on the mid-Quaternary basic and acidic dense rocks (basaltic andesites) exposed to preliminary tectonic and exogenous destruction. However, these are not lava covers. The landscapes on loose volcanic (pyroclastic) sediments are below 350–400 m of absolute height; the volcanic sediments washed by river streams are below 300 m, followed by volcanic-marine (below 200 m) and merely marine (below 100 m) sediments.

Volcanic ash multi-humus, acidic allophanic soils (andosols) are widespread on volcanogenic landscapes of the South Kuril Islands under the conditions of moderately warm and humid climate (Glazovskaya 1973). The silicate forms of aluminum compounds (allophans) typical of these soils are determined by the chemical composition of soil-forming pyroclastic material. The amorphous forms of iron, which are predominant in the soils of the Mendeleev Volcano, are composed mostly by amorphous humus compounds (Gladkova and Butovets 1988).

The andosols of South Kuril Islands, including Kunashir and the Mendeleev Volcano, have the following physicochemical characteristics (Glazovskaya 1973; Sokolov 1973; Rode 1973; Ivlev et al. 1987; Gladkova and Butovets 1988):

1. The predominance of coarse-dust and fine-grained sand with high water permeability and an equally high moisture capacity in fine-grained soil; under conditions of exceptional humidity, it ensures the washing regime of soils, while the predominance of coniferous vegetation ensures the alfehumus type of soil formation;
2. The high intensity of weathering of volcaniclastic sediments, which results in the dominance of allophan and amorphous hydrogen oxides as components of the newly formed minerals, as well as high absorption capacity and an equally low base saturation;
3. The low content of crystalline clay minerals favoring the intensive leaching of calcium and magnesium under conditions of high filtration capacity of soils and yielding acidic and unsaturated soils, respectively;
4. The weak argillification and ferrugination of thesoil profile and, on the contrary, its high humification;
5. The strong impact of organogenic factor (the high content of ash in the litter and tree waste) and deep impregnation of soil profile with humus (to 70 cm), resulting in a thick black humus layer (up to 15–20 cm and more, according to our observations) with a high content of humus (from 10–11% to 20%);
6. The deep illuviation of iron and aluminum sesquioxides, as well as fulvic acids, with the formation of the bright brown illuvial B horizon, 20–40 to 70–90 cm thick; under conditions of high atmospheric humidification, it favors a wide distribution of defluxion, mudflows and landslips on extrusive domes (Korotky and Skrylynik 2008);

7. The buried layers of humus, due to which the new illuvial horizon is quite often absent, while the contemporary accumulative-humus horizons lie on the black-brown buried layers of humus.

The Mendeleev Volcano is an elementary island-arc volcano of low maximum height (absolute height, ≤1000 m), and this height is determined by acidic magma that yields volcanic (extrusive) rocks with a high content of silica (49–74%) (Masurenkov 1991). The soils that are formed on the pyroclastic sediments of such composition have a high absorption capacity and an acidic reaction (saline pH = 4.2–5.5); therefore, the free Fe and AL hydrogen oxides are predominant, while fulvic acids (the hydrolyzed humus compounds) are represented mainly by the aggressive fraction 1a, which penetrates deep into the soil with the increase in its content (e.g., up to 50–60% on the Mendeleev Volcano). At the same time, continuous waterlogging and high acidity of soils slow down clay formation (Duchaufour 1968), providing the maintenance of their high water permeability, as well as deep migration of water-soluble organomineral substances.

The predominance of the mass of cereals (first of all, Kuril bamboo) and the leaves of trees and shrubs in the tree waste contributes to the development of not aggressive acidic but rather neutral humus – even in the belt of dark-coniferous forests where the Kuril bamboo cover is also rather thick (see below Figure 7.4). It originally favors soil formation processes developing by the pathway of brownification, which is probably preceded by the sod process corresponding to the first stage of succession – the meadow bamboo stage. The "relicts" of this stage were preserved as grassy heathland "windows" in forest cover on the slopes of the Mendeleev Volcano and the Mechnikov Mountain (see Figures 1.3 and 1.4). The bamboo stage followed by the dominance of bamboo tree waste are the factors of rapid accumulation of organogenic matter in the humus-accumulative A_1 horizon, which determines the rapid development of this horizon and its saturation with different fractions of humic acids and soil colloids.

The bamboo thicket impedes the surface fluid runoff in summer (on the contrary, in winter snow slides along them as a kind of avalanches). All moisture is transferred to the soil-ground runoff. Due to the high porosity of soil-ground substrate, moisture penetrates to a considerable depth and becomes unavailable for most of the plants. As a result of formation of B_{im} horizon in the soil, the ground water level increases, the watering of plant root system is improved, further processes of humus accumulation are activated, and soil productivity increases.

The apparently "peated" lower layer of the A_0 horizon and the A_1 horizon per se (Polivanov 1984), is more likely a humified litter layer with thin unstructured humus. This is a transitional stage from "mor" to "moder" (Duchaufour 1968). Bamboo is a light demander intolerant to soil waterlogging (Vorobyev 1963). At the sites with enhanced hydromorphicity, the bamboo ground cover is replaced by ferns and tall grasses.

The distinct phytogenic peculiarities of soil morphology (Ivlev et al. 1987) provide conditions for sufficiently accurate differentiation of soils by the groups of forest types.

1. *Ochreous brown soils* with argillification and ferruginization, with active accumulation of humus and its deep penetration into the soil profile, with weak illuviation of iron and aluminum sesquioxides, as well as fulvic acids, were

formed under coniferous/broad-leaved forests on marine terraces and piedmont plains.

2. *Humus-accumulative soils* with deep penetration of organic matter into the underlying soil layers, with the typical binomiality of the profile (the presence of buried humus layer) often resulting in the absence of illuvial layer, developed on the eluvium of pumices of mountain slopes and foothills, under dark-coniferous forests. The organogenic layers are well humified; humus accumulation is more marked that brown soil formation. The modern humus layers have a high content of iron and aluminum sesquioxides determined by the composition of scoria; however, illuviation processes are weak.
3. Finally, the thickets of cedar elfin woods, both predominant in the upper subgolets layer and sporadically distributed over the entire Mendeleev volcanogenic landscape, form peat-humus weakly illuvial-humus *soils* with typical "dry peat accumulation" as a surface humus layer, as well as with deep humification of soil profile (to the depth of 70 cm) and accumulation of Fe and Al sesquioxides. The formation of course-humus organogenic horizons and the accumulation of dry peat are interrupted by periodic ash falls.

In general, the volcanic landscapes of Kunashir, with regard to their basic physicochemical properties, are a certain *geographical "junction" of soil-forming processes* widespread on the insular and marginal-continental territories of the boreal and subpolar Northwestern Neo-Pacific – Kamchatka, Sakhalin and Hokkaido (Ivlev et al. 1987).

5.3. Landscape Structure of Mendeleev Ecoregion

Kunashir belongs to the mixed (volcanogenic, volcanogenic-denudation, denudation-accumulative) genetic group of landscapes of Kuril Islands (Ganzey 2010). The territory of experimental test site covers the two contrasting types of landscapes: subvolcanic of the mountain class (lava and pyroclastic stratovolcano) and the surrounding volcanogenic and alluvial fan pyroclastic class of plains (Melekestsev 1974; Grabkov and Khudyakov 1993). Let us consider this territory as part of the *Mendeleev lowland volcanogenic and denudation-accumulative landscape region* within the *Kunashir volcanic-tectonic boreal-forest landscape district.* The latter, in turn, is part of the Urup/Small-Kuril south forest region. For the purposes of stylistic simplicity, hereinafter we will use the term of "Mendeleev volcanogenic landscape" for our test site. It is a single-stratovolcanic low mountain region with coniferous/broad-leaved forests, elfin woods and moorlands. Here, the following two types of landscapes are predominant: (1) low-mountain extrusion-volcanic erosive-denudation and (2) piedmont plain pyroclastic transit-accumulative.

The independently ranked subregional typological units – *groups of microlandscapes* – have been distinguished for Mendeleevo experimental test site. Each of these groups is a particular paragenetic link of mesoscale landscape couplings – mesocatenas (Polynov 1956; Milykov 1981). These natural unities correspond to the taxonomic unit "group of urochishches" underlying the landscape structure of insular territories of less than 10 km^2 (Ivanov, 2014). According to the schemes of taxonomic and territorial division of mountain landscapes (Gvozdetsky 1963; Sochava 1978), this level corresponds to the category of subtypes, or kinds, of landscapes and is referred to the taxonomic rank of high-altitude terrain (highlands), complex stow, and mesogeochore.

The groups of microlandscapes were distinguished by the combination of two oro-climatic factors: (1) the vertical *landscape microstratification*, which includes a certain level of absolute heights and the predominant type of mesolocation, and (2) the *macro exposure* of slopes of the Mendeleev Volcano and the surrounding plains. In the landscape stratification of mountains, "... the vertical physico-geographical differentiation ... finds its most generalized complex manifestation" (Isachenko 1963 8). The four landscape microstrata have been distinguished: the piedmont-plain, as well as the lower, middle and upper low-mountain strata. Vertical stratification, as is known (Shchukin 1980; Kotlyakov and Komarova 2007), manifests itself in the successive change of particular combinations of the shapes of earth surface (the types of relief). The piedmont-plain stratum corresponds to the accumulative type of relief, while the upper strata correspond to the gently sloping/stepped, moderate- and high-steep, and flat-top low mountains (Korotky 2008).

The types of relief determine the structure of water and lithodynamic flows – dispersing, transit, and accumulative (Florensov 1978). The integral expression of this morpholitho-dynamics can be the "regional types of sites" (Barsh 1980), i.e., the background *types of mesosites*. Due to rather simple geometric construction of the Mendeleev volcanic cone, the above altitudinal microstrata correspond to the four types of mesosites, according to the classification proposed by Glazovskaya (1964): accumulative (A), transaccumulative (TA), transit–transeluvial (T–TE), and transeluvial (TE).

The factor of macro exposure of slopes in this ecoregion reflects the known bioclimatic asymmetry of island-arc systems of the Northwest Pacific, which is determined by the influence of warm and cold sea currents and by different properties of air masses coming from the Pacific Ocean and the Sea of Okhotsk, respectively (Zhuchkova et al. 1973; Ganzey 2010; Ivanov 2014). In accordance with the above-described regional climatic contrasts determined by the circulation exposure of macroslopes of the Mendeleev Volcano, we have divided the first three landscape strata into two parts: the Sea of Okhotsk (Okh, exposure sector SW–W–NW–N) and the Pacific (P/O, sector NE–E–SE–S). Thus, on the territory of the experimental test site there were 7 subregional geomorphological units, subsequently saturated with soil-phytocoenotic characteristics and denoted as *groups of microlandscapes* (Figure 5.10).

The map of microlandscape groups was plotted using the equation of multivariate logistic regression:

$$\mathrm{Ln}[P/(1\text{-}P)] = -\,0.002373\cdot(X\text{–}Y)/2^{1/2}{}_{-5.19} - 0.6940\cdot GA_{-4.32} + 0.02174\cdot Z_{+4.12} - \\ -\,0.2309\cdot F(35{,}180)_{-3.77} + 34448;\ \mathrm{R}^2 = 0.685;\ Degr = 15.5\%;\ \mathrm{P}<10^{-6}. \quad (5.2)$$

Here, P is the probability of finding microlandscapes of the Sea-of-Okhotsk sector (the rest of the territories are occupied by Pacific-Ocean sector microlandscapes). The predictors: $(X–Y)/2^{1/2}$ is the shortest distance to the southeastern coast; GA is the steepness of slopes; Z is the average height of the test site; $(–F(35,180)$ is the slope lighting from the south. The subscripts of each predictor are t-statistics; their module determines the significance of contribution of predictors to the spatial variability of temperature. R^2 is the coefficient of determination and P is the statistic characterizing the probability of error in the establishment of statistical dependence.

The model (5.1) describes nearly 69% of *P* variability in space by means of comprising factors. The model was successfully verified (tested) by Allen's cross-validation (Allen 1974): the empirical criterion (*Degr*< 50%) shows that the predictive strength of the model is sufficient for predictive mapping or calculating the characteristic at new points.

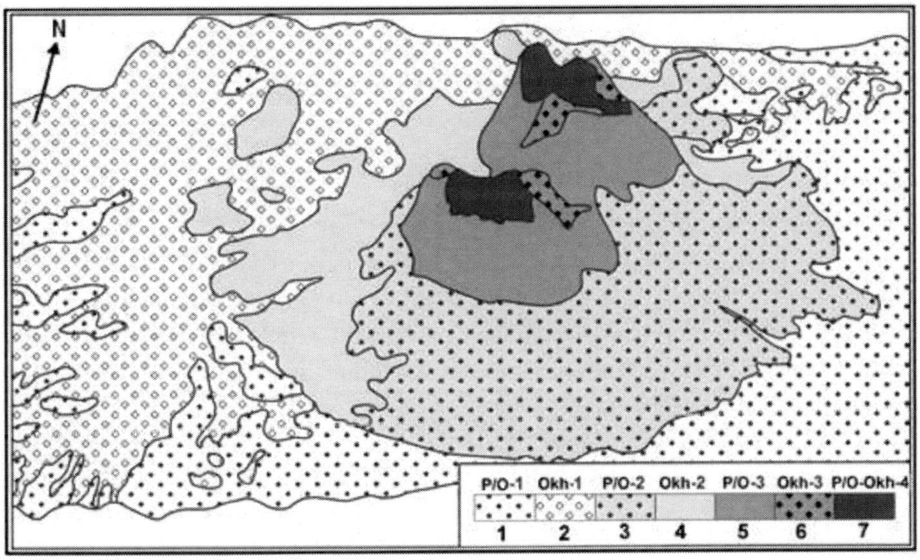

Figure 5.10. The map of microlandscape groups on the Mendeleev Volcano and in its surroundings. The map has been calculated and constructed according to Zalihanov et al. (2010). Legends, see in the text.

The types of water and geochemical regimes coupled with the selected groups of microlandscapes (Nikolaev 1978) impart certain environmental content to these natural complexes, and the landscape map per se becomes ecologically sound. We have previously constructed such landscape-ecological map for the territory of the Volga River basin (Kolomyts, 2005). It would be interesting to note that the analogous approach was used to create a small-scale map of hydrologic landscape classification of Primorye with runoff-forming, transit and accumulative landscapes (Tunegolovets et al. 2008).

Generally, the altitudinal bioclimatic zonality of the Mendeleev Volcano, with its relief (similar to the Golovnin caldera and most of the volcanoes of the Dokuchaev Ridge (see Figure 4.1, *a*) being within the low-mountain stratum, is a miniature repeat of the system of analogous belts of the Sikhote-Alin Range, which spread over all three geomorphological mountain strata – low mountains, middle mountains, and high mountains. Based on this analogy and on the landscape-forming principle of aligning the bioclimatic zonality of mountains with geomorphological zonality (Gvozdetsky 1963), we have distinguished the *microstrata* of the Mendeleev low mountains: the lower, the middle, and the upper.

Let us enumerate the distinguished groups of microlandscapes.

A. Piedmont-plain

Okh-1. The Sea-of-Okhotsk plain-hilly (marine terraces and pyroclastic plains, the absolute heights from 60–80 m to 230–250 m), accumulative (A), subhorizontal and gently sloping (at an angle of 0–8°), covered by coniferous/broad-leaved forests (fir/spruce/elm, fern/bamboo, liana, with phellodendron, yew, dimorphant, magnolia, Araliaceae), with humified brown-iron soils on the eluvo-deluvium of pumices.

P/O–1. The Pacific Ocean plain-hilly (marine terraces and pyroclastic plains, the absolute heightsfrom 60–80 m to 160–180 m), accumulative (A), gently sloping and mid-steep (from 0–5° to 12–14°), covered by fir/spruce nemoral herbaceous and dark-coniferous/broad-leaved tall-grass forests, fern/bamboo (with the admixtures of subtropical relicts), with humified brown-iron soils on the eluvo-delivium of pumices.

B. Low-mountain lower stratum

Okh–2. The Sea-of-Okhotsk lower stratum (from 230–250 m to 600–620 m), transaccumulative (TA), gently sloping and mid-steep (from 0–5° to 15–18°), covered by fir/spruce herb/bamboo/green-moss forests, with buffer stone birch communities at the upper boundary of spruce forests (at the altitudes of 580–610 m), with humus-accumulative, sometimes podsolized soils (with buried black-brown humus layers) on the eluvo-deluvium of pumices.

P/O–2. The Pacific Ocean lower stratum (from 160–180 m to 500–520 m), transaccumulative (TA), on mid- and high-steep slopes (from 8–12° to 30–34°), covered by fir/spruce herb/green-moss/bamboo forests, with humus-accumulative soils on the deluvium of pumices.

C. Low-mountain middle stratum

Okh–3. The Sea-of-Okhotsk middle stratum (from 600–620 m to 700–720 m), near-top transit (T), with a steepness up to 30–35°, and watershed transeluvial (TE), with a steepness of 0–5°, covered by stone birch crooked forests, herb-green moss and shrubby, often with cedar elfin wood in the undergrowth and with the admixture of spruce at the lower boundary, with buffer birch/cedar-elfin communities, on brown taiga illuvial-humus, sometimes podzolized soils.

P/O–3. The Pacific Ocean middle stratum (from 500–520 m to 670–690 m), near-top transit (T), up to 25–28° steep, and watershed (2–5°), covered by stone-birch/bamboo crooked forests with cedar elfin wood in the undergrowth, with the admixtures of spruce and fir, sometimes Japanese yew, at the southern boundary, on brown-taiga illuvial-humus, weakly podzolized soils.

D. Low-mountain upper stratum

OKh-P/O–4. The top and crest (above 670–720 m), transeluvial (TE), from steeply sloping to high-steep (from 1–5° to 28–32°), subgolets, with cedar elfin wood (green-moss/rhododendron and herbaceous), sporadically with the Japanese yew undergrowth, on peat-humified, weakly illuviated soils.

The total area of the Mendeleev volcanic landscape is about 200 km^2. According to the classification of (Sochava1974), such dimensionality corresponds to the topological level of geosystems; therefore, *the described groups of microlandscapes are topogeochores*, and the term "Mendeleev landscape" is used irrespective of its taxonomic content. Accordingly, the distinguished plant (bioclimatic) belts in the Mendeleev landscape should also be ranked to the topological level. As we can see, the altitudinal bioclimatic zonality, which is typical of the regional structures of mountain countries on the continent, emerges in the island-arc neo-Pacific at the local level, and this physico-geographical phenomenon will be discussed below.

Chapter 6

Structural Organization of Island-Arc Volcanic Landscape

Abstract

The novel methods of in-depth analysis of inter-component relationships using discrete empirical statistical modeling both at the regional level (insular physico-geographical provinces and regions) and at the scale of small natural complexes (microlandscapes) have been applied in this work to island-arc volcanic landscapes. Fundamentally new results have been obtained from the information and set-theoretical models of ecological connections between forest vegetation/soils and geomorphological/climatic factors on the Mendeleev volcanic landscape test site. The analysis and synthesis of phytocoenological, soil and microlandscape structures under particular geological-geomorphological and macroclimatic conditions on the basis of these models made it possible to determine the level of structural and functional unity or, on the contrary, disconformity of natural components and complexes formed on the morphostructures of different age and under the conditions of modern endogenous regime of passive volcanism, as well as different degrees of oceanicity of the climate. The factor of macro- exposure of slopes in this ecoregion reflects the known bioclimatic asymmetry of island-arc systems of the Northwest Pacific, which is determined by the influence of warm and cold sea currents and by different properties of air masses coming from the Pacific Ocean and the Sea of Okhotsk, respectively. The modeling results provided a much stricter and more streamlined insight into the known concept of "fuzziness/discontinuity" of altitudinal-zonal boundaries on Neo-Pacific islands and the "leveling" of these boundaries by local factors.

6.1. Basic Monosystemic Relationships in Boreal Forests

The Mendeleev Volcano is characterized by highly distinct and rather complete manifestation of altitudinal bioclimatic zonality, in spite of its relatively low altitude, and this phenomenon has already been mentioned in literature (Manyko and Rozenberg 1970). Altitudinal geobotanic zonation is expressed distinctly enough and, moreover, correlates well with great geological structures (see Figures 3.4 and 6.1). The volcano territory settles down in the general series of regular lowering of altitudinal-zonal boundaries along with the increasingoverall oceanicity of the climate and the decreasing height of mountains on the Kuril Islands (Vasilyev and Rozenberg 1977).

Figure 6.1. The common appearance of geological structures of the Mendeleev Volcano from the Sea of Okhotsk. On the right: the modern volcanic dome. On the left, in the background: the ridge of the second caldera (Mechnikov mountain). On the front: the first somma and the piedmont pyroclastic plane.

The materials of large-scale landscape-ecological survey at the Mendeleevo test site were used to construct a general information model describing the ordered system of chain reactions between the state characteristics of natural geo(eco)systems of island-arc Neo-Pacific (Figure 6.2). The model presents: the primary (morphostructural) factors of the abiotic environment and their localized geomorphological manifestations, lithoedaphic and hydrothermal conditions, and the floristic and phytocoenotic unities of forest cover. The methods of information model calculation and construction are described in Chapter 2.

First of all, the model clearly demonstrates that the considered intercomponental relations are organized into a highly determined system, with its core being the completely dominant factor of absolute height. It forms low-mountain microlandscapes of the volcanic cone (the first retransmitter with normalized contingency coefficient $C(A;B) = 0.466$) by its altitudinal strata (see above) and has a direct effect on floro- and phytocoenotic structures of forest cover, outlining the lithoedaphic and hydrothermal boundary conditions of their differentiation. The $C(A;B)$ coefficient of floristic phratries and groups of forest types with absolute heightand microlandscapes has rather high values (0.322 and 0.325, respectively), which forms the two main channels of intercomponental relations. In the mountains, the altitudinal-(stratum)-zonal factor is a complex geographical factor determining primarily the thermal conditions of surface air and soil (Gvozdetsky and Golubchikov, 1987).

Representing the vertical differentiation of regional climatic background, altitudinal zonation discloses the principal phenomenal feature of the bioclimatic system at the stage of primary island-arc biogeocoenogenesis. What is meant here is the direct subordination *of phytobiota to vertical hydrothermal gradients* and the formation of plant communities under climatic conditions similar to the "climate of free atmosphere" (Khromov and Mamontova, 1974), with the predominant influence of moist and unstable oceanic air masses (Blüthgen 1973) and with a relatively low refraction of climatic background by local geomorphological and edaphic factors.

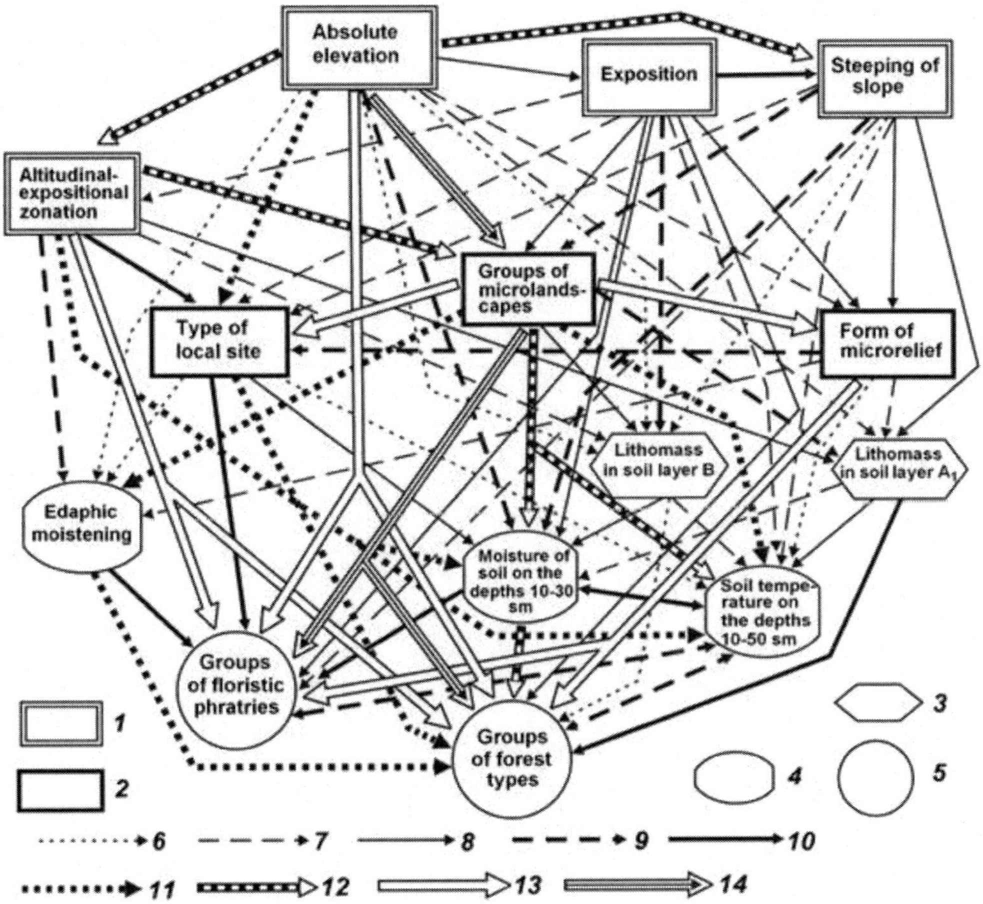

Figure 6.2. The general information-statistical model of geocomponent interrelations for low mountain ecoregion of Mendeleev landscape. Geocomponent blocks: 1, the initial abiotic factors; 2, the secondary abiotic factors – retransmitters; 3, the hydro- and thermoedaphic conditions; 5, the structural characteristics of forest communities. Normalized coefficient of interrelation *C(A;B)* between geocomponent attributes (Puzachenko, Skulkin, 1981): 6, 0.071–0.100; 7, 0.101–0.130; 8, 0.131–0.160; 9, 0.161–0.190; 10, 0.191–0.220; 11, 0.221–0.250; 12, 0.251–0.300; 13, 0.301–0.440; 14, 0.441– 0.720.

The South Kunashir low-mountain region has four distinct altitudinal bioclimatic belts (Figure 6.3, *a* and *b*): (1) coniferous/broad-leaved (Manchurian–Okhotian phratrie); (2) dark-coniferous fir/spruce (Okhotian); (3) spruce/stone-birch and complex spruce/cedar-elfin-wood/stone-birch (Okhotian–Kamchatian and Okhotian–Kamchatian–Beringian phratries, respectively); and (4) birch/cedar-elfin-wood (Kamchatian–Beringian).

The solar-circulation exposure (the second channel of relations) is represented by four octants of the four cardinal directions and neutral (subhorizontal) surfaces, and two pairs of octants were shown to be adequate to the two opposite sectors: the Sea of Okhotsk (315–45°

and 225–315°) and the Pacific Ocean (45–135° and 135–225°). Let us immediately note that the force of action of this factor on the floro- and phytocoenotic structure of plant cover of the volcanogenic-island ecosystem is more than three times weaker (*C(A;B)* = 0.109) compared to the factor of absolute height. These quantitative estimates argue against the current concept, which is based only on visual observations about the exceptionally high degree of disturbance of altitudinal zonation on the Kuril Islands by other abiotic factors, primarily the factor of exposure.

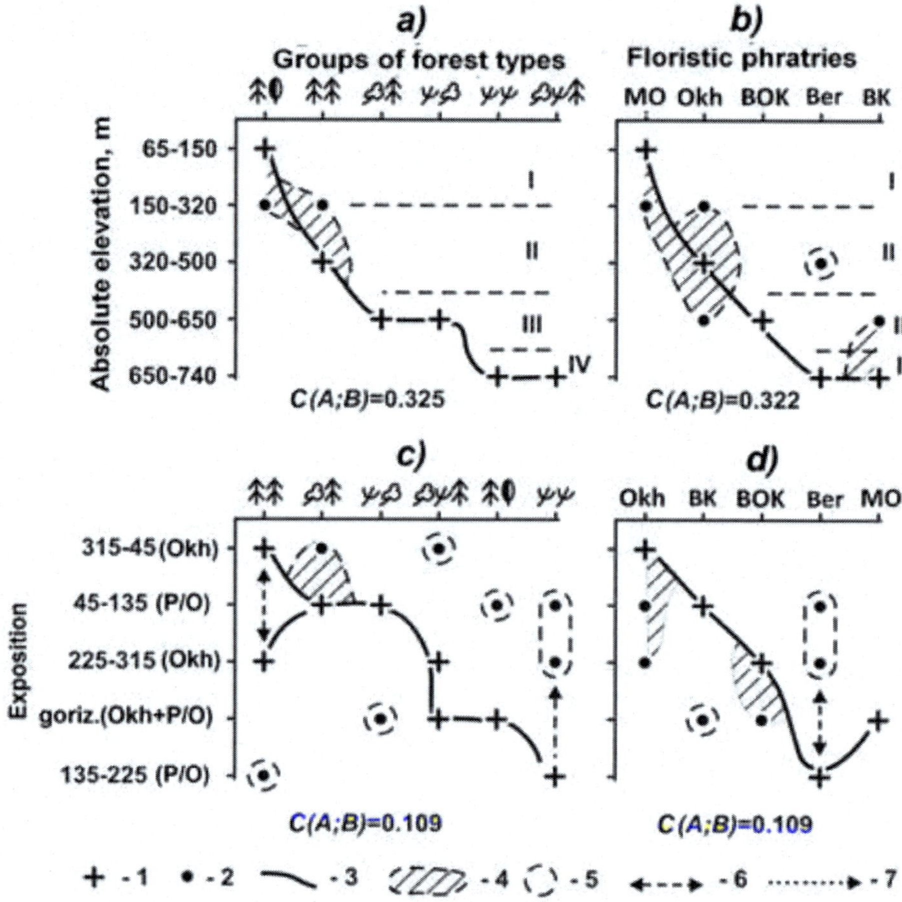

Figure 6.3. The binary ordination of forest type groups and floristic phratries by the absolute heightand exposure factors. 1 – the ecological dominant; 2 – the "fuzzy" part of ecological niche; 3 – the trajectory connecting ecological dominants; 4 – the space of ecological niche; 5 – the enclave; 6 – the altitudinal interval of fractionation of ecological dominants; 7 – the area of sporadic distribution of phenomenon within the given gradations of factor. I–IV, bioclimatic belts (see the text).

Nevertheless, there is no denying that the solar-exposure factor has some influence (see Figure 6.3 *c* and *d*). Coniferous/broad-leaved forests are widespread mainly on subhorizontal surfaces of piedmont plains and marine terraces and occur much less frequently in the lower low-mountain stratum on the NE–E–SE slopes (here, they gravitate rather toward the Pacific sector; however, this question is still open because of the lack of test sites). The fir/spruce forests of the Sea-of-Okhotsk flora are confined mainly (by 60%) to the northern and western quarters of the horizon, i.e., the Sea-of-Okhotsk sector, which is warmer and less humid; however, the graphs of relationships show more than 20% occurrence of enclaves belonging to the more humid sector (Pacific Ocean).

The northeastern, eastern and southeastern slopes facing the Pacific Ocean are the sphere of dominance of stone birch crooked forests, including those in combination with spruce and with cedar elfin wood in the undergrowth. It is more than 50% of the area of distribution of pure birch forests. Cedar elfin wood forming an independent altitudinal-zonal belt in the upper stratum of low mountains gravitates mainly (by 68%) to the SW–S–SE volcano slopes but is found also on the neutral slopes (western and eastern). As we can see, this typical representative of the Beringian flora has also its basic exposition niche in the Pacific Ocean sector; however, compared to stone birch, it prefers the warmer sunny habitats, which is evidence of its known relative thermophilicity. Finally, let us note the ambiguous exposure confinement of the most complex (ternary) buffer communities: stone birch/cedar elfin wood/spruce. They occur in the upper zone of the dark-coniferous forest belt (the absolute heights of 550–650 m) and mainly on the passes and mountain slopes facing both the Sea of Okhotsk and the Pacific Ocean.

The third channel, almost as powerful as the first one (*C(A;B)* = 321–0.350), goes through the altitudinal-exposure zonation. It is the second retransmitter taking into account the corrections for solar-circulation exposure in vertical zonality. It belongs to the category of dislocation zonality (Garzman 1971), which plays the key role in the bioclimatic system of the Pacific Ocean mega ecotone (Kolomyts 1987, 2015 a, b).

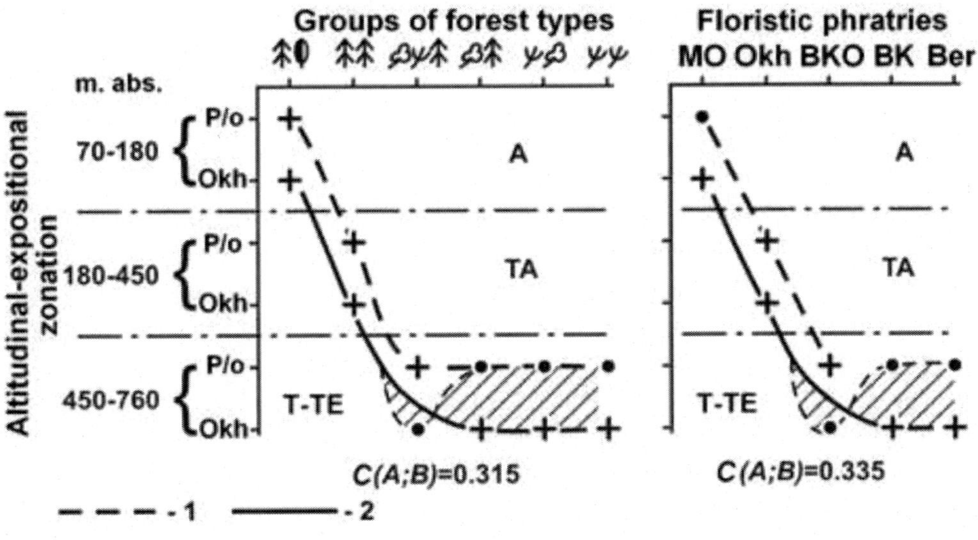

Figure 6.4. Distribution of groups of forest types and floristic phratries in space of dislocation altitudinal-expositional zonation. Sectors: 1 – Pacific Ocean (P/o); 2 – Sea of Okhotsk (Okh). A, TA, T–TE are the types of local sites. For other legends, see the text.

The factor of altitudinal-exposure zonation is slightly particularized by the regularities described above (Figure 6.4). It is obvious that the solar-circulation exposure makes only certain, generally insignificant corrections into the altitudinal-zonal distribution of floristic phratries and forest type groups. In general, the ecological dominants of TA coniferous/broad-leaved forests of the Manchurian–Okhotian flora, as well as the T–TE cedar elfin wood communities of Beringian flora and mixed birch/cedar-elfin-wood associations, are widespread almost to the same extent over both solar-circulation sectors but with the higher dominants (53–55%) in the Sea-of-Okhotsk sector. The fir/spruce forests of the Okhotian flora occupying TA mesosites at the lower stratum of low mountains (180–450 m) are also rather indifferent to the exposure factor. They occur on all slopes of the Mendeleev Volcano, with weakly marked dominants (by 36%) in both sectors.

Finally, the local representatives of floristic and coenotic complexes receive an additional weaker signal from the types of mesosites and forms of microrelief ($C(A;B)$ = 0.190– 0.250).

Thus, altitudinal bioclimatic zonality is the principal and even overwhelming factor of spatial subregional differentiation of forest cover (ranked as microlandscape groups). This fact contradicts the popular idea (Vorobyev 1963; Popov 1963; Ganzey 2010; etc.) of relatively weak manifestation of this factor on the Kuril Islands. The distribution of forest cover is directly associated with the regional altitudinal-zonal climatic system, with its very weak refraction by other subregional (exposure, lithogenic) and local geomorphological factors. It indicates that the geo(eco)systems of island-arc neo-Pacific are highly sensitive to the global and regional climate changes: the background hydrothermal signal it transmitted directly to phytobiota, almost without being considerably changed by intermediate local factors (retransmitters). It seems to be the cause of the known instability of island phytobiota, its climatogenic dynamicity.

The pressure of the macroclimatic factor on plant cover manifests itself not only vertically but also horizontally. The South Kuril Islands are characterized by exceptionally drastic submeridian changes of natural zones. At a little greater distance than 300 km (and about 170 km along the meridian), from the northern part of Iturup towards the southern part of Kunashir, the boreal stone-birch light forests with cedar elfin wood are replaced by larch and stone-birch/dark-coniferous forests, which then change into nemoral fir/spruce forests and dark-coniferous/broad-leaved forests with the abundance of thermophilic and evergreen species and lianas (Popov 1963). Thus, the *latitudinal-zonal range of phytobiota in the island boreal Neo-Pacific is found to be sharply narrowed, ecotonized throughout the entire three-dimensional space – both horizontally and vertically.* Accordingly, the phytocoenological gradients increase and the interaction between plant communities of different floristic phratries intensifies, followed by their interpenetration under suitable conditions of growth sites, *with the formation of mixed (buffer) communities,* as is clearly demonstrated by plant cover structure of the Mendeleev Volcano (Kolomyts and Surova 2018 a).

Such are the basic features of formation of forest geo(eco)systems of the Mendeleev Volcano, which reflect the monosystemic structural organization of landscapes of the South-Kuril ecoregion of island-arc Neo-Pacific. This information model is substantially different from the analogous model of the Lower-Amur marginal-continental Sub-Pacific situated in the same (boreal) belt (see below, Chapter 7). Contrary to the Priamurye Sub-Pacific, where altitudinal zonality is weakly manifested and smoothed over by other abiotic factors (exposure, lithogenic, etc.) under conditions of low mountains and the monsoon-continental climate, the altitudinal climatic zonality of a young island volcanic landscape (the Mendeleev volcano) formed under conditions of maritime climate is the major abiotic factor of low-mountain biogeocoenotic organization. This fact indicates that island biogeosystems, since the very beginning, being yet dominated by the ocean, are exposed to direct influence of the respective background climatic system.

The absence of the weak edificatory role of edaphic moistening as an intermediate converter of information signals is also the obvious contrast between the South Kuril Neo-Pacific and the Priamurye Sub-Pacific.

6.2. Ecological Role of Local Factors

Generally, it is strongly believed that extrazonal and local factors quite considerably manifest themselves in island landscape genesis. These factors are often supposed to smooth out the altitudinal zonality and to create the "fuzziness" and discontinuity of altitudinal belts, thereby determining an extremely mosaic pattern of the plant cover and small natural complexes of the islands in general (Korsunskaya 1958; Ivanov 2005, 2014; Ganzey 2010; etc.). Let us verify the actual efficiency of local factors under conditions of simplified morphotectonic structure of a single active Mendeleev Volcano. Let the *geotope*, i.e., the type of local site (Ramensky 1971; Krauklis 1979), and the respective *forms of microrelief* be the crosscutting features. These geomorphological features give synthetic expression to the sum of the most part of environment factors determining habitat conditions (Kolesnikov 1956). The distribution of floro- and phytocoenotic units by the types of habitats and microrelief in different strata of low mountains must demonstrate the geomorphological conditions of formation of the most part of representative local members of a particular altitudinal zone, as well as the forms of microrelief through which plant communities transgress beyond the limits of their altitudinal belt, thereby impairing the system of vertical zonality.

The binary ordination models show the following (Figure 6.5). To begin with, the types of sites are quite clearly differentiated by absolute height, i.e., stratification of mesorelief. The scattering types (the convex dry and sunny slopes) are located mainly in the middle TE stratum of the volcanic cone (in the height range of 500–650 m), though the EA anticlines – fragments of the ancient caldera of the Mendeleev Volcano – often occur here too. The latter become a dominant (in our sample of plots) in the above-lying near-top stratum, where there are also many TE and E geotopes. This is a level of the lower boundary of the stone-birch crooked forest belt; however, the communities of cedar elfin wood – a representative of Beringian flora, which is known to prefer sunny and heated spaces – also descend here along the above geotopes, especially along the southern slopes (at an angle of 135–225°, see aboveFigure 6-3*c*). Cedar elfin wood dominates on the scattering types of sites also in its own (subgoltsy) belt. At the same time, stone birch ascends here from below along the EA geotopes (near-peak anticlines) to form buffer cedar-birch communities at the lower boundary of the subgoltsy belt.

With descending to the lower stratum of low mountains, first T–TA and then accumulative types of local sites become dominant. Here the dark-coniferous taiga of the Okhotian flora is predominant under conditions of higher hydromorphicity (with a dominant of 64%); however, cedar-birch associations often descend to the heights of 550–600 m along the same geotopes (mainly by the slopes of the Sea-of-Okhotsk sector), and single communities of cedar elfin wood along the eluvial and transeluvial geotopes occur even lower – at the heights of 350–400 m. The most complex (ternary) cedar-birch-spruce complexes are also formed at the upper boundary of the dark-coniferous forest belt on EA geotopes (mainly in the Pacific Ocean sector).

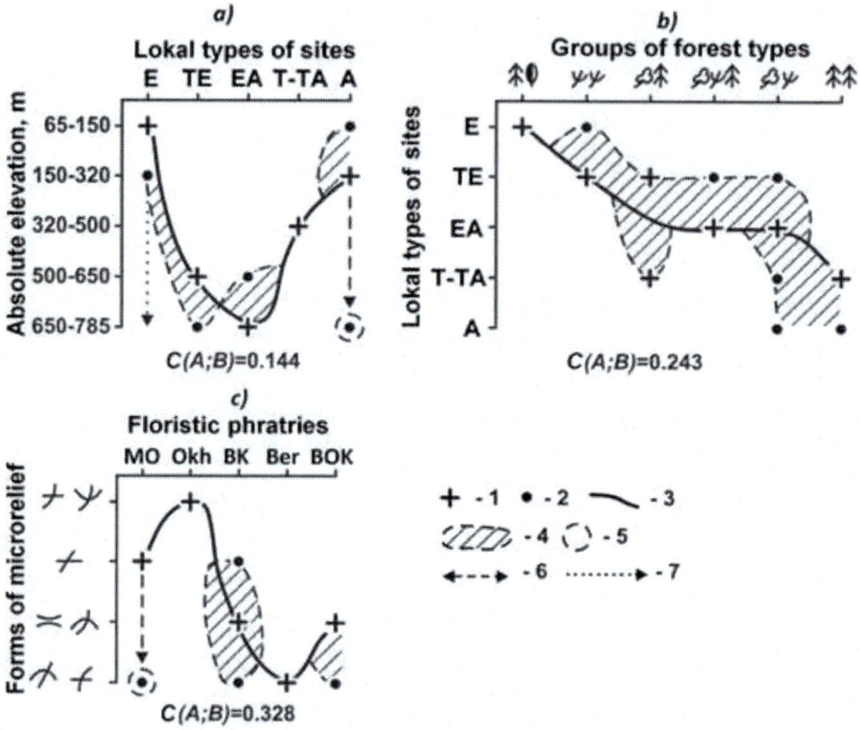

Figure 6.5. The binary connections for the groups of forest types and floristic phratries with absolute height, local types of sites and forms of microrelief. The symbols are the same as in Figure 6.3.

Finally, the eluvial (upland) sites with much less frequent occurrence of accumulative types are predominant on piedmont-hilly plain and marine terraces. Representatives of the zonal bioclimatic background of the southern half of Kunashir (coniferous/broad-leaved forests of the Manchurian–Okhotian flora, with a dominant probability of64%) are widespread at these geotopes.

These are the main results of binary ordination of floristic phratries and forest type groups describing local deviations from their altitudinal-zonal distribution in the territory of the Mendeleev Volcano test site. In general, they complicate the background pattern but do not "level" or "fuzzify" it, as is stated in the literature sources cited above.

Nevertheless, geotopes play a key role in establishment of regional natural-territorial structures by creating their local foci at the initial stages of subaerial landscape genesis (in the state of young volcanic islands). The most general morphological characteristics of soil (the thickness of genetic horizons) and the high-rank structural characteristics of forest cover (floristic phratries and forest type groups, floristic and phytocoenotic diversity of ground cover) are differentiated by the local geomorphological factors (the type of site and the form of microrelief) to no less and even higher extent compared to their changes caused by the territorially comparable altitudinal-zonal and exposure predictors (see Figures 5.8 and 5.9). The same can be referred to hydro-thermoedaphic characteristics and the main indices of forest ecosystem functioning: the total reserves of living phytomass, its productivity and biological cycle.

As is known (Krauklis 1974; Molchanov 1975), in the intracontinental ecoregions, which have already come a long way in subaerial development, the local soil-biotic contrasts, often higher than the regional (background) contrasts, are generally typical of topological geo(eco)systems; however, they concern only single representatives of the extrazonal physical-geographical background. During the formation of primary (volcanic) landscapes, the local structural and functional contrasts encompass the taxonomic norms of soil-biotic characteristics of a broad range of zonal-regional units, which are kind of focused in different forms of microrelief and types of sites.

Thus, in the *pioneer landscape genesis, local geomorphological conditions create the foci of emergence of various phytocoenological and soil structures of a higher (zonal-regional) level.* The prime cause of this process is the pronounced thermo- and hydroedahpic contrasts on local catenas. Such contrasts already quite definitely simulate the altitudinal-zonal and exposure differentiation of hydrothermal conditions of forest growth.

6.3. Forest Phytobiota in the Space of Humidity Factor

When exploring the mechanisms of spatial organization of the soil-biotic "core" of volcanogenic landscapes, we cannot ignore the most important ecological role of the heat and moisture ratio expressed by different of atmospheric humidity factor (F_{hum}). In this research we have used Vysotsky–Ivanov coefficient as a ratio of annual precipitation to annual potential evaporation (see Chapter 3, section 3.3).

The distribution of taxonomic norms of humidity factor by the groups of microlandscapes, floristic phratries and forest type groups are presented in Table 6.1. Information models, in contrast to continual statistic relationships (see Figure 3.3 *c*), are based on rather coarse quantization of a factor and a phenomenon and, therefore, give a more generalized pattern of discrete binary ordination, which does not always coincide with continual ordination. The piedmont-plain coniferous/broad-leaved forests of Mendeleev landscape were formed under conditions of the lowest atmospheric humidity (F_{hum} = 3.3–3.9), which slightly exceeds its zonal norm for the southern part of the Kunashir Island (see above Figure 3.2, Yuzhno-Kurilsk station).

On piedmont plains, as well as in the lower and middle strata, there are nosubstantial sectoral contrasts in F_{hum}. The interval of the minimum values of this parameter is the same: 3.3–3.9 for both sectors, and then generally increases to 4.4–4.8 with the increase in absolute height. There is an insignificant difference only in the upper stratum: the Sea-of-Okhotsk near-peak slopes are wetter than the Pacific Ocean slopes (the F_{hum} dominants are 5.0–5.2 and 4.8–5.0, respectively).

The broadest spatial variations of atmospheric humidity, with the minimum probability of ecological dominant (46%), are observed in the lower TA stratum of the Pacific Ocean sector (F_{hum} = 3.9–5.0). Here, the two contrasting solar exposures of equal dominance are widespread in the belt of dominance of mountain fir/spruce forests. The humidity factor on shady NW–NE slopes is higher by one grade compared to sunny SW–W slopes. As we can see, the dark-coniferous forests of the Mendeleev landscape grow within a sufficiently broad range of solar exposure.

The belt of mountain dark-coniferous forests and the belt of birch crooked forests are formed under comparatively similar conditions of atmospheric humidity, which gives grounds to consider them as a genetically uniform altitudinal-zonal formation. Through, stone birch forests are almost eurytopic by the F_{hum} values and have an enclave bringing them close to the subgoltsy belt of cedar elfin woods. The transitional communities of stone-birch/spruce forests are limited mainly by the relatively warm E–SE slopes of the Pacific Ocean sector with the minimum atmospheric humidity. They are much less frequent on the wetter shady NW–NE slopes of the sector.

Table 6.1. Taxonomic (average weighted) norms of Vysotsky–Ivanov's annual atmospheric humidity factor peculiar to different floristic phratries, groups of forest types and groups of microlandscapes

Divisions		F_{hum}	Divisions		F_{hum}
Floristic phratries	MO	3.91	Groups of microlands-capes	Okh–1	3.92
	Okh	4.52		P/o–1	3.92
	Bk	4.51		Okh–2	4.36
	Ber	5.08		P/o–2	4.66
	BKO	4.90		Okh–3	4.86
Groups of forest types		3.86		P/o–3	4.46
	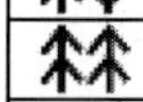	4.54		Okh–P/o–4	5.08
	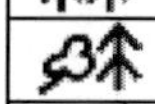	4.08			
	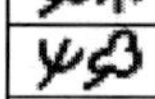	4.51			
	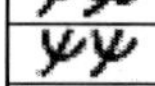	5.08			
	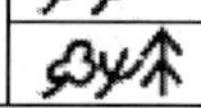	4.90			

Note. Conventional meanings see in the text.

The most structurally complex buffer cedarelfinwood/birch/spruce communities of the BKO phratrie, widespread in the upper stratum (650–750 m), actually in the subgoltsy belt, are characterized by a comparatively narrow range of F_{hum} = 4.8–5.0, which makes them more like the neighboring associations of pure cedar elfin wood. At the same time, these ternary buffer formations occur in the broad range of solar exposure and on the subhorizontal crests of watersheds.

The solar-circulation exposure introduces certain corrections into the vertical distribution of humidity factor. In general, the partial effect of exposure is weak, as is demonstrated by the low value of parameter *C(A;B)*, the low probabilities of the dominant areas (30–40%), and the substantial "fuzziness" of ecological niches, with enclaves and sporadic dispersion of F_{hum} by the gradations of exposure. Circulation exposure is the major contribution to atmospheric humidity of the Mendeleev landscape. This fact accounts for the "fuzziness" of ecological niches of F_{hum}. The SW–W slopes facing the Sea of Okhotsk are quite well heated and have a comparatively low F_{hum} (mostly 3.9–4.8), which brings them closer to the uplands of foothills and flat crests of the upper stratum. On the contrary, the SE–SW slopes of the Pacific Ocean sector with almost analogous sun exposure are characterized by maximal F_{hum} (5.0–5.2), with a 78% probability of the dominant. In this case, both components of the exposure interfere with different signs, with predominance of the circulation component.

The more sunny NE–SE slopes of the Pacific Ocean sector facing the Kuril-Kamchatka current have the same F_{hum} as the more shadowy NW–NE slopes of the Okhotsk Sea sector, where the branch of the warm Kuro-Sio current approaches (the compensatory effect). On piedmont plains, as well as in the lower and middle strata, there are no marked sector contrasts in F_{hum}. The most extensive spatial variations of atmospheric humidification are observed in the lower stratum of the Pacific Ocean sector. On the shadowy NW–NE slopes, F_{hum} is higher than on the sunny SW–W slopes. At the same time, the dark-coniferous forests of the Mendeleev landscape grow in a sufficiently broad range of solar exposure. *The belt of mountain dark-coniferous forests and the birch elfin wood band* are formed under the comparatively similar conditions of atmospheric humidification, which gives grounds for considering them as a *genetically uniform altitudinal-zonal formation.*

During the formation of young island-arc volcanic landscapes, the topological structural-functional contrasts encompass the taxonomic norms of soil-biotic characteristics of quite a number of zonal-regional units, which are as though *focused on various forms of lowland microrelief* (see Figure 6.5). Thus, at the initial stages of subaerial landscape genesis, the *local geomorphological conditions create the centers of origin of different phytocoenological and soil structures of a higher (zonal-regional) level.* The first cause is the pronounced thermo- and hydroedaphic contrasts formed on local catenas. Such contrasts quite definitely simulate the altitudinal-zonal and expositional differentiation of hydrothermal conditions of forest growth and soil formation.

6.4. Organization of the Soil System

6.4.1. Introductory Notes

It has already been mentioned that all microlandscapes of the Mendeleev Volcano region are rather young, the most ancient of them being about 35–40 thousand years old. At the same time, on the littoral marine terraces the soils (and landscapes) are somewhat more ancient; their age is a little more than 40 thousand years (Melekestsev et al. 1974). In particular, the terraces of 80–85 m in height are 20–45 thousand years old. The total period of formation of the Mendeleev stratovolcano is 200–230 thousand years and – since the beginning of the Sungacha time at the end of the Middle Pleistocene – Q_{II}^3, according to Korotky et al. (1999). The contemporary extrusive dome of the volcano is about 2500–2700 years old. Even this minimal period is quite sufficient for the development of dynamically mature soils (Targulyan 1984), all the more so because loose soil-forming rocks are predominant around here.

Indeed, all soil profiles of the test site have a complete set of genetic horizons which, according to Polivanov (1984), demonstrates a sufficiently high degree of organization of not only organic but also mineral soil matter in all bioclimatic belts. As is known (Glazovskaya 1953), a 0.5-m soil layer in the mountain dark-coniferous forests is formed during less than 10 thousand years even on the primary bedrock. At the same time, under meadow vegetation this process accelerates 3–4 times compared to forest territories. The age of soils of the Mendeleev Volcano and its surroundings is the Late Pleistocene, and the soils are really well developed. Out of 32 test sites, there was almost no buried soils as independent horizons A_I(bur.); more frequently there occurred only their dispersed traces, though often rather thick.

In light of the above, the statement (Ganzey 2010) that "underdeveloped soil profile" is a typical feature of volcanic landscapes of the Kuril Islands seems to be absolutely unsubstantiated.

The two characteristic events can be observed during soil development from fragmentary to "mature" (fully developed): (1) soil "continentality" is reduced (the soil temperature decreases and the soil moisture content increases as a result of fine grained soil accumulation, enhanced argillification of metamorphic hor. B_m, etc.); and (2) the largely neutral or alkaline soil reaction is replaced by more acidic one – due to emergence of fulvic acids contributing, in turn, to the precipitation of Fe, Al, and Mn sesquioxides in the lower illuvial hor. B_i (Polivanov 1984).

6.4.2. Slope Processes and Soil Formation

Each catena is created as a result of interaction primarily between the two processes – slope formation and soil formation, with mandatory involvement of the time factor (Gerrard 1981). The trajectory of elementary slope development includes a succession of four main stages (Rototaeva and Simonov, 1984): (1) the stage of denudation runoff slope, with bedrock outcrops; (2) the stage of the natural gradient of slope, with thin slope eluvium; (3) the stage of a slope covered with not only eluvial but also overlying specifically slope (transit) detritus – the result of mass wasting of loose masses: and (4) the stage of defluxion slope, with a well-marked cover horizon (soil hor. A_0 and A_1), with the underlying sloping and eluvial layers.

This generalized scheme can be simpler under real conditions. For example, in case of low tempos of soil development or too intensive denudation, a slope may have only the sloping and eluvial horizons, or the former. There is a developing denudation trend of pedo-lithogenesis, according to Targulyan (1982). With the weak development of mass wasting, the thickness of the transit layer sharply decreases and, in this case, the cover (soil) horizon can change directly into the eluvium of bedrocks, indicating the autochthonous type of pedo-lithogenesis. Finally, in case of accumulative trend, the lithogenic base of natural complex is made of slope detritus.

Thus, the balance of thicknesses and physicochemical properties of the major vertical strata of "slope" morpholithosystem at each local site represents the interaction between the three basic processes of exogenesis: weathering, soil formation, and slope denudation. The first two of them implement mainly the vertical movement of substance and, consequently, reflect the inter-component landscape relationships, while the third process causes the horizontal matter and energy transfer, thereby characterizing inter-complex interactions.

The special studies and experiments performed on the eastern slope of the Sikhote-Alin Range have shown (Karpukhin and Karyagin, 1984) that the cover of loose sediments on nearly all types of slopes can absorb any amount of atmospheric precipitation. At the same time, the level of moisture infiltration and accumulation in the loose cover significantly depends on vertical differentiation of the latter under the influence of weathering and soil formation processes. So, the filtration factor decreases by an order of magnitude upon transition from the sod and humus soil horizons to the underlying transit and then eluvial horizon and, therefore, the lower (water-proof) layers of loamy matter are characterized by the higher moisture and more viscous consistency. Under these conditions, on rather steep slopes (20-30° and more), even a not very thick interlayer of loamy matter (up to 2–3 cm) can cause mass wasting of the overlying loose cover. The overwetting of loamy soil-grounds often occurs not only in valley bottoms but also on the gentle parts of slopes, contributing to widespread loose matter movement without direct dependence on slope steepness.

6.4.3. "Maturity" of Soil System

The level of "maturity" of the upper organogenic zone of soil is determined primarily by the degree of organic matter "organization" in this zone, i.e., the ratio of organic matter represented by coarse plant debris, on the one hand, and humified matter with a high content of high-molecular organic acids of humate/fulvate type, on the other hand (Polivanov 1984). The indirect representation of such organic matter "organization" in soil may be the thickness ratio of forest litter (hor. A_0) to humus horizon A_1. It seems that the thickness of horizon A_1, which characterizes the period of soil development from the first tens to hundreds of years, according to Polivanov (1984), is rather representative per se. At the same time, the period of complete mineralization of tree waste under conditions of the temperature belt is 1–2 to a little more than 5 years (Osipov 2004), i.e., is much shorter. This is just the period of formation of horizon A_0, with attainment of a stable state.

Consequently, the thickness and chemical composition of hor. A_1 in the primary volcanic landscapes represents the features of the entire preceding soil-biotic cycle of soil development, while the thickness of forest litter (A_0) characterizes the results of the biological cycle of recent years and the tempos of the most rapid, according to Duchaufour (1968), phytocoenotic, as we call it (Kolomyts et al. 2015), cycle of organic matter turnover in a forest ecosystem.

The thickness and properties of illuvial B_i (or illuvial-metamorphic horizon B_{im}) reflect still greater characteristic time of soil formation, because illuvial processes are much slower than the degradation of primary minerals and organic matter (Glazovskaya 1981). Thus, the ordered series of genetic horizons $A_0 \rightarrow A1 \rightarrow B_i$ (B_m) as applied to volcanic landscapes can be interpreted as the age-related chain of morphological indication of evolutionary regularities of soil-forming processes, which have created the modern soil profile.

The most general spatial pattern of relationships between the basic elementary soil processes (organic matter degradation, mineralization, humification, illuviation, lessivage, and intrasoil weathering) in the territory of the Mendeleev volcanic landscape gives the averaged values of the above morphological parameters of soil profile by the groups of microlandscapes (Table 6.2), as well as a particular example by the topological profile (Figure 6.6). In general, the brown-ochreous soils of coastal-plain and piedmont microlandscapes proved to be better developed, as is shown by their maximum values of hor. Bi thickness and the A_1/Bi ratio, being indicative of a comparatively high level of intrasoil transformation of mineral mass.

At the same time, the more marked washing regime of soils in the excessively humid and cool Pacific Ocean sector intensifies the processes of illuviation and lessivage, which give the maximum thickness of hor. B_i. However, the processes of mineralization of decomposed organic matter are much slower here than in the warmer and drier Sea-of-Okhotsk sector; therefore, humus horizons reach the maximum thickness and the A_0/A_1 ratio is lower. The lowland and piedmont microlandscapes are also characterized by the thickest humus profile including up to two or three buried humus layers. On the steep slopes of the lower and middle strata of the volcanic dome, defluxion "erases" the buried layers and the total thickness of A_I + A_I(bur.) drastically decreases. It slightly increases again only on the peaks and crests of the ridges with more gentle slopes.

Table 6.2. The averaged morphological characteristics of soil in different groups of microlandscapes

Groups of microlandscapes	Morphological characteristics of soil					
	Thickness of the horizons, cm				Ratio	
	A_0	A_I	A_I+A_I (border)	B	A_0 /A_I	A_I/ B
1 – Okh-1	*4*	*6*	**37**	**58**	**0.78**	**0.51**
2 – P/O-1	*4*	**12**	**51**	**65**	*0.51*	**0.60**
3 – Okh-2	5	8	*13*	*36*	0.63	0.30
4 – P/O-2	*3*	*6*	*13*	41	**0.75**	*0.21*
5 – Okh-3	*3*	7	*11*	*36*	0.53	0.30
6 – P/O-3	*3*	7	*11*	43	0.57	*0.12*
7 – Okh– P/O-4	**9**	**16**	25	39	0.59	0.39

Note: The maximum values of the parameters are bold and underlined; the submaximum values are bold. The minimum and subminimum values are italic, underlined and not underlined, respectively. Other numbers show the mean values of the parameters.

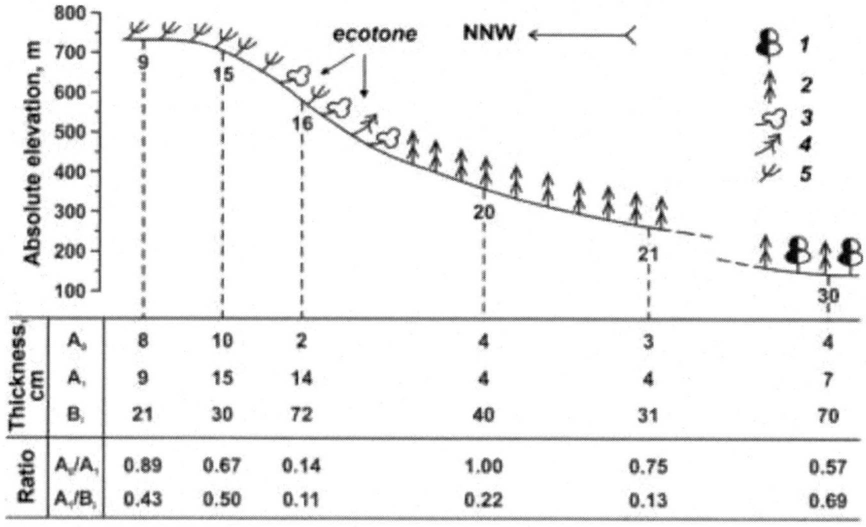

Figure 6.6. The changes in soil-morphological characteristics along the topological profile across the Pacific Ocean sector of the Mendeleev volcanic landscape. Forest-forming species: *1* – broad-leaved; *2* – fir/spruce; *3* – stone birch crooked forest; *4* – coniferous crooked forest; *5* – cedar elfin wood. 9, 15, 16, … – the numbers of test plots (see Figure 4.3).

The lower (TA), middle (T) and upper (TE) strata of the extrusive dome are characterized by the least-mature humus-accumulative and weakly illuvial-humus soils over the entire spectrum of altitudinal zonality (from mountain dark-coniferous taiga to the thickets of cedar elfin wood) as indicated by the minimum thickness of hor. B_i and the equally low A_I/B_i ratio (as concerns the abnormally high values of thickness of hor. B_i on test plot 16 in the belt of crooked birch forests, see below). The relatively weak degradation of dead organic matter also results in a less thick humus horizon. Positive anomaly is created only by the peaty-humus soils of cedar elfin wood, with "dry peat accumulation" in hor. A_0 and a thick peatified humus horizon.

There is a most marked spatial differentiation of the degree of soil maturity expressed by the thickness (*h*) of illuvial hor. B_i and its ratio to humus hor. A_1 (Tables 6.3 and 64, Figures 6.7 and 6.8). Illuvial-metamorphic processes are maximally developed in the ochreous brown earths of coniferous/broad-leaved forests on the plains and in the foothills and, to a lesser extent, in the humus-accumulative soils of fir/spruce forests in the lower TA altitudinal stratum (up to the height of 320 m). At the same time, they are slightly less marked in the Sea-of-Okhotsk sector compared to the Pacific Ocean sector: the gradation of the $h(B_i)$ dominant is 71–90 m in the former case and 55–71 cm in the latter case. In addition, the $h(B_i)$ dominant in the Pacific Ocean sector is particularly marked (with a probability of 80%), indicating the maximum development of sod process (Tsyganenko 1972). The warmer and drier Sea-of-Okhotsk sector is characterized by the active humification of substance, i.e., the sod process is much more developed, with twofold higher values of the A_1/B_i ratio compared to the Pacific Ocean sector.

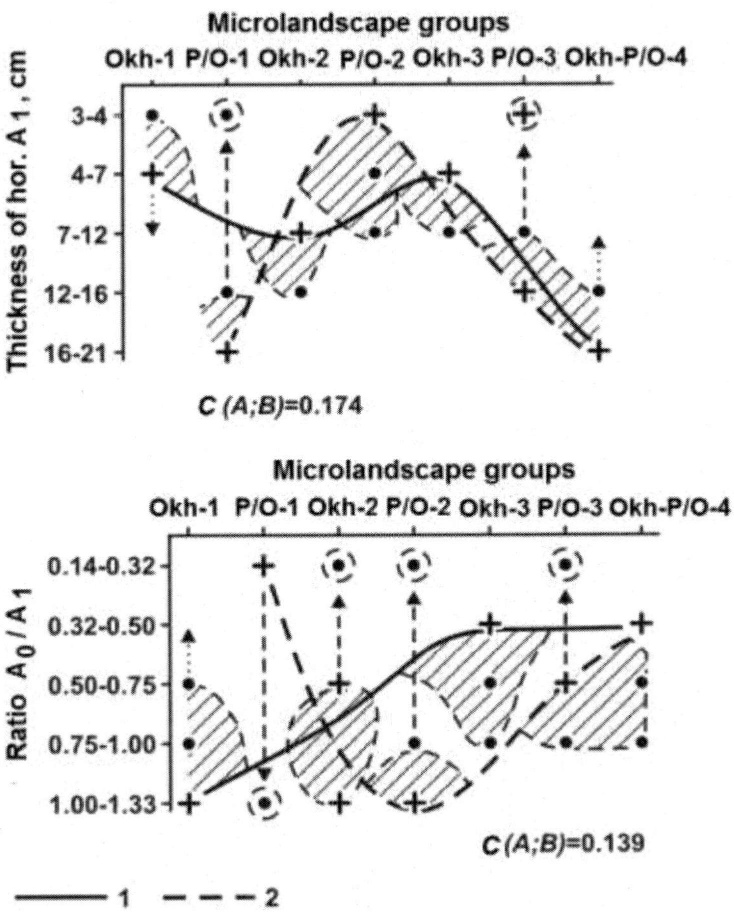

Figure 6.7. The distribution of characteristics of organogenic soil horizons by the groups of microlandscapes of the Sea of Okhotsk (1) and Pacific Ocean (2) sectors of the Mendeleev volcanic landscape. The symbols are the same as in Figure 6.3.

Table 6.3. Changes in the thickness of illuvial horizon B_i depending on the absolute height (the normalized partial coupling coefficient)

C(A;B) = 0.144)

Absolute elevation, m	Thickness of the horizon B_i				
	19–28	**28–40**	**40–55**	**55–71**	**71–90**
65–150	0.20	0.18		0.19	**0.52**
150–320		0.22		**0.47**	0.32
320–500		0.11	**0.60**	0.23	
500–650	**0.51**	0.11	0.12	0.12	0.16
650–785	0.29	**0.38**	0.28		

Note: The dominant areas of ecological niches are in bold.

Table 6.4. Distribution of thethickness of illuvial horizon Bi by altitudinal layers

C(A;B) = 0.133

Thickness of horizon B_i	Altitudinal layers					
	Lower (70–180 m)		**Middle (180–450 m)**		**Upper (450–780 m)**	
	T/o	Ox	T/o	Ox	T/o	Ox
19–28		0.14		0.20	0.24	**0.42**
28–40		0.14	0.36	0.20	**0.35**	0.14
40–55	0.20		**0.64**	**0.36**	0.21	0.12
55–71	**0.80**	**0.48**		0.24	0.21	0.32
71–90		0.24				

Note: The dominant areas of ecological niches are in bold.

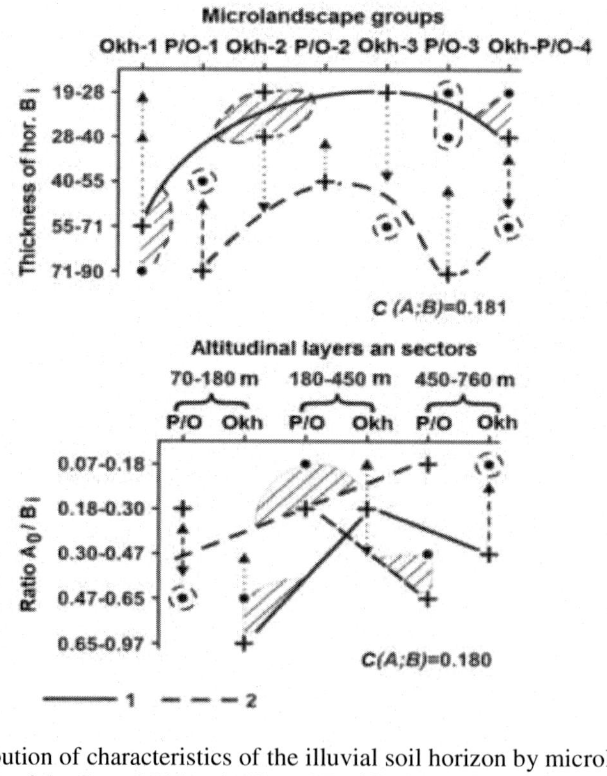

Figure 6.8. The distribution of characteristics of the illuvial soil horizon by microlandscapes, as well as by the altitudinal strata of the Sea of Okhotsk (1) and Pacific Ocean (2) sectors of the Mendeleev volcanic landscape. The symbols are the same as in Figure 6.3.

Such sectoral differences are generally maintained on the volcanic dome too. However, the hB_i parameter decreases upward the slopes everywhere, indicating that the soil becomes still younger and less developed on approaching the dome terminal. At the same time, the thickness of B_i on the slopes facing the Pacific Ocean remains the highest, while the A_l/B_i ratio is the lowest. The sectoral contrasts between these indices are considerably leveled in the middle altitudinal stratum (180–450 m) but increase again in the upper (top) stratum.

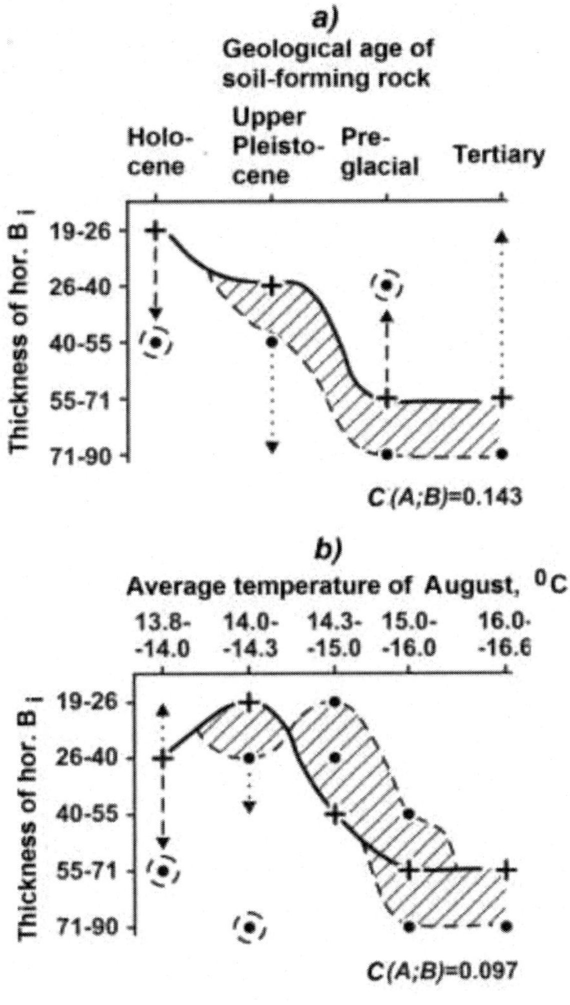

Figure 6.9. Distribution of the thicknesses of illuvial soil horizon in the space of soil-forming rocks of different age, as well as of the mean August air temperature.

The bottom-up decrease in soil maturity along the volcanic dome is obviously determined by the two factors: the geological age of pyroclastic soil-forming rocks and the thermal conditions (mainly in the warm season). Both factors interfere with the same sign and, therefore, it is very difficult to assess their partial role. Nevertheless, the values of the *K(A;B)* parameter and the binary ordination model (Figure 6.9) demonstrate the predominant role of the age of soil-forming rocks. In the first approximation, there are two contrasting levels of thickness of hor. B_i. The least-mature soils of the first level, with $h(B_i) \leq 40$ cm, belong almost exclusively to the youngest rocks: Holocene-aged in the central cone and Late Quaternary-aged in the second somma. Here there are also the lowest mean August air temperatures (≤14.0–14.5°). The most mature soils of the second level with thick illuvial horizons (55–90 cm) have developed on the tertiary sediments of piedmont plains and the first (preglacial) somma, where the temperatures are the highest (15.5–16.0°).

There are certain deviations against the described predominant background of morphology and age-related characteristics of soils. The comparatively low values $h(B_i)$ = 28–40 cm occur along the entire altitudinal profile of the Mendeleev Volcano. At the same time, the moderate thickness of illuvial horizon (40–55 cm) typical of the humus-accumulative soils of fir/spruce forests in the upper part of the middle stratum (320–500 m) often occur in the higher subgoltsy belt. Nevertheless, the peaty-humus soils of the subgoltsy belt with cedar elfin wood are the least developed.

One of the basic climatic factors of soil formation in the regions of contemporary volcanism is considered to be humidity or, on the contrary, aridity of the climate (Sokolov 1973). In the Mendeleev volcanic landscape, the humidity factor (F_{hum}) is rather closely related only tothe thickness of illuvial-metamorphic horizon $h(B_i)$. The relationship is linear and inverse:

$$h(B_i) = 200.11 - 33.4 \cdot F_{hum};\ R = 0.779. \tag{6.1}$$

As we can see, the soil-forming processes making hor. B_i in the volcanic landscape are rather sensitive to humidity of the climate; however, this factor is not the only one. The atmospheric humidity factor slightly increases (from 3,8 to 4,0 units) upwards the volcanic dome (to the absolute heightsof 700–800 m); however, intrasoil weathering and illuviation processes become much weaker, as indicated by a considerable decrease in hor. B_i thickness (from 70–85 cm to 25–40 cm). It seems that the key factor here was the age of soils, which drastically decreased with transition from tertiary and preglacial lowland plains and piedmonts to the Quaternary-Holocene volcanic dome. There was no significant relationship between the processes of dead organic matter mineralization and humification and the humidity factor either.

Thus, against the background of the general altitudinal-zonal decrease in the intensity of soil-forming processes, there are quite distinct sectoral contrasts between the latter, which are adequate to the known sectorality of natural conditions of the Kuril Island Arc. The soil cover of the colder and more humid Pacific Ocean sector of the Mendeleev landscape was formed under the predominant influence of podzol formation processes: illuviation and possibly lessivage, as well as intrasoil weathering. In the warmer and drier Sea-of-Okhotsk sector, the dominant significance in soil development was acquired by the sod process – humification resulting in the formation of the thickest humus horizon, with a relatively weak manifestation of illuvial-metamorphic processes.

The initial cause of the revealed sectoral contrasts in soil morphology (and the respective soil formation processes) should be apparently sought not only in the background climatic differences between these sectors (the effects of warm and cold sea currents, monsoon circulation of air masses, cyclone activity) but also in the *geothermal differences* associated with the spatial structure of the contemporary stage of island-arc volcanic activity – the stage of formation and intensive activity of fluid systems with fumarole manifestations (see below,

Chapter 8). Anyhow, the Sea-of-Okhotsk sector is a sphere of much more intensive and more extensive (by area) discharge of magmatic energy compared to the Pacific Ocean sector, which makes an indubitable contribution to the hydrothermal regime of soils, which becomes still more thermo-arid in the former case. Accordingly, the dominant soil-forming processes are developed – either more turfy or more podzolic.

The researchers note the apparent discrepancy between the low site quality of forests (the ratio of the average height of tree stands to their age) and rather high productivity of volcanic soils of island Neo-Pacific (Lashkov 1948; Gladkova andButovets 1988). Such disconformity is accounted for mainly by the influence of climatic conditions of the islands unfavorable for tree growth: first of all the strong, cold and moist winds and fogs, the negative effect of which becomes stronger as the trees grow up and reduces the tempos of their development and changes their life habit (right up to ugly "flag-like" shapes). The high content of aluminum compounds determined by the chemical composition of soil-forming pyroclastic material, the toxicity of which increases exactly under high acidic conditions typical of the soils of the Mendeleev Volcano, may also have a negative effect on plant growth (Kabata-Pendias and Pendias 1989). However, the negative role of this factor must decrease here due to the characteristic washing regime of pyroclastic soils and, accordingly, the intensive removal of amorphous free forms of aluminum into illuvial horizons, with a manifold decrease in its content in hor. A_1 (Gladkova andButovets 1988).

The soil-phytocoenotic disconformity of insular Neo-Pacific is an apparent manifestation of the key role of oceanic climatic background in the formation of forest cover on the islands, the evidence of considerable suppression of biogeocoenotic significance of the local geomorphological and edaphic factors by the climatic background. Hence, there is a marked direct dependence of latitudinal zonality and altitudinal zonation on the background hydrothermal gradients determining the drastically narrowed spectra of latitudinal zones and altitudinal belts.

6.4.4. Soil Moisture

The moisture content in the upper (organo-mineral) soil layer $W_{10–20}$ explicitly increases upwards the slope of the Mendeleev Volcano (Figure 6.10, *a*). The $W_{10–20}$ value is minimal on the piedmont plains and reaches its maximum on the tops and crests of the ridges. Since there is no relationship between the $W_{10–20}$ parameter and the soil and air temperatures, this moisture trend should be accounted for by the increase in precipitation with height. At the same time, soil moisture content decreases with depth at all altitudinal steps (Table 6.5), i.e., the upper organo-mineral soil layers have the highest water-retaining ability, while the illuvial horizon B_i is yet poorly developed and not sufficiently saturated with clay particles capable of retaining moisture.

The moisture content of soil is also positively distributed by the solar-circulation exposure and, accordingly, by the groups of microlandscapes, and exposure has a stronger effect on the hydrological regime of soils than on their temperature regime (Figure 6.10). Let us pay special attention to the upper organo-mineral soil horizons, which are most important for plant nutrition and biological cycle. Hydroedaphotopes have a differentiating effect on plant cover; in particular, on forest type groups (see Figure 6.10,*b*). Here, the dominants with a high probability (60–100%) are clearly distinguished in the system of hydroedaphic niches, in the absence of enclaves.

Table 6.5. Mendeleev landscape. Taxonomic standards of soil hydro-thermal characteristics in different groups of forest types

Groups of forestctypes (see in the text)	Soil temperature on the depths (cm)					Soil moisture (%) in the layers (см)		
	20	30	40	50	70	10–20	20–30	30–40
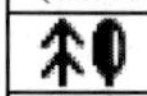	15.3	14.4	13.2	12.5	12.2	47	45	49
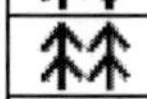	15.7	14.8	14.7	13.2	12.2	60	49	40
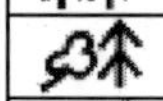	13.6	14.4	13.5	14.3	12.9	67	56	52
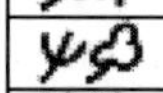	16.7	16.0	14.6	14.3	12.3	62	51	40
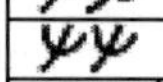	15.7	14.7	14.3	13.9	11.6	74	55	30
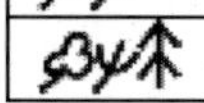	17.1	16.3	15.3	14.5	13.4	53	36	40

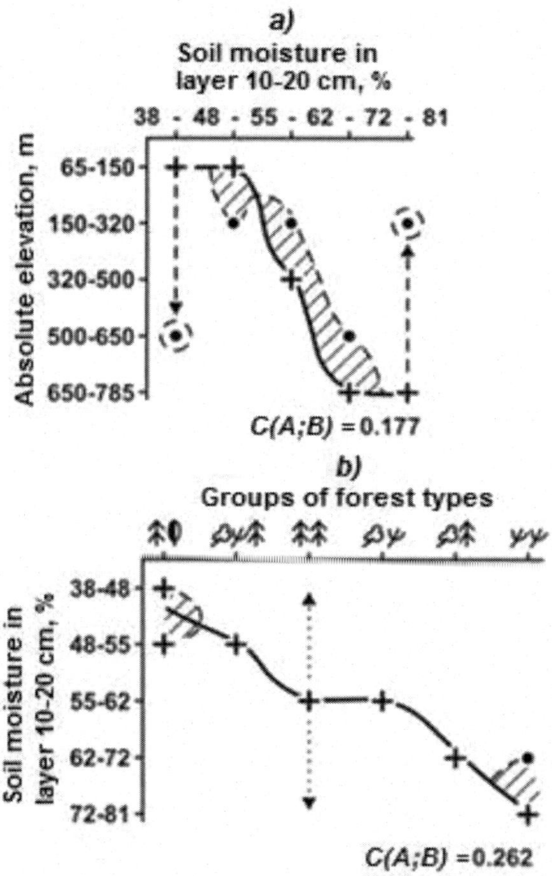

Figure 6.10. The distribution of soil moisture content by absolute heightand the groups of forest types in the Mendeleev landscape.

The soils with the minimum moisture content (W_{10-20} = 38–55%) are typical of subhorizontal sites and SW–W–NW–N slopes on the piedmont plains of the Sea-of-Okhotsk sector. As we know, this sector has also the highest soil temperatures. However, the known statement (Lashkov 1948) that the Sea-of-Okhotsk sector of the Kunashir Island is warmer and drier in the vegetation period, while the Pacific Ocean sector is colder and wetter, should be referred only to the piedmont plains of the Mendeleev volcanic landscape. The warming and, accordingly, aridizing effect of the Soya warm current on the Sea-of-Okhotsk sector is limited in the surface soil layers by coastal and piedmont territories with the predominant coniferous/broad-leaved forests of the Manchurian–Okhotian phratrie (see Figure 6.10, *a, b*). Even in the lower TA stratum, in the altitudinal range of 180–450 m, the soils of the northern and western slopes of the volcano are wetter than the soils of the southern and eastern slopes (W_{10-20} = 72–81% vs. 55–62%). In the middle transit stratum, their edaphic moistening becomes closer to each other, reaching the common maximum on the tops and crests of the ridges.

The middle-stratum dark-coniferous forests of the Okhotian flora and the spruce/birch forests replacing them in the upper stratum, as well as the Kamchatian–Beringian stone birch forests with cedar elfin wood in the undergrowth, grow under conditions of enhanced moisture content in the upper soil layers (the dominant W_{10-20} = 55–72%). Finally, the highest soil moisture content ((W_{10-20} = 62–81%) is typical of the soils under cedar elfin wood of the Beringian flora in the subgoltsy belt (see Table 6.5).

At the same time, mesorelief makes certain corrections in the distribution of moisture content in the upper soil horizons on mountain slopes. So, the topological profile across the Mechnikov Mountain (see Figure 5.5) clearly shows the maximum soil moisture content on the mountain slopes facing the Pacific Ocean, towards the moisture-bearing cyclones, and a sharp decrease in moisture content on the opposite slope under conditions of the "barrier shade." Let us add that the middle and lower soil layers in the Sea-of-Okhotsk sector are drier than in the Pacific Ocean sector not only at the foothills but also on all slopes of the Mendeleev volcanic massif.

Thus, *the moisture content in soils of the Mendeleev volcanic landscape is determined mainly by rainfall amounts, i.e., by the income component of the water balance, but not its outcome component* (evaporation and runoff). This conclusion is of no small importance for using soil moisture content in landscape-ecological forecasting as one of the transmission unit of the influence of climatic system on the functional and structural state of island-arc forest ecosystems. Mind that the global and regional climatic forecasts operate mainly with air temperature and atmospheric precipitation.

– «» –

The described models of structural organization of the Mendeleev volcanic landscape confirm the proposed (Pietsch et al. 2003) statement that the islands of the Kuril Archipelago are quite a representative natural laboratory for studying the origin, subsequent evolution and long-term maintenance of nonrelict island phytobiota. Our results substantially extend this statement, which is based only on the quantitative analysis of the lists of plant species, and fill it with the landscape-ecological contents. This statement certainly fits well into the elaborated *concept of island-arc biogeocoenosis* as the initial stage of formation of the next part of continental biosphere in the mobile belt of the planet.

6.5. Hierarchy of Characteristic Times of Landscape-Forming Processes

The buried humus horizons appeared in soil profiles as a result of periodic ashfalls during pyroclastic eruptions; the age of the lowest horizons is the Middle Holocene is 4000–6000 years (Razzhigaeva et al. 2011) and it apparently should be taken as the maximum absolute age of contemporary soils of this region.

The age of Holocene is generally typical, as is known (Bobrov 1980), of the contemporary vegetation in the South of the Far East. The age of vegetation on the Kuril Islands can be determined more exactly. According to the same radiocarbon dating (Razzhigaeva et al. 2011), nemoral broad-leaved forests were widespread in the southwest part of Kunashir in the subboreal period of Middle Holocene (3790 ± 70 years ago), when the average annual temperature was by 2° above the contemporary value. Then, after the cold snap in Late Holocene (2730 ± 60 years ago), they were replaced by birch forests and bamboo-grassy meadows. Further rearrangement of phytobiota occurred at the minor optimum of Holocene (1200–950 years ago). This temporal border can be considered as the beginning of formation of contemporary, steady-state plant communities which characterize, according to Zonov (1977), the fourth, final stage of the development of volcanogenic landscapes on the island.

Thus, as early as at the island-arc tectonic-volcanic stage of formation of subaerial landscapes there was a certain "...multi-speed "ladder" of characteristic times of ecosystem processes," which "...is obligatory for the development of ... any substantially multi-phase ("different-phase") ecosystem" (Targulyan, 198411), i.e., underlies the evolutionary biogeocoenotic process per se and provides the stability of the biosphere at each stage of this process (Puzachenko 1986).

In light of the above, it is necessary to touch upon the two important, in our opinion, landscape-ecological problems directly concerning the subject of this research. First, the concept that the geological-geomorphological factor is the leading factor at all hierarchic levels of geosystems in the regions of contemporary volcanism is now under discussion (Grabkov and Khudyakov 1993). What is meant here is that the existence of geosystem hierarchy in the biosphere is determined by different sensitivity of its different structural levels to the same landscape-forming factor with a particular oscillation frequency due to different frequencies of their own oscillations (Puzachenko 1986). It seems that one should consider not a single geological-geomorphological factor but a certain series of lithogenic factors of different spatial and temporal scales. It is a different matter if we speak about the catastrophic consequences of intensive volcanic activity for ecosystems. Volcanic eruptions periodically throw the affected surrounding natural complexes back to the initial stages of development and, eventually, exert a leveling effect on the continuous process of physical and geographical differentiation of a territory (Zhuchkova et al. 1973).

Second, there is no sense in searching for the correlations between even the youngest, 40-thousand-year-old basis of the Mendeleev volcanogenic landscape and its contemporary types of phytocoenotic structures aged no more than thousands of years. In addition, the studies of plant cover genesis in the south of the Far East (Urusov and Lauve 1980; Urusov, 1988, 1993; Urusov and Chipizubova 2008) assert the presence of a direct and synchronous relationship between tectonic and phytocoenological processes. In particular, it is proved that the establishment of contemporary floristic and phytocoenotic structures of the region is directly determined by tectonic subsidence of the margin of the Asian continent in the Oligocene–Quaternary period by more than 1.5 km; therefore, it is stated that "... the apparent stability of the upper forest boundary ... clearly contradicts the climate" (Urusov and Chipizubova 2008 211). It is also stated that "the assembly of communities in the most part of south of the region appeared as a result of tectonic subsidence of land, intrusion of nemoral elements coenotypes into taiga, and the thrusting of vegetation onto the relicts of cold stone deserts of the lowered uplands" (in the same place, 213).

Such "tectonic" hypothesis of the "evolutionary-biological effects" (Urusov 1993 17) can be seriously discussed with regard to rather long-term processes of species formation and florogenesis but seems to be quite ambiguous as applied to the much more dynamic "genesis of communities" (ibid.). First of all, this hypothesis contradicts the known palynological data. It has been shown that the types of plant cover in the continental part of the South of the Far East more than once underwent rather cardinal changes due to glacio-eustatic sea-level fluctuations and super-century climatic changes during Holocene only (10.5 thousand years): from birch/larch forests and light forests with frigid vegetation in the epochs of marine regressions and cold snaps to polydominant broad-leaved formations with the involvement of thermophilic species in the epochs of transgressions and climatic optima (Korotky et al. 1999; Korotky 2008). Though, the role of these factors (with a much higher frequency compared to geological processes) is also recognized by the authors of the "tectonic" hypothesis: on the Kuril Islands and Sakhalin, by the beginning of the latest marine transgression, the entire contemporary land (with a minor exception) ... was covered by taiga, subalpine prostrate shrubs and dwarf shrubs, alpine meadows and mountain tundras" (Urusov, 1988178).

Moreover, the concept of latitudinal shift of zonal bioclimatic boundaries on the plains and their lowering in the mountains as a result of land downwarping somehow extends also to the landscapes of the Kuril Island Arc (Urusov 1988; Osipov 2008). In the mean time, according to the theory of plate tectonics and orogeny, the northwestern margin of the Pacific Ocean was under conditions of tectonic compression and lithospheric uplift during a long geological period (throughout the Neogene), and it was a manifestation of ".... an obvious general tendency towards the buildup of continents here due to their accretion with fold belts" (Korzhuev 1990 119–120). The paleogeomorphological analysis has shown that throughout Late Pleistocene and Holocene there was a general tectonic uplift of the Kuril Islands, which proceeded at an average rate of 6.5–10 mm/yr and to date has reached 800–1200 m (Melekestsev et al. 1974).

Thus, the general evolutionary biogeocoenotic process on the Kuril Islands was predetermined by the ascending but not descending development of insular volcanic-tectonic morphostructure, with the extension of land area, the uplift of relief, and the buildup of a system of vertical bioclimatic zonality. This is precisely why we consider that the most ancient forest communities are nemoral-boreal types of the plain-piedmont stratum of the Mendeleev landscape and the youngest communities are subgolets crooked-forest/prostrate types of the upper stratum of low mountains (as opposed to the statements mentioned above). The relatively inert substrate geome of volcanogenic landscape at each stage of its development created the framework conditions for manifestation of much more high-frequency evolutionary biogenocoenotic processes associated with the super-century climatic fluctuations and with the respective changes in the ocean level.

Against the background of vigorous and steady tectonic uplift, the spread of glacio-eustatic sea-level fluctuations was comparatively small: from 120–140 m below the contemporary level during the Later Quaternary regression to 2–10 m above this level during the Atlantic transgression (Korotky et al. 1999; Pietschl et al. 2003), which corresponded to transition from the latest glacial epoch to the Holocene climatic optimum. Nevertheless, it is precisely this climatic trend that seemed to determine the transformation of supposedly original boreal-subalpine crooked-forest/prostrate phytocoenoses into fully developed nemoral-boreal forests represented in the contemporary forest belt of the Mendeleev landscape, as described in the work (Urusov 1988).

According to the data of spore-pollen analysis of the peat overlying the 80–85-m abrasive-accumulative terrace (Melekestsev et al. 1974), in the beginning of the Late Quaternary interstadial (31 thousand ± 100 years ago) the climate was slightly colder than today and the slopes of the Mendeleev and other mountains were occupied mostly by stone birch light forests with alder and tall grass. The sites of coniferous forests were preserved only in favorable habitats (narrow gullies and depressions of relief).

Based on the above, it may be supposed that the observed advance of coniferous and leaf forests into interfluvial tall-grass (bamboo) moors, as well as the transgression of dark-coniferous species to the belt of stone birch crooked forests, is evidence of contemporary (beginning from the second half of the XX century) warming of the climate. We also assume that the tall-grass woodless moors confined to the ridges between narrow gullies of southern, southwestern and other slopes of the Mendeleev–Mechnikov mountain massif are relicts of the previous, colder climatic epoch (XIV–XIX centuries). In that epoch, dark-coniferous forests were preserved only in the narrow gullies with greater snow accumulation and soil-ground moistening and better wind protection. Apparently, conservation of these forests in unfavorable climatic epochs was favored here by local geomorphological conditions, as well as the enhanced role of edaphic factor.

Chapter 7

FUNCTIONAL ORGANIZATION OF ISLAND-ARC FOREST ECOSYSTEMS

ABSTRACT

The most general features of the structure of functional phytocoenotic space of a young volcanogenic landscape are its high fragmentation and relative autonomy of metabolic processes in different groups of forest types, with the weak lateral interactions between separate fragments of phytocoenological mosaic. An exceptionally important backbone role of forest phytobiota in the establishment and development of young volcanic landscapes was demonstrated. The optimizing strategy program for the development of plant communities, which is aimed at the maximum biomass formation on forest island-arc landscapes, is much more marked compared to their zonal analogs on the neighboring continent. At the same time, an exceptionally high percentage of green mass in the structure of production results in the acceleration of biological cycle as a factor of stability of forest community. The energy bases of productivity of forest communities of the island-arc Neo-Pacific have been analyzed as a focus of local and regional diversity of terrestrial ecosystems, which is further developed on the continent. The two basic mega types of the heat balance structure in terrestrial ecosystems have been distinguished: continental and oceanic. These types significantly differ in the ratios of two main components of the expenditure part of the balance – evapotranspiration and turbulent heat exchange. The predominance of transpiration component of the heat balance in the oceanic type creates a high level of autotrophic biogenesis in forest island communities, providing their stability at the initial stage of subaeral landscape genesis.

7.1. GENERAL CHARACTERISTICS OF PRODUCTIVITY AND BIOLOGICAL CYCLE

Durgin the modeling, high emphasis was placed on the phytocoenotyic functional block – the parameters of the annual and perennial cycles of plantmatter turnover. It was shown that the shifts in the small biological cycle (phytomass production and degradation), including their climatogenic successions, were the main driving force for the evolutionary transformation of natural ecosystems. These changes occur in the first several years in the taiga zone and are completed within a year in the subzone of broad-leaved forests (Kolomyts 2012).

First of all, the map of total annual production of forest phytocoenosis (PC_{tot}) in the Mendeleev landscape was plotted (Figure 7.1). The map was calculatedusing the following empirical regression equation (Kolomyts et al. 2016; Kolomyts and Surova 2018 a):

$$PC_{tot} = -2.613 \cdot t_{30\,-3.60} - 0.4709 \cdot W(20\text{–}30)_{-3.18} + 0.2784 \cdot GA^{*}\cos(A_0)_{+2.48} + 84.07;$$
$$R^2 = 0.364;\ Degr = 16.7\%;\ P<10^{-3}. \qquad (7.1)$$

Here, t_{30} isthesoil temperature at the depth of 30 cm (^{0}C), $W(20\text{–}30)$ is thesoil moisture inthe layer of 20–30 cm (%),and$GA^{*}\cos(A_0)$ is thenorthern component of slope steepness.The model describes36% (R^2 = 0.364) of space changes forthetotal productivity of forest ecosystems bya comparatively low indicator of degradation. Onlyt_{30} and $W(20\text{–}30)$ are statistically significantpredictors in this equation. The productivity has a tendency to a decrease, if bothsoil temperature and soil moisture increase. Factor $GA^{*}\cos(A_0)$ reflects any increase in productivity on the steep slopes of northern exposure.

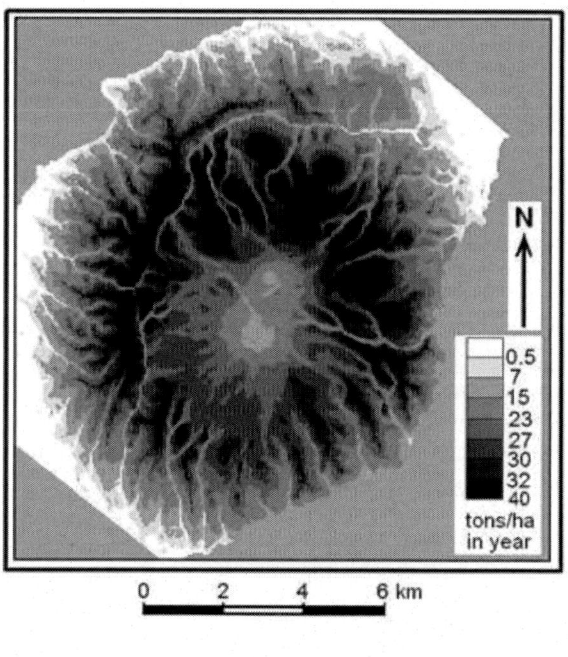

Figure 7.1. The raster map of total productivity of mixed and dark-coniferous forests, stone birch crooked forests and cedar elfin woods of the Mendeleev volcanic landscape.

The Mendeleev volcanic landscape, along with Kunashir Island and the entire South-Kuril Range, is a part of the Pacific marginal-continental and island-arc ring, differing from intracontinental territories by the generally higher indices of biological productivity and accumulation of living organic matter (Bazilevich 1981; Bazilevich and Titlyanova 2008). Our empirical data are in agreement with these concepts in regard to the boreal and subboreal forests (Figure 7.1; Tables 7.1 and 7.2). The coniferous/broadleaved forests of low-plain and piedmont Mendeleev microlandscapes have the reserves of above-ground living phytomass (the taxonomic norm $BL \geq 340$ tons/ha), which are considered maximal for the Pacific Northwest (about 300 tons/ha) and even exceed this maximum. For comparison, let us note that the most productive subboreal (pine-oak-lime) forests of thePrioksko-Terrasny Biosphere Reserve in the south of the subtaiga zone of the Russian Plain have a taxonomic norm ofBL = 376 tons/ha, while in cedar/broad-leaved forests of the boreal ecotone in the Lower by-Amur (Priamurye) this value hardly reaches 240 tons/ha (Kolomyts 2015 a). The fir/spruce forests of the lower altitudinal stratum (250–500 m) with BL = 215–230 tons/ha are not inferior to the dark-coniferous forests of Hokkaido, Sakhalin and Primorye with BL values varying from 130–150 to 230–280 tons/ha (Bazilevich 1981).

Table 7.1. The taxonomic norms of phytomass, productivity, and complex parameters of the biological cycle (CPBC) in the groups of forest types of the Mendeleev volcanic landscape

Phytomass, productivity, and CPBC *)		Groups of forest types					
		CB	DC	SbS	SbCe	CeW	CeSbS
BL		340.70	223.75	163.60	89.59	89.70	135.00
PG		11.16	14.95	11.75	8.62	4.18	14.19
PS_{tr-shr}		9.52	8.14	5.72	3.81	3.16	3.86
PV_{tr-shr}		3.99	3.21	2.41	1.83	1.62	1.94
$(PS+PV)_{tr-shr}$		13..51	11..36	8.13	5.64	4.78	5.00
PV_{tot}		15.15	18.16	14.16	10.45	5.80	16.13
PC_{tot}		24.67	26.31	19..88	14.26	8.96	20.00
CPBC	CR	0.044	0.036	0.087	0.117	0.065	0.119
	CA	0.467	0.536	0.520	0.523	0.488	0.566
	CY	1.68	0.54	1.31	0.46	5.95	0.46

*) Produictivity of: PS_{tr-shr} – wood mass of trees and shrubbery; PV_{tr-shr} – green mass of trees and shrubbery; $(PS+PV)_{tr-shr}$ – common ground of trees and shrubbery; PV_{tot} – total forest green mass; PC_{tot} – sum ground mass of the forest ($PG + PV_{tot}$). Rest meaninge see in the text of Chapter 2.

Table 7.2. The taxonomic norms of phytomass, productivity and complex parameters of the biological cycle (CPBC) in the groups of microlandscapes of Mendeleev volcanic landscape

Phytomass, productivity and CPBC *)	Groups of microlandscapes						
	Piedmont layer		Low layer		Middle layer		Upper layer
	Okh–1	P/O–1	Okh–2	P/O–2	Okh–3	P/O–3	
BL	355.58	228.5	230.8	217.9	114.2	103.4	89.6
PG	10.30	18.36	13.55	10.95	9.42	12.90	4.18
PS_{tr-shr}	16.25	15.44	14.90	12.20	8.38	7.83	6.50
PV_{tr-shr}	3.91	3.90	3.36	2.98	2.05	1.80	1.62
PC_{tr-shr}	20.16	19.34	18.26	15.18	10.43	9.63	8.12
PV_{tot}	14.21	22.26	16.91	13.93	11.47	14.70	5.80
PC_{tot}	30.46	37.70	31.81	26.13	19.85	22.53	12.1
CR	0.040	0.097	0.073	0.064	0.100	0.142	0.06
CA	0.452	0.572	0.520	0.489	0.516	0.580	0.39
CY	2.08	0.21	0.72	0.89	1.49	0.48	5.95

*) Conventional meanings see in Table 7.1.

The annual production of the phytobiota has still higher values. Let us emphasize *the altogether phenomenally high productivity of south Kunashir coniferous/broad-leaved and fir/spruce forests.* The taxonomic norm is $PC_{\underline{tot}}$> 24–26 tons/ha·yr, with the maximum values up to 5–40 tons/ha·yr (see Figure 7.1). This normal value exceeds the productivity of temperate coniferous and mixed forests of the Primorye, Manchuria and Northeastern China, with total production within a range of 11–18 tons/ha·yr, rarely reaching 21–24 tons/ha·yr. In the south taiga of West and East Siberia, these values are 2.5–7.1 tons/ha·yr (Bazilevich and Titlyanova 2008). In the coniferous/broadleaved forests of the Lower Priamurye, the mean values are PC_{vol} = 8.5–9 tons/ha per year in the boreal ecotone of the Volga River basin: about 11–12 tons/ha·yr. The productivity values similar to South Kunashir (from 7–10 to 20–25 tons/ha·yr on average) are observed only in the broadleaved and coniferous forests of Honshu and Hokkaido (Bazilevich 1981).

Thus, the maximal biomass production "programmed" by the developmental strategy of plant communities (Sochava, 1978 47) in the forest landscapes of insular Neo-Pacific is much more marked than in their zonal analogs on the neighboring continent. In other words, the boreal forests of Mendeleev volcanic landscape, in contrast to the analogous forests of Priamurye and Primorye, have more and less marked properties of patience and violence, respectively, according to L.G. Ramensky (1971). As regards structure and function, terrestrial ecosystems of the Pacific Fire Ring are "... a particular class of aggregates of natural formations" (Bazilevich 1981 208).

The implementation of the principle of maximal productivity, i.e., attainment of the maximum efficiency of phytocoenosis, is possible only under favorable conditions of the abiotic environment. This fact per se contradicts the widespread concept of existence of zonal types of plant formations on the northwestern Pacific islands with very limited heat and energy resources. Then, there must be a harmonic combination of photosynthetic and respiration functions for a plant community to reach the maximum productivity (Tooming 1984). Judging by the productivity values presented above, it would be right to suggest the presence of quite optimal functional interrelationships of forest communities in Mendeleev volcanic landscape. Accordingly, we can assume that these "climatically unjustified" terrestrial phytocoenoses (see below) have rather high level of adaptation to the conditions of extraordinary oceanic environment. It ensures their survival and increases their capacity for evolutionary transformations. As is known (Ashby 1956; Timofeeff-Ressovsky et al. 1969; Yablokov and Yusyfov 1976), adaptation underlies all forms of ecosystem evolution, and this mechanism has invariably worked in the critical epochs of history of the biosphere (Budyko 1977).

There are the two main characteristics of distribution of above-ground living phytomass reserves and forest productivity in the Mendeleev landscape (see Figures 7.1 and 7.2): (1) the general decrease in these values with absolute height and (2) the dominance of the Sea-of-Okhotsksector over the Pacific Ocean sector by the altitudinal-zonal values of *BL* and PC_{vol}. The former phenomenon can be logically explained by bioclimatic altitudinal zonality, while the latter seems to have more complex causal mechanisms. Here we can observe both the above-described circulation-exposure zonality, which is most favorable for the phytobiota of the Sea-of-Okhotsk sector, and the influence of volcanic geotherms on edaphotopes, which is predominant in the same sector.

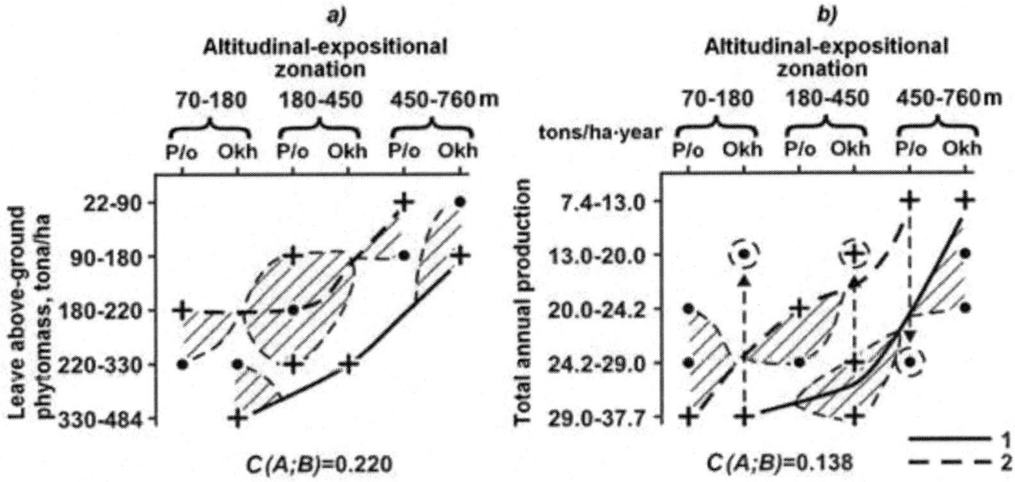

Figure 7.2. The distribution of phytomass reserves and forest productivity by three steps of absolute heights in the Sea-of-Okhotsk (1) and Pacific Ocean (2) sectors of the Mendeleev landscape. Symbols are the same as in Figure 5.8.

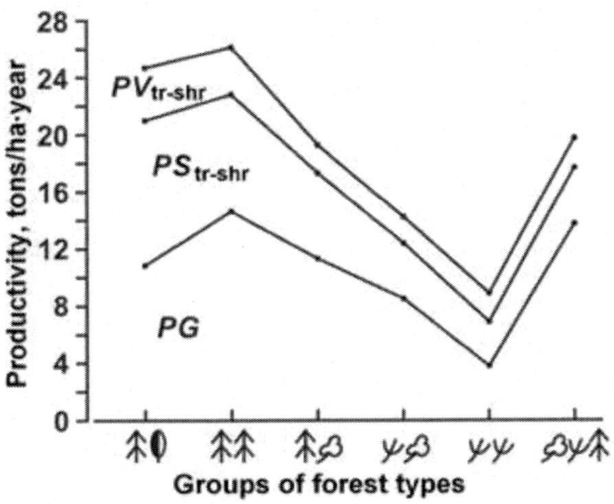

Figure 7.3. Distribution of the productivity of ground cover green mass (*PG*), tree and shrub skeletal mass (PS_{tr-shr}), tree and shrub green mass (PV_{tr-shr}) in different groups of forest types.

The subboreal and boreal forests of a young volcanogenic Neo-Pacific landscape dramatically differ from their zonal intracontinental analogs not only in total phytomass and productivity values but also in *production process structure* per se (Figure 7.3). The production of ground cover green mass (*PG*) is especially high on volcanic islands, with a considerable amount of tall grasses and Kuril bamboo thicket (Figure 7.4). In south Kunashir coniferous/broadleaved forests, the taxonomic norm of herbage green mass production (*PG*) is more than 11.0 tons/ha·yr, with a maximum up to 28.0 tons/ha·yr. It makes up about 44% of total above-ground phytomass produced, while the tree and shrub skeletal mass production (PS_{tr-shr}) is no more than 40%. Dark-coniferous forests are characterized by a still greater gap between these parameters (61% vs. 32%). In general, the portion of photosynthesizing green mass in boreal and subboreal forests of the Mendeleev landscape is 60–70% of the total above-ground productivity.

Figure 7.4. Kunashir Island. Mendeleev volcano. Kuril bamboo thickets.

By comparison, the middle and maximum *PG* values are within a range of 0.5–1.3 and up to 3.4 tons/ha·yr, respectively, in the marginal-continental Lower Priamurye vs. 5.5–7.5 and up to 10.5 tons/ha·yr, respectively, in the intracontinental Prioksko-Terrasny Biosphere Reserve. At the same time, the portion of total green mass is drastically reduced: from 33–36% to 42–47% in the former case and from 38–40% to 46–49% in the latter case, reaching 61% only in the pine-lime-oak forests.

As we can see, the autotrophic biogenesis in subboreal and boreal forests at the initial, island-arc stage of subaerial landscape genesis is directed mainly at production of green photosynthesizing mass that annually becomes tree waste. It gives an impetus to substantial acceleration of the biological cycle as a factor of functional stability of forest biogeocoenosis (Kolomyts and Sharaya 2015), which is confirmed by the values of complex parameters of metabolism (see Table 7.1). The subboreal and boreal forests of the Mendeleev island landscape have rather high values of the annual turnover of above-ground organic matter (normally, *CR* = 0.044–0.036). The normal values of this parameter in analogous forest biogeocoenoses are 0.023–0.040 in the marginal-continental Lower Priamurye and only 0.017–0.030 in the boreal ecotone of the Middle Povolzhye.Compared tothe East European boreal ecotone, the rate of phytomass turnover increases by 35–40% in south Kunashir coniferous/broad-leaved forests and by 40–50% in dark-coniferous forests.

The litter/tree-waste index *CY*, which is known (Dylis 1978) to be the major conductive channel for metabolic processes in a forest ecosystem, is also very representative. The boreal and subboreal forests of the Mendeleev landscape substantially differ from their intracontinental analogs in a much more optimal ratio between the income and consumption of dead organic matter in forest litter, which is demonstrated by the relatively low values of the *CY* parameter (the taxonomically normal values of 0.21–0.44 to 1.68–2.08 years vs. 6.42–7.95 years, e.g., in the Prioksko-Terrasny Biosphere Reserve). The capacity of this channel increases1.7–2 times in subboreal forests and 8–10 times in boreal forests. The marginal-continental Lower Priamurye is much closer to South Kunashir by the above values of the biological cycle, though it is also weaker here (see Tables 7.1 and 7.3).

Table 7.3. The taxonomic norms of complex parameters of the biological cycle in forest communities of intracontinental and marginal-continental ecoregions of Northern Eurasia

Ecoregions	Groups of forest types (types of local sites)	Complex parameters		
		CR	*CA*	*CY*
Middle Volga Region (By-Oka-Terrace Biosphere Reserve)	Pine forests with birch (E, TE)	0.027	0.437	3.7
	Spruce and pine-spruce forests (E, TE)	0.021	0.422	4.8
	Pine-lime-oak forests	0.026	0.428	3.0
	Lime-birch forests (TE, T)	0.029	0.441	3.5
	Spruce and pine-spruce forests (TA)	0.018	0.403	5.2
Lower b-Amur Region e (Komsomolsky State Nature Reserve) заповедник)	Broadleaf forests (E, Saq)	0.040	0.351	–
	Cedar-broadleaf forests (E, TE)	0.032	0.338	–
	Spruce-broadleaf forests (A, Saq)	0.032	0.313	–
	Silver-spruce forests (T–TA, A)	0.023	0.225	–
	Larch-tree forests леса (T–TA, TE)	0.029	0.289	–

Thus, extrazonal, i.e., going beyond the limits of intracontinental zonal norms, productivity and biological cycle indices in the subboreal and boreal belts of young island-arc landscapes demonstrate the *highly effective work of their soil-biotic core.* Intensification of metabolism is aimed at providing the functional stability and structural development of forest communities in the generally cold oceanic climate of Northwestern Pacific, which is extremely unfavorable for forest formation. Hence, the living above-ground phytomass can reach the values that approach the maximum in western intracontinental ecoregions. Such is, apparently, one of the most important regularities of the regional functional organization of forest cover at the initial, island-arc stage of subaerial landscape genesis.

The enhanced forest productivity and intensive metabolism inevitably cause the following two ecological consequences: the high sensitivity of forest communities to global climatic impacts (low resistance) and, on the contrary, the enhanced forest restoration potential (high elastic-plastic stability). In view of the above, *island-arc forest ecosystems must be functionally dynamic and, accordingly, have the higher tempos of structural transformations* at some stable climatic signal or another. Substantial reduction of organic deposit in tree-shrub skeletal mass can imply certain reduction of the regulatory role of forest cover in the carbon cycle and the respective weakening of its reverse effect on the carbon balance of atmosphere and on the climate. The obtained empirical conclusions directly concern the development of landscape-ecological predictions for the island-arc sector of the Pacific Ocean mega ecotone.

7.2. Altitudinal-Zonal Structure of Functional Phytocoenotic Space

The most common feature of this space is its high differentiation, the autonomy of metabolic processes in different groups of forest types. The orgraphs of inclusion relations (Figures 7.5–7.7) are represented by exceptionally weak connections in the absence of tolerance relations necessary for the formation of a topological core of a given system. The exceptionally high critical threshold of indistinguishability (Z_{crit} = 1) implies a very high lower limit of inclusion measures for elements of the system, when structural or functional unity of the latter is still preserved (Andreev 1980). For comparison, let us note that Z_{crit} = 0.975 in the Lower Priamurye (the marginal-continental sector of the Pacific Ocean mega ecotone), while on the boreal ecotone of the Volga River basin this parameter decreases to 0.942–0.937 (Kolomyts 2008, 2015 a). The functional phytocoenological space of a young island-arc volcanic landscape is still highly fragmented, with weakly marked lateral interactions between separate fragments of natural-territorial mosaic. The initial (island-arc) stage of subaerial landscape genesis differs from more mature (continental) stages in the common relatively low level of spatial organization of a regional geo(eco)system, with the absence in fact of its typological core that would provide its qualitative definiteness.

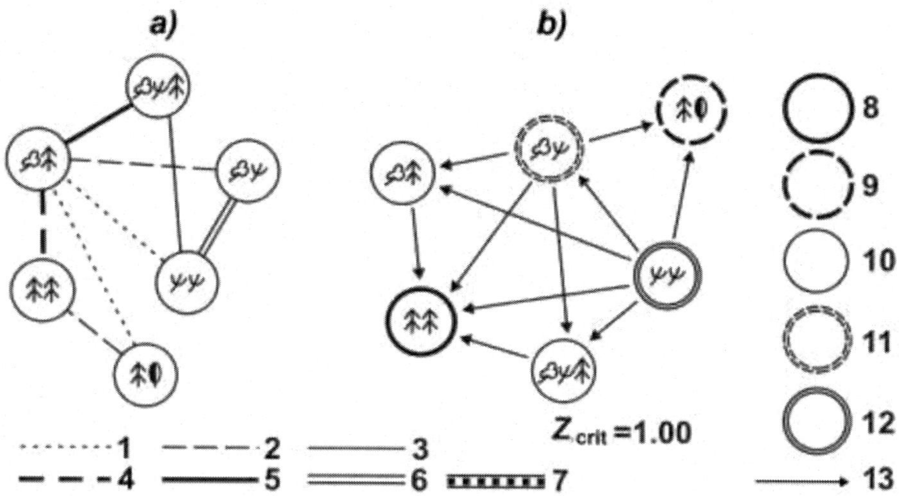

Figure 7.5. Mendeleev Volcano landscape. The graph of similarity relations (*a*) and the orgraph of inclusion relations (*b*) between groups of forest types by the phytomass supplies.

The measures of similar relation: 1 – 0.66–0.70; 2 – 0.71–0.75; 3 – 0.76–0.80; 4 – 0.81–0.85; 5 – 0.86–0.90; 6 – 0.91–0.95; 7 – 0.96–1.00. Ecotypes of forest communities: 8 – macrobolites; 9 – submacrobolites; 10 – mesobolites; 11 – submicrobolites; 12 – micriobolites. 13 – direction of the inclusion. Z_{crit} – the critical threshold of indistinguishability of objects.

Now let us proceed to particular functional analysis. With respect to above-ground phytomass reserves *BL*, piedmont-plain coniferous/broad-leaved forests are drastically different from all other forest formations of the Mendeleev landscape (Figure 7.5 *a*). The mean value of their similarity measures is N_{ik} = 0.58. Their weakly related positions of submacrobolites in the structural schemes of dominance with respect to not only phytomass but also total productivity is also characteristic (Figures 7.6,*b* and 7.7,*b*), demonstrating certain uniqueness of these subboreal forest formations, their functional isolation from the common altitudinal-zonal bioclimatic space of the Mendeleev landscape. Though the measure of similarity between coniferous/broad-leaved forests and the neighboring fir-spruce forests in *BL* and *PC* reaches 0.80 and 0.94, respectively, and their mutual inclusion relations are absent, i.e., they are not arranged into a subordinate series of autotrophic biogenesis.

Following the slopes of the volcanic dome upwards, the change in climate conditions (see Figure 3.3) is accompanied by the general decrease in *BL* and *PC*, as well as in the partial characteristics of productivity: skeletal wood-shrub, total green, and herbaceous cover (see Tables 7.1 and 7.2). Nevertheless, the boreal forest communities of the foothills, lower and middle layers of the Mendeleev Volcano low mountain country are much more similar in productivity (Figure 7.6 *a*) than in phytomass reserves.

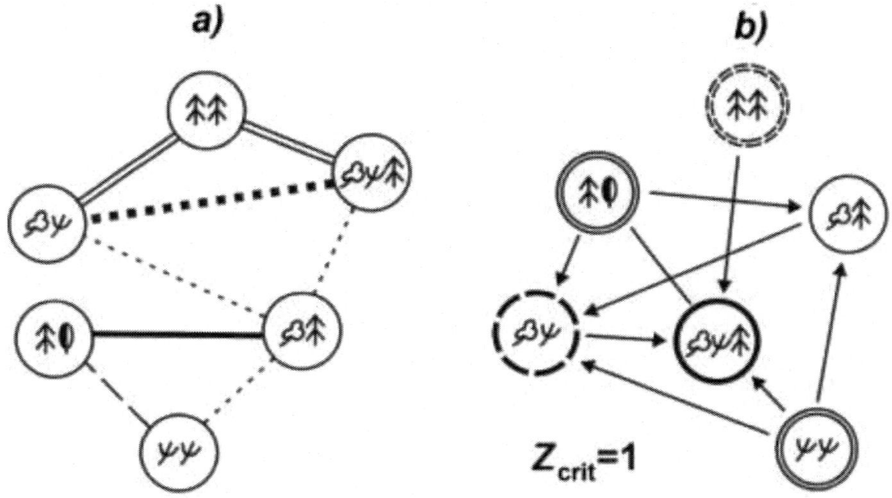

Figure 7.6. Mendeleev Volcano landscape. The graph of similarity relations (*a*) and the orgraph of inclusion relations (*b*) between the groups of forest types with respect to annual productivity. Symbols are the same as in Figure 7.5.

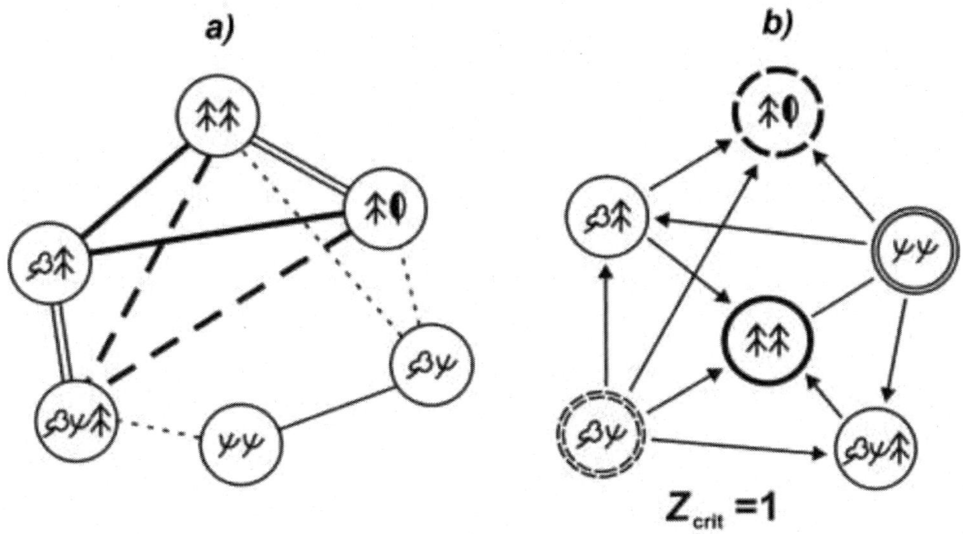

Figure 7.7. Mendeleev Volcano landscape. The graph of similarity relations (*a*) and the orgraph of inclusion relations (*b*) between the groups of forest types with respect to biological cycle. Symbols are the same as in Figure 7.5.

Being significantly inferior to subboreal forests in the reserves of living above-ground organic matter and insignificantly in the production of wood-shrub phytomasses (see Table 7.1), dark-coniferous communities produce the maximum amounts of soil cover green mass and generally take the leading position of *macrobolites* in total productivity. Judging by the values of metabolic index S^{m}_{k} and the measure of homeostaticity G_k, they maximally represent the altitudinal-zonal space of the Mendeleev volcanic landscape (Figure 7.5 *b*; Table 7.4) and their total production process is characterized by the maximum stability.

Table 7.4. Mendeleev landscape. The measures of homeostaticity (G_k) and the structural-metabolic indices ($S^m{}_k$) of different groups of forest types

Parameters		**Groups of forest types**					
					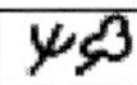	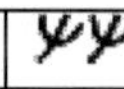	
Living phytomass	G_k	**0,324**	**0.343**	0.250	*0.180*	*0.108*	0.254
	$S^m{}_k$	**1.18**	**1.36**	1.08	*0.74*	*0.55*	1.09
Productivity	G_k	**0.327**	**0.333**	0.218	*0.167*	*0.103*	**0.249**
	$S^m{}_k$	**1.25**	**1.33**	1.00	*0.75*	*0582*	1.09
Biological cycle	G_k	*0.11*	*0.155*	0.231	**0.275**	*0.111*	**0.347**
	$S^m{}_k$	*0.70*	0.90	1.00	**1.20**	*0.70*	**1.50**

Note. The following metabolic ecotypes of forest communities were distinguished: (1) macrobolites and submacrobolites are in bold, underlined and not underlined, respectively; (2) microbolites and submicrobolites are in italic, underlined and not underlined, respectively. Mesobolites are not marked. Other symbols see in the text.

In the middle and upper strata of the Mendeleev volcanic dome, where dark-coniferous forests at the absolute heights of 500–600 m are replaced by stone birch with cedar prostrate shrubs, the living above-ground phytomass has approximately the same mean values of *BL* (90–110 t/ha) as in the piedmont zone of Central Kamchatka at the northern boundary of the boreal belt. The minimum reserves and annual production of above-ground phytomass are typical of the subgoletz zone of cedar prostrate shrubs (see Tables 7.1 and 7.2). The exceptionally high similarity in phytomass reserves between stone-birch elfin woodlands and cedar prostrate shrubs (N_{ik} = 0.98), as well as the fact that they leave the subboreal and boreal forests far behind in productivity (see Figure 6.5 *a*) demonstrate the functional unity of these phytocoenotic groups, supposedly forming the general Okhotian–Beringian bioclimatic system of the subgoletz zone.

Both of these phytocoenotic groups have a very weak similarity with the below-lying mountain fir-spruce forests (N_{ik} = 0.52–0.58); however, their higher similarity to stone birch/spruce forests (N_{ik} = 0.70–0.72) indicates the buffer nature of the latter, which is transient from the dark-coniferous zone to the subalpine zone. At the same time, there is a typical rather close relationship with respect to phytomass reserves between these binary buffer communities and the more complex (ternary) buffer group: cedar prostrate shrubs/birch/spruce (see Figure 7.5 *a*and 7.6, *a*), which is also formed at the upper boundary of the dark-coniferous forest zone.

In contrast to autotrophic biogenesis characterized by phytomass reserves and productivity, the small biological cycle has its own, substantially different regularities of the relations of similarity (isomorphism) and dominance of forest phytocoenoses. First of all, there is a much more contrasting distribution of metabolism among forest type groups, as is evidenced by the strongly rarified graph of the measures of similarity (Figure 7.7 *a*). Parameter N_{ik} varies from 0.41 to 0.99, with the base range of 0.56–0.92. It suggests the drastically marked heterogeneity of mechanisms of the biological cycle. At the same time, coniferous/broad-leaved and dark-coniferous forests are at different "poles" of this contrast: N_{ik} = 0.62, i.e., below the accepted level of minimum similarity, and their binary relations in the structural scheme of dominance (similar to productivity) are below the threshold of indistinguishability Z_{crit} (Figure 7.7 *b*). There is an obvious divergence between the mechanisms of metabolism of these two phytocoenotic groups comprising the functional "core" of the Mendeleev volcanic landscape.

The rate of the annual turnover of above-ground living phytomass substantially depends on the age τ of forest plantation. The following linear dependence of *CR* on τ has been obtained from the results of surveys in the Lower Priamurye:

$$CR = 0.0749 - 0.00037 \cdot \tau;\ R = 0.765. \quad (7.2)$$

The rate of annual phytomass turnover generally decreases as the major forest-forming species become older. Consequently, the *CR* value implicitly characterizes the degree of maturation of forest phytocoenosis, its closeness to the climax state. On this basis, the coniferous/broad-leaved forests of piedmont plaints of the Mendeleev landscape can be considered to be the oldest and, consequently, maximally approaching the climax state. At the same time, these subboreal forests with rather high productivity are characterized by the minimum values of parameters G_k and $S^m{}_k$, i.e., they have actually reached the limit of their functional development, already with the minimum representation of altitudinal-zonal diversity of the biological cycle, and thus belong to the most reduced ecotypes of forest communities of this ecoregion – microbolites. Accordingly, their biological cycle is minimally stable. The most productive piedmont and mountain fir-spruce forests comprising the microbolite category are related to them.

The subalpine communities of cedar prostrate shrubs, though being least productive, demonstrate a certain similarity to plain subboreal forests with respect to the biological cycle. This fact allows us to consider them to be rather mature, pseudoclimax phytocoenoses.

In contrast to the above-mentioned data, attention is drawn to the highly isomorphic triad of mountain fir-spruce forests with stone-birch elfin woodlands and the most complex buffer cedar prostrate shrub-birch-spruce communities (N_{ik} = 0.92). According to our observations, the ternary buffer SpsBS zone appeared near the upper boundary of the dark-coniferous forest zone as a result of spruce and fir transgression into stone-birch forests above the slope. It is confirmed, first, by the good condition of regeneration and undergrowth of coniferous species and, second, the presence of their satellites (yew, mountain ash, maple, lianas, ferns, Kuril bamboo) in the understory and soil cover. However, adult individuals have acquired the typically subalpine morphophysiological features: general stunting and often the prostrate form, tapering and crooked trunks and branches, dry-topped and wind (flat-topped) crowns. There is also an ample quantity of thin-stemmed brushwood of coniferous species. The above is the evidence of critical climatic conditions for the development of dark-coniferous species in these buffer phytocoenoses. There is an example of the phenomenon of obvious age-related decrease in the quality of locality of coniferous forest typical of the South Kuril Islands (Lashkov 1948). At the same time, these conditions are quite comfortable for stone birch: it almost does not creep but gives straight-trunk inclined life forms.

With regard to the biological cycle, SpsBS forests have a very high isomorphism with crooked birch forests (N_{ik} = 0.92). Obviously, after having acquired particular structural features of the dark-coniferous zone, these communities still preserved the functional properties of stone-birch elfin woodland. Being mesobolites in respect of productivity, they become dominant macrobolites in respect of the rate of biological turnover (see Figure 7.7 *b*). Within this range of phytocoenotic groups, these most complex buffer formations also seem to be the youngest, progressively developing ones, with the maximum functional stability as judged by the S^{m}_{k} and G_{k} values. Following the statement on the functional criteria of evolution (Timofeeff-Ressovsky et al. 1969; Ponomarenko 1998), one can suggest that the *buffer cedar prostrate shrub/birch/spruce forests* with the most intensive (and, in this sense, the most perfect) and stable biological cycle *are at the front of evolutionary progress* in the forest cover of the Mendeleev volcanic landscape.

Submacrobolites include the basal communities of stone-birch elfin woodland with cedar prostrate shrubs which, consequently, can also be considered as comparatively young and progressive associations with a rather high reserve of homeostaticity, i.e., stable metabolism.

Thus, the autotrophic biogenesis and the small biological cycle (in its highly simplified form under consideration) are the two qualitatively different aspects of functioning of island forest ecosystems with their metabolic parameters supplementing each other, which gives a more complete representation of the complex system of functional organization of forest cover in a young volcanic landscape.

7.3. Background Standards of Forest Ecosystem Functioning

The described regularities of autotrophic biogenesis and biological cycle in forest ecosystems of the Mendeleev landscape were implemented in the framework of particular background (altitudinal-zonal and sectoral) conditions of their functioning. Such framework conditions of metabolic processes can be revealed by using *flat-interfluve phytocoenoses* in each altitudinal layer and/or sector of low mountains. As is known, the plant cover of the plakor type of site (flat-interfluve geotope), being more or less "cleaned" from the refracting effect of local geomorphological and edaphic factors, maximally represents the zonal-regional bioclimatic norm of a given territory, thereby being an "intra-landscape core" of its *physical-geographical background* (Vysotsky 1960; Krauklis 1979). In the soil-geochemical respect, flat interfluves are eluvial, or autonomous, elementary landscapes (Polynov 1956; Glazovskaya 1964; Perelyman 1975). Although the concept of flat interfluves introduced by G.N. Vysotsky (1909) concerns plain territories, it can also be applied to fragmented mountain relief, because in each geomorphological layer or altitudinal zone there are elementary geo(eco)systems with soil-phytocoenotic characteristics corresponding to the given layer or altitudinal-zonal and exposure background. Due to the absence of the appropriate term, such natural complexes in Mendeleev landscape will be conditionally referred to as flat-interfluve. By analyzing natural complexes of the *flat-interfluve series,* we can gain a deeper insight into the *physical-geographical background:* the zonal, altitudinal-zonal, or sectoral geoecological space formed by the given regional bioclimatic system.

Six flat interfluves were distinguished in our statistical sample of test sites (t/s) (see Figure 4.3): two flat interfluves in each of piedmont layer A (t/s 8 and t/s 23 in the Sea-of-Okhotsk and Pacific-Ocean sectors, respectively) and the lower layer TA (t/s 32 and t/s 26 in the same sectors, respectively) and one flat interfluve in each of the middle T and upper TE layers (t/s 6 and t/s 9, respectively, confined to watersheds and therefore characterizing two layers at a time). Thus, the constructed flat-interfluve series of forest phytocoenoses made it possible to consider the background altitude-exposure structure of their functioning, though not completely. The sector differences in the functional parameters of forest ecosystems could be revealed only for the flat interfluves of plain-piedmont A and lower TA layers. The measures of nontransitive relations between the distinguished altitude-exposure units of physical-geographical background were calculated by three functional sets: the total above-ground productivity, the biological cycle, and the soil morphology as an indicator of the detritus branch of metabolism.

First of all, it has been shown that the above-described phytocoenological functional space of the Mendeleev landscape is characterized by organizational conformity with the background altitude-exposure ecological space, which is also represented by weak lateral connections and an extremely high value of the critical threshold of indistinguishability on the orgraphs of the measures of inclusion (see above). Both functional models are generally adequate with respect to both productivity and biological cycle, though there are certain clarifying differences. The averaged measures of similarity in autotrophic biogenesis in both cases are almost identical and equal to 0.68–0.74. The high production isomorphism of coniferous/broad-leaved and dark-coniferous forests (N_{ij} = 0.94) is determined by considerable closeness of analogous connections (N_{ij} = 0.86–0.90) between the lower TA layers of both sectors and the middle T layer (see Figures 7.6 *a* and 7.8 *a*). The conclusion about the affiliation of buffer stone birch-spruce and cedar prostrate shrub-birch-spruce communities with the upper near-boundary T belt of the dark-coniferous zone of low mountains and interpretation of the genesis of these buffer formations as a result of modern transgression of fir/spruce communities to the subalpine altitudinal zone has also been confirmed (see above, Section 7.2).

At the same time, the background bioclimatic model underlines particular differences between the Sea-of-Okhotsk and Pacific-Ocean sectors of plains and foothills both in production processes and in the biological cycle. First of all, one should note the more drastic decrease in autotrophic biogenesis with height and the weakening of phytobiota turnover in the colder and more humid Pacific-Ocean sector compared to the warmer and drier Sea-of-Okhotsk sector. Already in the lower TA altitudinal layer, the S^{m}_{k} and G_{k} values decrease to the level of submicrobolites in the former case but remain submacrobolites close to the macrobolite ecotypes of piedmont plains in the latter case (Table 7.5).

Subsequently, a certain degree of production isomorphism between zonal representatives of piedmont plains of the Pacific Ocean sector and the lower-layer slopes of the volcanic dome was determined (N_{ij} = 0.81). Accordingly, dark-coniferous forests are predominant in this sector, beginning from the coastal territories. With respect to the biological cycle, the Pacific piedmont plains are classified as submacrobolites (see Table 7.5), which is evidence of the relatively young age of flat-interfluve forests here.

On the piedmont plains of the Sea-of-Okhotsk sector, the zonal bioclimatic background has a highly marked metabolic abnormality – the very slow biological cycle, which determines the above-mentioned low values of S^{m}_{k} and G_{k} for subboreal mixed forests being prevalent here. It substantially differentiates the Sea-of-Okhotsk A microbolite (with respect to turnover) layer from the A macrobolite layer of the Pacific-Ocean sector (see Table 7.5) and, to the same extent, from the adjacent TA layer in its own sector (also macrobolite; N_{ij} = 0.64), already with prevalence of mountain fir/spruce forests with very stable functions.

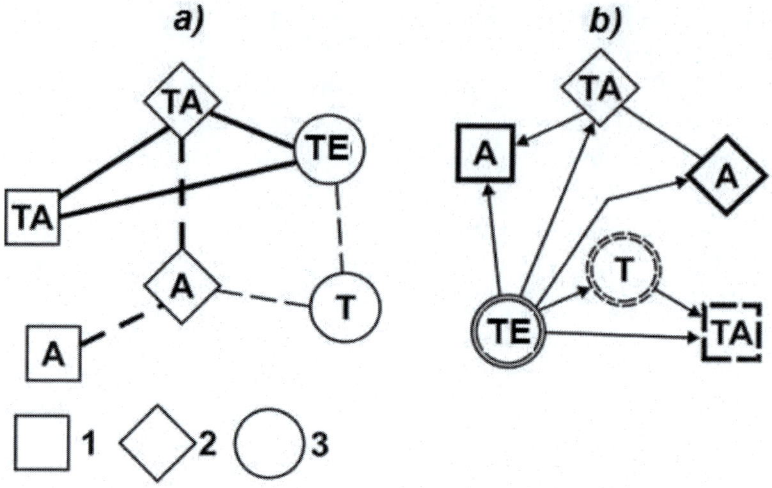

Figure 7.8. The schemes of nontransitive relations between flat-interfluve ecosystems in different altitudinal layers and sectors of the Mendeleev landscape in forest productivity: *a* – the graph of the relations of similarity; *b* – the orgraph of the relations of dominance. Altitudinal layers: A, plain-piedmont (70–160 m); TA, lower low-mountain (160–450 m); T–TE, middle and upper low-mountain (450–800 m). Sectors (for the first and second layers): 1 – Sea-of-Okhotsk; 2 – Pacific Ocean. 3 – the middle and upper layers without distinguishing the sectors. The symbols for the measures of similarity (*a*) and ecotypes (*b*) see in Fig. 7.5.

Table 7.5. The measures of homeostaticity (G_k) and structural-functional indices ($S^m{}_k$) of flat-interfluve forest communities in different altitudinal layers and sectors of the Mendeleev landscape

Parameters		Altitudinal layers an sectors					
		Piedmond, A		Lower, TA		Middle, T	Upper, TЭ
		Sea-of- Okhotsk	Pacific-ocean	Sea-of-Okhotsk	Pacific-ocean		
Productivity	G_k	**0.352**	**0.352**	**0.339**	*0.200*	*0.188*	*0.116*
	$S^m{}_k$	**1.30**	**1.30**	**1.20**	*0.90*	*0.80*	*0.50*
Biological cycle	G_k	*0.144*	**0.457**	**0.434**	*0.161*	*0.161*	*0.161*
	$S^m{}_k$	*0.60*	**1.40**	**1.60**	*0.80*	*0.80*	*0.80*
Soil	G_k	**0.750**	**0.750**	*0.250*	**0.750**	*0.250*	*0.250*
Блок	$S^m{}_k$	**1.25**	**1.50**	*0.75*	**0.25**	*0.75*	*0.50*

Note. Conventional meanings see in Table 7.4.

The most exotic buffer communities (cedar prostrate shrub-birch-spruce) function within the scope of the altitudinal-zonal bioclimatic system of the middle T layer providing a low (submicrobolite) background standard for both productivity and biological cycle (see Figures 7.8 *b* and 7.9 *b*). Here, flat-interfluve ecosystems are in the states of functional submicrobolites and microbolites. In the meantime, the SpsBS buffer communities, as it has already been mentioned, are characterized by higher productivity and maximum intensity of the biological cycle (see Figures 7.6 *b* and 7.7 *b*). With respect to the levels of these parameters, they are very similar to fir-spruce forests of the lower TA layer of the Sea-of-Okhotsk sector (N_{ij} = 0.83–0.92) with the most stable biological cycle (G_k = 0.434) belonging to the highest metabolic ecotype ($S^m{}_k$ = 1.60). At the same time, the bioclimatic background of forest transit layer drastically differs from the background of the subgoletz transeluvial layer, especially in the biological cycle (N_{ij} = 0.56).

Thus, the comparison of background and local parameters of metabolism of the buffer SpsBS communities once again confirms, as it has been mentioned above, that they belong to the dark-coniferous forest but not subalpine zone. However, these communities may have a high productivity and will be rather stable only as *extrazonal formations* – in the local, most favorable sites (on the sunny slopes with highly trophic, adequately moistened substrate), which are to some extent simulate the background zonal conditions of fir-spruce forests below the slope. Due to this reason, the buffer SpsBS associations occurred sporadically, and their tree layer on flat interfluves per se was highly suppressed (see above).

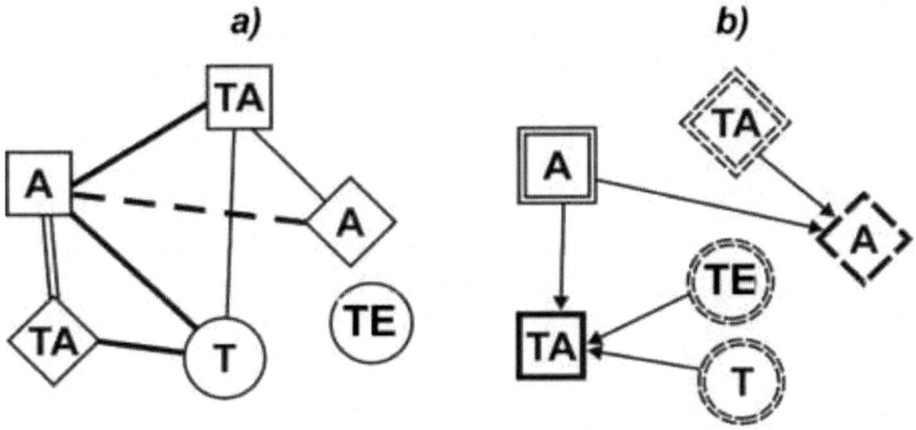

Figure 7.9. The scheme of nontransitive relations between flat-interfluve forest ecosystems in different altitudinal layers and sectors of the Mendeleev landscape with respect to the biological cycle: *a* – the graph of relations of similarity; *b* – the orgrpah of relations of dominance. Symbols are the same as in Figure 7.8.

The birch-spruce forests (binary buffer formations) near the upper boundary of the dark-coniferous zone function at the level of mesobolites (see Table 7.1; Figures 6.6 and 7.7 *b*). With respect to their slow biological cycle, these buffer communities approach the coniferous/broad-leaved forests of piedmont plains (N_{ij} = 0.89), which is in agreement with a rather high degree of isomorphism of background bioclimatic systems in transitive and accumulative altitudinal zones of the Mendeleev landscape (see Figure 6-9 *a*). The productivity of stone birch-spruce forests, being similar to the productivity of SpsBS communities (N_{ij} = 0.93), also corresponds to the submicrobolite ecotype of productivity typical of the background bioclimatic system of the middle transit layer of the volcanic dome (see Table 7.2, Figure 7.9 *b*). Thus, in contrast to the ternary SpsBS buffer formations characterized, as has been mentioned above, by extrazonal functioning, the transitional communities between dark-coniferous and stone birch forests are rather zonal. They are quite in agreement with the background bioclimatic system of the middle transit layer and function steadily on flat-interfluve geotopes.

7.4. Functional Phytocoenotic Contrasts between the Sea-of-Okhotsk and Pacific-Ocean Sectors

Let us consider some substrate functional characteristics of the height/exposure geospace of the Mendeleev landscape. The dominant areas of productivity of coniferous/broad-leaved forests in plain microlandscapes (the piedmont layer) are approximately the same in both the Sea-of-Okhotsk and Pacific-Ocean sectors (see Figures 7.1 and 7.2), though both sectors are not quite isomorphic by the mean statistical value of productivity (N_{ij} = 0.82, see Figure 7.7 *a*). There is also a significant spread of PC_{vol} towards a decrease. The zonal geoecological space of production processes in both sectors takes the position of leader macrobolites in the structural scheme of dominance (see Figure 7.7, *b*). Thus, the above-mentioned *high level of organic matter production typical of the plains and foothills of the Mendeleev volcanic landscape is determined by the regional bioclimatic system of the South Kuril Islands.*

On the other hand, the total reserves of phytomasses on plains in the warmer and drier Sea-of-Okhotsk sector, with respect to their dominant, are by 150–240 tons/ha greater than in the colder and more humid Pacific-Ocean sector. In the Sea-of-Okhotsk sector, the range of the lowest values BL = 22–180 tons/ha falls on the middle (T) and upper (TE) layers of low mountains; in the Pacific-Ocean sector, it occurs already in the lower TA layer.

It can be supposed that such "brackets" in the sectoral differentiation of PC_{vol} and BL on plains and in foothills are determined by two factors: (1) the increased floristic and phytocoenotic diversity of forest communities in the less oceanic Okhotian sector and (2) the increased lifetime of each generation of the major forest-forming species under the same conditions. The indirect evidence of significance of the former factor is the measures of floristic and phytocoenotic diversity of herbage: 3.77 and 1.78 bit (binary units) in the former case and only 3.14 and 1.59 bit, respectively, in the latter case.

With regard to the lower TA and middle T–TE altitudinal zones of the volcanic dome, the Sea-of-Okhotsk sector differs from the Pacific-Ocean sector not only in the high reserves of living phytomass in dark-coniferous forests but also in their enhanced total productivity (by 4–9 tons/ha·yr on average). However, even in the transaccumulative zone, the sectoral contrasts in productivity are substantially smoothed: N_{ij} = 0.90 (see Figure 7.8 *a*). Still higher production isomorphism seems to be typical of the sectors of middle and upper zones of the volcanic dome, though the difference in phytomass reserves in the zone of stone-birch elfin woodland and subgoletz cedar prostrate shrubs persists (by 70–90 tons/ha with respect to the dominants).

It has already been shown (see Chapter3) that more favorable climatic conditions of forest growth are typical of the Sea-of-Okhotsk sector over nearly the entire altitudinal-zonal range of the Mendeleev landscape; therefore, the upper forest boundary also rises here to higher absolute values compared to the PacificOcean sector (see Table 7.1; Figure 7.5). As we can see now, these climatic contrasts manifest themselves not only in the structure of forest communities but also in their function.

Thus, the negative forest-growth effect of cold sea currents under the conditions of extremely oceanic climate is quite clearly manifested already at the early, island-arc stage of subaerial landscape genesis, within a comparatively small territory of an island with low-mountain relief.

With regard to allometric coefficient *CA* (see below, Chapter 8), the low- and middle-layer mountain fir-spruce forests of the Mendeleev landscape correspond quite well to the zonal standard for south taiga ecosystems (see Table 7.1), which demonstrates the closeness of these minimally disturbed forests to the climax state. The dark-coniferous forests of piedmont plains of the Sea-of-Okhotsk sector, which are partially exposed to clearance, are also rather close to the latter ($CA \approx 0.450$). It cannot be referred to the coniferous/broad-leaved forest cover of foothills of the PacificOcean sector with drastically reduced values of *CA* (to 0.220–290), in spite of the relatively high total and partial productivity. It seems to be due to the maximum disturbance of these forests by cutting and fires, i.e., their youngest age and the respective remoteness from the climax state (they seem to be most productive exactly because they are the youngest).

– «» –

Thus, the results of the functional analysis of forest cover at the Mendeleev Volcano test site disclose the two phenomena of its structural-functional organization. First, there is an obvious *exceptionally high system-forming role of forest phytobiota in the establishment and development of young volcanic landscapes.* The Neo-Pacific forest communities in the southern part of the boreal belt are characterized by the annual turnover of plant matter (see Table 7.1) classified in the category of intensive metabolism and, with regard to zonal measures, is typical of subtropical forests only. The buffer communities are especially representative in this respect: cedar prostrate shrubs-birch-spruce in the middle zone of the Mendeleev Volcano and stone birch-spruce at the upper boundary of dark-coniferous forests (KY = 0.47–0.77 years). Here, one cannot but see *the origin of progressive evolutionary tendencies in the buffer forest formations that are so typical of the Pacific Ocean mega ecotone of North Eurasia* (Kolomyts 2015 b). As is known (Timofeeff-Ressovsky et al. 1969; Shwarts 1980; Ponomarenko 1998), the increase in intensity of the function of biogeocoenoses and the improvement of biological cycles underlie the evolutionary progress and are the main essence of evolution. Such evolutionary tendencies in the plant cover originate as early as at the initial, volcanogenic stage of formation of continental biosphere.

Second, the intensive annual turnover of above-ground phytobiota and the high rate of the entire biological turnover in southern Kunashir landscapes are the major factors of their stability under the conditions of cold oceanic climate unfavorable for forest formation and, together with geothermal wells (see below), contributes to the formation of *subboreal "climatically unjustified forests"* (see below Chapter 8). This is the second *phenomenon of organization of boreal forest ecosystems at the initial, island-arc stage of subaerial landscape genesis.* With transition into the state of marginal-continental and then intracontinental formations, forest ecosystems generally reduce the intensity of their functioning: the annual turnover of above-ground phytomass of indigenous forests drops 2–3 times, and their biological cycle decreases to the moderate and even slow category ($CY \approx$ 6.5–8.0 years).

7.5. Background Standards of Soil System

Finally, let us touch upon the background bioclimatic standards of the soil unit of forest ecosystems of the Mendeleev Volcano. It should be recalled that the thickness and ratios of soil horizons A_0, A_1 and B_i (B_m) allow us to assess at a first approximation the detritus branch of metabolism, i.e., the processes of mineralization, humification and intrasoil transformation of mineral mass (Duchaufour 1970; Glazovskaya 1981; Polivanov 1984; etc). These processes are elucidated in Section 5.3, Chapter 5.

The soil-forming rocks of the Mendeleev Volcano as an isolated Strombolian-type volcano are mostly loose pyroclastic volcanites, and their age gradually increases from the top to the foot of the volcano. Consequently, the distribution of soil morphological characteristics by altitudinal zones must characterize the age trajectory of soil-forming processes that have resulted in the formation of modern soil profiles with different degrees of soil maturity, its closeness to the climax state.

The background soil-morphological space of young volcanic landscape differs from the same phytocoenological space in the lower value of Z_{crit} on the orgraphs of relations of inclusion (Figure 7.10 *b*). Obviously, the soil systems of different altitudinal zones and sectors that we consider are spatially interrelated (and mutually determined) more closely than phytocoenological systems. It means that the lateral exogenous processes, i.e., the surface, ground and lithodynamic runoff, influence soil formation more effectively than forest formation per se. Due to this fact, *the soil cover of the extrusive dome and its surroundings acquires the higher territorial uniformity compared to the forest cover.*

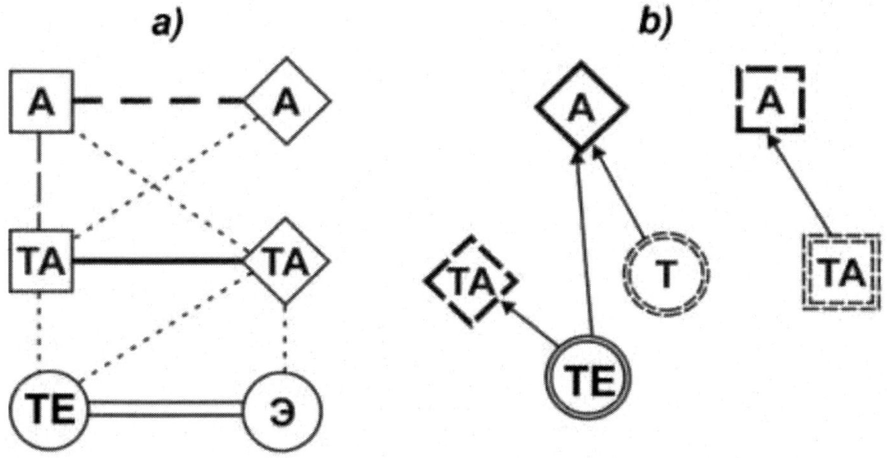

Figure 7.10. The schemes of nontransitive relations between flat-interfluve forest ecosystems in different altitudinal zones and sectors of the Mendeleev landscape with respect to the degree of soil "maturity": *a* – the graph of the relations of similarity; *b* – the orgraph of the relations of dominance. Symbols are the same as in Figure 7.8.

The models of morphological similarity and relations of inclusion for the soils of flat-interfluve geotopes of the Mendeleev landscape show (see Figure 7-10) that the forest ecosystems of both sectors are rather clearly differentiated into three altitudinal belts with respect to the degree of soil development (maturity): the lower (corresponding to the A piedmont-plain zone), middle (TA zone of the lower part of the volcanic dome), and upper (including the middle T and upper TE zones). The zones differ primarily in the age of soils, as well as in the intensity of soil-biotic processes. Bottom-up the volcano slopes, soils become younger and younger and the intensity of soil-forming processes decreases. Consequently, these factors interfere with the same sign, which intensifies their total soil-forming effect.

Indeed, the isomorphism of the Sea-of-Okhotsk and Pacific-Ocean sectors with respect to the degree of maturity of soil background state reaches the maximum values in the upper T–TE zone and decreases on transition to the transaccumulative and then accumulative zones (see Figure 7.10 *a*). On the piedmont plains of the Mendeleev landscape, sectoral contrasts in the development of flat-interfluve (zonal) soils manifest themselves to the maximum extent. The soils of A and TA layers from different sectors are found here even in different fragments of the divided orgraph of the relations of inclusion (see Figure 7.10 *b*). The trajectories of their development diverged very substantially, being in agreement with the fact that the sectoral bioclimatic contrasts are maximally manifested exactly on the piedmont plains of the Mendeleev landscape (see above).

It should be noted that the zonal (flat-interfluve) soils of relatively narrow piedmont plains of the Pacific-Ocean sector have proved to be maximally developed (macrobolite ecotypes), which does not agree with the younger age of its erosion relief compared to the vast terraced plains of the Sea-of-Okhotsk sector, where the relief is erosion-accumulative and generally has a "more ancient appearance" (Korsunskaya 1958). In this case, the age of zonal soils does not correspond to the age of exogenous relief, which seems to contradict the known theory of paedogeomorphological systems (Gerrard 1981).

However, with the dominant landscape-forming role of volcanic-tectonic processes (Grabkov and Khudyakov 1993), the age of soils in young island-arc landscapes is determined primarily by the age of volcanites as soil-forming rocks, which are generally more ancient in the Pacific-Ocean sector of piedmont plains. Here, soils were formed on a tertiary foundation: terrigeneous–pyroclastic sediments (Markhinin 1956; Gorshkov 1967); therefore, they have attained the maximum metabolic index and the highest functional stability (see Table 7.5). At the same time, at the foothills of the Sea-of-Okhotsk sector, the ancient tertiary foundation occupies a comparatively narrow littoral zone. Here, the main soil-forming rocks are much younger pyroclastic sediments that emerged in the preglacial period during the giant eruptions of the volcano and created its first somma (see Figure 3-3). Accordingly, the piedmont zonal soils of the Sea-of-Okhotsk sector are less developed, being in the state of the submacrobolite ecotype.

Quite clear sectoral differences are observed also in the altitudinal-zonal age-specific gradients of soil-forming rocks, inevitably affecting the age-related state of soils. In the PacificOcean sector, the soils of piedmont plains are drastically different from the soils of the T and TE layers in age and, accordingly, in the degree of maturity. Mountain soils were formed here on the pyroclastics of the second somma (see Figure 3.3) that appeared in the epoch of the latest (Wurm) interglacial period (about 39000 years ago). That is why they are yet in the state of submicrobolites and microbolites, at a much earlier stage of development compared to the soils of foothills on tertiary rocks. In the Sea-of-Okhotsk sector, the inter-layer age-related differences between soil-forming rocks are not so pronounced (below, the early Pleistocene; above, the latest interglacial period). Under these conditions, the measures of mutual inclusions of morphological characteristics of zonal soils between the A and T layers remain below the threshold of indistinguishability and the background altitudinal-zonal structure of soil cover is less contrasted.

7.6. Insular Neo-Pacific Heat Balance Structure

The performed analysis of interrelationship between the productivity of forest communities of the Kuril Island Arc and the thermal balance gave us grounds for *seriously doubting the existing concepts of energy basis for the functional organization of island biogeosystems*.

As has been mentioned, one of the characteristic geographical regularities of insular extratropical Neo-Pacific is a general southward shift of the boundaries of climatic belts and natural zones compared to the neighboring continents, with the simultaneous meridian extension of belts and zones per se, which is accounted for primarily by heat and energy deficiency (Budyko 1977; Markov 1980;Ivanov 2005, 2014). The values in Table 7.6 generally confirm this statement; however, it is caused by much more complex landscape-geophysical processes. The matter is that the temperature of the vegetation period on islands decreases to a much greater extent than the incoming solar radiation and radiation balance.

Table 7.6. Comparison of the radiation-thermal parameters of intracontinental, marginal-continental and insular territories of the Asian continent in the southern part of the temperate zone

Region (meteorological station)	Energy parameters, MJ/m²· month, and ⁰C		
	Q	B	$t_{VII-VIII}$
South Transbaikalia (Borzya)	515	297	20.0
Central Transbaikalia (Chita)	540	268	18.8
Middle Priamurye (Chabarovsk)	633	318	21.1
Sakhalin (Yuzhno-Sakhalinsk)	451	310	16.9
Kunashir (Yuzhno-Kurilsk)	450	169	15.8

Note: Q and B are the total radiation and the radiation balance, respectively. The parameters were taken for horizontal surface under the conditions of average cloudiness and calculated for May–September. Parameter $t_{VII-VIII}$ is the mean temperature of the warmest month (⁰C).

Thus, the total radiation balance of Kunashir for May–September decreases by 4–11% and the warmest month temperature decreases by 16–21% compared to the southtaiga and foreststeppe regions of Transbaikalia. In the southern Sakhalin, the parameter B value is nearly the same as in the Middle Priamurye and much higher than in Central and South Transbaikalia; nevertheless, the summer temperature on the island decreases by 4.2°and 3.1°, respectively, compared to these regions. In general, with the same radiation balance of the vegetation period (B_{wp}), the difference in the mean temperature of the warmest month (t_{wm}) between the marginal-continental (as well as intracontinental) sector (1) and the insular sector (2) of the Pacific Ocean mega ecotone is about 3.0–3.5° (Figure 7.11). This difference is rather distinct and expressed by the following linear regression equations:

$$t_{wm} = 8.14 + 0.039 \cdot B_{wp}; \qquad R^2 = 0.78; \qquad (7.3)$$

$$t_{wm} = 0.21 + 0.056 \cdot B_{wp}; \qquad R^2 = 0.79. \qquad (7.4)$$

Such a contrast between the radiation and thermal parameters is determined by sectoral differences in the structure of the expenditure part of heat balance – mainly in the expenditure of residual radiation on evapotranspiration and turbulent heat exchange (the latter exactly determines the temperature of the near-surface air layer. As is known, with sufficient and excessive atmospheric humidification (F_{hum}> 1), the total evaporation from the surface of closed canopy is represented mostly by transpiration (Budyko 1971), which is a necessary condition for the primary bioproductivity and biomass accumulation (Sochava 1978). In the mixed-forest and taiga zones of the Russian Plain, the share of transpiration in the total evaporation of deciduous and coniferous forests is 65–70% (Rauner 1972). On oceanic islands of the boreal belt, this value should be apparently increased due to high air humidity.

As has been mentioned, the insular and coastal ecosystems of the Pacific Fire Ring differ from their zonal analogs in the intracontinental regions in enhanced values of primary productivity. In addition, the extremely high values of annual production and accumulation of living phytomass have been observed in the forest communities of the Mendeleev volcanic landscape. Naturally, the higher productivity must be provided by more considerable heat expenditure on transpiration.

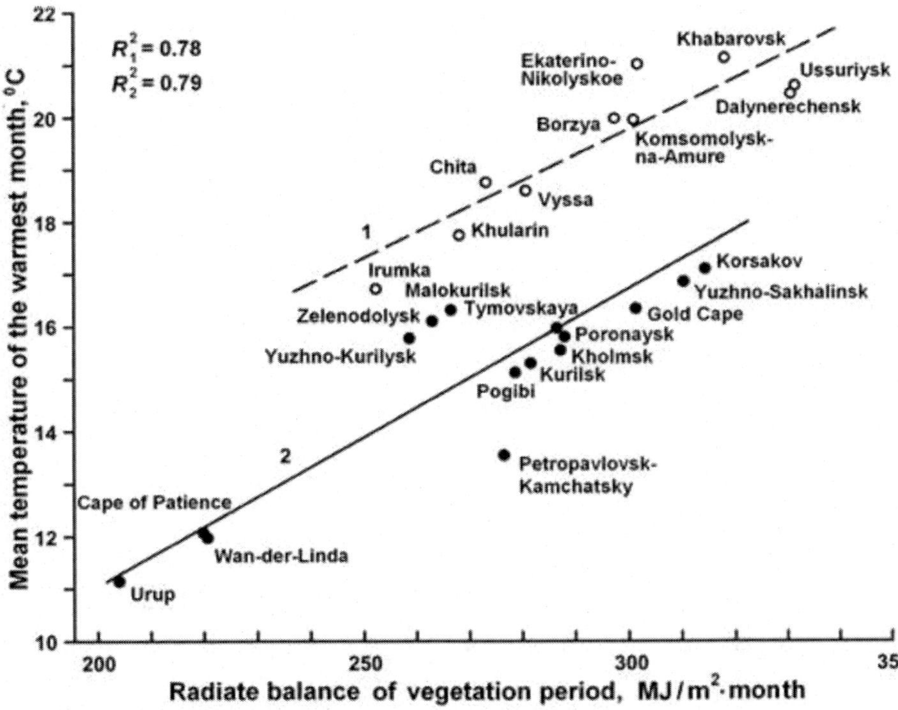

Figure 7.11. The correlations between the mean temperature of the warmest month and the radiation balance of the vegetation period. Longitudinal sectors of the Pacific Ocean mega ecotone: 1 – marginal-continental (Sub-Pacific – Lower Priamurye, Primorye) and intracontinental (Paleo-Pacific – Central and South Transbaikalia, Middle Priamurye); 2 – insular and peninsular (Neo-Pacific – Kuril Islands, South Sakhalin, South Kamchatka). The names of meteorological stations are indicated.

This expenditure was calculatedfor the two ecoregions (the Lower Priamurye and the south of Kunashir) by the comparable forest plain-piedmont communities: coniferous/broad-leaved and dark-coniferous forests (Table 7.7). With a generally insignificant difference in the radiation balance, the contrasts in heat expenditure on transpiration between analogous communities and between ecoregions are merely striking. In the Lower Priamurye, these expenditures on the total production and generation of total green mass are 26–28% and 17–19% of radiation balance of the vegetation period for mixed (cedar- and spruce-broadleaved) forests, respectively, and decrease to 14% and 7% for fir/spruce forests. In mixed and coniferous forests of the boreal ecotone of the Volga River basin, the value of productive transpiration is 260–300 mm (Kolomyts 2012), with an energy equivalent of 650–750 MJ/m^2, which is rather close to the value of the Lower Priamurye.

Table 7.7. The radiation budget in forest communities of marginal-continental and island-arc regions for transpiration during the formation of herbage mass, total green mass, and total annual production

Energy parameter, MJ/m^2		**Ecoregion (test ground) and group of forest types**				
		Lower Priamurye (Komsomolsky Nature Reserve)			**Kunashir Island (Mendeleev Volcano)**	
		Cedar-broadleaf	Spruce-broadleaf	Fir-spruce	Dark coniferous-broadleaf	Fir-spruce
Total monthly radiation balance for vegetation period		2357	2357	2422	2031	2077
Radiation budget for production of	herbage mass	34	42	53	672	864
	total green mass	447	404	269	1184	1095
	total phytomass	670	610	348	1440	1520

Note. The following values were used for the calculations: the data of actinometric observations at the Khabarovsk and Yuzhno-Kurilsk weather stations, the conversion factors of radiation balance during the transition from field (weather station site) to forest (Rauner 1972), and the transpiration coefficients for different tree species (Rudnev 1984).

In the Mendeleev island landscape, the transpiration share of radiation balance for both groups of boreal forests (mixed and dark-coniferous) is up to 71–73% for total production and 52–58% for green mass. The contrasts in heat expenditure on the production of ground cover green mass are still greater (33–42% vs. 1–2%), reflecting the known phenomenon of gigantism in the herbage of insular and coastal-marine Neo-Pacific ecosystems. In Kunashir, these are bamboo thicket and tall grasses, the development of which takes no less than one third of the heat balance of the vegetative season (Figure 7.12).

With such abnormally high expenditure of residual radiation energy for transpiration, the share of turbulent heat exchange between the earth surface and the atmosphere substantially decreases, causing a decrease in the temperature of the vegetative season, with the respective change in the zonal hydrothermal parameters in the Neo-Pacific sector of the Pacific Ocean mega ecotone and with the above-mentioned southward shift of zonal boundaries. However, the "common" concept of reduced latitudinal-zonal rates of heat and energy resources in the Neo-Pacific Northwest, which is based on temperature indices, seems to require certain correction with due regard to the structure of expenditure parts of the heat balance. In the intracontinental and even marginal-continental territories, this structure is essentially different from insular oceanic territories. Otherwise, it is impossible to explain the seemingly paradoxical phenomenon: the abnormally high productivity of natural ecosystems of insular

Neo-Pacific at a relatively low temperature of the vegetative season, which is set up for the index of "heat and energy resources." Such an index should be undoubtedly the radiation balance per se.

Figure 7.12. Piedmont landscape of Tyatya volcano. Dark-coniferous forests with open lots of tall herbaceous vegetation and with "small island" of thermophilic nemoral-boreal vegetation caused by the local warming influence of modern geotherms from this volcano.

Thus, we can distinguish the *two main megatypes of heat balance structure in terrestrial ecosystems: continental and oceanic* (with probable intermediate types). These types rather dramatically differ in the ratio of the two basic components of expenditure part of the balance, namely, evapotranspiration and turbulent heat exchange. The predominance of transpiration component in the oceanic type of heat balance is due to the immanent strategy of the development of plant communities of insular Neo-Pacific by the principle of maximal productivity (see above), which provides the survival of young terrestrial (especially forest) ecosystems under unfavorable conditions of extra-oceanic climate at the initial stage of subaerial landscape genesis. Having an adverse effect on expenditure parts of the heat balance, forest ecosystems, even at the given stage of continental biosphere development, are a sufficiently powerful regulator of atmospheric hydrothermal regime via their biochemical cycle and, first of all, autotrophic biogenesis. Later on, this regulating and stabilizing mechanism of forest cover becomes global, as has been quite comprehensively described in the works (Sukachev 1972; Moiseev et al. 1985; Tarko 2005).

Different combinations of the conditions of radiation and thermal regimes entail significant ecological consequences (Budyko 1977). Essential reduction of residual radiation energy expended for the turbulent heat exchange in Northwestern Pacific islands leads to a general decrease in the underlying surface and ground air temperatures, which is a background for the appearance of critical thermoedaphotopes for particular plant communities. The abiotic environment becomes more contrasted territorially. Accordingly, certain structural and functional diversity of the regional and local geo(eco)systems is formed under the same background (zonal) conditions. By the example of mountain sub-Arctic landscapes of South Chukotka it has been shown that, due to the contrasts between ecological regimes, there may be "... a wide coexistence of heterogeneous elements within the confines of the same flora" (Yurtsev, 1978 33) and, we may add, the same geographical zone.

Such biogeocoenotic diversity is determined by the fact that the local geomorphological and edaphic factors differently deflect the structure of expenditure part of the heat balance: reduce and enhance the heat expenditure on turbulent heat exchange and evaporation (and partially soil heating), respectively. Consequently, the summer air temperatures decrease. In the localized types of sites, they decrease down to the level critical for a given zonal plant formation. It is just under such critical conditions that the new intrazonal types of ecological systems appear (Budyko 1977). It seems to be the energy basis of the *initial role of insular Neo-Pacific as a focus of local and regional diversity of terrestrial geo(eco)systems,* which is further developed on the continent. The local contrasts of hydrothermal conditions determine also the spatial-temporal variability of island geo(eco)systems, their high susceptibility to climatic changes and capacity for evolutionary transformations.

7.7. Some Inferences

The following two phenomena of structural-functional organization of the Mendeleev volcanic landscape have been revealed. First, there is an *exceptionally high system-forming role of forest phytobiota in the formation and development of young volcanic landscapes.* The forest communities of insular Neo-Pacific in the south of the boreal belt are characterized by the annual turnover of plant matter falling into the category of intensive metabolism and being typical only of subtropical forests by the zonal standards.

Then, the intensive annual turnover of above-ground phytobiota and the high rate of the entire biological cycle in South Kunashir landscapes are the main factors of their stability under the conditions of relatively low temperatures and, together with magmatic geotherms (see below), contribute to the formation of *subboreal "climatically unjustified" forests.* It is the second *phenomenon of organization of forest ecosystems at the initial, island-arc stage of subaerial landscape genesis.* Here, one cannot but see the emergence of progressive evolutionary tendencies in buffer forest communities, so typical of the Pacific Ocean mega ecotone of Northern Eurasia. The known proposition of evolutionary theory of the biosphere (Schwartz 1973) has been confirmed, stating that *the enhanced intensity of biogeocoenoses and the improvement of biological cycles underlie evolutionary progress and are the main contents of evolution. Such evolutionary tendencies in plant cover emerge already at the initial, volcanogenic stage of formation of the continental biosphere.*

Chapter 8

Effect of Geothermal Energy on Forest Formation in Island Volcanic Landscapes

Abstract

The chapter presents for the first time the quantitative estimates of the influence of magmatic energy of a passively active volcano on the thermal regime of the soil and forest-forming processes in the island-arc volcanic landscape as a representative of the Neo-Pacific sector of the Pacific Ocean mega ecotone of Northern Eurasia. The investigation was carried out by the example of degassingthe Mendeleev Volcano. The climatically determined soil temperatures have been calculated and to compare there with the real temperature. The conducting channels of magma-based geothermal systems causing abnormal increases in soil temperature have been found in the island volcano landscape. This process was favorable for transgression of more southern species and entire communities to this borealforest ecoregion over a long period of time. The "climatically unjustified" but sufficiently stable subboreal forest ecosystems were formed, as well as the early stages of buffer forest communities. Geothermal energy induces rearrangement of the plant cover towards the increase in its floro- and phytocoenotic diversity to the detriment of biomass formation rate. The largest possible (under the given climatic conditions) diversity of floristic and phytocoenological structures creating terrestrial vegetation in the transitional zone from continent to ocean originates just from island volcanic landscapes.

8.1. Prefatory Remarks

The landscape, geobotanic and soil studies in volcanic regions concern mostly the catastrophic impacts of volcanic eruptions on the surrounding soil-plant cover and the subsequent restorative successions of meadow/forest formation process, with formation of a new stable biogeocoenotic structure (Sokolov 1973; Zhuchkova et al. 1973; Manyko 1974; Grishin 2003; Pietschl et al., 2003; Ganzey 2014; Vyatkina et al., 2014). These studies were carried out almost solely in the paleogeographical aspect of postvolcanic vegetation dynamics, because they were based on the analysis of the Holocene spore-pollen spectra. They occasionally mention some influence of hot springs on the plant and soil cover of modern landscapes of the South Kuril Islands (Lashkov 1948; Urusov 1993). There are some casual remarksthat in the coniferous/broad-leaved and dark-coniferous forests of Kunashir, for example, "...the life forms of subtropical coenoelements ... preserved trees only close to the hot springs" (Urusov 1988, 102).

The geological literature considers the problems of thermal relief of volcanoes and the influence of steam-and-gas streams on the state of surface waters and near-ground atmosphere close to active volcanoes (Droznin et al. 2004; Kozlov, Zharkov 2012); however, we do not know any studies of the effect of geothermal energy on soil temperature regime and on the structural and functional characteristics of plant cover of volcanic landscapes. The studies of geocryological conditions on the territory of the Klyuchevskaya group of active volcanoes of the Kamchatka Peninsula (Abramov, Gilichinsky, 2008) are a rare exception, but they do not dwell on the plant cover either.

Quantitative assessment of the effects of magmatic energy of active volcano on the thermal regimes of soil and near-surface air layer, as well as on the water balance, forest formation and soil processes in an island volcanic landscape, has been made for the first time in given investigation. Geothermic energy has important influenced on the structure and functioning of forest ecosystems. The empirically substantiated concept of geothermal energy of passively erupting volcanoes as a factor of accelerated evolution of island phytobiota has been set forth as a generalizing result.

8.2. Geothermal Energy of Active Volcano

In the regions of active tectonic activity of the Pacific Ocean (mid-ocean and volcanic ridges, rift zones), the ascending flows of heated and deconsolidated mantle substance create ubiquitous heat fluxes, up to 2–4 microcalories per 1 cm^2 per second, which cause active hydrothermal activity in these regions (Galerkin et al. 1982). In transitional areas of the northwestern part of the Pacific Ocean (the Sea-of-Okhotsk and Sea-of-Japan depressions), heat fluxes decrease from 2.2 to 0.8 on average in the direction from island arcs to deep-sea trenches.

Volcanoes carry out the intraterrestrial energy by two heat fluxes: conductive, due to the thermal conductivity of rocks, and convective, via fluid systems and hydrothermal vents formed by these systems (Sugrobov and Yanovsky 1991). By the example of South Kamchatka it has been shown that the total heat efflux by unloading hydrothermal vents in young (Pleistocene) volcanic belts is 923 MWt, with a heat flux of 63.8 mWt/m^2, i.e., twice higher than its conductive efflux. Thus, the hydrothermal vents of the Mendeleev Volcano can be considered the major geothermal factor of soil-ground temperature regime. On the slopes of the active Shiveluch Volcano (the north of Central Kamchatka), the rate of gas efflux in fumaroles varies from 10 to 150 m/s at a temperature of 90–130°, with mineralization of a condensate (sulfur, chlorine, boron and ammonium) from 1 to 10 g/l and with a heat efflux of about $28.8 \cdot 10^7$ J/s on average (Melekestsev et al. 1991).

The detailed seismic investigations have shown (Zlobin et al. 1997) that under the Mendeleev Volcano there is a transitional type of crust (from oceanic to continental) with a thickness of 22–27 km. It has a multilayered structure and consists of three major layers: volcanogenic-sedimentary, upper granite-gneissic and lower "basaltic" (see Figures 4.4 and

4.5). The deep structure and the internal heterogeneities of the volcano are associated with its magma foci and zones of magma formation.

It should be noted that not all volcanic formations give rise to geothermal vents. It has been shown that the hydrothermal systems of the Kuril-Kamchatka Island Arc "... are confined to the continuously developing negative tectonic structures – calderas and volcanic depressions with persistent active volcanism" (Safonov et al. 1998 182). The necessary condition of emergence and development of geothermal vents is *caldera-forming eruptions with shallow magma foci.* Such conditions are maintained on Kunashir Island (see Chapter 4, Section 4.3). Calderas of different ages appeared on the Mendeleev Volcano throughout its eruptive history (see Figure 4.2), while the active Tyatya Volcano has a clearly marked modern vast caldera.

The modern fumarolic/solfataric activity of the Mendeleev Volcano is determined by the fact that the areas of lithospheric fusion and "erosion" of the boundaries between the "basaltic" and granite-gneissic layers directly approach the upper volcanogenic sedimentary layer. At these "erosion" sites, the overlying structural boundaries disappear too and the lower roof of the volcanogenic-sedimentary layer is sagging. Such depressions are the most probable location of steam/hydrothermal vents, which are an extension of fluids – gaseous and liquid solutions that carry magmatic substances from lower to upper layers of the lithosphere (Paffengolyts 1973; Letnikov 2000). "The regular, almost ubiquitous spatial confinedness of the distinguished ... areas to active volcanoes has been established" (Zlobin et al. 1997, 84). These are the areas of "... heating ... and intrusion of additional volumes of fused magma. Such situation is most clearly observed, for example, under the active Tyatya Volcano" (ibid.). Undoubtedly, the conduits of steam-and-gas hydrothermal vents heating more or less the entire body of the volcanic dome and its foothills appear in these areas. For example, in the Klyuchevskaya Group of volcanoes, the talik zones in permafrost being the manifestation of volcanism are confined to the terminal parts of volcanic constructions (Abramov, Gilichinsky 2008).

There are two types of geothermal vents confined to different morphotectonic structures and, accordingly, two chemical types of steam-and-gas flows and waters (Markhinin 1956; Baskov and Surikov 1975; Markhinin and Stratula 1977). In the Mendeleev landscape, the first type is widespread along its periphery, mainly in the Sea-of-Okhotsk sector, in sedimentary-neogenic masses. Its fumaroles are connected with volcano-tectonic deformations, mostly with the northward course (see Figure 4.2). These are the weakly alkaline and neutral sodium chloride thermal springs with the mean outlet temperature of 60° and more. The vitally active thermophilic microflora of these springs is an indicator of their dilution by infiltrated atmospheric waters, which makes these geothermal springs less toxic for the root systems of terrestrial plants. The second type of springs is confined to the volcanic dome itself, at a height of 350–400 m, and is associated directly with its formation. These are acidic and ultra acidic sulfate and chloride steam-and-gas flows (pH = 2.5–3.5) of deep formation "... concentrating the components of magmatic emanations" (Safonov et al. 1998 182). They are characterized by the high content of hydrogen sulfide, the outlet temperature in brooks and lakes from 35–40° to 90–95°, and the maximum toxicity for phytobiota. Their fumaroles are confined to ring faults and fractures delimiting the central cone (see Figure 4.3).

The data of thermometric observations in deep drilling holes in the Kluchevskaya Group of volcanoes have shown that each magmatic dike giving rise to steam-and-gas geothermal vents with fluid processes in the pyroclastic mass (Kotlyakov and Komarova 2007) creates a very limited field of temperature anomalies on the volcanic dome in the mass of pyroclastic sediments (from 50–60 m to several hundreds of meters crosswise), which does not exceed the 10-fold size of the dike (Connor et al. 1997). The thermometric survey performed in September, 1954, on the Goryachiy Plyazh t.ermal field near the northeastern foothill of the Mendeleev Volcano has shown that the total area of the heated site with the ground temperatures of 30° to 101° at a depth of 15 cm is about 5000 m^2 (Markhinin, Stratula, 1977). It is also characterized by the thermal spatial heterogeneity andmultidimensionality of heat flow (Abramov, Gilichinsky, 2008). However, judging by the fact that the thermal anomalies of soils on the Mendeleev Volcano dome are in the form of rather large foci (see below), such localized individual influence of fluid dikes seems to be summed up by their massive intrusion within each focus.

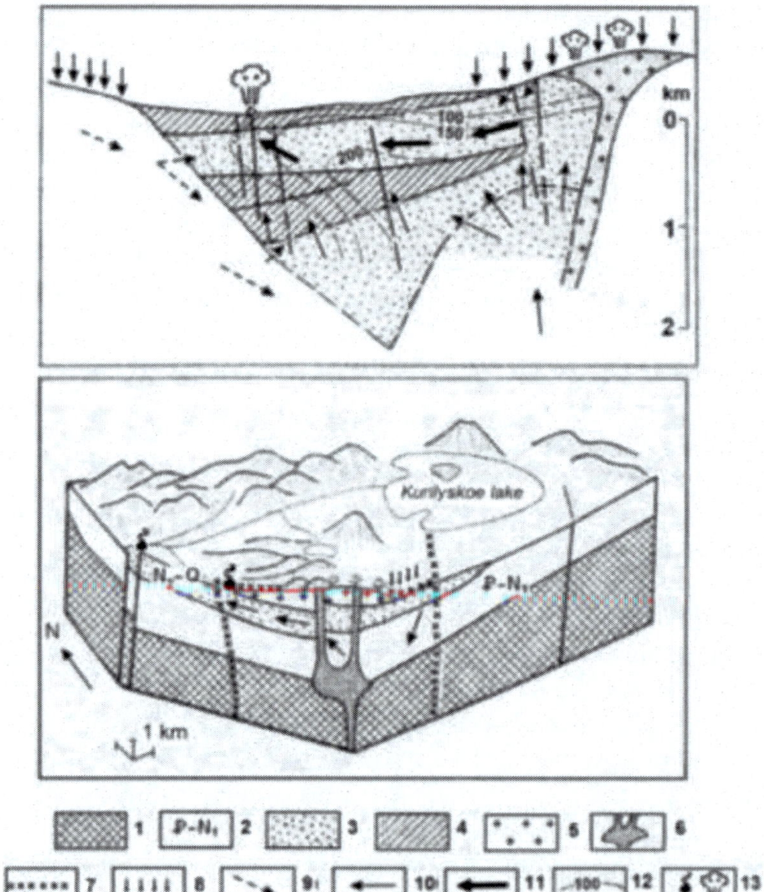

Figure 8.1. The model of the Pauzhet hydrothermal system in Kamchatka (Sugrobov, Yanovsky, 1991). 1 – the foundation; 2 – the Paleogene–Neogene sediments; 3 and 4 – the permeable and waterproof sediments; 5 – the volcanites; 6– the magma focus; 7 – the permeable zones of the ridge; 8 – the infiltration nutrition zones; 9 – the cold underground water movement; 10 and 11 – the ascending and lateral flows of hydrothermal vents; 12 – the isotherms; 13 – the boiling springs and steam jets.

It is known that fluid mass transfer occurs mainly along the sheeting planes of the rocks. The most part of thermal water does not reach the earth surface, being filtered into loose gravel-pebble and sandy-loam deposits (Markhinin and Stratula 1977). All the above extends many times the area of distribution of geothermal vents (Letnikov, 2000) and results in massive geochemical transformation of soils. A brilliant example can be the model of the Pauzhet hydrothermal system in Kamchatka (Figure 8.1) with the well-marked lateral flow through permeable rocks. Here, steam is generated at the depths of 120–250 m, and steam condensation near the thermal field surface causes kaolinization of the rocks (Baskov and Surikov, 1975). At the same time, it is characteristic that geothermal vents are more confined not to the central part of the volcanic dome with the least permeable rocks but to its periphery (the same is observed on the Mendeleev Volcano).

In addition, the steam-and-gas effects on the soil-plant cover have the meandering contours (Abramov and Gilichinsky 2008): the fumarolic activity is ceased in some areas and revived in other areas (Korsunskaya 1946). Finally, there is a significant example of the Zavaritsky caldera complex on Simushir Island, which appeared after the strong eruption in 1957. Here, steam-and-gas hydrothermal vents have not ceased their activity in the volcano body thus far and heat the surface waters of crater lakes up to 65^0, though the previously exposed hot springs have already dried up (Kozlov 2013).

The geothermal vents creating, e.g., volcanogenic taliks in the permafrost rocks of the Klyuchevskaya Group of volcanoes (Abramov 2009) are confined mainly to the negative forms of mountain meso- and microrelief. On the Mendeleev Volcano, these are mostly atrio hollows and the upper reaches of small rivers (Lesnoy, Kisly, Lechebny, Chetverikova Brooks), exactly where all four major solfatare fields are located (see Figure 4.2). At the foot of the Tyatya and Roruy Volcanoes in the northeast of Kunashir, geothermal vents cover the upper reaches of the Tyatyin River basin (see below).

As is known, the global geothermal heat flows ascending to the earth surface give 0.6 Wt/m^2, which is negligibly small as compared to absorbed solar radiation (about 168 Wt/m^2). Such are the heat and energy conditions of existence and functioning of intracontinental natural landscapes. A completely different situation is observed for primary landscape genesis in the mobile belt of the planet. For example, the narrow zone adjacent to the Kuril Island Arc shows a 6–7-fold higher conductive heat flux compared to the average planetary level (Gorshkov 1967).

It has also been shown that the power of geothermal process is inversely proportional to the area of explosion or the hydrothermal region as a whole (Masurenkov 1991). We have used the formula presented in this work to calculate the specific power of geothermal vents of the Mendeleev Volcano for two areals: (1) the total geothermal area (44.16 km^2) contoured with a 300-m isohypse, above which the soil temperature anomalies (see below) are manifestation and (2) the area of the ancient caldera (9.6 km^2) as an indicator of the most powerful manifestation of explosion in the Quaternary period (see Figure 4.2). The specific power of hydrothermal flow was 2.84 Wt/m^2 in the former case and above 13 Wt/m^2 in the latter case.

Thus, in the area of the Mendeleev Volcano that encompasses nearly all of its altitudinal layers, the power of geothermal flow exceeds almost 5-fold its average planetary values though does not reach 2% of the power of solar radiation. At the same time, the power of geothermal vents in the Mendeleev landscape region is at a relatively low level compared to the entire Kuril-Kamchatka volcanic arc, and this level seems to correspond to the state of passive volcanoes. According to the data (Kotlyakov et al. 2009), the power of geothermal flow for Kunashir Island in general is within the range of 90–110 mWt/m^2, and only the Tyatya Volcano region has a power of about 120 mWt/m^2. On the northwestern coast of the neighboring Iturup Island, the heat flow power is up to 150 mWt/m^2. In the territories of active volcanoes of Kamchatka, the average power of hydrothermal flows is about 42 Wt/m^2, while the maximum values are more than 208 Wt/m^2 (Masurenkov 1991).

8.3. Soil Cover Temperature Field

Single stratovolcanoes are elementary cells of the volcano-tectonic system forming characteristic ring structures. The hydrothermal fields of these structures correspond to the very high density of endogenous flow with concentric thermal zonality: the temperature of the central part of the flow can be several degrees higher compared to its periphery (Masurenkov 1991). This pattern is clearly exhibited also in abnormal temperature distribution along the soil profiles on the Mendeleev Volcano and in its surroundings.

Such abnormalities were revealed by the landscape-ecological survey of this territory. The measurings of soil temperature on different depths and that have permitted to carry out the temperature mapping of soil. Based on empirical data, we mapped soil temperature parameters using the new methods of geomorphometry (Shary et al. 2002) – the science of relief modeling. The multi-aspect properties of relief are expressed by more than 20 characteristics calculated from the matrix of heights of the earth surface. The calculation and construction of large-scale maps of soil temperatures and temperature anomalies (see below) were made on the basis of NASA satellite data on the relief: the 90-m resolution matrices of heights obtained in eth framework of the USA–Italy–Germany Shuttle Radar Topography Mission (SRTM).*Thematrix maps* describing in detail the distribution of temperature parameters of the Mendeleev landscape were calculated (Figures 8.2 and 8.7). The image of a matrix map was formed in the GIS program by the appropriate legend, with the procedures of *interpolation and extrapolation of the measured data by the matrices of factors* with the closest statistical relationships revealed by multiple regression analysis.

The soil temperature map for the depth of 40 cm (t_{40}, Figure 8.2) was constructed by the equation (Kolomyts et al. 2016):

$$t_{40} = -0.0005623 \cdot (X - Y) / 2^{1/2}{}_{-5.84} - 0.0005660 \cdot D(\mathrm{MECHN})_{-5.27} -$$
$$- 1.079 \cdot FORMS\_4_{-4.79} + 0.2745 \cdot rot^{S}{}_{+3.00} + 8179;$$
$$R^2 = 0.768;\ Degr = 12.8\%;\ P < 10^{-6}. \qquad (8.1)$$

Here, the predictor $(X–Y)/2^{1/2}$ is the shortest distance to the southeastern coast, *D* (MECHN)is the minimal distance from the Mechnikov Peak (see Figure 5.5; here we should mention that the most of sample areas were on the slopes and foothills of this mountain). *FORMS*_4 are the forms of relief; the 4 types of local forms rank from 1 to 4: depressions, medium concave cols, medium convex cols, and hills. The subscripts of each predictor are *t*-statistics; their module determines the significance of contribution of predictors to the spatial variability of temperature. R^2 is the coefficient of determination and *P* is the statistic characterizing the probability of error in the establishment of statistical dependence. The "minus" predictor describes the temperature decrease in this series. The parameter of relief dissection rot^S shows that soil temperature tends to increase on its strongly dissected forms with clockwise geoflows.

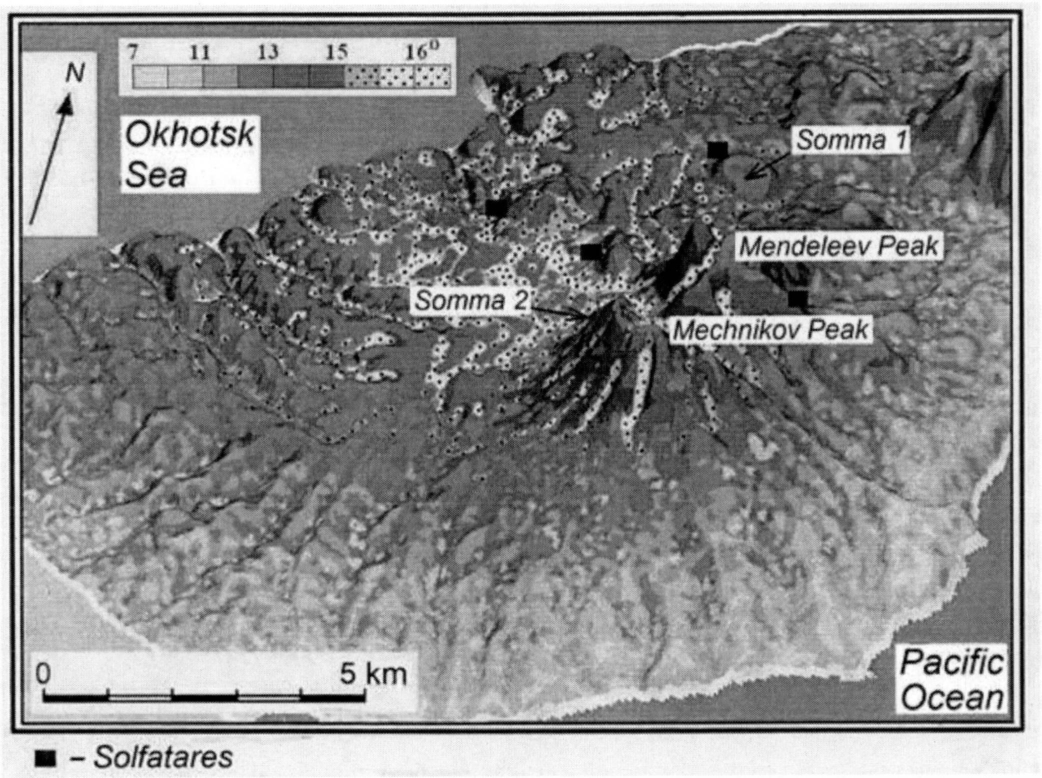

Figure 8.2. Soil temperature distribution at a depth of 40 cm over the territory of the Mendeleev volcanic landscape. The map has been calculated and created by L.S. Sharaya.

The cardinal temperature changes take place along the geographical axis (*X–Y*) responsible for climatic events and processes occurring in summer in this part of the island: the entry of cold air from the east and southeast. The "minus" sign of the predictor means that the lower soil temperatures are characteristic of the southeastern (Pacific) sector of the Mendeleev landscape. The temperature decrease seems to be due to both the cold sea current and the less marked effects of geotherms. North-westward (over the territory of the Sea-of-Okhotsk sector), the effects of both the Soya warm current and geothermal fluxes are enhanced (see above) and, hence, soil temperature increases. The distance from the Mechnikov Peak *D*(MECHN) with the "minus" sign describes the decrease in soil temperature downwards the mountain (from top to bottom) due to the weakened effect of gas-steam hot springs.

The model (8.1) describes 77% of t_{40} variability in space by means of comprising factors. The model was successfully verified (tested) by Allen's cross-validation (Allen, 1974): the empirical criterion (*Degr*< 50%) shows that the predictive strength of the model is sufficient for predictive mapping or calculating the characteristic at new points.

The altitudinal-zonal temperature field of the Mendeleev Volcano is highly mosaic, as is demonstrated by numerous enclaves distant from the main niche and its dominants on the graphs of partial connections. Nevertheless, judging by the ecological dominants, there are certain patterns of soil temperature distribution by absolute height (Figures 8.3 and 8.4). In the plain–piedmont layer, under coniferous/broad-leaved and dark-coniferous forests, the soils have both higher and lower temperatures in an almost equal proportionwith the two dominants: t_{30} = 15.5–16.8° and 11.4–14.3°. In the places where fir/spruce forests come down to the coastline, soil temperature decreases by 2.5–5.0°.

The total decrease in soil temperature (along its entire vertical soil profile) occurs in the lower TA–T layer of low mountains (above 150 m), where the altitudinal zone of dark-coniferous forests begins (t_{30} = 11.4–13.3°). Indeed, the dominants of these temperature fields are weakly marked (with a probability no more than 32–38%); however, here and there low temperatures can be followed up to the absolute heights of 500 m, i.e., almost to the upper boundary of the zone of mountain fir/spruce forests.

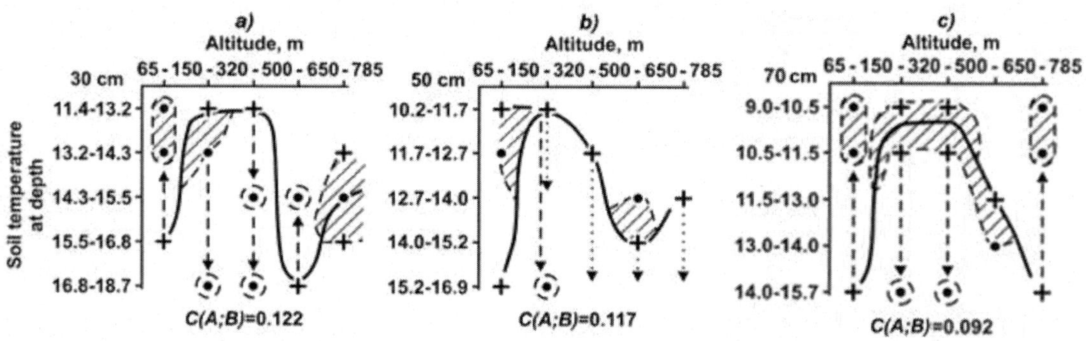

Figure 8.3. Distribution of soil temperature at the depthof40 cm by altitudinal layers and sectors, as well as by floristic phratries and groups of forest types. Legends are as in Figure 6.3.

The transition to the middle T–TE layer of low mountains, which occurs already in the dark-coniferous forest altitudinal zone, is signified by an equally abrupt increase in soil temperature at all depths measured (down to 70 cm). Here, soil has an equal and even higher temperature compared to piedmont mixed forests (the dominant t_{30} = 16.8–18.7°). Sporadic temperature growth begins already after 320 m (see the enclaves on the graphs) and, beginning from the 500-m level, this growth becomes total, as is demonstrated by the nearly complete absence of low-temperature enclaves.

The "normal" (climatically determined) trend of soil temperature decrease with altitude (e.g., the dominant t_{30} = 15.5–16.8° →11.4–13.3°) is observed only at the foothills and in the lower TA layer of the Mendeleev volcano (Figure 8.3). The transition to the middle T layer of low mountains, already in the dark-coniferous forest zone, is represented by a drastic increase in soil temperature at all depths measured (down to 70 cm). Here, the soil was shown to have the same and even higher temperature values as in the piedmont coniferous/broad-leaved forests (the dominant t_{30} = 16.8–18.7°). Sporadic temperature growth begins already after the 320-m altitude (see the enclaves in Figure 8.3, *a,b*), and this growth becomes total beginning from the 500-m level.

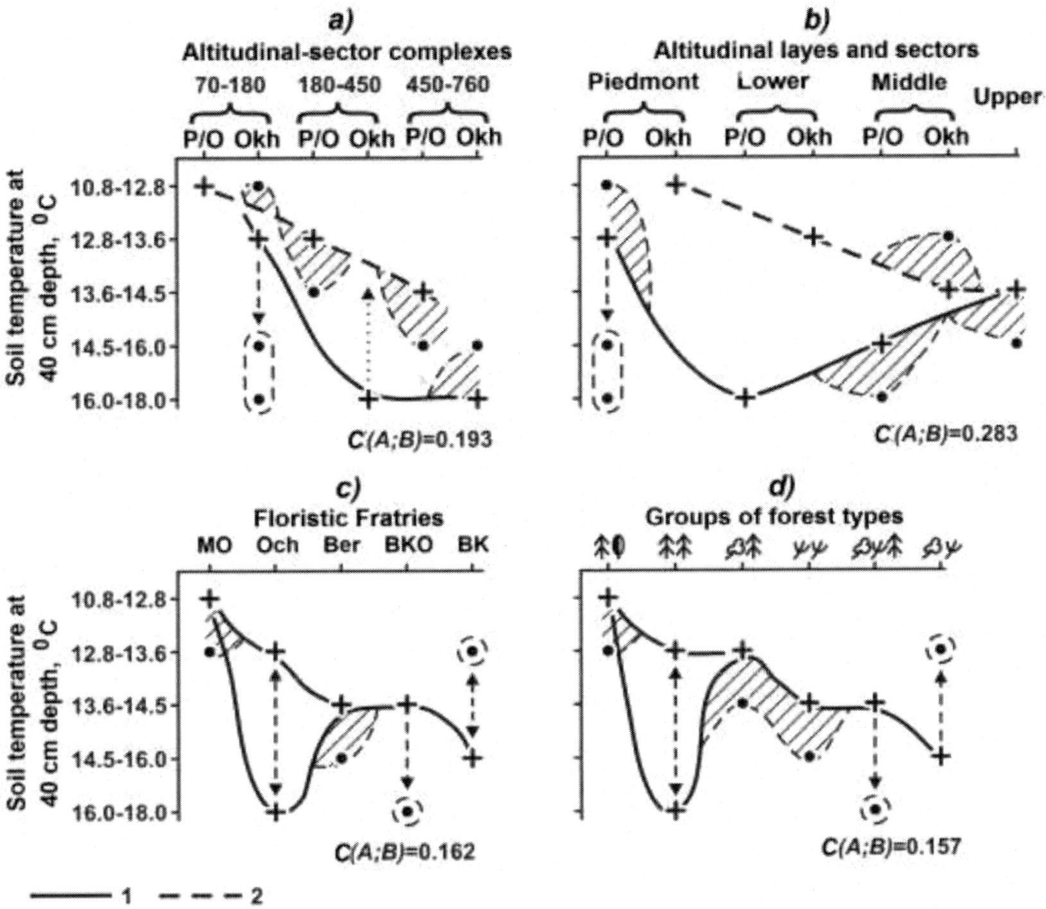

Figure 8.4. Distribution of soil temperature atthe depth of40 cm by altitudinal layers and sectors, as well as by floristic phratries and groups of forest types.

The topological profile across the Mechnikov Mountain (Figure 8.5) clearly shows two temperature leaps. The first, stronger leap is caused by approaching the tectonic fault on the mountain slopes, according to the scheme of E.K. Markhinin (1956). The second soil temperature rise occurs in the atrio hollow, which separates the ancient somma from the Mendeleev Volcano conduit with fumarolic vents. It is also influenced by the closeness of the southeastern fumarole fields stuck on the ellipse of the faults (see Figure 4.2).

Thus, not only on the five above-mentioned fumarole fields but also along the entire ellipse of the faults encompassing the middle and upper altitudinal layers of low mountains, one can find the conduits of magma hydrothermal vents located close to the day surface, which cause *the phenomenal event on volcano slopes – the abnormal increase in the temperature of bedrocks and soils, as well as surface waters.* The thermal imaging of the lakes located in the caldera of the neighboring Golovnin Volcano shows (Kozlovand Zharikov 2012; Kozlov 2013) the water temperature of 90°C in three powerful steam hydrothermal vents and up to 35° on the lake surface. This fact suggests the effective heating of surface waters and soil-grounds around such vents.

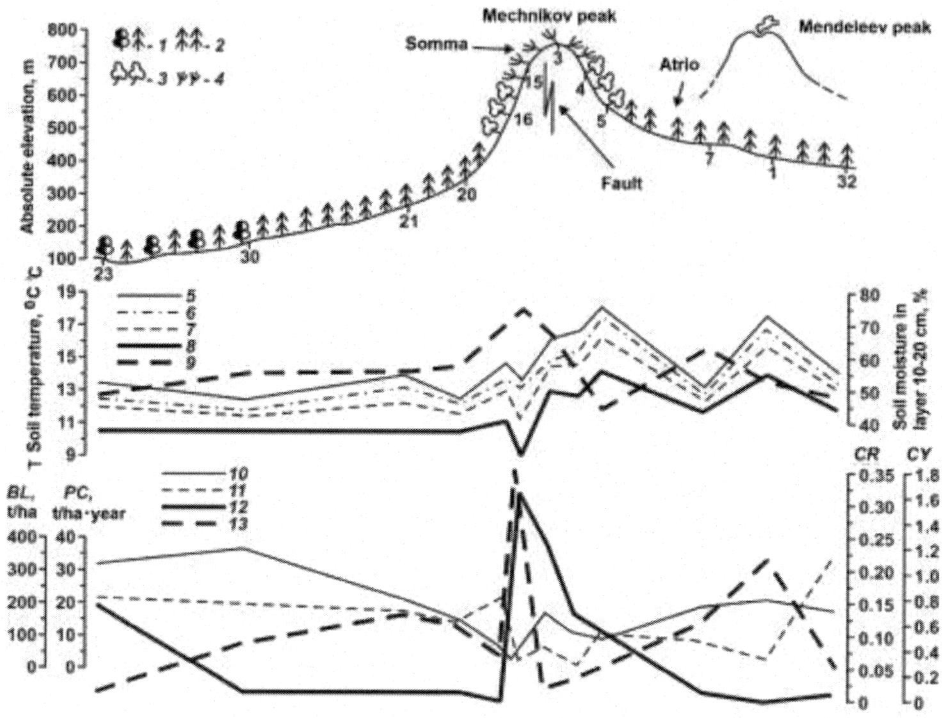

Figure 8.5. Distribution of the groups of forest types, soil temperature and moisture content, as well as the productivity parameters of forest communities, by the topological profile across the Mechnikov Mountain. Plant communities: ***1*** – coniferous-broadleaf forests; ***2*** – fir-spruce forests; ***3*** – birch crooked forests (including those with cedar prostrate shrubs); ***4*** – cedar prostrate shrubs. Soil temperature at the depths of: ***5*** – 30 cm; ***6*** – 40 cm; ***7*** – 50 cm; ***8*** – 70 cm. ***9*** – Soil moisture content in the layer of 10–20 cm. Functional parameters: ***10*** – the total reserves of living terrestrial phytomasses (*BL*); ***11*** – the total productivity of forest phytocoenosis (*PC*); ***12*** – the coefficient of annual phytomass turnover (*KR*); ***13*** – the coefficient of annual destruction of terrestrial phytomasses (*KY*), i.e., the litter-fall index. 23, 30, etc. – the numbers of test plots (see Figure 4.2).

The distribution of geothermal fields is determined by volcanic foci (Safonov et al., 1998), and the power of hydrothermal systems directly depends on the age of foci giving rise to them (Leonov 1991). The main feature of the history of tectonic development of the Kuril Island Arc system, beginning from Late Oligocene, is its discrete displacement and successive rejuvenation towards the Pacific Ocean (Avdeyko et al. 2001). Consequently, the relatively gentle and long Sea-of-Okhotsk macroslope of the Mendeleev Volcano differs from the steep and short Pacific-Ocean macroslope (see Figure 4.2) in the longer history of morphostructural development, which is represented by the geomorphological west-east asymmetry of the morphosculpture of this volcanic massif (see Chapter 3, section 3.3). Accordingly, the deep geotherms of the Sea-of-Okhotsk sector with the longest period of development must also have the maximum effect, with respect to both intensity and area, on the soil-plant cover of the Mendeleev landscape.

These points are confirmed by the data of our thermometric survey. The Sea-of-Okhotsk slopes of the Mendeleev volcanic massif, which include the most ancient first and second sommas (see Figure 4-2) and are outlined with residual crests of the respective calderas (Safonov et al. 1998), are characterized by the most frequent and considerable soil temperature anomalies, though the modern solfataric fields are present here only on the northwest. On the Pacific-Ocean macroslope with the highest fumarolic activity, the general soil temperature anomalies on the sites outsides the solfatares are much less marked.

The Sea-of-Okhotsk sector of the Mendeleev landscape differs from the Pacific sector not only in the higher soil temperatures in the piedmont-plain layer but also in the much more pronounced influence of hydrotherms on soil temperature in the lower and middle low-mountain layers of the volcanic massif (see Figure 8.4 *a, b*). The difference in temperature anomalies between the Sea-of-Okhotsk and Pacific-Ocean sectors in the lower and middle layers of low mountains (beginning from the absolute heights of 180–200 m) by their dominants is up to 3.5–5.0°, though the strong "fuzziness" of the niche of thermal anomalies reaching 50–60% of occurrence frequency and often their mutual overlapping demonstrate the broad distribution of sites with similar manifestations of geothermal effect in both sectors.

Such situation seems to be typical of the Large Kuril Islands in general, because the volcanic belt here is confined to the western, gentler slope of the ridges (Aprodov 1982). The major fumarolic fields stuck on the ellipse of the faults (three out of four fields) are situated exactly in the Sea-of-Okhotsk sector of the Mendeleev landscape. Only in the upper layer, on the tops and crests of the ancient somma (the Mechnikov Mountain), where the influence of hydrotherms becomes weaker, soil temperatures on the W–NW and E–SE slopes are closer to each other. At the same time, while the transition to the top mountain layer in the upper organomineral soil layers (to 30–50 cm deep) is characterized by a slight temperature decrease (see Figure 8.3 *a, b*), the temperature in the lower illuvial-metamorphic horizon (70–100 cm) continues to increase in some places. The heating effects of magmatic geotherms on soil-forming rocks and lower soil layers can be traced up to the peaks and crests of the ridges.

It should be noted that soil temperature at all test plots decreases with depth and no reverse thermal stratification has been found anywhere. It means the following. First, the main temperature background of the soil cover of volcanic landscape is created by the vertical flows of radiation heat coming from the day surface, while the geothermal flow coming from the entrails of the Earth makes only certain and, moreover, focal corrections in soil-ground temperatures. The low thermal conduction of pyroclastic material (Abramov 2009) equally weakens both radiation and geothermal flows without disturbing the climatogenic temperature stratification. Second, it is confirmed that the geothermal flow comes bottom-upwards along separate conduits, from where it diffuses laterally, increasing the total heat-and-energy level of soil-ground mass.

8.4. Interference of Radiation Andgeothermal Waves

The structural ordination of soil thermal niches by solar exposure at different depths (Figure 8.6) revealed the pattern of distribution and interference of heat waves coming from two alternative sources: solar radiation and gas-steam geotherms. We discovered the regions with the conduits where geothermal heat comes from, as well as the areals that lack such ascending flows and demonstrate the radiation-thermal soil regime closely related to slope exposure. At the depths of ≥70 cm including the illuvial layer B, the predominant temperature values shift to the geothermal field with a high probability of 65–75%. At the depth of 40 cm, both (radiation and geothermal) heat flows demonstrate a mutual lateral overlap; above 30 cm, the geothermal flow occurs only sporadically.

The fumarole fields are objects of thermal relief of the gas-steam and magmatic activities of the volcano, associated with the productive geothermal reservoirs (Droznin et al., 2004). These are the areas of "... heating ... and intrusion of an additional volume of molten magma" (Zlobin et al. 1997 84). These areas are where the conduits of gas-steam geotherms appear, which more or less heat the entire body of the volcanic dome, as well as its foothills. However, the conduits of magmatic geotherms close to the day surface are observed not only on the four fumarole fields of the Mendeleev volcano, but also over the entire ellipse of fractures encompassing the middle and upper altitudinal layers, and cause the *phenomenal event on volcanoslopes: the abnormally high temperatures of bedrocks and soil, as well as surface water.*

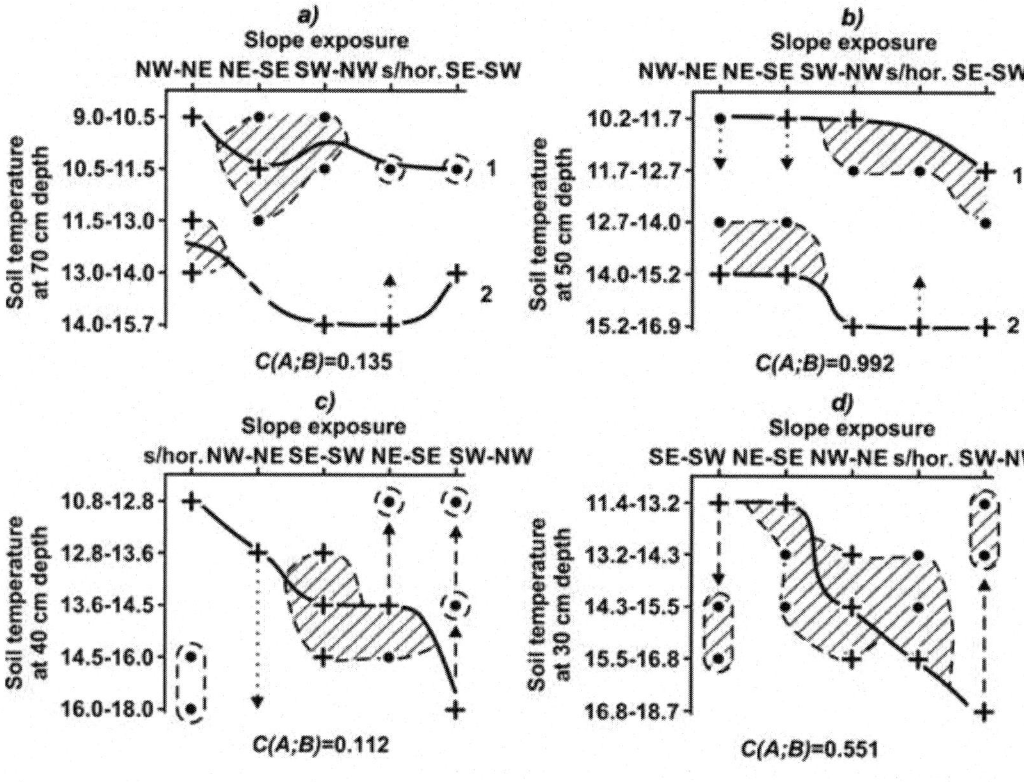

Figure 8.6. Soil temperature distribution at different depths over the slopes of different solar exposure. Areas of the temperature field: 1 – in the absence of the effect of geothermal flow; 2 – in the presence of this effect. Legends are as in Figure 6.3.

In general, the *inversion high-zonal distribution of summer soil temperature* is beginning to emerge, with the respective values for floristic phratries and groups of forest types (see Figure 8-3; *c* and *d*). Soil temperatures usually do not decrease but increase bottom-up along the altitudinal bioclimatic belts. The coldest soils are those of plain-piedmont coniferous/broad-leaved forests (t_{40} = 10.5–13.5°) and the warmest soils (t_{40} = 16.0–18.0°) are those of the mountainous fir/spruce belt (often up to its upper boundary with the buffer CeSbS communities). At the same time, the Sea-of-Okhotsk sector of the Mendeleev landscape is different from the Pacific Ocean sector not only in the higher soil temperatures of the piedmont-plain layer but also in the much more marked influence of hydrotherms on soil temperature in the lower and middle low-mountain layers of the volcanic massif (see Figures 8.2 and 8.7). The difference in temperature anomalies between the Sea-of-Okhotsk and Pacific-Ocean sectors in the lower and middle layers of low mountains (beginning from the absolute heights of 180–200 m) reaches 3.5–5.0° by their dominants.

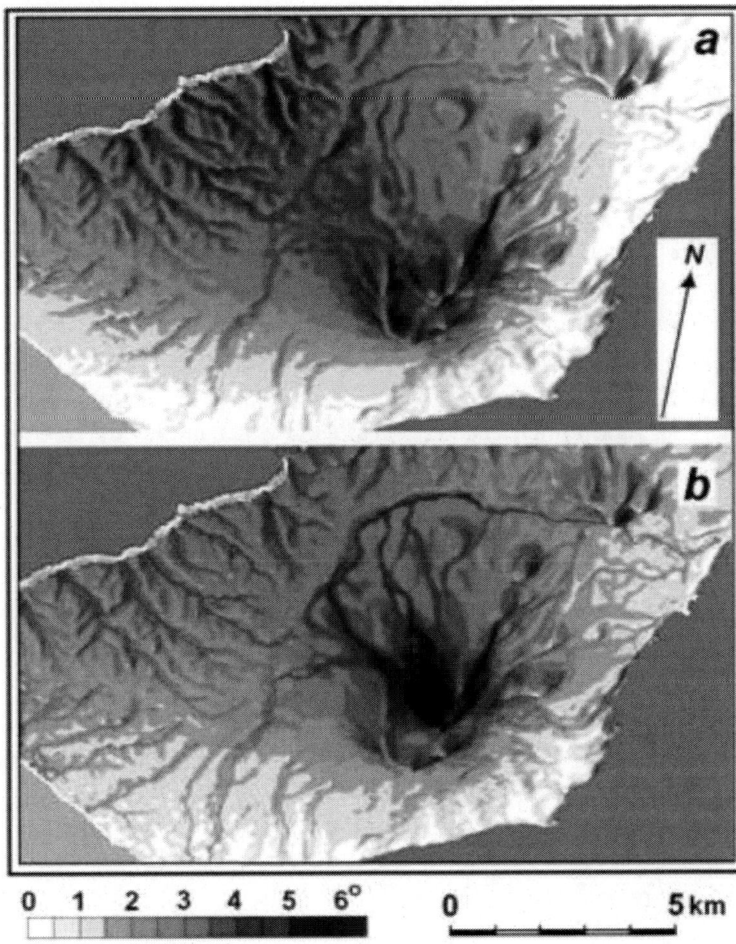

Figure 8.7. Distribution of soil temperature anomalies at the depths of 40 (*a*) and 70 cm (*b*) over the territory of the Mendeleev volcanic landscape (Kolomyts et al., 2016).

The power of geothermal flow is inversely proportional to the area of explosion or the hydrothermal region as a whole (Masurenkov 1991). We have used the formula presented in this work to calculate the specific power of the Mendeleev volcano geotherms for the two areals: (1) the total geotherm area (44.16 km^2) outlined with a 300-m isohypse, above which there begin to appear soil temperature anomalies, and (2) the areas (9.6 km^2) of the ancient caldera as an indicator of the most powerful manifestation of explosion in the Quaternary period. The specific power of geothermal flow was 2.84 Wt/m^2 in the former case and more than 13 Wt/m^2 in the latter case. Thus, the power of geothermal flow on the volcanic landscape area including actually all of its altitudinal layers is nearly 5-fold higher than its average planetary values, though does not reach the 2% power of the solar radiation flux. It should be noted that the power of the Mendeleev landscape geotherm is at a relatively low level along the entire Kuril-Kamchatka volcanic arc, and this level seems to correspond to the state of volcanoes. According to Kotlykov et al. (2009), the power of geothermal flow is in the range of 90–110 mWt/m^2 for the Kunashir Island in general and up to 120–150 mWt/m^2 only near the Tyatya volcano and on the western coast of the neighboring Iturup Island.

8.5. Calculations and Mapping of Soil Temperature Anomalies

For revealing the role of geothermal factor in the soil thermal regime and forest-forming processes, it was necessary to calculate the climatically determined soil temperature and to compare it with the real temperature. First and foremost, an air temperature matrix map was obtained using the WorldClim data containing global surfaces on climate for terrestrial parts of the biosphere. The Mendeleev landscape matrix contained the elements of 600 m×600 m in size. Then, the multiyear normal values of the mean August air temperatures were reduced to those of 1985, when our thermometric survey of the soil was conducted.

Under oceanic climate conditions, diurnal variations of the air (and soil) temperatures are very weak, which is especially noticeable in August. The diurnal amplitude of air temperature at the Yuzhno-Kurilsk station situated at the bottom of the Mendeleev Volcano is 2.0°. The long-term normal value of the daily mean air temperature is t_{air}m = 15.8^0; the temperature values in the periods of 11–15 hours, when we measured factual soil temperatures on the testing areas, were within a range of 17.4–17.8°. The difference is 1.8° on average.

As is known (Dimo 1967), the daily mean soil temperature at a depth of 20 cmt_{20}(av) measured by elbow thermometers (i.e., for bare soil) is very close to the daily average air temperature t_{air}(av). The following linear regression equation with a very high correlation coefficient R was obtained for the Yuzhno-Kurilsk station on the basis of long-term monthly norms for July–September and the data forparticularyears:

$$t_{20}(av) = 0.754\ t_{air}(av) + 4.187;\ R^2 = 0.86. \quad (8.2)$$

This formula was used to calculate the climatically determined (virtual, as we will conventionally call them) mean August temperatures of bare soil at the depth of 20 cm. Then it was established that, first, soil temperatures for the period of 11–15 hours were very close to its maximal temperatures t_{20}max(1). Second, for the Yuzhno-Kurilsk station, t_{20}(av) = 16.4° and t_{20}max(1) = 16.5–16.7°. The difference is t_{20}max(1) – t_{20}(av) ≈ 0.2°, i.e., almost an order of magnitude less than the analogous difference for the air temperature. The control at the Aleksandrovsk-Sakhalinsky station showed nearly the same (and even lesser) difference.

The meteorological data of the Kurilsk station were used for the transition from t_{20}max(1) of bare soil to t_{20}max(2) of grass-covered (meadow) grounds, i.e., from the readings of elbow thermometers to the readings of stretch thermometers:

$$t_{20}max(2) = 0.87\ t_{20}max(1) + 1.54;\ R^2 = 0.97. \quad (8.3)$$

At the next stage, it was necessary to turn from meadow to forest, i.e., to take into account the decrease in soil temperature under forest canopy. The correction factors $\Delta t_{20}\max = -3.0 \div 3.5°$, which are characteristic of the subtaiga fir/spruce forests of South Sakhalin, were used for the CB and DC forests (Protopopov 1975). According to other works (Molchanov, 1961; Tarankov, 1974), the temperature corrections for SbS forests, buffer CeSbS communities, as well as CeW and SbCe formations (see Chapter 5) were –2.5°, –1.8°, –2.0°, and –1.5°, respectively.

Thereafter, the data of our temperature measurements on the test ground were used to calculate the values of virtual (climatically determined) soil temperature at the depths of 30, 40, 50, and 70 cm (here, $t_{20}\max(2) \equiv t_{20}$):

$$t_{30} = 0.838 \quad t_{20} + 1.78;\ R^2 = 0.94; \tag{8.4}$$

$$t_{40} = 0.732 \quad t_{20} + 2.73;\ R^2 = 0.83; \tag{8.5}$$

$$t_{50} = 0.737 \cdot \quad t_{20} + 1.9;\ R^2 = 0.80; \tag{8.6}$$

$$t_{70} = 0.654 \quad t_{20} + 1.7;\ R^2 = 0.57. \tag{8.7}$$

Finally, the real (measured) soil temperatures were compared for the above layers with the virtual (calculated) temperatures to find out the values of temperature abnormalities determined by the influence of gas-steam-thermal springs on the thermal field of each of the 32 sample areas. On the basis of proportional relationships between the real soil temperatures and the real air temperatures, we obtained the so-called *equivalent mean August air temperatures.* The latter reflect the total effect of climatic and geothermal factors, assuming that the geothermal factor is completely replaced by the climatic one (Table 8.1).

Table 8.1. Taxonomic (weighted average) norms of soil temperature anomalies for different groups of forest types of the Mendeleev volcanic landscape, as well as the equivalent values of the mean August air temperature corresponding to these anomalies

Groups of forest types	Mean August air temperature *)	Temperature anomalies at the depths, cm					Equivalent values of August air temperature and its increase	
		30	40	50	70	mean value	mean values temperature	increase of mean temperature
	16.1	2.5	2.0	2.4	2.4	2.3	19.2	3.1
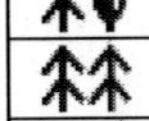	14.9	3.8	3.5	3.1	3.0	3.4	19.7	4.8
	14.4	1.7	2.3	3.0	3.3	2.6	18.3	3.9
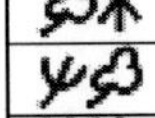	14.3	3.7	3.0	2.9	2.1	2.9	18.9	4.6
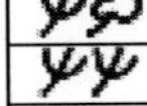	13.9	3.1	3.0	2.3	2.1	2.6	18.1	4.2
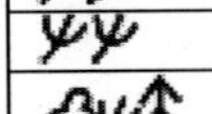	14.4	4.5	4.0	4.0	3.6	4.0	19.0	4.6

*) The mean August air temperature was determined by the temperature surface of the South- Kuril ecoregion on the basis of WorldClim, using the data on 24542 points. Interpolation was realized with the ANUSPLIN software package according to (Hutchinson 2004). The algorithm takes into account the latitude, longitude and altitude as independent variables. Later on, the WorldClim matrices were transformed into the Gauss-Kruger projection (zone 25) with a resolution of 600 m.

The map ofsoil temperature anomalies at a depth of 40 cm (Δt_{40}, Figure 87) was constructed by the multiple regression equation in the form of:

$$\Delta t_{40} = -\,0.0005186 \cdot (X\!-\!Y) / 2^{1/2}{}_{-4.51} - 0.0005194 \cdot D(\mathrm{MECHN})_{-4.16} +$$
$$+\,0.3572 \cdot rot^{S}{}_{+3.38} - 0.4395 \cdot H^{P}{}_{-2.36} + 7531;$$
$$R^2 = 0.660; \qquad Degr = 17.3\%;\; P < 10^{-5}. \qquad (8.8)$$

Here, H^P is the characteristic of the degree of concavity of relief forms (other symbols see above). The analogous equation was derived for mapping the temperature anomalies at a depth of 70 cm.

Judging by the parameter $(X\!-\!Y)/2^{1/2}$, the marked temperature anomalies are observed mainly in the Sea-of-Okhotsk sector of volcanic landscape (see Figure 8.7), which is accounted for by the most efficient manifestation here of steam and gas conduits. The geothermal effect intensifies also up the slopes towards the top of the Mechnikov Peak. Finally, temperature anomalies increase with the enhancement of erosional dissection of the relief and on its concave forms (the closer to depressions, ravines and valleys are relief forms, the higher are anomalies).

Gas-and-steam hot springs have a certain influence on the components of the expenditure part of water balance of the territory. Piedmont coniferous/broad-leaved forests are characterized by an increase in the total evaporation and a more significant decrease in the runoff and surplus summer productive moisture reserves in soil. On the slopes of the volcanic dome, both in the belt of dark-coniferous forests and in the sub-goltzy belt, evapotranspiration is kept to the same level, the runoff increases, and the excessive soil moisture content is preserved.

8.6. Volcanogenic Landscape with "Climatically Unjustified" Forests

The Mendeleev volcanogenic landscape located in the south of the boreal belt, similar to all landscapes of the South Kuril Islands, belongs to the subtaiga (transitional to subboreal) type and the Far-East subtype (Isachenko 1985). Recall that the analogs of Mendeleev coniferous/broad-leaved and dark-coniferous forests are widespread in Sakhalin, in the Lower and Middle Priamurye, as well as in the Northern and Middle Sikhote-Alin (Sochava and Bayborodin 1977). However, south-Kuril subtaiga landscapes drastically differ from their continental-insular and continental analogs both in the much lower supply of warmth and especially in the greater precipitation, yielding exceptionally high atmospheric moisture quotients in the Pacific sector of the Pacific Ocean mega ecotone (see Table 3.3), essentially incomparable with the quotients on the Priamurye sub-Pacific territories. Due to the high oceanicity of climate on the Kuril Islands, the southern boundary of the boreal belt shifts far southwards on their plain areas (Markov 1980) and the boundaries of altitudinal belts in the mountains descend (Nikolyskaya and Khomentovsky 1967).

First of all, there are essential differences in the warmest month temperature and in the accumulated biologically active temperatures ($\sum t \geq 10^0$): in the generally accepted values of warmth supply and the basic criteria of attributingthis plant formation to a particular bioclimatic belt. In the plain and piedmont broad-leaved and coniferous/broad-leaved tall-grass forests of the Middle Amur and Ussuri Riverbasins (Dalynerechensk, Neran, Zhuravlyevkastations), the mean air temperature of the warmest month is 19.5–21.0°, i.e., by 3.5–5.0° higher than in analogous formations in the foothills of the Mendeleev Volcano (see Table 3.3). The subtaiga forests of southern Kunashir, with considerable addition of nemoral elements and even subtropical relicts, by $\sum t \geq 10° = 1140–1610°$ areunderthe conditions of northern dark-coniferous taiga of Eurasia, according to the climaarealsof natural ecosystems (Bazilevicn et al. 1986), i.e., they must be located mostly near the northern boundary of the boreal belt.

The summer surface atmosphere in the low-mountain nemoral-grass fir/spruce forests of the Mendeleev landscape is by 1.5–2.5° colder than their mid-mountain phytocoenotic analogs in the Bureinsky and Badzhalsky ridges on the left bank of the Middle AmurRiver (Ulunga and Irumka stations, see Table 3.3). The sum of biologically active temperatures (1000–1430°) also cast out the Mendeleev fir/spruce forests into the climaareals of north-taiga spruce forests. Finally, the communities of subalpine cedar elfin woods in the sub-goltsy altitudinal belt of the Mendeleev volcano by $\sum t \geq 10° = 950–1070°$ approach the plain Eurasian foresttundra but are at a lower heat and energy level compared to their high-mountain subalpine analogs in the Dusse-Alin Mountain Range, in the upper reaches of the Bureya River ($\sum t \geq 1260°$, Sofiyskii Priisk station).

On the other hand, Vysotsky-Ivanov's annual atmospheric humidity factor (F_{hum}) is also very representative. This complex landscape-geophysical parameter is determined as a ratio of annual precipitation to annual evaporativity (potential evaporation). Kunashir substantially differs in its rather high F_{hum}values not only from the above marginal-continental regions with analogous forest formations but also from the neighboring insular territories, e.g., boreal Southern Sakhalin and subboreal Hokkaido (in Sapporo, $F_{hum} = 1.44$). The similar level of atmospheric moisture is observed on the southern subtropical islands of Japan ($F_{hum} = 2.9–3.6$) and the Black-Sea subtropics of Transcaucasia (in Batumi, $F_{hum} = 3.14$).

Thus, the atmospheric hydrothermal conditions of coniferous/broad-leaved and dark-coniferous forests of the Mendeleev volcanic landscape do not correspond to the climatic norms typical of their marginal-continental formation analogs. These types of forest communities are kind of *"climatically unjustified," extrazonal*, coming from the more southern natural zones with the higher heat and energy level and lower moisture quotients. One of the causes of their sustainable existence and development under these unfavorable climatic conditions should evidently be the positive compensatory effect of thermo-edaphotopes rather strongly affected by magmatic geotherms.

The proposed hypothesis was verified by calculating the virtual (climatically determined) soil temperature at the depths of 30, 40, 50, and 70 cm by the methodical technique presented above (see above). Then they were compared with real soil temperatures, and the so-called *equivalent air temperatures* were obtained on the basis of their proportional relationships with the real air temperature. The equivalent air temperatures represent the combined effect of climatic and geothermal factors, assuming that the geothermal factor has been completely replaced by the climatic one. The difference between equivalent and real air temperatures is, in essence, the measure of influence of steam-gas thermal springs of the Mendeleev Volcano on the heat-and-energy level of volcanic landscapes (see Table 6.1). This measure proved to be rather considerable and reached its maximum (4.6–4.8°), as expected, in the belt of mountain dark-coniferous forests and on the ternary ecotones of their transition into stone birch crooked forests with the maximum geothermal activity. In the foothills of the volcano and on piedmont plains, where geothermal effects are weaker, the equivalent temperatures are lower, though the temperature increase is still rather high (more than 3°).

Thus, the calculations show that *forest communities of Mendeleev volcanic landscape have been formed and function under the same heat-and-energy conditions as their zonal marginal-continental analogs.* This conclusion leads to the following ecological generalizations. Obviously, the zonal types of geo(eco)systems are predetermined not only by the *atmospheric factor,* i.e., the heat and moisture ratio at a given radiation balance (Grigoryev 1966; Budyko 1977), but also by the *lithoedaphic factor,* i.e., i.e., the chemical composition of total and specific uptake of soil nutrient fluxes by plants (Bazilivich and Titlyanova 2008). These two system-forming principles interact via the primary biological productivity and the biological cycle (Perelman 1975; Glazovskaya 1988). Accordingly, certain ecobiomorphs are formed (Reymers 1990), i.e., the arrays of life forms of the phytobiota as representatives of a particular zonal type of geographical environment.

What are the causal mechanisms of influence of soil temperature increase on biogeochemical processes that can realize the functional and structural changes in a forest community, up to the zonal level? First, the higher heat-and-energy level of edaphotope increases the ability of plant roots to take up water and nutrients, due to the enhanced permeability of the cytoplasm (Kulytiasov 1982). Under high moisture supply, it gives an impetus to the intensification of autotrophic biogenesis, including the translocation of substances from roots to above-ground phytomass, as well as the acceleration of the biological cycle (Abaturov, Lopatin 2002). Reproductive processes in the phytobiota become strained; seed germination and shoot elongation are accelerated (Larkher 1976). Accordingly, the development of each new generation of forest stands is accelerated. For the temperate forests, such thermoedaphic signal, if it is sufficiently appreciable and durable, must induce a metabolic shift of the given zonal type of forest community towards the metabolism of more southern types.

Second, soil temperature increase leads to the changes in the geochemical structure of *NPP* (net primary production), i.e., the ratio of basic elements of plant nutrition (N, P, K, Ca) and their involvement in metabolic reactions. The ratio of specific uptake of these nutrition elements by roots determines the chemism of dominant species comprising particular zonal types of plant communities and determining the life cycles of woody plants.

In coniferous forests of the boreal belt, at a relatively low total uptake of nutrients, the leading element is nitrogen. With the transition to subboreal broadleaved forests, the total uptake substantially increases, potassium becomes more important, and calciphilic communities become predominant (Bazilevich and Titlyanova 2008). At the same time, the coefficient of biogeochemical activity (the ratio of the biological uptake of nutrients to their removal by the surface and subsurface runoff) increases dozens of times, which means a drastic increase in the biogeochemical stability of subboreal communities compared to boreal ones (Glazovskaya 1988).

Since ecobiomorphs are material manifestations of the changes in *NPP* structure, it would be logical to think that the subboreal transformation of dark-coniferous forests of Mendeleev landscape described above indicates that the geothermal energy from the depths of active volcanoes causes irreversible landscape-geochemical shifts of boreal forest ecosystems towards subboreal ecosystems. Such shifts make a particular contribution to their evolutionary transformations. At the same time, the higher total value of consumed substances, the greater role of calcium in the structure of specific uptake of typomorphic elements and the higher coefficient of biological activity must contribute to the stability of the formed "climatically unjustified" forests and preservation of not only new but also relict species.

Additional geothermal heat allows the forest communities of Mendeleev landscape to attain their normal zonal and altitudinal-zonal hydrothermal values. The active volcano compensates for climatic resource deficiency and ensures steady functioning and structural design of these "climatically unjustified" island-arc ecosystems. Here, one cannot but see one of the manifestations of the Dokuchaev's law of interchangeability of environmental factors (Dokuchaev 1953; Vysotsky 1960).

The long-sustained total heat-and-energy level of soils, which is higher compared to the partial climatically determined level, essentially activates one of the most important functions of soil habitat (Dobrovolysky and Nikitin, 1990): to be an accumulator and source of energy for structural complication of the phytobiota. As is known (Odum 1971), deterioration of climatic conditions results first of all in the mineral malnutrition of plants in phytocoenoses. Elaborating upon this important proposition of ecology, we can say that heat (or moisture) deficiency under extreme (for a given type of the phytobiota) hydrothermal conditions can be to a certain extent compensated by the influx of necessary nutrients from the soil, according to the above-mentioned law of interchangeability of environmental factors. In this respect, soil unit acts as a major homeostatic stabilizer and as a factor of adaptive plasticity of zonal types of natural ecosystems (Dobrovolysky and Nikitin, 1990) and, in addition, as a certain stimulator of their structural and functional development. The striking example is the conditions and results of formation of island-arc altitudinal-zonal types of forest vegetation in Mendeleev volcanic landscape, which we have considered.

The volcano stimulates the development of surrounding landscapes not only by geothermal heat but also by the properties of volcanic soil-forming rocks favorable for the phytobiota. The processes of initial biogenic land development and preparation for further hypergenic transformation are substantially accelerated due to the mineral composition of original lithological material. Young volcanic soils, especially those developed on fine-grained pyroclastic deposits, have high water permeabilityand a considerable fertilizing potential. The lava andash/slag/gas products of volcanic eruptions contain chemical elements, which are most important for the organic synthesis: hydrogen, organic carbon, nitrogen, phosphorus and sulfur, as well as trace elements forming prebiological structures. All of them often exist as salts easily soluble in water, actually ready to be assimilated by living organisms (Manyko 1974; Markhinin 1980). The ashes also contain a great number of microbial cells, providing the development of blue-green and diatomic algae – the pioneers of soil cover formation on a young land (Markov 1980).

The Mendeleev volcano is not the only example of effective geothermal influence on the soil-plant cover of an island-arc landscape. The even brighter forest-forming example is the active Tyatya volcano in the extreme northeast of the Kunashir Island. Here, in the sea of dark-coniferous forests in the southwest foothill of the volcano with hot spring outputs (Korsunskaya 1958), there is a large areal of fern/moss lianoid fir/spruce forests (Koltlykov et al, 2009; see also Figures 5.1 and 7.12) with nemoral elements (oak) and even subtropical relicts (*Kalopanax septemlobus*, *Magnolia abovata, Taxus cuspidata*). This “small island” of thermophilic nemoral-boreal vegetation caused by the local warming influence of modern geotherms of the Tyatya Volcano and, partially, the neighboring passive Ruruy Volcano, is quite similar to the forest cover of far-away piedmont plains of the southwestern part of the island (the surroundings of the Mendeleev Volcano and the Golovnin Caldera), though, according to some data (Koltlyakov et al. 2009; Kozlov and Zharikov 2012), it is separated from this part by high-rank bioclimatic and geobotanic borders (Vorobyev 1963; Krestov 2006).

The analogous extrazonal forest communities are also observed in the central part of the more northern Iturup Island (Figures 8.8 and 8.9). Here, the zonal types of vegetation are stone birch (*Betula ermanii*) forests and crooked forests (herbaceous, elfin wood and bamboo). The piedmont fir/spruce (*Picea yezoensis, Abies sachalinensis*) forests, with stone birch and bamboo, adjoin the warmer Sea-of-Okhotsk coast only in the extreme southwestern part of the island (Kotluyakov et al., 2009).

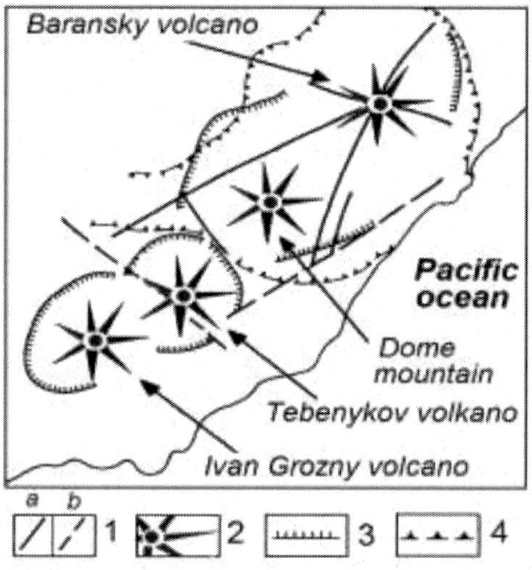

Figure 8.8. The structural scheme of geothermal fields at the Grozny Mountain range by Pacific Ocean shore in the middle part of Iturup (Safonov et al., 1998). 1 – The faultsdetermined on the basis of geophysical data (*a*) and hypothetical (*b*); 2 – volcano craters and domes; 3 – caldera ridges; 4 – the boundaries of geothermal fields.

Against this background, the central broader part of the island is an evident floro-phytocoenotic anomaly (Figures 8.10 and 8.11). Here, the Grozny Ridge with a chain of active volcanoes runs along the southeastern coast (Safonov et al. 1998). Fumarole fields, mud pools and boiling rivers are widespread on the northwestern slopes and at the foot of the ridge. Xerophilic larch (*Larix kurilensis*) forests with stone birch and oak (*Quercus crispula*), with bamboo and cedar elfin wood here and there, are widespread here. On volcano slopes, there is a sharp transition from subgoltsy elfin-wood/bamboo thicket and crooked birch forests to the belt of these apparently xero-thermophilic mountain forests.

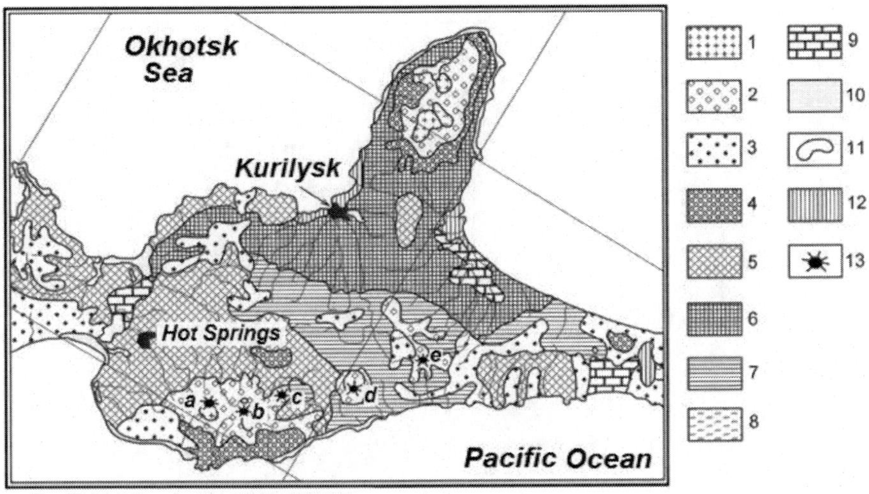

Figure 8.9. The map of vegetation in the central part of Iturup (Kollyakov et al., 2009). Mountain tundras, meadows and elfin woods: 1 – heath tundras, heaths and cubalpine meadows; 2 – cedar elfin woods, on upper bound in complex with the heaths. Forests, light and crooked forests: 3 – elfin wood/bamboo thickets, here and there crooked forests, with herbs, elfin woods and bamboo; 4 – stone birch forests, here and there crooked forests, with elfin woods and bamboo; 5 – larch-tree forests and light forests, with stone birch and oak, often with bamboo, here and there with cedar elfin woods; 6 – oak forests, with larch-tree and stone birch, here and there with undergrowth from ilex; 7 – stone-birch forests with broadleaf species; 8 – partial and almost closed vegetation (of different succession stages) at old and ancient effusieve relief. The vegetation of falls and sea shores: 9 – low scrub, forb and herbs meadows and swamps as well; 10 – shore complexes of rocks and beaches. Anothere meanings; 11 – lakes; 12 – agriculture and residential areas; 13 – volcanos of Grozny mountain range (*a* – Ivan Grozny volcano; *b* – Tebenykov volcano; *c* – Dome mountain; *d* – Baransky volcano; *e* – Burnt mountain, see Figure 8.8).

Figure 8.10. Iturup island. Phreatic eruption Ivan Grozny volcano at the Grozny Ridge in August 2012. The dark coniferous forest with deformeted top of a tree is on the front plane.

Figure 8.11. Iturup island, the Grozny Ridge. River Boiling near south-west foot of Baransky volcano.

The described phenomenon of geothermally determined extrazonality of forest ecosystem is as surprising as not something special in nature. Let us give a striking example. In the tundra zone of eastern Chukotka, in the Amguyema River valley, the ground waters of large river valleys also have a warming effect on soils and vegetation; therefore, the flood plains here are occupied by high relict Chosenia/poplar forests and shrubs, which drastically contrast with the surrounding dwarf-shrub, moss tundras – the zonal type of vegetation of this region (Reutt 1970).

The second example describes the specific peculiarities of high-mountain geoecology. In the alpine zone of the Central Alps, the temperatures of surface and upper soil layers due to strong insolation are higher than the temperature of the near-surface air layer, which results in a longer vegetation period (Valter 1968) and, accordingly, the emergence and development of more thermophilic plants. The analogous microbioclimatic conditions are also formed in the high mountains of Greater Caucasus. It suggests that mountains in general are a source of heat for the adjacent layers of the atmosphere (Shikhlinsky 1969).

8.7. Structural and Functional Changes in Forest Communities

The energy of gas-steam geothermal reservoirs has rather contradictory effects on the structural and functional state of boreal forests of Mendeleev volcanic landscape. Due to geothermal heat flows through the soil, the coniferous forest enriches oneself by nemoral species. The floristic diversity of forest stand increases: the mixed buffer communities with broad-leavedEast Asian species (*Quercus crispula, Q. dentate, Acer tschonoskii, A. japonicum, Magnolia hypoleuca, Ulmus laciniata, U. propinqua,* etc.) are widespread in the lower piedmont altitudinal belt with the zonal type of boreal dark-coniferous (spruce/fir) forests. Simultaneously,the lifetime of each age generation of basic forest-forming species increases,leading to the higher forest density and to the increase in live aboutgroundphytomass (Figure 8.12, *a*, *b*). The highly productive lower-layer fir/spruce forests and piedmont coniferous/broad-leaved forests, where geotherms have the strongest influence on soil temperatures, have undergone the greatest structural changes.

The analysis of floristic and geobotanic data for South Kuril Islands (Urusov, Lurye, 1980; Urusov, 1993) and, in particular, for the Mendeleev Volcano (Manyko and Rozenberg 1970; Vasilyev and Rozenberg 1977) confirms our conclusions. These forests, "…where the conditions of growth of dark-coniferous species are far from extreme" (Manyko and Rozenberg 1970 67), are characterized by large-stem stands and park-like appearance (see above Figure 5.3). At the same time, "the signs of suppression of broad-leaved species at the upper boundary of their dissemination are absent" (Urusov and Chipizubova, 2008 216). It is also stressed that these forest communities are saturated with thermophilic subtropical relict species: *Kalopanax septemlobus* (Figures 8.13 and 8.14), *Taxus cuspidate, Magnolia abovata, Ilex crenata, Actinidia colomicta, Skimmia repens, Onoclea sensibitis*, *Hydrangea anomala petuiolaris, Toxicodendron orientals*.

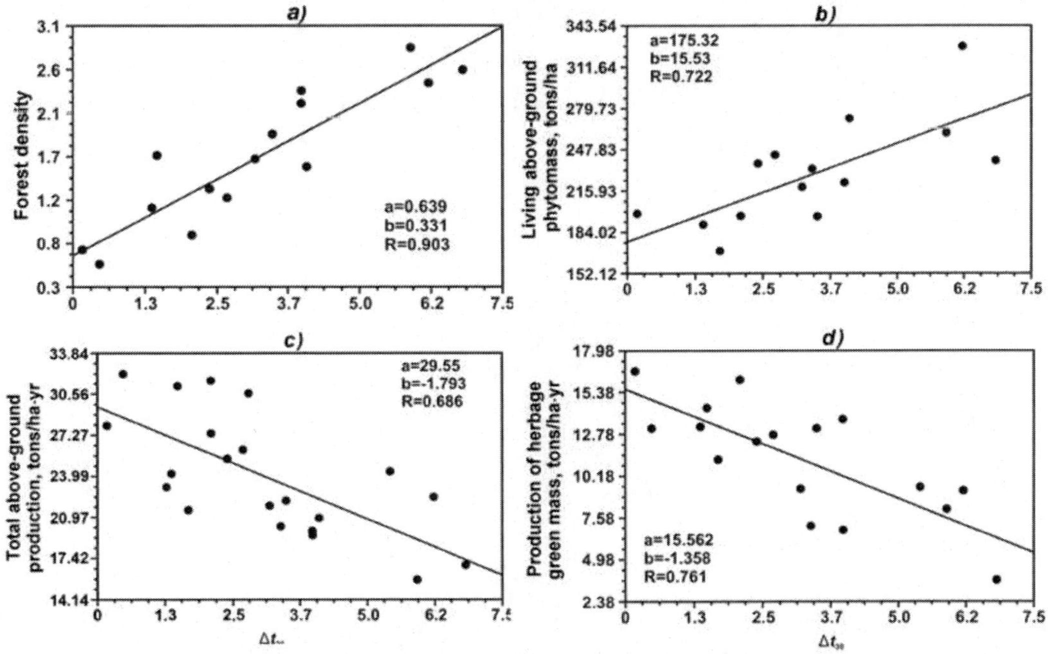

Figure 8.12. The correlations of different parameters of autotrophic biogenesis with soil temperature anomalies at a depth of 30 cm (Δt_{30}).The linear regression equations: $Y = a + b{\cdot}X$; R is thecorrelation coefficient.

Figure 8.13. Dimorphant (*Kalopanax septemlobus*) – the "white nut" of the ginseng family. The relict subtropical species in South Kuril broad-leaved and mixed forests.

Figure 8.14. White-leaf Japanese magnolia, or inversely heart-shaped silver magnolia (*Magnolia abovata*) – the subtropical relict tree in South Kuril boreal and mixed forests.

Apparently, it was just the geothermal factor that contributed to the “combined flora” (Urusov 1988) in the Northwest Pacific, i.e., the combination of genetically and ecologically heterogeneous elements into mixed (buffer) communities, followed by the adaptive transformation of ecological niches of these elements. This pendulum process, manifesting itself most actively in the warm climatic epochs and decaying in the epochs of cooling, can be considered as an *accelerated evolution of original (Holocene) island ecosystems,* as is mentioned in (Urusov and Chipizubova 2008). The major factor of acceleration of this evolution, in our opinion, could be the deep magma energy of active volcanoes.

The obvious warming effect of gas-steam geotherms is also observed in the overlying belts of birch crooked forest and cedar elfin wood. Spruce forest stands with viable undergrowth as well as Manchurian and Japanese species (Japanese yew, wrinkled holly) are frequent here. In the upper layer with the maximum soil temperature anomalies (4.0–4.5^0 on average), the most complex cedar elfin wood/birch/spruce buffer communities are well developed.

Thus, *geothermal energy exerts a positive effect on the structure of forest communities of volcanic landscape*. It does not concern the functional parameters, first of all, the common forest productivity which, on the contrary, is negatively influenced by gas-and-steam geotherms (see Figure 8.12, *c*). The gradient of total above-ground productivity ΔPL = –(1.47÷2.00) t/ha ·yr per 1° increase in t_{20} of the soil in the mountain dark-coniferous forests and –1.33 t/ha ·yr in the lower TA layer. However, this negative effect is nevertheless an order of magnitude weaker compared to the positive effect of solar energy (Tables 8.2 and 8.3).

Table 8.2. The comparison of total above-ground production (PC_{tot}, tons/ha per year) in mountain dark-coniferous forests of the Mendeleev Volcano, as well as its changes (ΔPC_{tot}) caused by the increase in the mean August air temperature (*a*) and the analogous increase in soil temperature at a depth of 20 cm (*b*)

a)

Parameters of the productivity	Average August temperature of the air			Change of the temperature	
	13.7	**14.6**	**15.2**	**0.9**	**1,5**
PC_{tot}	10.5	25.0	28.6	–	–
ΔPC_{tot}	–	–	–	+14.5	+18.1

b)

Parameters of the productivity	Average soil temperature at depth 20 cm			Change of the temperature	
	14.3	**15.0**	**16.2**	**0.7**	**1.9**
PC_{tot}	26.2	24.8	23.4	–	–
ΔPC_{tot}	–	–	–	–1.55	–1.87

Table 8.3. The comparison of the changes in total above-ground production PC_{tot} (tons/ha per year) in the plain coniferous/broad-leaved and dark-coniferous forests of Mendeleev landscape caused by the increase in the mean August air temperature and the analogous increase in soil temperature at a depth of 20 cm

Temperature characteristics			Parameters of the productivitu	
			Meaning PC_{tot}	Change PC_{tot}
Temperature of the air	Average temperuture of August	15.8	30.1	–
		16.6	19.4	–
	Change of the temperature	0.8	–	– 10.8
Soil temperatrure at depth 20 cm	Average temperature	17.1	22.2	–
		19.3	18.3	–
	Change of the temperature	2.0	–	– 1.9

It should be noted that the *PL* parameter decreases mainly due to the drastically reduced production of herbage green mass, making up to 45% and above 60% of the total above-ground production in coniferous/broad-leaved and dark-coniferous forests, respectively. The productivity of ground cover: tall grasses and Kuril bamboo (*Sasa kurilensis*), – decreases most abruptly (see Figure 8.12 *d*). At the same time, the production of skeletal and green tree-bushes phytomass increase, although it douse not compensate the common decrease of forest productivity.

The increase in soil temperature under the conditions of humid climate and highly acidic pyroclastic rocks causes (Sokolov 1973) quite a number of changes in soil formation processes: the rapid decomposition of primary minerals, the higher content of amorphous aluminum toxic for plants, the weaker removal of organic matter, and the stepwise increase in mineralization of buried organic matter (thickness of humus profile decreases from 50–70 cm to 10–20 cm). The total acceleration of soil-forming processes in the foci of geotherms leads to the more intensive transformation of pyroclastic materials of different age (with the buried humus layers) into a simpler soil profile, with its secondary differentiation. As a result, the thickness of the illuvial layer decreases (from 60–85 cm to 25–40 cm) and the information measure of soil profile complexity is reduced, which is indicative of soil "rejuvenation." At the same time, the soil and groundwater are saturated with nitrogen-carbon and carbon-nitrogen vapors and gases, as well as with hydrogen sulfide and sulfur dioxide gas (Sokolov 1973; Markhinin 1980). These changes are suppressing for the aqueous and mineral nutrition of plants (Kabata-Pendias and Pendias 1986), which may be an additional factor of reduced forest productivity on volcanic landscapes.

The functional changes in forest communities determined by the effects of gas-steam thermal springs are also represented by a composite index of autotrophic biogenesis – the allometric coefficient *CA*: the ratio of the logarithm of productivity to the logarithm of biomass. The indicative role of this parameter at the topological level of geo(eco)systems is considered in (Perelyman 1975). The island-arc boreal Neo-Pacific differs from its intra- and marginal-continental analogs in the lower general level of values of this coefficient (Figure 8.15, *a*), which corresponds to the above-mentioned specificity of its bioclimatic system. Against this back ground, one can clearly see a considerable decrease in the *CA* parameter, with the increased soil temperature anomalies. Consequently, the volcanogenic aridization of soils causes a certain decrease in the ecological efficiency of production in forest communities. These are the peculiar features of the ascending branch of the biological cycle in "climatically unjustified" subboreal forests of island volcanic landscape.

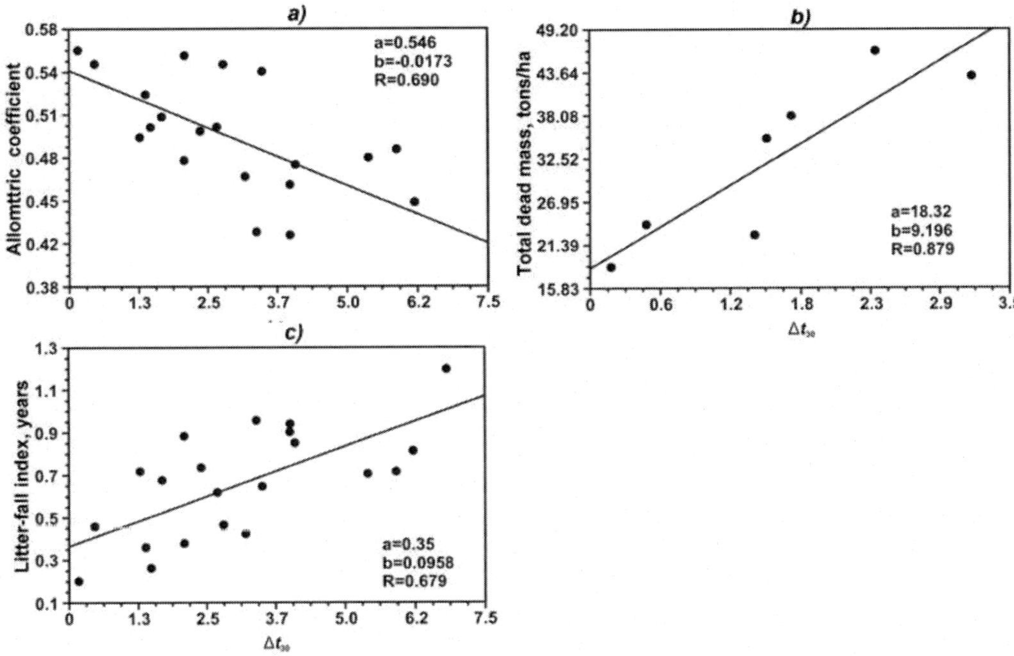

Note: Litter-fall index – the ratio of forest litter mass to the mass of green substance produced annually. The rest conventions see in the text and in Figure 8.12.

Figure 8.15. The correlations of different parameters of biological cycle with soil temperature anomalies at a depth of 30 cm (Δt_{30}).

The detritus branch of biological cycle in the South-Kunashir boreal forests is also somehow exposed to geothermal heat. The rise in soil temperature increases total evaporation and reduces, though negligibly, soil moisture. Against this background, the coolest and most humid Pacific Ocean sector, with a comparatively low geothermally induced rise of soil temperatures ($\Delta t_{30} \leq 3.5°$), shows a threefold increase in total dead phytomass – debris and forest litter (Figure 8.15, *b*). The Sea-of-Okhotsk sector, in any case warmer and drier, is much more influenced by strong geothermal flows($\Delta t_{30} \rightarrow$ 6.5–7.0°) intensifying the processes of decomposition of dead organic matter, resulting in a noticeable decrease in its total mass.

The forest litter mass (up to 20–30% of total mortmass) increases under the influence of geotherms in both sectors of the Mendeleev landscape. It is obvious that mineralization processes do not manage the amount of annual tree waste (+ tree waste). Accordingly, the known litter-fall index (Figure 8.15, *c*) also increases. Thus, gas-and-steam geothermsdo not accelerate but slow down the small biological cycle, stabilizing the state of boreal forest community as a detritus ecosystem (Odum 1971). Hence, it is less probable that forest-covered territories will be transformed into meadows or invaded by Kuril bamboo thickets, which is usual for the places of fires and cuttings.

As it has already been mentioned, the atmospheric hydrothermal conditions of boreal forests of the Mendeleev volcanic landscape do not correspond to the climatic norms typical of their marginal-continental analogs. One of the causes of sustained existence and development of these climatically unjustified forest communities seems to be the positive compensatory influence of thermoedaphotopes exposed to rather powerful influence of magmatic geotherms. The calculations of virtual (climatically determined) soil temperature at the depths of 30, 40, 50, and 70 cm, as well as the equivalent mean August air temperature (see above), have revealed a rather great measure of influence of gas-steam geotherms on the heat-energy level of the Mendeleev volcanic landscape. This measure reaches the maximum (4.6–4.8°, an increase from 14.3–14.9°to 19.0–19.7°) in the belt of mountain dark-coniferous forests and in the band of their transition to stone birch crooked forest with the maximum activity of geotherms. In the foothills of the volcano and on the piedmont plains, where geothermal effects are weaker, the equivalent temperatures are lower, though the temperature increases by more than 3°.

Thus, one can state that forest communities of the Mendeleev volcanic landscape have been formed and function under the same heat-and-energy conditions as their marginal-continental analogs. Being supplied with more geothermal heat, they reach their zonal and altitudinal-zonal hydrothermal norm. The active volcano compensates for climatic resource deficiency, providing stable operation and structural design of these “climatically unjustified” island-arc ecosystems.

8.8. Changes in Soil-Forming Processes and Soil Morphology

The heat-and-energy effects of geothermal reservoirs in volcanic landscapes undoubtedly influence soil formation processes as the terminal element of the descending branch of the biological cycle. Under the conditions of humid climate, temperature increase causes the following changes in geochemical processes in volcanic soils (Sokolov 1973): (1) the destruction of primary minerals is accelerated; (2) the mineralization of organic matter is intensified and the accumulation of amorphous organomineal compounds containing allophanes (alumosilicates) and “free” hydrous ferric oxides, aluminum hydroxide and hydrated silica increases; and (3) the scavenging of organic matter is impeded, resulting in accumulation of large humus stocks in the soil profile.

The transformation of amorphous minerals (allophans and "free" hydrous ferric oxides, aluminum hydroxide and hydrated silica) into crystallized minerals weakens the illuvial Al-Fe-humus process. The degree of sesquioxide saturation in the infiltrated humic acids increases, and their removal to the lower horizons is reduced. Soil horizon B_i can become less marked not only illuvially but also metamorphically, which is facilitated by the growth of crystallized minerals during intrasoil weathering. Finally, the temperature increase intensifies the transformation of buried humus horizons, with the formation of organomineral compounds of the same composition as the products consolidating here as a result of modern illuvial-humus processes.

The soils of the extrusive dome of the Mendeleev Volcano contain high amounts of amorphous aluminum (more than 58% of the gross amount), where the amorphous forms as free Al components may be up to 100% (Gladkova and Butovets 1988). The excess of amorphous aluminum with high biological availability makes it toxic for plants, especially in acidic soils and at the higher contents of Fe, Mn, and other heavy metals. As a result, metabolism is disturbed and productivity of the phytobiota decreases (Kabata-Pendias and Pendias 1986). It seems to be one of the causes of the above considered decrease in soil cover productivity in the forests of the Mendeleev volcanic landscape under the influence of geothermal wells. This factor partially neutralizes the positive effect of geothermal increase in soil temperature on the productivity of trees and shrubs.

The steam-and-gas geothermal wells of active volcanoes have not only thermal but also geochemical effects on soil. The major components of fluids in the lithosphere are the gases of the C-H-O system, as well as the acidic compounds of Fe, Cl and S, with the involvement of nitrogen and noble gas compounds (Letnikov 2000). The analysis of the gases from fumaroles and the extracts from fresh ashes showed (Markhinin 1980) that the Klyuchevskaya group and Kuril Ridge stratovolcanoes (the final stage of continental mantle degassing) are characterized by the highly acidic gas composition of magma (saline pH = 2.5–5) with the high content of H_2, N_2, HCl, HF, H_2S, CO, SO_2, as well as boric and arsenous acids, metal chlorides and fluorides. In the SouthKamchatka geothermal province, with the youngest (quaternary) volcanoes, the steam and gas outflows are of diverse chemical composition: from hydrocarbonate sodium to sulfate sodium waters, with nitrogen-carbon dioxide and carbon dioxide-nitrogen gas composition (Sugrobov and Yanovsky 1991).

The gas streams of fumaroles onthe Mendeleev Volcano contain, in addition to water vapor, hydrogen sulfide and sulfur dioxide gas (Markhinin 1956). The solidified sulfur flows are formed at the outlets of fumaroles, and vegetation is absent over tens of meters around these outlets and acidic streams (see above Figure 4.4).

Due to the prevalent interstratal mass transfer (Letnikov 2000), as well as the exceptionally high porosity and water permeability of pyroclastic soil-forming rocks, the above-listed substances, which are for the most part toxic for plant root systems, have an opportunity of broad lateral distribution from outlet geothermal channels (Sugrobov and Yanovsky 1991). Thereby they suppress the aqueous and mineral nutrition of plants, and this suppression, as it has already been mentioned, can reduce the productivity of forests in volcanic landscapes. In the meantime, the known works on the soil of volcanic regions neither address this issue nor present any data on the soil content of toxic substances. It has only been reported of the abnormally high acidity of volcanic soils, probably due to not only the abundance of silica in pyroclastic material (Sokolov 1973) but also the additional acidifying effects of steam-and-gas geothermal wells.

Near the fumaroles, pyroclastic rock is in the plastic state because of high temperature (Korsunskaya 1946). The upper part of soil-forming rocks in the vicinity of the Mendeleev Volcano (up to 60 m thick) is a zone of active coalinization and argillitization, with high-salinity chloride sodium waters (mineralization from 1–2 to 10–25 g/kg) with a considerable content of calcium (Baskov and Surikov 1975). The abundance of hot water springs in this region determines the development of ferrous iron forms in the illuvial soil layer (Gladkova asnd Butovets 1988). Obviously, the fluids of steam hydrothermal wells, being a result of phreatic eruptions, cause further hydromorphization and allitization of soils. Piercing through fine-dispersed material, they cause its fluidity (Kotlyakov and Komarova 2007) and thereby contribute to the development of defluxion, landslides and landfalls on steep mountain slopes – the abnormal slope processes so typical of the Kuril Islands (Korotky and Skrylynik 2008).

According to our observations on the Mendeleev Volcano, the accumulation of silt and ortsands in hor. B_i makes this horizon waterproof, and in all of the above-lying horizons on steep mountain slopes there is an intensive intrasoil runoff (water often oozes over the walls of soil pits and road trenches). The waterlogged fine-dispersed material of the illuvial horizon per se becomes viscous (the state of thixotropy) and serves a good "lubricant" for the sliding of large volcanic debris.

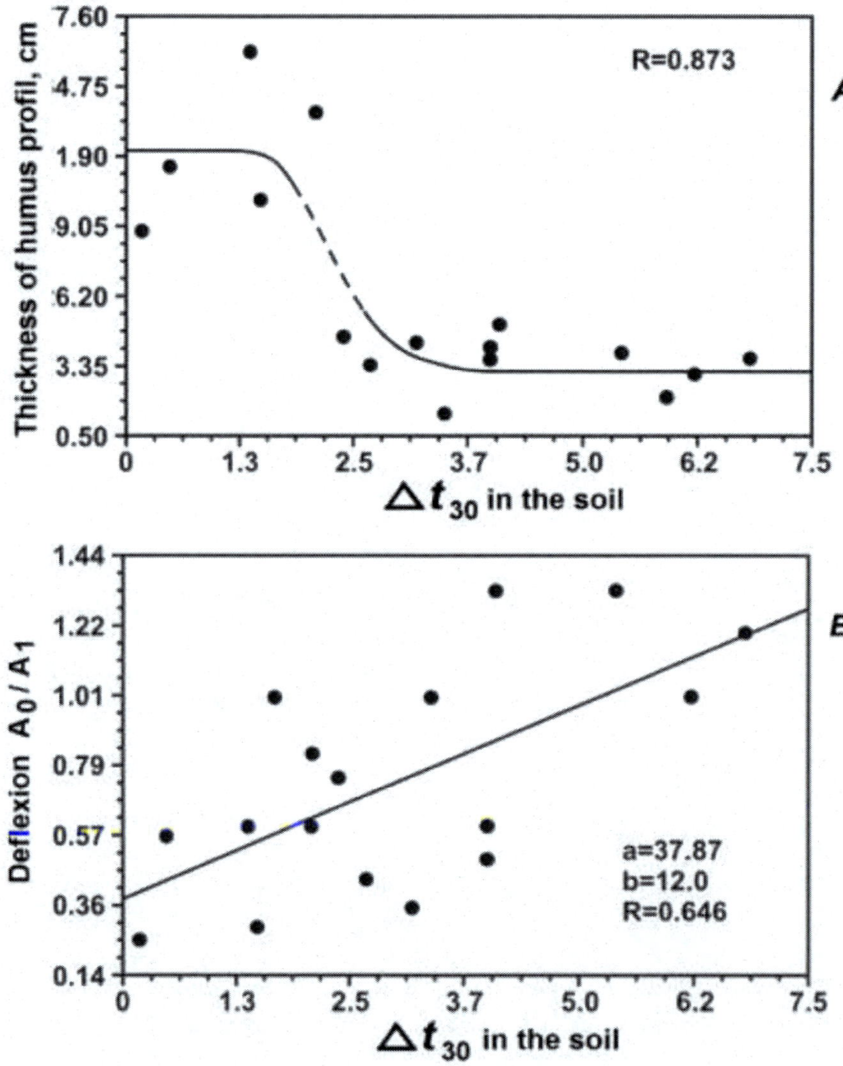

Figure 8.16. The effects of temperature anomalies in soil at a depth of 30 cm (Δt_{30}) on the morphology of the upper organomineral part of soil profile in mixed and dark-coniferous forests of Mendeleev landscape. The linear regression equations: $Y = a + b{\cdot}X$. R is the correlation coefficient.

The described regularities of geothermally determined shifts in soil formation influence the dynamics of morphological characteristics of soil of Mendeleev landscape. It is most pronounced in the complex of coniferous/broad-leaved and dark-coniferous forests on piedmont plains and in the lower layer of the volcanic dome (Figures 8.16 and 8.17).

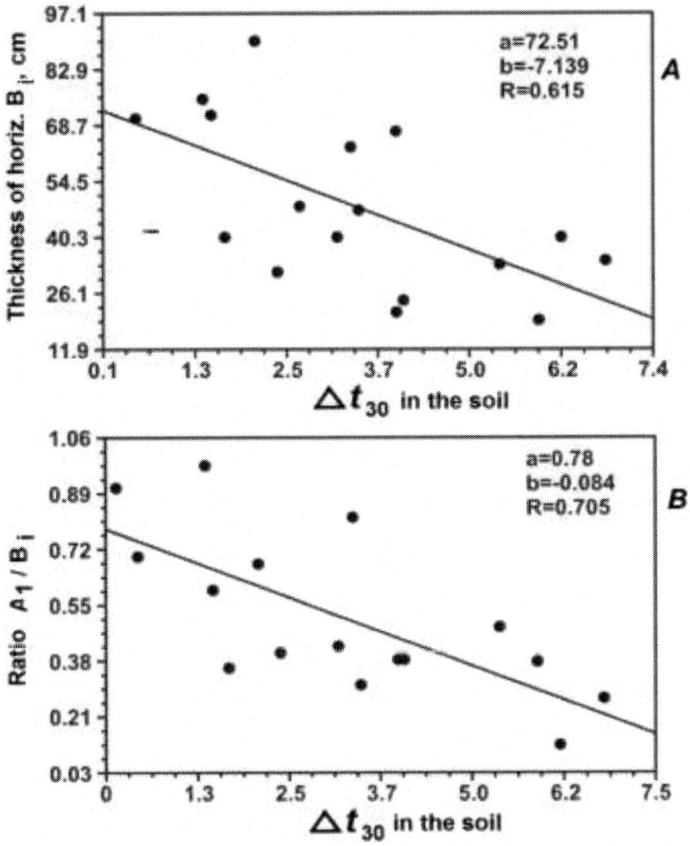

Figure 8.17. The connections between the morphology of soil hor. B and the temperature anomalies in soil at the depth of 30 cm (Δt_{30}). The regression equations: $Y = a + b{\cdot}X$. R is the correlation coefficient.

First of all, substantial changes occur here in the structure of the humus profile including A_I+ A_I(bound.). The graph of correlation connections (see Figure 8.16 *A*) clearly shows the two broken areals, within which the connections per se are almost absent. The buried humus horizons persist at soil temperatures close to the climate normal but are abruptly "dissolved" when the anomalies get over 2–2.5°. It occurs on the piedmont plains of the volcanic landscape (see above Figure 4.3, test plot 21) with relatively "old" soils that have gone the longest way ofgeothermal transformation. As a result, the thickness of the entire humus profile decreases from 50–70 cm and more to 10–25 cm and then almost does not vary, in spite of the increasing influence of geothermal wells. Thus, the geothermal increase in the energy level of soils leads to the abruptly intensified mineralization of buried organic matter.

Simultaneously, mineralization of organic matter is activated also in the modern humus layer, as is demonstrated by the increase in the A_0/A_I ratio (Figure 8.16 *B*). It occurs in spite of a certain decrease in the replenishment of tree waste with the green mass and drain due to decreased $PV_{d\text{–}k}$ and *PG*.

The general acceleration of soil-forming processes in geothermal foci leads to more intensive transformation of the vertical stratification of pyroclastic materials of different age (with numerous horizons, not only A_I but also B_i) into a simpler soil profile with secondary differentiation into genetic horizons. As a result, we can observe a decrease in the thickness of the iluvial horizon (see Figure 8.17 *A*), as well as in the information measure of complexity of the entire soil profile (H_{pp}). This measure has a reverse linear connection with temperature anomalies:

$$H_{pp} = 2.606 - 0.125 \ \Delta t_{30}; \ R = 0.706. \qquad (8.9)$$

According to some data (Duchaufour 1965; Polivanov 1984), the decrease in the $h(B_i)$ and H_{pp} values must indicate a general decrease in soil maturity, in other words, its "rejuvenation." The decrease in the A_1/B_i ratio (see Figure 8.17 *B*) demonstrates that the intrasoil transformation of mineral mass in this process begins to prevail over humification.

The positive soil-temperature anomalies reach their maximum in the zone of birch elfin woodlands at the absolute heights of 500–700 m; here, they cause highly peculiar changes in soil-forming processes influencing the altitudinal-zonal distribution of characteristics of the vertical soil profile (see above Figure 6.6). In contrast to the peak subgoletz zone with cedar prostrate shrubs and the underlying zone of dark-coniferous taiga, the near-pear layer of stone birch forests (see test plot 16 in Figure 6.6) shows noticeable activation of the processes of both mineralization of dead organic matter ($A_0 \rightarrow$ min) and humification (the thickness of hor. A_1 increases, while $A_0/A_1 \rightarrow$ min). Moreover, the export of organomineral compounds to hor. B_i is intensified; as a result, its thickness increases up to 70 cm and more, becoming equal to the illuvial horizons of the most developed soils of piedmont-plain coniferous/broad-leaved forests, where steam-and-gas geothermal wells are less marked. Undoubtedly, the inflow of geothermal heat to the soil of the stone birch elfin woodland zone not only accelerates the small biological cycle but also substantially increases the degree of soil maturity. However, in contrast to the plain mixed forests, in the mountain stone birch forests it is achieved due to obvious prevalence of illuvial processes over humification processes ($A_1/B_i \rightarrow$ min). In the mixed forests, these processes are relatively balanced ($A_1/B_i = 0.69$).

8.9. Some Inferences

The empirically substantiated concept of geothermal energy of passively erupting volcanoes as a factor of accelerated evolution of island phytobiota has been set forth as a generalizing result. The quantitative assessment of the effects of magmatic energy of active volcano on the thermal regimes of soil and near-surface air layer, as well as on the water balance, forest formation and soil processes in an island volcanic landscape, has been performed for the first time. The geothermal factor has a positive effect on the floro- and phytocoenotic structure of island natural complexes and makes a certain contribution to the heat-and-energy basis of their formation and development, which is a phenomenon of the primary subaerial landscape genesis. The conducting channels of magma-based geothermal systems causing abnormal increases in soil temperature have been foundin the degassing region of the Mendeleev Volcano. Geothermal energy has a significant effect on forest ecosystems.

The obvious warming effect of gas-and-steam geotherms is also observed in the overlying belts of birch crooked forest and cedar elfin wood. Spruce forest stands with viable undergrowth as well as Manchurian and Japanese species (Japanese yew, wrinkled holly) are frequent here. The most complex cedar elfin wood/birch/spruce buffer communities are well developed in the upper layer with the maximum soil temperature anomalies (4.0–4.5° on average).

The geothermal heating of edaphotopes on the slopes and near the tops of volcanoes of the Kuril Range is not an exceptional but a mass phenomenon in the hydrothermal regime of forest island-arc ecosystems. Gas-and-steam flows existing in the bodies of even ancient (badly destroyed) eruptive cones and being widespread in the Northwest sub-Pacific had a substantial influence on the thermoedaphic conditions of forest growth throughout the entire Pleistocene Epoch and maintain these conditions at present. This influence promotes the sustained development of near-volcano boreal forest communities accompanied by their enrichment with subboreal species and subtropical relicts and certain reduction of their functional parameters (green mass productivity and biological cycle).

Magmatic geotherms (steam/hydrotherms) were favorable for transgression of more southern species and entire communities to this boreal-forest ecoregion over a long period of time, and these species and communities remain here as intrazonal and relict formations. The "climatically unjustified" but sufficiently stable subboreal forest ecosystems have been formed, as well as the early stages of buffer forest communities.

At the same time, the steam-and-gas flows of heat and chemical solutions change the water balance and soil geochemical processes, reducing the productivity of forest communities and slowing down the biological cycle in general – mainly due to the rapid decrease ingrass cover production.

Geothermal energy was a factor of accelerated evolution of the island phytobiota. It induces rearrangement of the plant cover towards the increase in its floro- and phytocoenotic diversity to the detriment of biomass formation rate. The largest possible (under the given climatic conditions) diversity of floristic and phytocoenological structures creating terrestrial vegetation in the transitional zone between the continent to the ocean originates precisely from island volcanic landscapes.

The Neo-Pacific volcanic island-arc landscape, which are in the range of influenceof underground gas-and-steam hot springs and solfataras, *become a kind of a cradle for buffer forest communities* (on the South Kuril Islands, these are Manchurian–Okhotian and Okhotian–Kamchatian–Beringian). The latter are characterized by further development and more complex spatial differentiation when the territory changes from the insular state of Neo-Pacific to the marginal-continental state of Sub-Pacific, with the generally weaker oceanicity of the climate and the complicated mountain-valley relief.

PART III: EVOLUTIONARY ECOLOGY OF MARGINAL-CONTINENTAL BOREAL LANDSCAPE

Chapter 9

FLORA, VEGETATION AND MICROLANDSCAPES OF EXPERIMENTAL TESTING GROUND

ABSTRACT

The territorial object of this research is the Far-East branch of the sublatitudinal boreal bioclimatic ecotone of Eurasia. Here, the boreal ecotone encompasses the mountain and valley systems of the Lower Priamurye. The large-scale landscape-ecological survey was carried out by the technique that we had developed in the territory of the Komsomolsky State Nature Reserve. The ecology of boreal forest landscape of the continental margin characterizes the mature stage of evolution of continental biosphere. The structural "core" of this part of the boreal ecotone is represented by the subtaiga zone. Here there are four basic floristic phratries (Manchurian mesophilic and xerophilic, Okhotian, and Angaridian) and two buffer phratries (Manchurian–Okhotian and Manchurian–Angaridian). The following groups of forest types were distinguished: (1) broad-leaved; (2) cedar/broad-leaved; (3) spruce/broad-leaved; (4) fir/spruce; (5) larch-tree; (6) forest bogs represented by larch-tree mari. We have also considered the microlandscapes consisting of paragenetic spatial series (catenas) of elementary natural complexes – *landscape facies*. Seven elements (groups of microlandscapes) were distinguished: (1) mountain transeluvial; (2) mountain transit and transaccumulative of shadowy slopes; (3) mountain transit of sunny and neutral slopes; (4) piedmont transaccumulative sunny and neutral; (5) valley eluvial; (6) valley accumulative; (7) valley superaqual.

9.1. INTRODUCTION

The Pacific Ocean mega ecotone of Eurasia is a three-dimensional formation. It was created by three processes of ecotonization of the biosphere: latitudinal-zonal (macro-climatic), longitudinal-sectoral (macro-orographic), and altitudinal-belt (meso-oroclimatic), which is clearly reflected in the formation structure of plant cover (Sochava 1977). Distinct regional and local contrasts between phytocoenological and landscape structures appear on the continental margin under the conditions of high meridian hydrothermal gradients complicated by the mountain-valley relief (Krištofovič 1946). It creates the high mosaicity of natural complexes, as well as the frequent zonal and altitudinal-zonal inversions favoring the formation of *ecotone (buffer) biotic communities* that have become widespread here (Sochava 1977).

Staying for a long time under the conditions of monsoon climate, the plant cover of the temperature belt on the mega ecotone developed continuously. The transition of flora and vegetation from the tertiary to quaternary period was relatively smooth (Krištofovič 1946; Sochava 1946), with migration of individual elements of the tertiary flora from north to south and back and with theintrusion of representatives of cold climate (Nikolyskaya and Khomentovsky 1967).

This specific and long-term evolutionary process resulted in gradual transformation of tertiary thermophilic sub-oceanic formations into more temperate formations adapted to drastic seasonal hydrothermal contrasts.

Due to the above, the mixed cedar/broad-leaved and, to a lesser extent, coniferous forests have a multispecies composition and ecological/biological diversity of phytocoenotic components in all layers, with numerous tertiary relics. They are characterized by the complexity and intricacy of interspecies relationships and the diversity of forest-forming process, which is composed of age-related, secular, and restorative changes of forest communities creating certain genetic rows of plant groups (Kolesnikov 1956).

9.2. Geographical Location of Testing Ground

The Far-East branch of the boreal bioclimatic ecotone of Eurasia (Bazilevich et al. 1986; Kolomyts 1995) embraces the marginal-continental mountain and valley systems of the Lower Priamurye being under specific monsoon-continental climatic conditions (Gartman 1971). The territory of the Lower Priamurye is part of the Mesozoic morpho-texture and "The Systems of Mountains and Intermontane Valleys of the Middle and Lower Amur River" physical-geographical country (Nikolyskaya et al. 1969). V.B. Sochava (2005) includes this territory into the larger Amur-and-Sakhalin physico-geographical area. It is characterized by two longitudinal-sectoral rows of latitudinal zonality: the eastern one, most completely reflecting the influenceof the Pacific monsoon, and the western one with more marked continental features. The mostly mountain relief of this territory predetermined rather high levels of habitat diversity, habitat isolation and often relative antiquity, the accelerated evolutionary rate in bionts (Krištofovič 1949; Panfilov 2005) and their proper species saturation. In 1931, the well-known biogeographer D.N. Kashkarov wrote that "mountains are a smith's shop where novel species are molded by the environment and developed" [cit. by: Milykov, 1981 159)].

The structural "core" of this part of the boreal ecotone is represented by the subtaiga zone (Sochava 1979). Here, the large-scale landscape-ecological survey was carried out by the technique that we had developed in the territory of the Komsomolsky State Nature Reserve (see Figure 1.1) in 2011, 2012 and 2014. The data on 76 qualitative and quantitative characteristics of structural and functional states of forest ecosystems were obtained on 55 plots of the test ground.

The model territory of the Komsomolsky State Nature Reserve characterizes only the low-mountain/plain part of the Priamurye boreal ecotone (with absolute heightsbelow 1000 m, Figure 9.1). The Nature Reserve is situated at the zonal boundary between boreal and subboreal landscapes of the South of the Far East, in the Amur–Primorye physico-geographical country, the Lower Amur region and the province of the same name (Nikonov 1975). In the territory of the Nature Reserve there is one of the large stands of dark-coniferous and coniferous/broad-leaved forests preserved in the Lower Priamurye. The widespread type of soils under forest communities is brownified soils: brown-taiga (including illuvial-humus) soils in the south taiga and subtaiga forests and brown (including brown mountain-forest) soils in subnemoral forests.

Figure 9.1. Lower Priamurye. Territory of Komsomolsky State Nature Reserve. The valley of Gorin river, left tributary of Amur river, and surrounding low mountain.

The following two categories of phytocoenological formations were studied: (1) *classes of plant formations – floristic phratries* (Sochava 1946) as the highest regional typological unit (phytocoenomere) combining the classes of plant formations and corresponding to the rank of two phytocoenochores – the regions with latitudinal zonality and altitudinal zonation (Sochava 1979); and (2) *groups of forest types* (Kolesnikov 1956) as a total of timber species, which is a certain stage of forest-forming process typical for the given conditions of forest growth.

The four basic phratries were distinguished: Manchurian mesophilic (Mm), Manchurian xerophilic (Mx), Okhotian (Okh), and Angaridian (An). The mixing of favorable habitats results in the stability of buffer (transient) phratries(Sochava 1946; Nakamura and Krestov 2005). We have distinguished two buffer phratries on the territory of Priamurye sub-Pacific: Manchurian–Okhotian (MO) and Manchurian–Angaridian (MA). The buffer Okhotian–Angaridian phratrie has not been distinguished as it has no independent forest-typological significance (Vasilyev et al. 1976).

The following groups of forest types were also distinguished: (1) broad-leaved (Br); (2) cedar/broad-leaved (CrBr, see Figure 1.3); (3) spruce/broad-leaved (SBr, see Figure 1.4); (4)

fir/spruce (FS, see Figure 1.5); (5) larch-tree (Lar); and (6) forest bogs (FB) represented by larch-tree mari. The several structure characteristics of forests for different floristic phratries are showed on Table 9.1.

We consider also *microlandscapes*, consisting of paragenetic spatial series (catenas) of elementary natural complexes – *landscape facies (see Chapter 5).* Seven elements (*groups of microlandscapes*) have been distinguished: (1) mountain transeluvial, TEm (mount), sublithomorphic; (2) mountain transit and transaccumulative of shadowy slopes, T^{m}_{shad}–TA (mount/shadow), litho-hydromorphic; (3) mountain transit of sunny and neutral slopes, T^{m}_{sun} (mount/sun-neutral), lithomorphic; (4) piedmont transaccumulative sunny and neutral, TA (piedmont/sun-neutral), sublithomorphic; (5) valley eluvial, Ev (valley), sublithomorphic; (6) valley accumulative, Av (valley), subhydromorphic; (7) valley superaqual, Saq (valley), hydromorphic.

9.3. Floristic Phratries

Being strongly influenced by the ocean, under conditions of year-round or seasonal (summer) high moisture content of the atmosphere, the phytobiota of insular and marginal-continental land is represented by forest vegetation of mainly mesophyte appearance along nearly the entire meridional length of the Pacific Ocean mega ecotone. Previously it was believed that forest cover had remained stable here, with minor and short-term fluctuations (floristic rises and falls), at least since the end of the Cretaceous period, i.e., throughout the past 50–60 million years (Krištofovič 1946). However, later it was revealed that the climate fluctuations that occurred in the Far East in Pleistocene–Holocene were no less considerable than those in the intracontinental regions, causing a significant shift in the zonal types of vegetation both along the meridian and in the spectra of altitudinal zonality (Bobrov 1980; Korotky et al. 1999). In the Lower Priamurye, mesophilic broad-leaved forests with relicts of the Tertiary flora were predominant on the plain in the warm interglacial epochs and replaced by low-bush forest tundra and larch mari in the glacial epochs. The species composition of the forests was depleted after each new cold snap, with gradual disappearance of subtropical relicts. The contemporary plant cover in the Primorye, Lower Priamurye, Kamchatka and Kuril Islands is rather young (the Holocene Epoch) (Bobrov 1980).

Once, V.L. Komarov (1897) distinguished the four botanic-geographical regions for the south of the Far East: Manchurian, East-Siberian, Daurian–Mongolian, and Okhotian, with the respective plant species composition. Following Komarov, botanic florists distinguished three types of forest floras in the territory of Primorye and Priamurye (Sochava 1946, 1958; Kolesnikov 1969 a): Manchurian, Okhotian (Okhotian–Kamchatian), and Angaridian (East-Siberian). Each of them forms its own "zonal complex of forest formations" (Kolesnikov, 1956 140) including both aboriginal and derived forest communities of a given florogenetic type. This highest regional typological unit (phytocoenomer) combines the classes of plant formations and is referred to as *phratrie*, which is ranked as two phytocoenochores – the regions with latitudinal zonality and altitudinal zonality (Sochava 1979). The forest phratries of the South of the Fat East are integrated from the two lower classification units: (1) the geomorphological complex of forest types (e.g., "mountain" and "valley" cedar forests) and

(2) the climatic complex of forest types ("typical," "northern," and "southern–hornbeam" cedar forests). The latter are composed by forest type groups (Kolesnikov 1956).

The most prominent representatives of sub-Pacific forest cover in Priamurye and Primorye (both in the mountains and on the plains) are forest formations of the *Manchurian phratrie*; among them there is a highly diverse group of formations of *cedar/broad-leaved forests* that has been formed historically over many millennia and generally possessing the ecological and physiological characteristics from sub-oceanic to sub-continental (Sochava 1946; Kolesnikov 1956, 1969 a; Tsymek et al. 1969; Akhmetyeva 1977). Following B.P. Kolesnikov, for the sake of brevity, hereafter we will refer to this group of forest formations as cedar forests. The biogeographical features peculiar to these formations are usually called Manchurian, and they more or less represent the Pacific Ocean influence on forest cover of the region (Sochava 1946, 1980). This phratrie was also proposed to be referred to as *Priamurskaya* – as consistent with its actual distribution (Kurentsova 1973), but this term did not come to stay.

The long-term co-habitation of coniferous and deciduous species created a typical plurodominance of the tree layer of Manchurian mixed forest; V.L. Komarov supposed that coniferous species were the initial basis of this class of forest formations [cit. by (Sochava 1946)]. Due to the recent expansion of Korean cedar, this complex is considered to be relatively young, though its nemoral elements are very ancient and throw back to the Tertiary pra-Manchurian/north-Japanese phratrie (Sochava 1946). However, Korean cedar per se is not a heritage of the Tertiary Period. Its dissemination over the south of the Far East began only in the ancient Holocene and reached the Lower Priamurye in the Holocene Climatic Optimum, i.e., 5–4.9 thousand years ago (Neyshtadt 1952, 1957). Therefore, the contemporary complex of Manchurian mixed forest is rather young. A proof of this fact can be the negative coenotic relationship between Korean cedar and the components of broad-leaved forests (Udra 1975). However, similar to the more thermophilic plants, cedar could be preserved in the epochs of cooling in the refugiums distant from its contemporary habitats, in the periphery of the East-Manchurian Mountains and on the Sea-of-Japan slope of Sikhote-Alin, and then spread from these places (Urusov1988).

The latest wave of formation of contemporary Manchurian forests in the south of the Far East seemed to occur in the Holocene Climatic Optimum, 7.5–4.9 thousand years ago (Korotky et al. 1999), which coincided not only with the sea level rise up to 45–50 m since the beginning of Holocene, but also with the activation of warm sea currents. It resulted in extensive transgression of nemoral elements into boreal forests, with destruction of previously stable ecological niches in the latter and development of introgressive hybridization (hybrid mixing) of the dominants (Bobrov 1980). Eventually, both transitional (buffer) communities and aboriginal phytocoenoses of contemporary Manchurian mixed forest were formed. All of them are characterized by the typical "assemblage" of the flora, where the boreal elements proved to be more ancient than the nemoral ones and were preserved due to their "biological resistance" (Bobrov 1980; Urusov andLauve 1980).

The broad-leaved forests of the Manchurian phratrie are the most direct heritage of the Turgay flora (Krištofovič 1958). Their florogenetic basis is the thermophilic relicts of mixed post-Tertiary (beginning from Oligocene–Miocene) forests of the Turgay type (Sochava 1946; Vasilyev 1958): Manchurian walnut (*Juglans mandshurica*), Amur, Taque and Manchurian lime (*Tilia amurensis* Rupr., *T. taquetii* C., Sneid, *T. mandshurica* Rupr.), Amur cork tree (*Phellodendron amurense* Rupr.), Mongolian oak (*Quercus mongolica* Fisch.),

Manchurian ash (*Fraxinus mandshurica* Rupr.), white-bark and mountain elm (*Ulmus propinqua* Koidz., *U. laciniata* Mayr.), maples, Manchurian aralia (*Aralia mandshurica* Rupr. et Maxim.), Chinese magnolia vine (*Schizandra chinensis* (Turcz.) Baill.), spiny eleutherococcus (*Eleutherococcus senticosus* Max.), and Amur vine (*Vitis amurensis* Rupr). Among coniferous species, there are Korean cedar (*Pinus koraiensis* Sieb. et Zucc.) in the Amur River basin and needle (Manchurian) fir (*Abies holophylla* Max.) in the south of Primorye. On the Sakhalin and Kuril Islands, the disappearance of cedar is followed by the development of Ajan spruce/broad-leaved forests represented by the Ajan spruce (*Picea ajanensis* Fisch.), Sakhalin and Mayra fir (*Abies sachalinensis* Mast., *A. mayriana* Miyabe et Kudo). All the above species can be considered as the most ancient "core" of contemporary mixed forests in the South of the Far East (Kurentsova 1973).

It will be expedient to divide the Manchurian floristic phratrie into the two major types conventionally named as "mesophilic" (Mm) and "xerophilic" (Mx) (Kolesnikov 1969 a). For the sake of simplicity, we will refer to each type as a separate phratrie. Mesophilic cedar/broad-leaved and broad-leaved forests are characterized by a predominance of the relict species listed above, rather insignificant occurrence of Korean cedar (no more than 30–50% of total forest stand, according to our data) and more significant occurrence of Mongolian oak (a maximum of 40–60%). The forests grow mainly on gentle mountain slopes and in river valleys with nutrient-rich and constantly moist (but not bogged) soils, i.e., under conditions of flow humidification (Krestov 1993). They are characterized by the highest diversity of forest stand and the vigorous development of undergrowth and soil cover, which often smoothes over the vertical structural differentiation of forest communities. These forests are also characterized by the highest productivity for the Far East: more than 10 t/ha · yr (Sochava 1977). The communities with elm, ash tree and maple signify further nemoralization of the forest cover of the region and therein are the youngest (Vasilyev 1958).

The western boundary of distribution of this type of the Manchurian phratrie goes along the Lesser Khingan Range and approximately coincides with the bounds of the most marked effect of summer oceanic monsoon (Sochava 2005). From south to north, the Manchurian flora gradually drops out – beginning from 50° N (to the south from the Komsomolsk-on-Amur – Blagoveshchensk line), i.e., in the northern part of the Middle-Amur Plain and along the northern boundary of the Middle Sikhote-Alin Range.

The xerophilic type of the Manchurian phratrie differs from the mesophilic type in a higher abundance of cedar (up to 65% of total forest stand), as well as Mongolian oak (up to 60–85%). The latter is a "continental-xerophyte" species and its usual habitats are mainly the SE–S–SW rocky mountain slopes with thin soils. Mongolian oak is one of the most drought- and frost-resistant species, which sustains frosts down to –60°(Tsymek et al. 1969). Its ecological-physiological group began to form no later than in the Pliocene as a result of xerophytization of the climate, with strengthening of the Asian anticyclone (Vasilyev 1958), and became a dominant in cedar/broad-leaved forests in the dry Pleistocene and Holocene Epochs (Neyshtadt 1957). At present, oak forests are often after-fire derivatives in the place of mixed and even dark-coniferous forests (Kurentsova 1973). The western border of the area of distribution of mixed forests with oak goes along the upper reach of the Amur River and then along the Greater Khingan Range (Sochava 2005) being the western boundary of the Pacific Ocean monsoon climatic area (Alisov 1956). Thus, the xerophilic type of the Manchurian flora in the regional aspect characterizes the western, most continental sub-Pacific sector in the Amur River basin.

In the Lower Priamurye,the Mongolian oak is somewhat involved (up to 10%) in the formation of forest communities of the Okhotian floristic phratrie and still more involved (up to 30–50%) in forests of the buffer Manchurian–Okhotian complex (see below).

The second predominant flora, which reflects the Pacific Ocean's influence on the boreal forest cover of the South of the Far East, is of taiga *Okhotian* (OKh) type, with a predominance of the Ajan spruce and whitebark fir (*Abies nephrolepis* Maxim.), with the involvement of stone birch (*Betula ermanii* Cham.) and, to a much lesser extent, the tertiary endemic species. Though, the phylocoenogenetic relationship between dark-coniferous and Manchurian mixed forests proves that the Ajan fir/spruce forests belong to the Manchurian rather than Okhotian phratrie, the more so that both phratries share a common origin in Turgay forests (Vasuilyev 1944 a; Sochava 1944). The "Okhotian" taiga is even considered as a depleted derivative of Manchurian mixed forest (Sochava 1946). However, the Turgay flora per se was not homogeneous over such a vast territory as the South of the Far East (Krištofovič1946). The species that often occur in the dark-coniferous forests of Priamurye are yellow maple (*Acer ukurunduense* Trautv. et Mey.), Actinidia kolomicta (*Actinidia kolomicta* Maxim.), spiny eleutherococcus and other species, which are relatively cold-resistant but fastidious (like the Ajan spruce) in regard to enhanced air humidity. These species in the geological past were coupled not with spruce forests but with thermophilic Turgay deciduous forests (Kurentsova 1973).

V.B. Sochava (2005) distinguishes three concentric borders of (a) continuous, (b) considerable (in the mountains only) and (c) sporadic distribution of the Ajan dark-coniferous forests, the core of which is situated in the Lower Amur, on the Sakhalin and Hokkaido Islands. The members of the Okhotian flora are admixed with cedar or substitute for it and create mixed forests in cold intermontane valleys and on mountain slopes above 400–500 m. At the same time, many thermophilic broad-leaved species drop out, while the role of ferns in soil cover increases. These communities are structurally similar to the neo-Pacific Ajan spruce/broad-leaved forests of the Sakhalin and Kuril Islands (Kolesnikov1969a).

The large taiga territories of Priamurye sub-Pacific (on shady mountain slopes and valley pediments under conditions of climatic inversion) are covered with forests of the *Angaridian taiga phratrie* (An) succeeding from the psychrophytic continental flora of the most ancient land massif, Angarida, situated in place of contemporary East Siberia (Sochava 1944 b; Yurtsev 1974). It is typically represented by *larch forests*. Their ancestors were part of Turgay forests; the contemporary species originated no earlier than the Pleistocene due to cold snap and habitat isolation (Kurentsova 1973). There are three groups of larch forests (Kolesnikov 1956; Vasilyev 1958): (1) rhododendron, cowberry-ledum and herbaceous in the mountains on rocky ledges; (2) fern-hazel, with the involvement of cedar, spruce and oak, on the hills (uplands, according to our terminology, see below); and (3) sedge-small reed and sphagnum-heather on waterlogged low-lying plains and in river valleys. The larch mari widespread in the Northern Priamurye are of postglacial origin (Kolesnikov 1956). They are distributed up to the Zeisk-Bureinsk Plain, which is considered to be the western outpost of Pacific influence with regard to this feature (Sochava 1980).

Finally, we should note the presence of representatives of the *Beringian flora* as *cedar and alder elfin woods*, forming peculiar creeping forests in the subgoltsy belt of the mountains, on the downwind northern slopes of Sikhote-Alin and the Bureinsk Range system (Kolesnikov 1969 b). Emergence of the groups of cedar elfin wood (*Pinus pumila* Rgl.) and stone-birch crooked forests among fir-spruce forest stands in the mountains indicates the lower boundary of high-mountain vegetation.

Thus, the forests of the Lower-Amur sub-Pacific are characterized by substantial floristic richness: primarily in the species compositions of timber stand and undergrowth. The mixed nature of these forests is a basis of their structural peculiarities (Sochava 1946). The natural phytocoenotic mosaic of forest cover in this continental margin is highly dynamic (on not only a geological but also a historical timescale), and it concerns first of all the northern formations of cedar/broad-leaved forests situated exactly in the Lower Priamurye. The component forest-forming species (*Pinus korajensis*, *Quercus mongolica*, *Picea jezoensis*) belong to the type of tertiaryrelicts with progressive distribution due to alteration of their ecological properties (Sochava 1946). As is noted by researchers, in the postglacial period and especially in the past few centuries cedar trees and forests in the subtaiga zone were pushed to the south and down along the mountain slopes by the taiga coniferous species of the Okhotian flora and the fir/spruce forests formed by these species, with century-long transition stages: from cedar forests with fir and spruce to fir/spruce forests with and without cedar (Ivashkevich 1933; Vasilyev 1939; etc).

In the territory of the Komsomolsk State Nature Reserve, the members of the Ajan flora: the Ajan spruce (*Picea ajanensis* Fisch.), whitebark fir (*Abies nephrolepis* Maxim.) and stone birch (*Betula ermanii* Cham.), are present in nearly all florogenetic formations (phratries), even in the Angaridian one. This share is comparatively low (no more than 15–35% of timber stand) in the forests of Manchurian phratries and increases (up to 60%) in fir-spruce forests. However, spruce and fir reach their maximum relative abundance (up to 80–90%) in forests of the buffer Manchurian–Angaridian formation. At the same time, the share of larch (*Larix amurensis* Kolesn. ex Dul.) as a representative of Angaridian flora is no more than 30–55%. The apparent predominance of dark-coniferous timber stands is indicative of suboceanic features of this buffer formation on the Priamurye boreal ecotone.

The replacement of cedar/broad-leaved forests by fir/spruce forests is associated with the increasing habitat humidity, which, in turn, is accounted for both by global climate changes and (much more convincingly) by endoecogenetic successions: the local transformation of phytoclimate and soil hydrothermal regime during the centuries-long helical-cyclic stage-age successions of forest communities, eventually resulting in the substitution of one forest type for another (Kolesnikov 1956). As cedar dies in the age of 200–240 years, the entire step-by-step process of such successions extends to more than 1200–1400 years.

In other cases, cedar/broad-leaved forests acquire an admixture of the Dahurian larch (*Larix daurica* Turcz.) or the Cajander larch (*L. cajanderi* Mayr), with white birch (*Betula platyphilla* Suk) and aspen – representatives of permafrost-taiga Siberian flora, which is indicative of the enhanced continentality and severity of these habitats approaching the bioclimatic conditions of the neighboring Paleo-Pacific (the Upper Priamurye).

There is also a process of mixing of the Okhotian and Angaridian floras. However, larch communities usually form a series of reconstructive stages of indigenous fir/spruce forests, and these stages are not longer than the age of one generation of larch (Vasilyev 1946; Vasilyev et al. 1976). Insomehabitats of the lower mountain belt, larch is not always completely replaced by spruce. The emerging transitional Okhotian–Angaridian associations occupy the gentle parts of slopes adjoining river flood plains. We conventionally classify them as larch forests with a share of spruce in the timber stand up to 30% (Vasilyev et al. 1976).

In light of the above, it is obvious thatextensive migration and regional differentiation (fragmentation) of the tertiary pra-Manchurian/North-Japanese formation (Sochava 1946) resulting in the isolation (segregation) of the above basic floras were also accompanied by their *integration* – certain mixing in favorable habitats, with the formation of stable buffer (transitional) phratries. In the territory of Priamurye sub-Pacific, we have distinguished *two buffer phratries – Manchurian–Okhotian* (MO) and *Manchurian–Angaridian* (MA). Buffer formations are very typical of forest cover of the Amur River basin (Sochava, 1980). In light of the above, the mixed Okhotian–Angaridian phratrie is not distinguished.

Table 9.1. The taxonomic (weighted average) base norms for the structural characteristics of different floristic phratries

Parameters		Floristic phratries					
		Mm	Mx	MO	Okh	MA	An
Average height of forest stand, m		18.6	16.9	22.9	20.5	13.6	11.7
Average stem diameter, cm		22	20	21	19	8	13
Average height of undergrowth, m		1.3	1.9	3.1	1.5	1.4	1.6
Average height of understory, m		1.8	0.9	1.3	0.9	0.8	1.0
Crown density, %		59	51	44	54	85	31
Quality of locality – bonitet *)		III (V)	IV (II. III)	I, II	III (II. IV)	I, IV	V (I)
Measure of floristic diversity, in (bits)	stand	2.30	4.01	3.95	2.94	3.95	3.04
	undegrowth	2.24	1.97	2.12	2.04	1.87	0.82
	herbage	3.29	1.71	2.29	1.97	1.40	0.75

*) The regions of dominance are presented; the enclaves on the graphs of binary ordination are in brackets.

The forest communities of both basic and buffer phratries are rather clearly differentiated by their major structural characteristics, including the generalized structural quality expressed by quality classes (Table 9.1). The buffer Manchurian–Okhotian phratrie is characterized by forests of the highest quality (class I–II). The Okh and Mm phratries have forests of quality class IV. Class V is typical solely of forest phytocoenoses of the Angaridian floristic phratrie, though the forests of the An and MA floras may also be of the highest class.

9.4. Forest Type Groups

Flora, as a genetically separate complex of species, is inseparable from phytocoenoses made up of its components (Sochava 1946). Each test plot of 20×20 m in size characterized a particular *forest phytocoenosis* as a stable system of elements of the phytobiota, which coexist on a certain area of land, and the phytocoenotic environment created by these elements and their precursors (Bykov 1973). Phytocoenosis is characterized by specific composition and structure, a particular biological cycle and productivity. It is the lowest-level topological category of the phytosphere, which is determined as a plot, or allotment, of plant association (Valter, 1968; Sochava 1979). Forest phytocoenoses similar in species composition and life forms are grouped into *associations* (Rabotnov 1978; Reymers 1990), which can be considered, according to G.F. Morozov (1949), typologically as *types of plantation* and spatially as microphytocoenochores (Sochava 1979). It is also an elementary but, to a certain extent, generalized unit of plant cover. The type of plantation corresponds to the concept of the type of forest biogeocoenosis (Sukachev 1972), which was the basis for including this operation unit as the lowest-level phytocoenotic category into our geo(eco)system analysis. The following seven groups of the types of plantation were distinguished for the territory of the Komsomolsky State Nature Reserve: broad-leaved, cedar/broad-leaved, spruce/broad-leaved, spruce/fir, larch and small-leaved forests; larch sphagnum mari form a special group of types of plantation.

The composition and structure of leaf canopy and other layers, which are used for differentiation between the types of plantation, are rather dynamic and change especially dramatically during short-term restorative (century-long) successions (Kolesnikov 1956); therefore, together with the types of aboriginal and conventionally aboriginal (relatively stable) associations, we have distinguished a group of types of small-leaf forests combining various transient restorative stages of vegetation.

The phytocoenological category of the next rank – *forest type* – was used for the deeper landscape ecological survey of plant cover of the Nature Reserve and especially for the development of its predictive climate scenarios. In the hierarchic series of phytocoenomers, forest type is adequate for a group and/or class of associations (Sochava 1979). Each forest type includes both its aboriginal phytocoenotic structure and its different variable states – the stages of short-term restorative successions, all of them varying during the evolution of forest cover. In this respect, forest type is an invariant concept drawing closer to the concept of epi-association (Sochava 1979). It is important to emphasize that each forest type conforms to its own type of habitat, which is determined by terrain and the terrain-related complex of factors of plant life.

As is defined by B.P. Kolesnikov (1956), *forest type* is an ensemble of tree species, which is a qualitative element in the genetic series of forest vegetation development in some landscape-geographical area or another, a stage of particular forest-forming process typical of the given forest growth conditions. Forest type combines the plots with different stages of age-related and short-term restorative successions, i.e., aboriginal and derivative communities. It is characterized by the commonality of the principal and admixed tree species, as well as the types of undergrowth and soil cover. Finally, each forest type is characterized by its own productivity expressed in terms of quality classes.

In forest typology, forest types are usually differentiated by the composition of herbage or moss cover. According to the concepts of G.F. Morozov and A. Cayander (Sukachev 1972), it is exactly soil cover that characterizes soil-ground conditions and biologically equivalent habitats as a basis for classification of plantations. However, the species composition of undergrowth is a more important indicator of forest type for the Far East forests. Herbage has much lesser diversity with a predominance of boreal species (Sochava 1946) and is furthermore characterized by high variability of composition at different stages of forest community development (Kolesnikov 1956). Due to the known phenomenon of interchangeability of ecological factors (Dokuchaev 1949), each forest type can be represented by different interchanging variants (climatic, altitudinal, exposure, inversion, edaphic) arranged in certain ecological replacement series (Kolesnikov 1956).

Six forest type groups were distinguished in the territory of the Komsomolsky State Nature Reserve:

(1) broad-leaved (**Bl**); (2) cedar/broad-leaved (**CB**); (3) fir/spruce/broad-leaved (**SB**); (4) spruce/fir (**SF**); 5) larch (**Lar**); (6) forest marsh (**FM**) represented by larch mari.

The main forest-forming species of the first and second forest type groups belong to the Manchurian flora; those of the forth group belong to the Okhotian flora; and those of the fifth and sixth groups belong to the Manchurian–Angaridian and Angaridian, respectively. The first three groups of forest types are represented in the ecological-phytocoenotical scheme of V.B. Sochava (1946) concerning the formations of Manchurian mixed forest.

The Lower Priamurye is an area of distribution of south-taiga coniferous (psychromesophilic) forests, as well as the northern climatic complex of cedar/broad-leaved forests, with the impoverished composition of nemoral species and considerable content of taiga species: the Ajan spruce and Khingam (whitebark) fir. Here, the concomitant species of cedar are as follows: the Mongolian oak (*Quercus mongolica* Fisch.), small leaf maple (*Acer mono* Maxim.), the Amur cork tree (*Phellodendron amurense* Rupr.), the Manchurian ash (*Fraxinus mandshurica* Rupr.), the Taque-type ash, whitebark and mountain elms (*Ulmus propinqua* Koidz.; *U. lociniata* Mayr.); the shrubs are the Manchurian hazelnut (*Corylus mandshurica* Max.), spiny eleutherococcus (*Eleutherococcus* senticosus Max.), mock orange (*Philadelphus tenuifollus* Rupr.); the lianas are Chinese magnolia vine (*Schizandra chinensis* (Turcz.), the Amur vine (*Vitis amurensis* Rupr), and Actinidia kolomikta (*Actinidia kolomicta* Max.). Ferns are typical for grass canopy.

The region of the Komsomolsky State Nature Reserve is situated close to the northern boundary of the natural habitat of Korean cedar. This is an "area of diffused state of cedar forests" (Kolesnikov 1956), which have no continuous distribution but occur as separate (insular) stands in the ocean of coniferous taiga forests, which especially accentuates their relict nature. Here, cedar forests are represented mainly by mock-orange/Actinidia associations, with yellow birch (*Betula schmidta* Rgl.), lime tree (*Tilia mandshurica* Rupr.) and ferns. They are confined to the well-drained and warm habitats most favorable for cedar, on steep rocky slopes up to the absolute heights of 200–250 m (occasionally up to 400 m, mountain cedar forests) and on the high terraces of broad river valleys (valley cedar forests confined to their lower altitudinal boundary). Between separate cedar stands there are widespread long- and stable-derivative deciduous forests with a predominance of oak and linden, which appeared as a result of frequent fires and economic activity. There are almost no cedar forests on the Sea-of-Japan coast continuously exposed to maritime fogs. "Siberian continental ecotones" (Sochava 1980) with strong winter cooling and permafrost-taiga soils are formed in the intermontane valleys with a marked climatic inversion; here, the forests with Manchurian flora disappear.

The relationships between forest types and types of plantations have been revealed using the information measures of their binary relations (see Chapter 2). There are two pleiads of forests types: (1) Bl + CB and (2) FS + SB, within which the same type of plantations may be part of both the former and latter element of this pleiad. In addition, the ecological dominant of the type of plantations for Bl and CB forests corresponds to the forest type of the same name. Moreover, almost pure broad-leaved plantations are often included in CB forest types, if they have a considerable undergrowth of Korean cedar. On the contrary, overmature cedar forests with a thick undergrowth of broad-leaved species are included in the broad-leaved forest type per se. All the above is determined, first, by the reciprocal florogenetic similarity of these forest communities and, second, by the age and century cycles of dynamic successions of these representatives of Manchurian flora (Kolesnikov 1956).

In the second pleiad, the mixed SB plantations proved to be principal for discrimination of not only this buffer association but also of the basal type of fir/spruce forests. Thus, the Ajan fir/spruce forest type (Okhotian flora) includes dark-coniferous plantations with a considerable admixture of broad-leaved species. Boreal species are predominant in the soil cover of these forests, but with substantial involvement of nemoral species.

In the larch forest type, in addition to the Daurian and Cajander larch, there is a considerable share of fir/spruce communities with larch.

9.5. FOREST FORMATION PROCESSES

When thoroughly describing these processes in the south of the Far East, B.P. Kolesnikov characterized the forests of this vast ecoregion "... as the forests of mountain valley landscape and monsoon climate, ... highly variable in space and extremely dynamic, i.e., as exclusively multiform, with a versatile appearance" (1956 136). It is exactly *multiformicity* and *dynamicity* that make the forest formations of boreal and nemoral Eurasian sub-Pacific different from their intracontinental plain and mountain analogs on the Russian Plain and in West Siberia, in the Ural and Altai Mountains, in the Sayan Mountains and Transbaikalia, in the Carpathian Mountains and in the Alps. These analogs are characterized by a relatively poor and ecologically homogenous set of phytocoenotic components and rather structurally uniform forest types.

In contrast to Western and Eastern Europe, as well as West Siberia, there was no continuous glacial cover throughout the entire Quaternary Period in the South of the Far East, and it had absolutely no effect on the Priamurye plains (Nilolyskaya 1969); therefore, the plant cover here had no developmental hiatus. Beginning from Paleogene/Neogene, the Amur River valley played the role of a vast refugium of forest, steppe and floodplain meadow plants of highly different ecology; it also proved to be the arena of their migration pathways (Titov 1952). According to the spore-pollen data of coal sediments in the Middle Bikinskaya depression (Khudyakov 1972), the mixed coniferous/broad-leaved forests grew as early as in the Miocene period (about 25 million years ago), with the involvement of both subtropical species and representatives of moderately thermophilic (subboreal) and even moderately psychrophilic (boreal) forms.

Thus, the highly contrasting floras in Primorye and Priamurye were mixed long before the Quaternary glaciations, undoubtedly as a result of influence of the already existing monsoon climate. Previously it was believed (Korotky 1972; Akhmetyeva 1977) that the climate in Primorye and Priamurye always remained humid, which must have facilitated the preservation of floristic richness of forests. However, the later studies (Korotky et al. 1999; Korotky 2008) showed substantial variations not only in temperature but also in annual precipitation of this region in Pleistocene and Holocene, which could play a progressive role in forest formation. The varying moisture conditions extremely stimulate morphogenesis (Sochava, 1946), which should contribute to the increase in phytocoenological diversity of forests.

The plant cover of Priamurye and Primorye continuously developed while staying under the conditions of monsoon climate for a long time. At a scale of the entire south of the Far East, the transition of flora and vegetation from the Tertiary to Quaternary Period was relatively smooth though wave-like (Krištofovič1946; Sochava 1946), with migration of single elements of the Tertiary flora from north to south and back and with intrusion of cold climate species (Kolesnikov 1969 a). It resulted in the "assemblage" of flora typical of the Manchurian floristic region (Urusov 1988).

The formation of a relief-substrate basis under conditions of frequent and dramatic climate changes in the Quaternary Period resulted in the enhanced plasticity of landscapes, while the common trend of continentalization determined still more and more increasing landscape-forming role of climatic processes (Korotky et al. 2011). There was "... a successive formation of plant cover as surf waves from north and northeast, with sedimentation ... of the products of floristic input" (Krištofovič1946 64). Therefore, although the Manchurian coniferous/broad-leaved forests are rather young, they have been found to be depositories of many ancient elements of the flora (*Alnus japonica*, *Ulmus laciniata*, *Quercus mongolica*, etc.), the generic groups of which ascend to the Cretaceous period (Vasilyev 1958).

As is known (Tolmachev 1974; Urusov and Lauve 1980), plant migration in geological history was especially efficient during changes in the land to ocean ratio, with the appearance and disappearance of bridges as a result of tectonic and glacioeustatic processes. The continental margin of Northern Eurasia proved to be the intersection of migration pathways (Yurtsev 1974). Such specific and long-term forest-forming processes resulted in gradual metamorphosis of tertiary thermophilic suboceanic formations into more temperate formations adapted to the marked seasonal hydrothermal contrasts. Even representatives of the contemporary permafrost-taiga flora, "...larch forests during their evolution combined tertiary species and their descendants different in genetic relations and centers of origin" (Kurentsova 1973 173). According to the figurative expression of A.N. Krištofovič, the richness of flora in the South of the Far East "... is merely a result of its saturation to the climatic limit" (cited by: Kolesnikov 1969 218).

The most important factor of these forest-forming processes was the latest tectonic uplifts in the area of the Sikhote-Alin, Badzhalsky and Bureinsky Ridges, with the emergence of altitudinal bioclimatic zonality and numerous trenches – the foci of formation of continental air masses (Sochava, 1950).

In the Quaternary Period, the dominant process was steady subsidence of trenches in the Far East seas, with the involvement of continental margins (Khudyakov 1977). Against this background, the higher-frequency glacioeustatic fluctuations of sea level in the Late Pleistocene–Holocene created the transitions (extremely constrained by mountains) from the subtropical belt via taiga to the subalpine altitudinal zone (Osipov et al. 2008).

The known metachronicity of over-century temperature and precipitation fluctuations in the Pleistocene (Markov et al. 1965) seemed to be one of the causes of transgression of the members of cold humid Okhotian flora (first of all, the Ajan spruce), which composed the coniferous taiga zone as early as in the Miocene, into cedar/broad-leaved forests (Krištofovič 1946). One of the latest transgressions could be such transgression, e.g., in the phase of cedar/spruce/broad-leaved forests at the interface of the Atlantic and subboreal periods of the Holocene (4.7–4.1 thousand years ago), during a cold snap at a still high moisture content (Korotky et al. 1999). This forest-forming process, as a result of Pacific Ocean influence, continues in the Priamurye sub-Pacific at present. Transgression of the members of Angaridian continental flora (in particular, the Dahurian and Cajander larch) proved to be a more localized event associated with climatic inversions under conditions of mountain and valley relief. Nevertheless, the larch area extended during century-long successions, with its intrusion into mixed forests (Kurentsova 1973).

The result of the described forest-forming processes was the known floristic richness and coenotic diversity of forests in the south of the Far East. First of all, the mixed cedar/broad-leaved and, to a lesser extent, coniferous forests here are characterized by the "assemblage of flora" (Urusov and Lauve, 1980), i.e., the multispecies composition and ecological-biological diversity of phytocoenotic components at all layers. It is associated with the complexity of inter-species relationships and the diversity of forest-forming processes including the age-related, century-long and restorative successions of forest communities. These relationships result in: the vertical density (monolithic character) of plant layers, the arrangement of individuals into groups, the presence of many generations of stands in a forest community, the periodical pattern of tree growth in general and any generation in particular, the fierce competition between the species and, as a consequence, the general decrease in the quality class and productivity of forests, which does not correspond to the potentialities of habitats (Kolesnikov 1956).

On the other hand, due to multispecies composition, the age-specific successions of forest stands occur in different habitats by the common helical scheme, with no less than 2–2.5-century cycles (Kolesnikov 1956). The helical (unclosed) pattern of these cycles is manifested in long-term restorative and even stably-irreversible successions and causes a general tendency to replacement of cedar in the stand structure by deciduous species in Primorye and coniferous species (mainly the Ajan spruce) in the Middle and Lower Priamurye. The emerging long-term restorative and even stable naturally-irreversible forest communities consist of the members of other forest floras. For Priamurye, it is the replacement of Manchurian flora by either Okhotian or Angaridian flora. The nonuniform age structure is a characteristic feature of virgin broad-leaved/cedar forests (Nakamura and Krestov 2005).

The dramatic spatial variability of forests is determined by the change of habitat conditions, i.e., meso- and microscale geomorphological factors and territorial contrasts of oro-climatic interactions, while their multispecies composition and structural diversity result from superposition (interpenetration) of age, century-long and restorative (demutational) successions of forest communities. At the same time, age successions cause gradual qualitative transformations of habitat conditions from generation to generation, primarily the accumulation of organic matter, the changes in the hydrothermal parameters of edaphotopes and soil trophicity. Such environment-forming activities of some components of the phytobiota (Isakov et al. 1986) create preconditions for irreversible century-long alternations of basic forest-forming species. Each of these alternations signifies the natural completion of a given climax, with mineral resource depletion in the edaphotope and subsequent succession of the phytocoenosis aimed at restoration of soil resources favorable for the functioning of a new climax (Kerzhentsev 2006).

On the Far East boreal ecotone, with its floristic richness of mixed forests and with their complex multilayer structure of different ages, this process has become especially widespread, a vivid example of which is the above-mentioned replacement of Korean cedar by the Ajan spruce, with the development of cedar-spruce and spruce/broad-leaved communities. The multiyear dynamics of cedar/broad-leaved forests, according to the scheme of Ivashkevich–Kolesnikov, is described in detail in the work (Nakamura and Krestov 2005).

In other cases, especially after fires, the process of age-related and century-long alterations of cedar/broad-leaved forests is accompanied by the replacement of cedar by small-leaved species (yellow and Amur birch and aspen), with broad-leaved species (oak, ash, linden) moved towards the middle ground and with mesophilic Manchurian plant species replaced by psychromesophilic taiga species (Kolesnikov 1956). Such pyrogenic alterations of forest cover can be classified as *explerents* (Mirkin and Rozenberg 1983). They are typical of many areas of the Komsomolsk State Nature Reserve, especially in its northeastern, flattest region.

Thus, the century-long phytocoenotic successions, eventually resulting in substitution of a forest formation for another one, are composed of a number of stages and, therefore, can be presented as a certain *genetic series of forest groups* (Kolesnikov 1956). Consequently, the forest type groups distinguished for the ecoregion of the Komsomolsk State Nature Reserve reflect not only the zonal-geographical, altitudinal-zonal and local spatial differentiation of plant cover in this territory but also certain stages of endoecogenetic successions (Sukachev 1972) in the framework of particular oro-climatic conditions. This is the undoubted complexity of factorial-dynamic ordination of Priamurye sub-Pacific forests, since it is necessary to differentiate between the phytocoenological structures determined by external (climatic, geomorphological and lithoedaphic) factors, on the one hand, and by the processes of autochthonous development of forest communities, on the other hand; the latter can significantly "blur" the role of some external factor or another, reducing the results of gradient factor analysis of forest type groups and forest biogeocoenoses in general.

9.6. CATENAS AND MICROLANDSCAPES

In mountain countries, all natural-territorial structures are somehow oro-climatic (Kotlyakov and Komarova 2007). The superposition (overlap) of two initial factors of mountain landscape formation (erosion-tectonic relief and zonal-regional climate) creates a particular *oro-climatic system.* The latter is interpreted as a spatially ordered interacting set of different-order abiotic factors creating certain boundary conditions for the formation of a soil-biotic "core" of natural complexes of the respective taxonomic rank. The oro-climatic system comprises the two basic hierarchical levels of the geosphere: regional and local (topological). As applied to mountain countries, the regional-level system includes three major elements: altitudinal-layer (geomorphological), longitudinal-sectoral, and circulation-exposure (climatic). The local oro-climatic system of mountains consists of the following elements: altitudinal-zonal (intra-layer, bioclimatic), solar-exposure, and lithoedaphic.

The oro-climatic system of mountain country, even by its definition and landscape appearance (Kotlyakov and Komarova 2007), should be referred primarily to the zonal (isopotential) type of geographical environment, as is manifested in its altitudinal-zonal bioclimatic structure. In addition, the very system of geomorphological stratification of mountain territories has pronounced climatic altitudinal-zonal features (Isachenko 1963) suggesting the combination of bioclimatic zonality and vertical geomorphological layers in the mountains (Gvozdetsky 1963).

The large-scale landscape-ecological survey of the low-mountain territory of the Komsomolsk State Nature Reserve elucidates the role of orographic ecotones of different scale in the spatial phytocoenological and landscape organization. We will consider the *two hierarchical levels of oro-ecotones: local and subregional.* The first level is represented by a collecting catenary system of local types of sites (through the entire model low-mountain/valley territory). Within the range of the testing ground, all forest biogeocoenoses (landscape facies) were distributed over six local types of sites (S) forming a particular *catena* – a vector topological system of landscape couplings. As it has been shown (Polynov 1956; Glazovskaya 1964), there are the following S types (in the direction from the local drainage divide to the local basis of denudation): evuvial (E), transeluvial (TE), transit (T), trans-accumulative (TA), accumulative (A), and superaqual + eluvial-accumulative (Saq). The *flat interfluve groups of forest geo(eco)systems* formed at eluvial sites are subregional and local representatives of a given zonal-regional bioclimatic background (Vysotsky 1909, 1960). Other geotopes and microlandscapes are considered as *extrazonal* (Sprygin 1986), i.e., representing the background conditions of other, often very distant, regions.

The second level of oro-ecotones is formed by a vector system of *microlandscapes* (Glazovskaya 1964) consisting of the paragenetic spatial series (catenas) of elementary natural complexes – landscape facies. According to the schemes of taxonomic and territorial division of mountain landscapes (Gerenchuk et al. 1963; Fedina 1971; Gvozdetsky 1972; Miller 1974; Sochava 1978), this level corresponds to the category of subtypes, or kinds, of landscapes and belongs to the taxonomic rank of high-altitude (mountain) terrain, complicated stow, as well as mesogeochore.

Groups of microlandscapes: (1) mountain eluvial–transeluvial E–TE(m.), sublithomorphic; (2) mountain transit and transaccumulative of shadowy slopes T–TA(m-shad), litho-hydromorphic; (3) mountain transit of sunny and neutral slopes T(m-s-n), lithomorphic; (4) piedmont transaccumulative sunny and neutral TA(p/m-s-n), sublithomorphic; (5) valley eluvial E(val), sublithomorphic; (6) valley accumulative A(val), subhydromorphic; and (7) valley superaqual Saq(valley), hydromorphic.

In accordance with B.B. Polynov's concept (1956) suggesting a "through" (trans-level) system of landscape couplings, the territorial unities determined for the second level (microlandscapes) were analogous to the unities of the first level. Microlandscapes were distinguished by the combination of two geomorphological characters of forest biogeocoenoses: the types of their meso-scale sites and the solar exposure of slopes (for the transit types of sites).These associations are elements of a complex mountain and valley system of meso-scale landscape couplings (mesocatenas). They characterize the differentiation of thermo- and hydroedaphic conditions, the ratio of mineralization and humification as two antagonistic soil-forming processes.

There are seven such elements – *groups of microlandscapes* (Figure 9.2):

1) E–TE (m), low-mountain; dividing ridges and near-peak slopes with cedar/broad-leaved dry forests, on stony brown-podzol soils, sublithomorphic;
2) T–TA (m-shad), low-mountain; shadowy slopes of different steepness and their bottoms, with fir-spruce, spruce/broad-leaved and larch forests (moist) on brown-podzol and humus-illuvial stony soils, litho-hydromorphic;

3) T(m.-s–n.), low-mountain; sunny and neutral slopes of different steepness, with cedar/broad-leaved and spruce/broad-leaved forests and oak forests, with ash (moist), on brown, on high gravel content, weakly podzolic soils, lithomorphic;
4) TA (p/m-s–n), piedmont; sunny and neutral slopes of different steepness with cedar/broad-leaved and spruce/broad-leaved forests and oak forests, with ash (moist), on brown gravelly, weakly podzolic soils, sublithomorphic;
5) E(v), valley, upland; ridge medium-steep and gently sloping watersheds with oak-linden, as well as cedar/broad-leaved and spruce/broad-leaved forests (most recent) on brown-podzolic slightly rocky soils, sublithomorphic;
6) A(v), valley; the bottoms of mountain river valleys and piedmont trains with broad-leaved and larch forests (wet) on alluvial podzolic- and peaty-gley, here and there permafrost, soils, subhydromorphic;
7) Saq, closed potholes on plains and in river valleys with waterlogged sphagnum light larch forests (mari, wet) on peaty-clumpy long-seasonally frozen soils, hydromorphic.

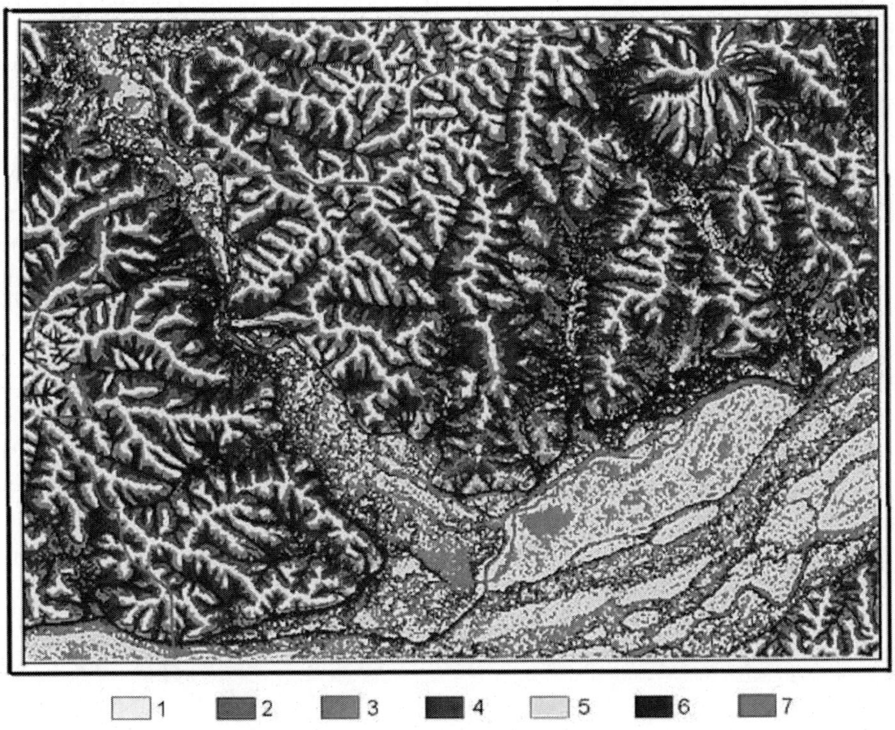

Figure 9.2. Map of microlandscape groups for experimental test site on the territory of Komsomolsky Nature Reserve (Lower Priamurye, see Figure 1). Map have been created by L.S. Sharaya and P.S. Petrenko. The boundaries of reserve is shown by red line. The rest of conventions meaning see in the text.

The microlandscapes of sunny and neutral mountain slopes were combined into a single group on the basis of mutual similarity of these slopes in total radiation and in the radiation balance of vegetation period for the Priamurye–Primorye ecoregion (Vygodskaya 1981).

Designations for landscape groups are as follows: 1 – TEm; 2 – T^{m}_{shad}; 3 – T^{m}_{sun}; 4 – TA; 5 – Ev; 6 – Av; 7 – Saq.

Table 9.2 gives the distribution of floristic phratries, forest type groups and their zonal-geographical groups over the groups of microlandscapes (based on the results of binary ordination, see Chapter 2).

Table 9.2. The Komsomolsk State Nature Reserve. The affiliation of floristic phratries, forest type groups, and their zonal-geographical groups with different groups of microlandscapes (according to the matrices of partial coupling coefficients, see Chapter 2)

Microlandscape groups	Floristic phratries (*C(A;B)* = 0.170)	Groups of forest types (*C(A;B)* = 0.169)	Zonal groups of forest (*C(A;B)* = 0.212)
1 – TEm (mount)	Mx	**CdBr** + Br	**Boreal-nemoral** + boreal and nemoral (equally)
2 – T^{m}_{shad}–TA (mount/shadow)	**Och** + MA	**PiBr** + **SiPi**	Nemoral-boreal
3 – T^{m}_{sun} (mount/sun-neutral)	MO	**SiPi** + CB + SB	**Boreal-nemoral** + boreal
4 – TA (piedmont/sun-neutral)	MA	**Larch** + SiPi + CdBr	Nemoral-boreal
5 – Ev (valley)	**Mm** + **Mx** + **MO**	**PiBr** + Br + CdBr	Boreal and nemoral (equally)
6–7 – Av, Saq (valley)	**Mm** + **An**	**Larch** + Br	**Boreal** + nemoral-boreal

Footnote: ecological dominants are in bold. Abbreviations are expanded in the text.

The territory of the Komsomolsk State Nature Reserve is the southwestern outpost of the Lower-Amur orographic group (Nikolyskaya et al. 1969), with the predominant low-mountain relief and with a wedge-shaped northeastern area of the Middle-Amur Plain (along the left bank of the Amur River and the lower reaches of its tributary, the Gorin River). The morphotectonic sub-Pacific representativeness of this territory is as follows: it is a Mesozoic epigeosynclinal orogenic structure, which progressively developed as early as in the Paleogene and Neogene as a border zone between the Pacific Ocean and the continental areas of Asia, with a geosynclinal complex of rocks lying on the oceanic crust on the one side and on the continental crust on the other side (Khudyakov 1977). As a result of geomorphostructural integration of depressions and elevations into integral passive geoblocks (segmented plates) in Cenozoic, the Nature Reserve territory was in the Amur–Ussuri suture zone dividing two such plates: Mongolian–Okhotian and Sikhote-Alin. The morphotectonic peculiarity of the suture zone is the active "polarization" of the two major geomorphostructures: simultaneous downwarping of intermontane depressions and uplift of mountain ranges, with localization of valley pediments and with "overwhelming" of outlier uplands with a mantle of coarse-grained material (Khudyakov 1977). Under such morphotectonic conditions, the paragenetic system of meso-scale landscape couplings must be especially marked, as has been represented by the respective spatial differentiation of microlandscapes over the Reserve territory.

Chapter 10

STRUCTURAL ORGANIZATION OF FOREST GEO(ECO)SYSTEMS

ABSTRACT

The general information model of monosystemic organization of plain-lowland subregional and local geo(eco)systems of the Lower Amur Sub-Pacific has been used to characterize the most common regularities of formation of phytocoenological and landscape structures under the conditions of epigeosynclinal mountainous country of Mesozoic age and boreal bioclimatic ecotone. The causal mechanisms of appearance of the known floristic richness and phytocoenotic diversity of forest cover typical for marginal-continental lands of the southern Far East have been established. The more detailed mechanisms of these biospheric processes have been revealed using discrete partial models of the binary ordination of geocomponent characters. The polysystem space of the Priamurye (by-Amur) boreal ecotone has a high degree of intercomplex relationships. It was represented by mutual penetration of floras and phytocoenotic structures and resulted in the formation of buffer forest formations – a phenomenal event in the temperature belt of the Asian sub-Pacific marginal continental land. The mechanisms of appearance of buffer forest communities, including spruce/broad-leaved and nemoral fir/spruce forests of the Manchurian–Okhotian phratrie, have been revealed. It has been shown that the forests of this phenomenal buffer flora differ from the forests of basic floras in the higher indices of structural and functional development and have reached the state most similar to the evolutionary climax. The earlier concepts of the Pacific Ocean mega ecotone of Northern Eurasia as a focus of evolutionary processes in the continental biosphere have been confirmed.

10.1. THE GENERAL STATISTICAL INFORMATION MODEL

The basic regularities of monosystemic organization of the geographical environment in the Priamurye sub-Pacific at the subregional-topological level are presented in the general information model (Figure 10.1). This model characterizes the plain/low-mountain system of intercomponent relationships in the boreal ecotone in the marginal-continental sector of Eurasia, with the relatively young (Mesozoic) orogenic morphostructure and active monsoon-like climatic interaction between the continent and the ocean. The topological-level natural complexes are much younger. The relative age of low-mountain/valley morphosculpture, the main features of modern climate and the floristic structure of plant cover in this territory can be assessed as 4.5–5 thousand years, since the end of the Holocene climatic optimum, compared both to the boreal period (Khotinsky 1977) and to the Atlantic period (Korotky et al. 1999).

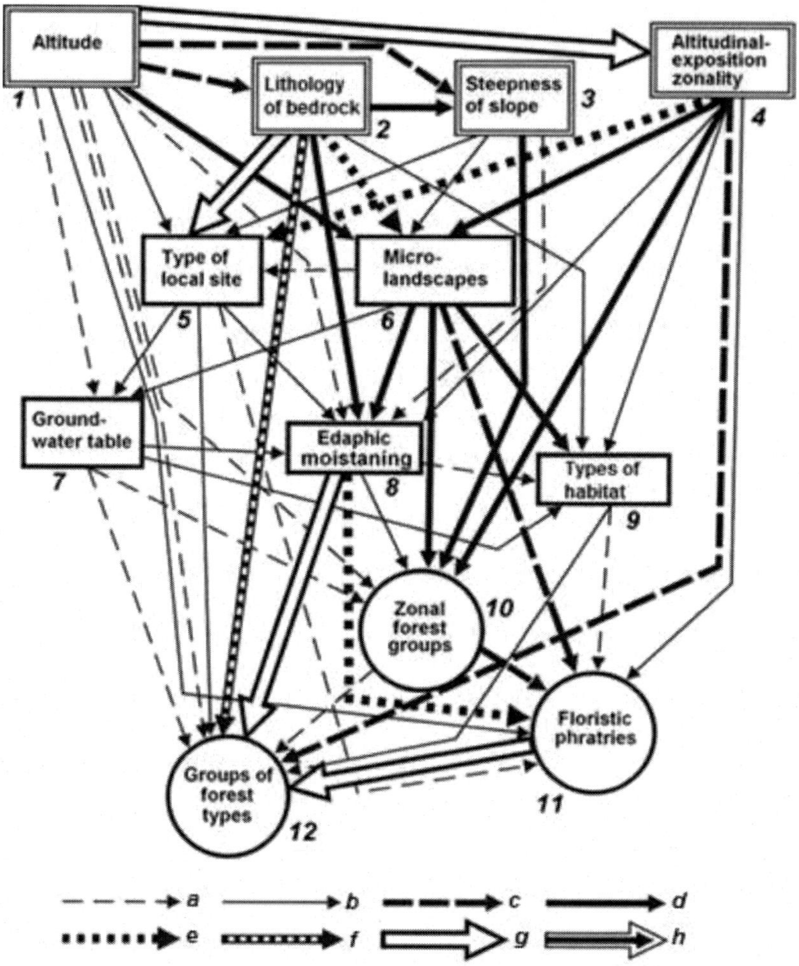

Figure 10.1. The general information-statistical model of geocomponent interrelations for the regional ecosystem in Priamurye sub-Pacific (Komsomolsky Nature Reserve). The normalized coefficient of interrelation *C(A;B)* between geocomponent attributes: *a*, 0.101–0.130; *b*, 0.131–0.160; *c*, 0.161–0.190; *d*, 0.191–0.220; *e*, 0.221–0.250; *f*, 0.251–0.300; *g*, 0.301–0.440; *h*, 0.441 – 0.720.

Following the cold and dry Postglacial (Preboreal) stage of total elevation and downcutting, the warm and wet Holocene optimum was characterized by land subsidence, marine transgression, and the predominance of lateral cutting over deep cutting with the formation of a bench of the first fluvial terrace above the floodplain and the emergence of modern floodplain, with the conversion of loose sediments and the development of peat deposits in river valleys and lake basins (Nikolyskaya 1969). Finally, then there was the beginning of alternation of the phases of cedar/spruce/broad-leaved, cedar/broad-leaved and coniferous/broad-leaved forests with larch, which created the modern plant cover of the Lower Priamurye.

For comparison, we will mention the information models of some ecoregions in the boreal ecotone of the Volga River basin (Kolomyts 2008) remaining under the conditions of long-term (beginning from the Precambrien) platform geotectonic regime of the Russian platform, with senile topography (Figure 10.2). This region is a part of the Atlantic-continental forest climatic area (Alisov 1956) and gravitates towards the passive Atlantic margin of the continent of Eurasia (Khain 1973; Rice 1978), ecological systems of regions under consideration. It can be readily used to follow the main directions of transmission of both internal and external signals and to assess the relative rates of their propagation with the predominant westerlies of air masses. The nature reserve is situated on a lowland plain built of horizontal strata covered with coniferous, broad-leaved and mixed forests.

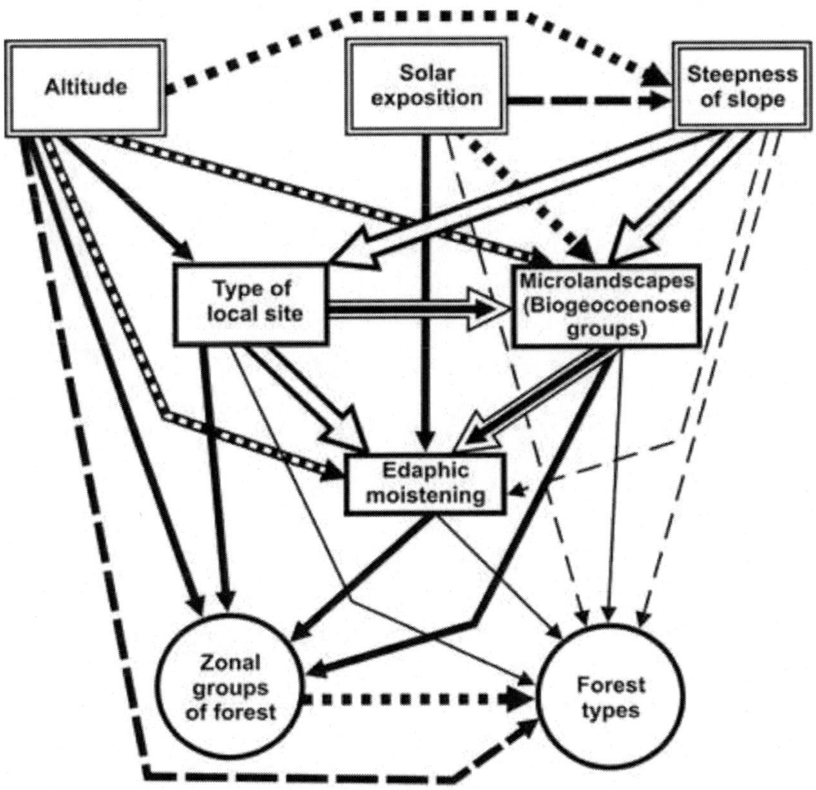

Figure 10.2. The general information statistical model of geocomponent interrelations for the regional ecosystem on the high-plain province of the Middle Volga Region (in the vicinity of Nizhny Novgorod). Legends are as in Figure 10.1.

The general information model reflects the ordered hierarchy of mutual influence of natural components and, in essence, is a *model of chain reactions* in the local landscape-ecological systems of the regions under consideration. It can be readily used to follow the main directions of transmission of both internal and external signals and to assess the relative rates of their propagation along the chain of componental interrelations, using the values of the normalized contingency coefficient $C(A;B)$ of phenomenon A with factor B.

The system of local component interconnections in Priamurye Sub-Pacific (see Figure 10-1) is formed by the first (upper) echelon consisting of the four initial and relatively independent geological-geomorphological factors: absolute height, altitudinal exposure or dislocation (Gartsman 1971), zonality (see below), bedrock lithology, and slope steepness. This is a morpholithogenic "memory" of the system (Solntsev 1974), its most conservative part that creates a stabilizing mechanism of inter-component relationships. The bedrock lithology of the test ground is rather diverse, which determines the efficiency of its landscape-forming significance in the Mesozoic orogenic morphosystem. Lithology includes: (1) granodiorites, diorites and dioritic porphyrites (GDr); (2) conglomerates, breccias, gravelites (CBr); (3) aleurolites (siltstones), sandstones, clay states (SCs); (4) river lake-alluvial deposits – pebbles, sands, clays (RLd).

Microlandscapes (as subregional groups of biogeocoenoses) and types of local sites (LS, geotopes) make the second echelon of partially dependent geomorphological variables. Each of them is represented, as has already been mentioned above (see Chapter 2), by a landscape coupling (catena) consisting of four subdivisions of meso- and microsites, respectively. The sites determine the hydrological regime of soil-grounds, which is involved in the exchange-transit part of geo(eco)system (Krauklis 1979) and combines their inert and mobile parts into an integral whole. This is the third echelon of inter-component relationships. It includes, in this case, the depth to groundwater (Z_{GW}), the moistening of edaphotopes represented by ecological groups of soil cover (Ramensky 1971), as well as the types of habitats by the scale of P.S. Pogrebnyak (1955). The following ecological groups of herbage have been distinguished: (1) MX (xerophytes + xeromesophytes); (2) XM (mesophytes + xeromesophytes); (3) mesophytes (M); (4) MH (hygrophytes + mesohygrophytes); (5) hygrophytes (H). The types of habitats (litho-hydroedaphotopes) are given as a combination of soil trophicity (A – sands; B – fine sandy loams; C – light and medium loam soils; D – heavy loam soils and loams) and moisture content (1 – dry; 2 – fresh; 3 –moist; 4 – wet; 5 – waterlogged).

Finally, the fourth, terminal echelon of indicators is formed by zonal groups of forest, floristic phratries and groups of forest types. There are four zonal groups determined by the floristic composition of herbage: (1) boreal; (2) nemoral-boreal; (3) mixed boreal + nemoral; (4) boreal-nemoral. The floristic phratries and groups of forest types have been characterized above (see Chapter 2).

The graph information models of the Prioksko-Terrasny Biosphere Reserve test ground have an analogous structure.

In general, the monosystemic organization of local natural complexes of the low-mountain Lower Amur sub-Pacific is formed by moderate information couplings ($C(A;B)$ = 0.160–0.220), from the four initial abiotic factors to phytocoenological characteristics (see Figure 10.1). The predominant relatively low contingency coefficients ($\leq$ 0.130) not only create certain information "noise" in the channeled system of the dominating chain reactions but also make this system more diversified.

The low mountain landscape structures of the marginal-continental sector in the south of the Far East are characterized by a comparatively dense network of moderate and weak geocomponent interrelations, which makes them drastically different from intracontinental landscape structures that have passed through long-term subaeration development (Kolomyts 2008). There are numerous, almost equivalent main pathways of information interactions, with the primary and secondary communication channels for most characters and with the presence of doubling connections. It implies, first, the yet insufficiently developed hierarchic structure of monosystemic landscape organization in the given ecotone and, second, the possibility of creating similar (and even analogous) phytocoenological formations at different combinations of geological-geomorphological and hydroedaphic factors. It leads to the essential diffuse "fuzziness" and fractionation of ecological niches of phytobiota, often the appearance of several ecological dominants in a niche, the formation of numerous *enclaves* distant from the main niche under the influence of local factors and, finally, the emergence of mixed (buffer) florogenetic and forest-typological formations. The diagrams of partial connections (see Figure 10.3, 10.4, etc.) show diffuse "blurs" and breaks of the ecological niches of phytobiota, appearance of several ecological dominants per niche, formation of enclaves and, finally, mixed (buffer) florogenetic and forest-typological communities.

According to the known concept of plant cover evolution (Krištofovič 1946; Vasilyev 1946), the above characters demonstrate that the given *monosystem landscape structure of the boreal sub-Pacific type* could be created only due to the *migration biogeocoenotic processes implementing this evolution*, which have occurred in the recent past and continue at present. The necessary condition of manifestation of evolutionary processes has been, and still is, the clearly marked seasonal hydrothermal contrasts associated with the monsoon circulation of air masses.

All of the mentioned biospheric processes recorded in discrete form in the model of partial coupling coefficients (see below) have two important ecological consequences. First, they have created and still continue to recreate the known florogenetic richness and phytocoenotic diversity of forest cover of the plain and mountain landscapes of Priamurye sub-Pacific. Such biodiversity, with a mixture of various floras (from subtropical to moderately cold) have existed here since the preglacial (Palaeogene–Miocene) time (Khudyakov 1977), when the climate seemed to be already monsoon-like.

Second, the complicated structure of intercomponent relationships, with the presence of numerous doubling connections and fractionation or, on the contrary, generalization of information signals at certain stages of their transduction, must ensure rather high stability of phytocoenological and landscape structures during multiyear and supercentury climatic fluctuations. The latter manifested themselves mainly in temperature variations (January, July and annual temperatures, by not more than 8–13°), in the absence of drought warm epochs.

In the warm climatic epochs, the climate was sufficiently humid, which was an important factor in preserving the mixed Manchurian–Okhotian floras. In the cold and more arid epochs, on the contrary, the Manchurian flora was mixed with the Angaridian flora. The low mountain regions and adjacent valleys of the Middle Priamurye (in the area of the Komsomolsky Nature Reserve) were covered with cedar/broad-leaved forests in the Riss–Würm Interglacial epoch (130–75 thousand years ago) and with open larch and birch forests in the climatic minimum of Late Würm (20–18 thousand years ago), in the absence of periglacial tundras and forest tundras (Gerasimov and Velichko 1982; Korotky et al. 1999). By analogy with the Quaternary paleogeography of Beringia (Yurtsev 1974) it may be supposed that the Manchurian and Okhotian floras were mixed in the warm climatic epochs with marine transgressions, while the Manchurian and Angaridian floras were mixed in the cold snap epochs with marine regression. This process was facilitated by the periodic weakening of temperature gradients: latitudinal-zonal in the former case and longitudinal-sectoral in the latter case.

The transition from the Pacific Ocean mega ecotone to the plain areas of intracontinental paleo-Pacifics (on the Russian Plain) that have had quite a long path of subaerial development (see Figure 10.2) is accompanied by disintegration of the weakest connections and dramatic intensification of the most significant connections. It has two important ecological consequences. First, the more developed hierarchic structure of monosystemic landscape organization was formed. It became more clearly echeloned, with reduction of the information "noise" in the main communication channels and enhancement of the role of intermediate repeater nodes. The rigidity of mutual determination of geomorphological and hydroedaphic characteristics increased, with the reduction and even disappearance of doubling connections. For instance, the complementarity of the types of sites and habitats, hydroedaphotopes and biogeocoenoses sharply increased (*C(A;B)* = 0.674–0.722). Second, the ecological niches of indicator characteristics (types of forest and biogeocoenoses) narrowed, with more distinct manifestation of their dominant areas.

As a result of intracontinental natural complexes we have: (1) a decrease in bio- and geodiversity and (2) a potential decrease in their potential resistance to global climatic changes. The latter is proved by the known extraordinary dynamics of climate and plant cover of the Russian Plain in Pleistocene and Holocene (Gerasimov and Velichko 1982; Velichko et al. 1991). In the culmination phase of Mikulino Interglacial (about 120 thousand years ago), the mixed spruce/broad-leaved forests grew on the territory of the Middle and Upper Povolzhye; in the Middle-Quaternary glaciation epoch, the region was completely covered with a glacier; in the epoch of the latest, Valdai Ice Age (35–24 thousand years ago), forest tundra and periglacial (cryogenic) forest-steppe were predominant here. The amplitude of oscillation of the mean January temperature between the extreme climatic epochs was 22–26°, and that of the annual precipitation was 300–400 mm (Velichko 1999).

In the low-mountain Priamurye sub-Pacific, the absolute height does not play a determining bioclimatic role in itself: the high-zone factor here is weakly manifested in the distribution of floristic phratries and forest types (see below Figure 10-4). The altitudinal-exposure zonality, with the combined influence of circulatory and solar exposure (see below), and bedrock lithology (with a partial effect of slope steepness) are much more effective. Here, the *C(A;B)* values reach 0.239–247 and the relationship between the local sites and lithology is up to 0.414. Moreover, under the conditions of relatively young highlands, with rather great morpho- and lithodiversity, and aggravated circulatory climate-forming processes, there is an obvious direct effect of both these abiotic factors, bypassing intermediate transducers, on the phytocoenological structure of forest cover, which is represented in this case by zonal groups and forest types (the *C(A;B)* parameter is 0.181–0.196 and 0.270, respectively).

Typically, the two intermediate characteristics: the type of local site (geotope) and the depth of groundwater (Z_{GW}), are apart from the main network of chain reactions. At the same time, the type of local site under the given morphotectonic conditions proves to be a kind of dead end of information flows and, therefore, cannot be considered, as it is accepted in traditional landscape science (Isachenko 1965; D.L. Armand 1975; Sochava 1974; etc.), as a governing factor of spatial organization of geo(eco)systems of topological dimensionality. Under the conditions of "senile" denudation-accumulative relief of the ancient platform, geotope proves to the one of the most important data communication centers, which corresponds to the known canons of landscape science. Here, the role of groundwater depth and, accordingly, soil-ground moistening drastically increases; together with the mechanical composition of soil-forming rocks, the latter forms the second important information center: a type of habitat (see Figure 10.1).

The latter, as we can see, is not such a node under the conditions of low-mountain Priamurye Sub-Pacific, where the bedrocks lying close to the day surface provide rather high general moderately high fertility of soils. Here, the type of habitat is the second blind alley of information signals.

10.2. The Main Channels of Component Interconnections

Let us consider in more detail the general information model of monosystem landscape organization of the Lower Amur Sub-Pacific. First of all, it is confirmed (Isachenko 1965; Milykov 1981) th.t absolute height per se plays no determinative bioclimatic role under low mountain conditions: here, the altitudinal-zonal factor is weakly manifested in the distribution of floristic phratries and forest types (see below Figure 10.7, *a, b*). The altitudinal-exposure zonality, with the joint effect of circulatory and solar exposure (see below) and bedrock lithology (with the partial effect of slope steepness) are much more marked. The *C(A;B)* values here are 0.239–0.247, while the relationship between the types of local sites and lithology is up to 0.414. Moreover, under the conditions of relatively young mountain country with sufficiently high morpho- and litho-diversity and with acute circulatory climate-forming processes, there is a clear direct influence of these two abiotic factors on the phytocoenological structure of forest cover, which is represented in this case by zonal groups and types of forest (*K(A;B)* 0.181–0.196 and 0.270, respectively).

In the general information model, there are two major *intermediate information nodes*: microlandscapes and edaphic moistening represented by the ecological groups of soil cover (see Figure 10.1). The first node "multiplies" information signals towards the third and fourth echelons (the multivalued effect) and the second one substantially reduces them (the effect of selectivity of information flows) but, simultaneously, drastically amplifies the selected integrated signal – towards the types of forest (*K(A;B)* = 0.331) and floristic phratries (*C(A;B)* = 0.241). The analogous intensification of integrated signal (up to *C(A;B)* = 0.313) occurs in the binary connective of forest types with floristic phratries.

Let us mention several important statistical regularities in the system of componental interconnections in the Lower Amur boreal ecotone. First, the types of forest are rather autonomous (weakly dependent) categories relative to its zonal groups, i.e., this model is dramatically different from the model of boreal ecotone of the Volga River basin. In the low-mountain Lower Priamurye, the following two factors have an incomparably more substantial phytocoenological effect: edaphic humidification and floristic composition. They smooth over the direct influence of geographical zonality on forest structure and perform territorial fractionation of its zonal groups by microlandscapes and hydroedaphotopes.

Second, the type of habitat traditionally represented as a combination of trophism and soil humidity (Pogrebnyak 1955) has no such considerable effect here on forest types (*C(A;B)* = 0.166) as it does under the platform conditions of the Russian Plain (*C(A;B)* = 0.304). The bedrocks lying close to the day surface provide rather high general trophicity of soils; therefore, the type of habitat becomes a dead end of information signals. As has already been mentioned, the direct influence of original oro-climatic factors (lithogenic and dislocation-zonal) and the associated microlandscapes on the florogenetic and coenotic forest structure is much more significant.

Third, edaphic humidification, in contrast to the compared ecoregions of the Volga River basin, is rather weakly connected with groundwater depth and local type of site ($C(A;B)$ = 0.137 and 0.149, respectively). It also much more, and directly, depends on geological and geomorphological factors ($C(A;B)$ = 0.211–0.218).

The described regularities are an important distinctive feature of monosystemic organization relative to the young low-mountain landscapes on the Priamurye boreal ecotone of the sub-Pacific Lower Amur – the mountain country of epigeosynclinal origin. The feebly marked echelonment of intercomponental relationships and the disordered hierarchy of geocomponental characters are the evidence of yet insufficient development of the hierarchic system of landscape organization, which is known (Puzachenko 1986) to be one of the most important indicators of geosystem stability.

Thus, the presented information model of monosystem organization of plain/low-mountain subregional and local geo(eco)systems of the Lower Amur Sub-Pacific characterizes the most common patterns of formation of phytocoenological and landscape structures under conditions of the epigeosynclinal orogen of Mesozoic age and the boreal bioclimatic ecotone. This model already discloses the causal mechanisms of the known florogenetic diversity of forest cover typical of marginal-continental land in the south of the Russian Far East.

10.3. Altitudinal-Exposure Zonality of Forest Distribution and Emergence of Buffer Communities

In highlands, natural zonality is much more diverse than on plains, as is represented in the further development of Dokuchaev's theory of geographical zonality (Gartsman 1971). Superposition of the two principles of regional differentiation of a mountain territory (morphotectonic and macroclimatic) predetermines the entire gamut of zonal phenomena: latitudinal, altitudinal, longitudinal- and exposure-circulatory, barrier, and solar exposure. Being combined, they yield the complex forms of the so-called *dislocation zonality*. In this context, zonality is closely related to the concept of vector (gradient) and is defined as a regular, spatially ordered trend of changes in the properties of natural complexes in general and their individual components.

Let us consider the manifestation of dislocation zonality in forest vegetation of the low-mountain boreal ecotone of Priamurye. To this end, let us imagine the oroclimatic system of the Komsomolsky Nature Reserve as a superposition of two major abiotic factors: absolute height(A) and slope exposure (B). The following three altitudinal bands (and, accordingly, three gradations of this factor) can be distinguished in the sample of our testing areas: A_1 (20–110 m abs.), A_2 (110–310 m), and A_3 (310–680 m). The first band includes the plain and piedmont territories with river valleys, as well as the lower parts of mountain slopes; the second band includes the middle, most part of slopes and peaks of bald mountains; and the third band includes near-peak slopes and crests of the highest ridges.

In accordance with the basic processes of climate formation in the Priamurye part of the Far East (Vitvirsky 1969), all mountain slopes of the model territory were divided into the two opposite exposure categories actually combining the circulatory and solar exposures. The first category B_1 is formed by the coldest and leas hydrated slopes of the NW–N–NE–E orientation. The second category B_2 is formed by the SE–S–SW–W slopes with the warmest and humid habitats. The separate subhorizontal surfaces B_0 correspond mainly to piedmont plains and small river valleys (A_4), where the role of solar exposure is neutralized and, simultaneously, the negative effect of winter climatic inversions is displayed.

The binary ordination of the above phytocoenological and landscape indicators by the factor of dislocation zonality (Figures 10.3), with rather high *C(A;B)*, demonstrates that the altitudinal bioclimatic zonality of low mountains under the conditions of monsoon-continental climate is very selectively manifested and significantly disturbed by exposure contrasts. The measures of partial contingency *C(A;B)* of floristic phratries and forest types with absolute height are 0.126 and 0.102, respectively. For solar slopes, they are even higher: 0.174 and 0.157, respectively. The role of sectorality is demonstrated first of all in the distribution of larch-tree forests (representatives of the cryogenic-taiga Angaridian flora) along the river valleys and mountain slopes of NW–N–NE–E exposure in the middle belt of low mountains. Complex larch-tree forests with broad-leaved species appeared on the same slopes as a result of the Angaridian flora mixing with the xerophilic Manchurian flora. They belong to the buffer Manchurian–Angaridian phratrie.

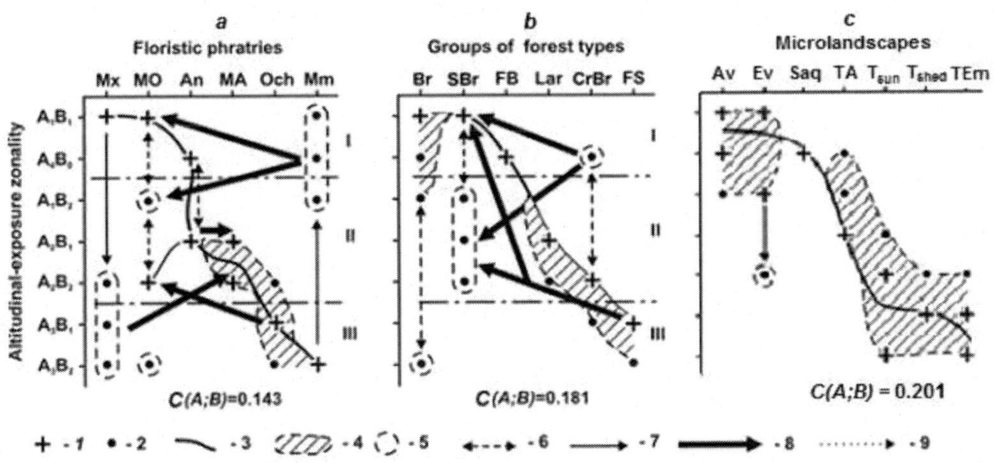

Figure 10.3. The binary ordination of floristic phratries and forest type groups by the factor of altitudinal-exposure (dislocation) zonality. Symbols: 1 – the ecological dominant; 2 – the "fuzzy" part of ecological niche; 3 – the trajectory connecting ecological dominants; 4 – the space of ecological niche; 5 – the enclave; 6 – the altitudinal interval of fragmentation of ecological dominants; 7 – the direction of possible transgression of a given phenomenon from its ecological dominant; 8 – the direction of transgression of basal phratries (or forest type groups), with their mutual displacement and formation of buffer phratries (forest type groups); 9 – the direction of sporadic distribution of a phenomenon with the given gradations of a factor. I, II, and III – the upper, middle and lower altitudinal layers of low mountains, respectively.

The deflection of altitudinal zonality by the complex factor of circulatory-solar exposure is determined, as one may believe, by the duality of regional bioclimatic system of the southern Far East. What is meant here is the commensurate expression of sub-oceanic latitudinal zonality (broad-leaved forests ↔ mixed forests ↔ dark-coniferous forests) and eastern marginal-continental longitudinal sectorality. Sectorality manifests itself first of all in the intrusion of plain and mountain larch forests (representatives of the taiga Angaridian flora) into river valleys and the NW–N–NE–E mountain slopes of the middle band of low mountains, respectively. As a result of mixing the latter with the xerophilic Manchurian flora, the complex larch forests with broad-leaved species belonging to the buffer Manchurian–Angaridian phratrie appeared in the same band and on the slopes of the same exposures.

In addition, there is no unambiguous altitudinal-zonal correspondence of Manchurian floristic phratries as the primary phytocoenological representatives of Priamurye sub-Pacific. The broad-leaved and cedar/broad-leaved forests occur almost over the entire altitudinal profile of low mountains (even though sporadically) and, in addition, have two ecological dominants: (1) xerophytic on the plains and in piedmonts and (2) mesophytic on the southern and southwestern slopes in the upper band of ridges. The predominant area of fir-spruce forests appears in the same band but on the northeastern slopes; it is the type of forest formations represented by the Okhotian flora. As we can see here, the altitudinal zonality though begins to manifest itself but rather selectively: along the slopes of the most favorable exposure, according to the known Alekhin–Walter "rule of landscape precedence" (cited by: Milykov 1981).

The transgression of Mx and Ok floras from the upper mountain slopes along sunny slopes to the middle band, followed by their mixing, resulted in the formation of spruce/broad-leaved forests. However, the forests of the same buffer phratrie are no less widespread in the foothills, with an enclave on the southern slopes of the lower mountainous belt.

Table 10.1. Distribution of the groups of forest types by floristic phratries (the normalized partial coupling coefficients

C(A;B) = 0.313)

Groups of forest types	Floristic phratries					
	Mm	**Mx**	**MO**	**Okh**	**MA**	**An**
Br	**0.71**	0.67				
CrBr	0.29	0.33	0.22		0.35	
PBr			**0.40**	0.44	0.21	
SE			**0.38**	**0.56**		
Lar					**0.45**	0.34
FB						**0.59**

Footnote. Dominant fields of ecological niches are showed by half-rich print.

The typical ecological consequence of dislocation geographical zonality in the low mountains of Priamurye sub-Pacific is the mutual penetration of representatives of different floras into some or other forest type selectively by altitudinal belts and slope exposures. The combination of the absolute heightand solar-circulation exposition created certain "ecological corridors" for the *migration flows* of phytobiota and the habitats for *selection* of novel plant communities. Thereby, the conditions were created for evolutionary transformation processes in forest cover in the south of the Far East, according to the concept (Komarov 1897; Krištofovič 1946; Vasilyev 1946).

The correlation between forest type groups with the forming floristic phratries is exclusively high (Table 10.1; the measure of $C(A;B)$ = 0.313). It demonstrates that oro-climatic conditions predetermine certain stability of florogenesis creating a particular type of forest. The stability of florogenesis is also confirmed by the fact that the ecological niches of floristic phratries and forest type groups are very discrete in the space of dislocation zonality (see. Figure 10.3). They are "fuzzied" very little and represented mainly by their dominant areas and numerous enclaves. There was territorial differentiation of forest-formation process by the elements of rather complex dislocation zonality, with the formation of the respective phytocoenological structures. It seems to be one of the mechanisms of origination of the known floristic richness and species abundance in the forests of the Primorye–Priamurye sub-Pacific.

The correlation between the groups of forest types and the floristic phratries they are formed of is exclusively high (Table 10.1), demonstrating the oroclimatically predetermined stability of florogenesis that creates different types of forest. This stability is also confirmed by the fact that the ecological niches of floristic phratries and groups of forest types in the space of dislocation zonality (see Figure 10.3) are highly discrete. They are very little "fuzzy" and represented mainly by their dominant regions and numerous enclaves. There was territorial fractionation of florogenesis by the combination of absolute heights and solar-circulatory exposure of the slopes, with formation of the respective phytocoenological structures. It seems to be one of the mechanisms of emergence of the known floristic richness and species abundance of the forests of Priamurye sub-Pacific.

The broad-leaved forests consist mainly of the species of Manchurian flora (up to 80–100% in forest stand), equally for the Mm and Mx types (by 52% and 48%, respectively). It cannot be referred to the cedar/broad-leaved forests, which have the richest floristic composition of forest stand and undergrowth (see Table 10.1). They have the approximately equal portions of both Mm and Mx forest communities (21% and 25%, respectively), as well as the forests transitional to the Ohkotian flora, with up to 15% of fir/spruce stands. However, the dominants (with up to 32% probability) here there are the buffer MA associations with no less than 30% of the Dahurian larch and *Larix cajanderi* in their forest stand (the domination region is 55–80%).

Thus, there is direct evidence of significant intrusion of representatives of the continental Angaridian flora (first of all, larch) and the sub-oceanic Ajan dark-coniferous forests into northern cedar forests of the Lower Priamurye. It is a characteristic feature of florogenetic structure of the basic zonal forest formation of the Priamurye boreal ecotone. On the other hand, larch forests, being the "legal" representatives of the Angaridian phratrie, have an even more significant (up to 43%) admixture of the buffer MA communities. The Manchurian–Angaridian flora occurs also in the forests of the Okhotian flora, i.e., larch forests transgressed (by 20%) also into these mixed forests of the Lower Priamurye.

The cedar/broad-leaved forests are more mesophilic than the broad-leaved forests (mesoxerophilic, with oak). Therefore, the buffer Manchurian–Okhotian forest communities are formed when the immigrants from the Ajan fir/spruce forests are mixed with the representatives of cedar forests as more similar to them in the general level of humidity. This forest-forming process associated with the well-known phenomenon of substitution of the

Ajan spruce for the Korean cedar pine (Kolesnikov 1956), which results in the presence of two equivalent floristic domination regions in the spruce/broad-leaved forests. They remain the buffer Manchurian–Okhotian type in some cases and move into the category of Okhotian phratrie in other cases.

The theater of local contacts between the primary phytocoenological representatives of sub-Pacific and paleo-Pacific corresponds to lowland plains and foothills on the Priamurye boreal ecotone, both of them encountering mainly on subhorizontal surfaces: plakors. On mountainside, their ecological dominants are dispersed over different solar exposures: sunny and neutral for broad-leaved forests of the Manchurian mesophytic floristic phratrie and shadowy for larch forests of the Angaridian phratrie.

The dominants of the two buffer floristic phratries of the forests (Manchurian–Okhotian and Manchurian–Angaridian) were formed in the territorially close altitudinal belts (within an interval of 110–310 m). The mixing of Okhotian and Manchurian floras proceeded from the upper band of low mountains, with the "focus" of the predominant Ajan fir/spruce forests, down the slopes – towards the dominant of the ecological niche of mountain mesophilic cedar/broad-leaved forests. The formation of spruce/broad-leaved forests, i.e., the intrusion of the Ajan spruce into ash-tree/oak/linden communities, occurred mainly on the neutral and, less frequently, shadowy slopes, to which the ecological dominant of this buffer MO phratrie was shifted. In addition, its dominant proved to be 100–200 m lower relative to the dominant of the mountainous Mm phratrie.

The SBr and FS forests, on the one hand, and CrBr forests, on the other hand, are well differentiated by the *solar-circulatory exposure* (Figure 10.4). The Ajan fir/spruce forests are widespread on all neutral slopes and also cover the northern slopes. However, the ecological dominant of forests of the Ok flora is on the slopes of western rhumbs facing moisture-laden air. At the same time, the domination region of broad-leaved forests of the basic Mk flora is on the southern (sunny) and eastern (the least humid) slopes. The transgression of mountainous broad-leaved forests from eastern slopes into fir/spruce forests of western slopes resulted in the buffer MO formation (spruce/broad-leaved forests) on these slopes. For its part, the distribution of fir/spruce forests along river valleys and their intrusion into Mm broad-leaved forests resulted in the formation of valley enclaves of this buffer phratrie.

Thus, the two opposite "poles" of the floras (Mm and An phratries) have their own domination regions at nearly the same absolute heights (from foothills to crests of ridges) under the conditions of close exposure (neutral but with the predominance of sunny areas for the Mm flora and shadowy areas for the An flora) and with the same plain meso- and microrelief (the steepness of 0–16°). However, the buffer flora (MA) is widespread both on valley uplands (plakors) and on the crests and peaks of ridges. In both cases, the Manchurian mesophilic and Angaridian continental floras merge here, with intrusion of larch into cedar/broad-leaved forests. However, the broad-leaved forests with oak (Mk flora) are also widespread here as a consequence of simultaneous aridization of mesophilic Manchurian forests.

As we can see, the really existing mosaic of floras under consideration in the Lower Amur sub-Pacific low mountains is not strictly related to the geomorphological framework but formed by the bioclimatic system: the century and supercentury climatic fluctuations, as well as the ectogenetic successions of forest vegetation.

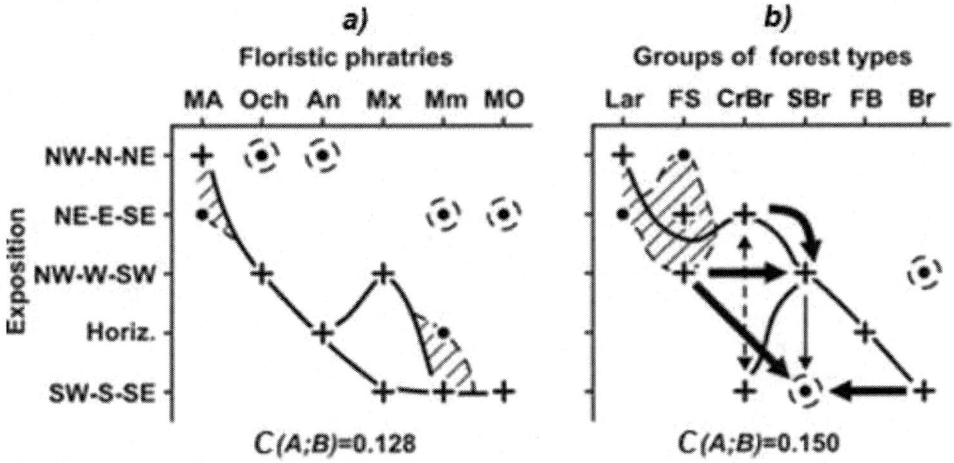

Figure 10.4. The binary ordination of floristic phratries and groups of forest types by the factor of slop solar-circulatory exposure. Legends are as in Figure 10.3.

10.4. Exposure Factor of Migration Forest-Forming Processes

The pattern of altitudinal zonality of the Lower Amur low mountains is intervened, as has already been mentioned above, with the *solar and wind exposure of slopes*. The slopes of southern exposure receive almost twice more radiation heat in summer compared to the subhorizontal surfaces of valley bottoms. The wind exposure manifests itself most distinctly on the crests of watersheds and the upper parts of northern and northwestern mountain slopes exposed to severe winds of winter monsoons. It leads to the emergence here of psychrophilic groups of plants, which decrease the lower boundary of the respective altitudinal zone.

The broad-leaved, cedar/broad-leaved, fir/spruce and larch forests, i.e., the communities of all four basic floristic phratries, have very "fuzzy," or with enclaves, partial ecological niches with respect to the solar exposure of slopes. All four groups of forest types occur on the slopes of nearly all exposures (the only exception is the absence of cedar forests on shady mountain slopes and fir-spruce forests on sunny slopes), though the areas of their dominance are azimuthally dispersed. The dominants are on SE–SW slopes for broad-leaved forests (with a probability of 38%), on neutral slopes for cedar/broad-leaved forests (55%), on NW–NE slopes for fir-spruce forests (49%) and, finally, on flat-interfluve territories for larch forests (38%).

Such a high tolerance of basic plant formations of the Lower Priamurye to the solar exposure of low mountain slopes is undoubtedly determined by the generally suboceanic climate of Priamurye Sub-Pacific. As is known (Blüthgen 1966; Alisov, Poltoraus 1974), the solar-exposure bioclimatic contrasts are most marked under conditions of strongly continental (radiation-type) climate and much weaker in the advection-type suboceanic climate. The obvious examples of local bioclimatic contrasts are Paleo-Pacific regions: the Upper Priamurye (Kolomyts 1977) and southeastern Transbaikalia.

The generally suboceanic pattern of forest plant cover of Priamurye Sub-Pacific manifests itself, in particular, in the coenotic groups of soil cover. The meadow species of herbage are widespread nearly in all floristic phratries. The forest coenotic groups are predominant only in xerophytic Manchurian forests (with oak). They are also predominant in the Ajan fir/spruce forests. The mixed conenoses are widespread too.

The SB and FS forests, on the one hand, and CB forests, on the other hand, are well differentiated with respect to *solar-circulation exposure* (see Figure 10.4). The Ajan fir/spruce forests are widespread on all neutral slopes and also cover the slopes of northern orientation. However, the ecological dominant of forests of the Okh flora occurs on the slopes of western points oriented towards moisture-laden air masses. The area of dominance of broad-leaved forests of the basal Mc flora is on the slopes of southern (sunny) and eastern (least moistened) points. The transgression of mountain broad-leaved forests from eastern slopes to fir/spruce forests of western slopes resulted in the buffer MO formation on these slopes: spruce/broad-leaved forests. In turn, the spread of fir/spruce forests along river valleys and their penetration into Mm broad-leaved forests resulted in the formation of valley enclaves of this buffer phratrie.

Thus, the two opposite "poles" of the floras: Mm and An phratries – have their own areas of dominance at nearly the same absolute heights (from bottoms to crests of the ridges) under conditions of similar solar exposure (neutral, but with predominance of sunny slopes for the Mm flora and shady slopes for the An flora) and with the same plain meso- and microrelief (the steepness of 0–16°). However, the buffer flora (MA) is widespread mainly on the E and TE types of local sites, both on flat interfluves of the valleys and on the crests and peaks of the ridges. Here, in both cases there is the mixing of Manchurian mesophilic and Angaridian continental floras, with larch penetration into cedar/broad-leaved forests. However, here there are also the widespread broad-leaved forests with oak (Mc flora) – the consequence of simultaneous aridization of mesophilic Manchurian forests. As we can see, the really existing mosaic of the floras under consideration in the Lower-Amur low mountain Sub-Pacific is not rigidly bound to the geomorphological framework and is formed by the bioclimatic system – century-long and super century-long climatic fluctuations, as well as ectogenetic successions of forest vegetation per se. As is known (Vasilevich 1983), vegetation does not passively follows the changes in the environment but transforms them through the mechanisms of inter- and intraspecies competition; therefore, entirely different communities occur under the same climatic conditions, which is especially typical of the Priamurye boreal ecotone.

10.5. Ecological and Landscape-Forming Role of Lithological Factor

The lithological composition of bedrocks has moderate influence on the distribution of floristic phratries ($C A;B) = 0.120$) and much stronger influence on forest types ($C(A;B) = 0.270$). These relationships quite definitely duplicate the ordination of forest geo(eco)systems by absolute heightdue to the ordered altitudinal deposition of bedrocks (Table 10.2; Figure 10.5), which provides for clearly differentiation of high-altitude terrains (microlandscapes).

Table 10.2. The distribution of lithological bedrock composition by absolute heights (the normalized partial coupling coefficients

C(A;B) = 0.163)

True altitude, m	Bedrock Granodiorites, diorites, diorite porphyrites , GeDr	Conglomerates, breccia, gravelites, CoDr	Aleurolites, sandstones, clay slates, PSL	Lake-alluvial deposits (sands, clays, gravels), LaAl
20–40			0.10	**0.50**
40–110		0.18	0.18	0.19
110–190		0.21	**0.21**	0.11
190–310		0.21	0.17	0.21
310–450		**0.41**	**0.21**	
450–680	**1.00**		0.14	

Footote: The domination fields of ecological niches are showed by half-rich print.

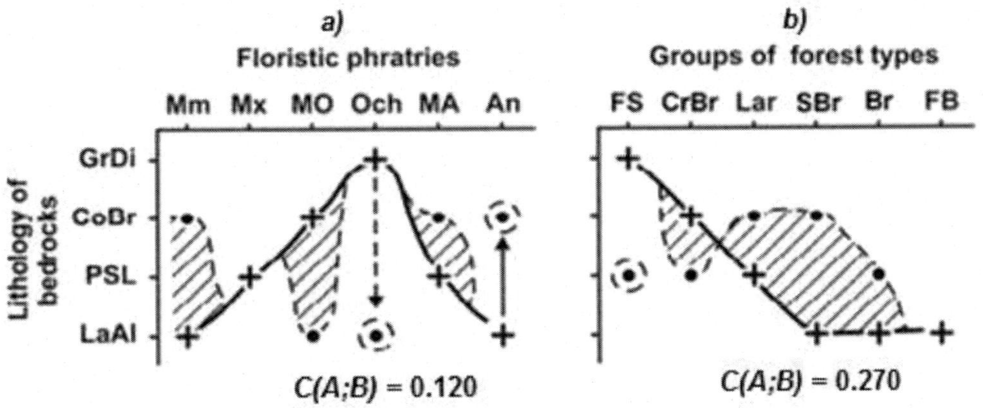

Figure 10.5. The binary ordination of floristic phratries and groups of forest types by lithological factor. Legends are as in Figure 10.3 and Table 10.2.

The fir/spruce forests of the Okhotian phratrie are predominant in the transit shadowy microlandscapes of the upper band of low mountains with the outcropping massive crystalline rocks (diorites, granodiorites, etc.). The mountain xerophytic cedar/broad-leaved forests (with oak) of eluvial–transeluvial and transit sunny microlandscapes grow on sedimentary/metamorphic rocks, mainly rudaceous (conglomerates and breccia) and less frequently fine (sandstones, clay shales, and aleurolites). The ecological dominant of buffer spruce/broad-leaved forests of sunny-transit microlandscapes also corresponds to rudaceous rocks, while the fine sedimentary-metamorphic rocks create the domination region of mountain larch forests of the Manchurian–Angaridian phratrie in the microlandscapes of transaccumulative piedmonts. Finally, the loose lake-alluvial sediments of eluvial (plakors), accumulative and superaqual microlandscapes create conditions for the growth of mesophytic broad-leaved forests, valley larch and fir/spruce forests, as well as larch mari.

Characteristically, most forest types have their domination regions on the most disperse and least dense bedrocks, while the "fuzzy" parts of niches are spread over the more rudaceous and less dense rocks.

10.6. Zonal Groups of Forests and Mechanisms of Formation of Buffer Communities

The zonal-geographical structure of phratries is represented by the ground cover as an indicator of *biologically equivalent habitats*, according to A. Cajander (cited by: Sukachev 1972). The Manchurian phratries are most interesting in this respect, as they were a basis of forest cover formation in the Priamurye sub-Pacific. The ground cover in the forests of the Mm phratrie consists by 40% of a complex of boreal + nemoral species (dominants), with a considerable enclave of merely boreal species, which gives a total of 74%. In the meantime, in the Manchurian xerophytic flora (the broad-leaved forest with oak) this sum is not more than 19% and the boreal species are completely absent. The domination region is represented by the boreal-nemoral species with a probability of 62%, while the "fuzzy" part of the niche equally encompasses both the mixture of boreal + nemoral species and nemoral-boreal spacies.

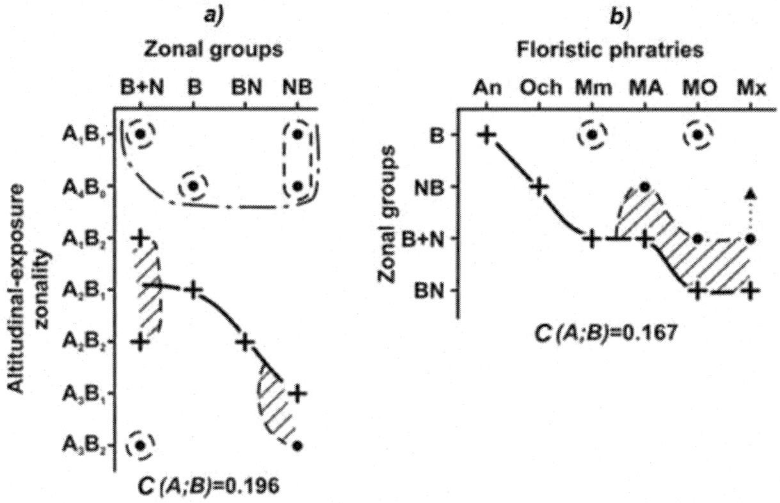

Figure 10.6. The ecological niches of zonal groups of forests in the spaces of altitudinal-exposure zonality and floristic phratries. Legends: High-altitude layers: I – low-mountain; II – piedmont-plain. Zonal-geographical groups: B – boreal; NB – nemoral-boreal; B+N – boreal and nemoral (equally); BN – boreal-nemoral. Other legends see in the text and in Figure 10.3.

The boreal-nemoral group of cedar/broad-leaved forests of the Manchurian mesophytic flora is predominant on the crests of ridges and ridge-top sunny slopes. The nemoral-boreal group of the Ajan fir/spruce forests also occurs on transeluvial and transit shadowy and neutral sites (zonality types A_3B_1 and A_3B_2, see above). The buffer spruce/broad-leaved communities with boreal and nemoral herbage species in approximately equal proportions appear closer to the bottom of ridges. However, boreal species become predominant in TA microlandscapes due to formation of the nemoral-boreal group of larch forests of the buffer Manchurian–Angaridian phratrie.

The zonal structure of the buffer floristic phratries sheds some light on the mechanisms of their formation (Figure 10.6). In the buffer Manchurian–Okhotian flora, the boreal-nemoral herbage species are predominant (the probability of 47%), though merely boreal species occur up to 20% (the heritage of the basal Okhotian phratrie where they make 100%). Since the boreal-nemoral species are absent in the Manchurian mesophytic flora, it may be assumed that the buffer MO phratrie originated from the Okhotian flora mixed with the Manchurian mesophytic but not xerophytic flora. Then, the buffer Manchurian–Angaridian phratrie was formed by mixing the Angaridian flora with the same Manchurian mesophytic flora.

It is demonstrated, first, by the dominant position of boreal + nemoral species both in the Mm flora (with a probability of 40%) and in the MA flora (54%) and, second, by the substantial presence of boreal species in the Mm flora (34%), while their occurrence in the An flora is 73%.

Thus, both buffer phratries (Manchurian–Okhotian and Manchurian–Angaridian) were formed on the basis of the same Manchurian mesophytic phratrie, which proved to be much closer to other basal phratries (MO and An) by the factor of biologically equivalent habitats compared to the xerophytic variant of Manchurian flora.

10.7. Geotopology of Forest Ecosystems in the Lower-Amur Sub-Pacific

10.7.1. Geotopes as the Intra-Landscape "Core" of Ecological Connections

The distribution of floristic phratries and groups of forest types by local types of sites (Figure 10.7, *a* and *b*) is a rather clear manifestation of floro- and coenogenesis at the level of microcatenas – the coupled system of elementary forms of mountain-valley relief (geotopes). The profiles of the areas of dominance of floristic and phytocoenotic characteristics represent a certain collective microcatena. Within each microlandscape, its floristic phratries and groups of forest types are distinctively distributed over this catenary system consisting of three pleiads of geotopes. The first and third pleiads combine representatives of the most contrasting and therefore most competitive floras. The first, theautomorphic pleiad of E and TE types of local sites is occupied by cedar/broad-leavedcommunities of the Manchurian xerophytic phratrie and, to a considerable extent, the complex Manchurian–Angaridian larch forests. Cedar forests, over 67% of their areals, are concentrated on flat interfluves, while the area of dominance of larch forests (with a much lower probability of 39%) is on the upper convex parts of slopes. The mixing of Manchurian floras with the Angaridian one is also sporadically manifested on flat interfluves and in the lower elements of microcatenas.

The second, most extensive pleiad comprising the middle and lower parts of slopes (T-TA and A geotopes) is the area of dominance of Okhotian and Manchurian–Okhotian floristic complexes, as well as the mountain part of the Angaridian flora. Finally, the third accumulative–superaqual pleiad of geotopes is occupied by mesophytic broad-leaved forests, on the one hand, and permafrost/taiga larch mari (the lowland-kettle variant of Angaridian flora), on the other hand.

Thus, occurring actually at the same absolute heights (from 20–80 m to 400–450 мm), the two types of Manchurian flora are spaced by the microcatena profile: the forests of the Mc phratrie are confined to E and, more seldom, TE geotopes, while those of Mm are confined mainly to superaqual geotopes. The communities of Okhotian and Manchurian–Okhotian floristic complexes are formed in the vicinity of the latter (at the Ta–A geotopes), in the middle zone of low mountains.

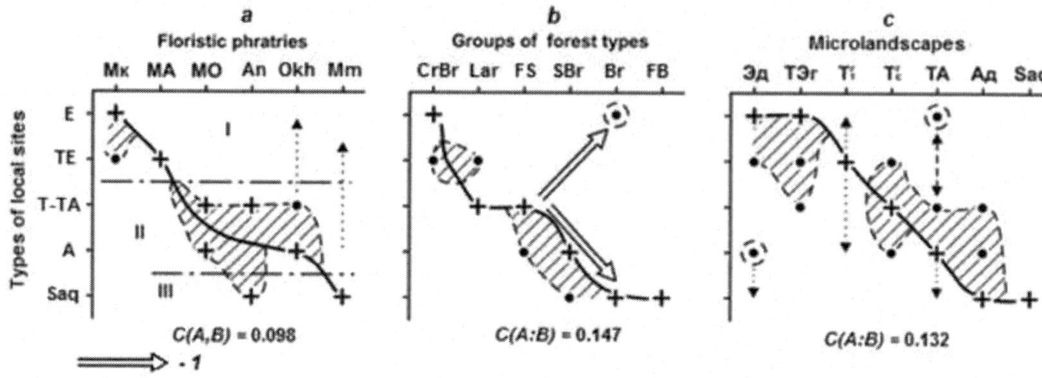

Figure 10.7. The binary ordination of floristic phratries, groups of forest types and microlandscapes by the factor of local site. I–III – Correlation pleiads (see in the text). *1* – Directions of local transgression of silver-spruce forests to areals of broad-leaved and cedar/broad-leaved forests. Other legends are as in Figure 10.3.

The mixing of Manchurian flora with Okhotian flora, on the one hand, and Angaridian flora, on the other hand, was quite clearly differentiated not only with respect to slope exposure (see above, Chapter 9) but also with respect to geotopes. On the gentle-concave geotopes of sunny slopes, the Ajan spruce and the accompanying taiga elements penetrated into mesophytic cedar/broad-leaved forests. At the same time, larch penetrated into xerophytic (with oak) cedar forests widespread on the crests of bald mountains and ridge-top convex slopes. With respect to height, the areals of forests of the two formed buffer phratries (Manchurian–Okhotian and Manchurian–Angaridian) overlap very rarely. The former occupies mostly the middle and upper zones of low mountains and the latter occupies the lower zone and piedmont plains.

Let us trace back the distribution of local types of sites over microlandscapes (see Figure 10.7, *c*). It is actually the relationship between micro- and mesocatenas in the given statistical sample. The relationship is quite significant and very illustrative. Ecological dominants show that the local types of sites, which we have distinguished by the groups of forest biogeocoenoses, generally represent the adequate higher-rank territorial unities – microlandscapes. At the same time, in the lower elements of the catena, the "fuzzy" fields of conformity comprise the geotopes more drained relative to the dominants. The opposite pattern is typical of the E and TEm geotopes. It is noteworthy that the second (T–TA and A) pleiad of geotopes forms a rather vast group of microlandscapes including their T^{m}_{sun}, TA and Ad types.

The resultant connections yield the structural and functional characteristics of microlandscapes by the groups of forest types, which belong to some type of local sites or another. Each of these groups includes mainly forest communities that may characterize the adequate type of site of the higher hierarchic level – mesocatena elements in the mountain–valley system. Such inter-rank connections and relationships at the qualitative level have been considered previously by B.P. Kolesnikov (1956). He distinguished the following three zones at the scale of mountain-valley relief: drifting, transit and accumulation of groundmass with a consistent rise of the groundwater table. Against this general background of three coupled mountain-valley types of growth sites, there are widespread local catenas (with local erosion bases) reproducing in miniature the mesolevel of landscape couplings.

10.7.2. Geomorphology of Low Mountain Microlandscapes

The highly distinct ordination of microlandscapes by altitudinal-exposure zonality and by absolute height per se has been established (see Figure 10.3 *c*). At the same time, in the sample under consideration, the test areas (geographical points) on eluvial–transeluvial and transit sites cover almost the entire altitudinal range of low-mountain ridges, beginning from 200 m. However, the transit groups of biogeocoenoses on shady and neutral slopes are predominant (with more than 40% probability) in the highest zone of low mountains (above 450 m) transitional to middle mountains. The sunny transit sites are predominant (with the same probability) at the lower absolute heights: in the middle low-mountain zone (below 200 m).

The graphs of partial coefficients of connections clearly show two altitudinal pleiads of microlandscapes – piedmont-plain and low-mountain. The latter pleiad is characterized by the highest geomorphological diversity. It especially concerns the peaks and crests of bald mountains, as well as the adjoining upper parts of mountain slopes. Transaccumulative microlandscapes occupying the margins of morphotectonic depressions are transitional between the two pleiads.

The types of sites are directly related to the important "direct-acting," according to L.G. Ramensky (1971), landscape-forming attribute – the degree of lithomorphism–hydromorphism. As it has been mentioned (Rakovskaya et al. 1984), introduction of the characteristic of edaphic moistening as a classification criterion at the lowest stages of the hierarchy of regional natural complexes makes it possible to bring together the landscape-forming effects of zonal and azonal factors.

The alternative manifestations of litho- and hydromorphism are determined by the competitive relationship between the two properties of edaphotopes: the mechanical composition of soil-forming rocks and the degree of soil-ground moistening, which is related to drainage intensity. In this context, these properties were characterized by the content of lithomass (bedrock fragments) in soil horizon B and the moisture content of horizon A_1.

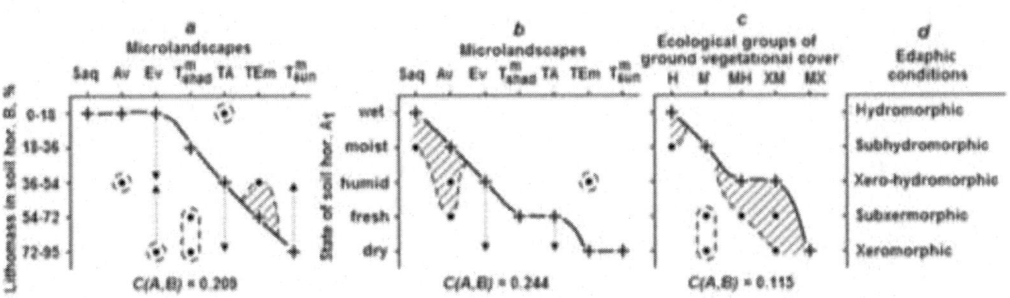

Figure 10.8. The binary ordination showing the lithomorphity–hydromorphity of habitat conditions. Legends are as in Figure 10.3.

On the binary ordination graphs (Figure 10.8, *a, b*), the distinguished microlandscapes form two ecological gradients characterizing the mutually exclusive manifestations of the processes of litho- and hydromorphization. It is appropriately expressed in the ecological groups of soil cover (see Figure 10.8, *c*). Finally, there is a certain inversely proportional relationship between the soil moisture and rubble content of soil (Table 10.3): in general, the lower the rubble content, the higher the moisture content.

Table 10.3. Dependence of moisture content in soil hor. A_1 from lithomass content in hor. B (partial association coefficients)

Content of the lithomass in hor. B, %	State of soil hor. A_1				
	wet	**moist**	**humid**	**fresh**	**dry**
0–18	**1.00**	0.40	0.05	0.24	0.24
18–36			**0.62**		
36–54		**0.60**	0.08	0.24	0.24
54–72			0.25	0.19	0.20
72–95				**0.34**	**0.31**

Footnote. The domination fields of ecological niches are showed by half-rich print.

The transition from low-mountain and elevated mesosites to lowland-valley sites is accompanied by worsened drainage and the respective activation of hydromorphization processes, with the weaker role of lithomorphism. The process of lithomorphization manifests itself mainly on well-drained fragmented uplands with the removal of fine-grained sediment, while hydromorphization is typical of lowland-valley sites with the accumulation of fine-dispersed material and the overload of surface and ground waters. Under mountain relief conditions, the solar exposure of slopes makes certain corrections in this scheme: sunny slopes become merely lithomorphic (microlandscape $T^{m}{}_{sun}$), while shady slopes shift towards hydromorphism (litho-hydromorphic microlandscape $T^{m}{}_{shad}$).

It would be interesting to note that eluvial (flat-interfluve) microlandscapes of the elevated parts of plains, which represent the zonal-regional bioclimatic norm of the Priamurye boreal ecotone, are in sharp contrast to the neighboring lowland Ad subhydromorphic landscapes and occur in the sublithomorphic group – together with mountain TE_m and piedmont TA natural complexes. The valley flat interfluves have a very broad dispersion of the degree of soil rubble content and an equally extended range of moisture content in hor. A_1 (from dry dominant to moderately humid enclave). Due to such contrasting litho-hydroedaphic conditions, the sites become an area of distribution of different floras and, accordingly, are occupied by forest communities of different types.

10.7.3. Distribution of Forest Communities over Microlandscapes

The system of ecological niches of floristic phratries and groups of forest types in the space of microlandscapes clearly demonstrates two geomorphological complexes of forests: mountain and piedmont-plain (Figure 10.9). The obtained models considerably particularize this phenomenon described by B.P. Kolesnikov (1956) for cedar/broad-leaved forests of Priamurye and by V.A. Rosenberg (1963) for fir/spruce forests of the Lower Priamurye.

The eluvial (flat-interfluve) microlandscapes of piedmont plains and broad river valleys (E_D) are confined by 80% of their areas to the sunny slopes of moderate and even enhanced steepness (from 3–5^0 to 10–15^0) providing their good drainage, which is necessary, as is known (Vysotsky 1909), for the formation of exactly flat-interfluve types of plant cover.

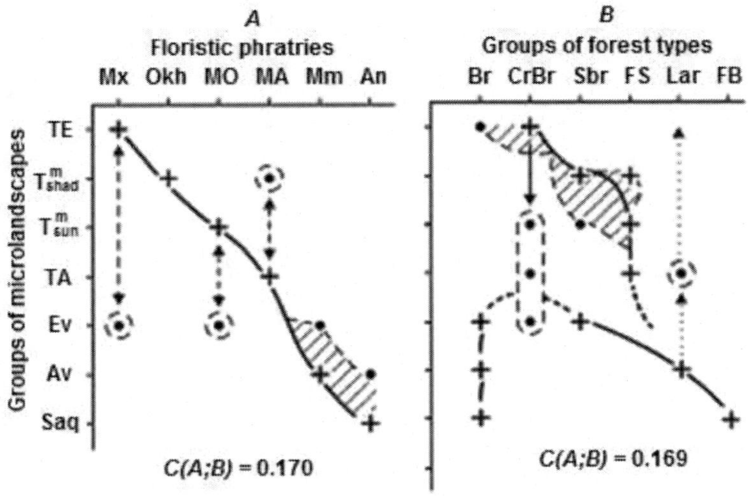

Figure 10.9. The ecological niches of floristic phratries and groups of forest types in the paragenetic system of microlandscapes. Legends are as in Figure 10.3.

It is significant that none of the floristic phratries (basic or buffer) and none of the forest types are represented by their dominants in flat-interfluve microlandscapes. This fact suggests that each of these florogenetic and phytocoenological units cannot be the only zonal representative of the Priamurye boreal ecotone. They are dispersed by their dominant areals over different elements of mesocatenas and over the sites with different solar-circulation exposure. It can only be noted that mainly the forests of three floristic phratries (Mx, Mm and MO) occur on flat interfluves as subdominants. At the same time, Manchurian flora per se, being initial basic one for the general appearance of forest cover of the plain-lowland Lower Amur Sub-Pacific in this ecoregion, is localized enough and represented by disconnected areas.

In the spatially ordered (by microlandscapes) series of floristic phratries, the two varieties of Manchurian flora (mesophilic and xerophilic) are "scattered" over different poles of mesocatena, which has also been observed for the broad-leaved/coniferous zone of the Middle Sihkote-Alin (Krestov 1993). Cedar forests with oak are confined mostly to ridge crests and ridge-top slopes, and lime forests are confined to lowland plains and river valleys. The Ajan fir/spruce forests inhabited shady mountain slopes, while the Manchurian mesophytic and Okhotian floras were mixed on sunny slopes. The transgression of larch forests of the An permafrost-taiga flora from the bottoms of small valleys up the slopes resulted in the formation of forests of the MA buffer phratrie in the lower zone of low mountains.

These territories are occupied by forests of highly variegated floristic composition and coenotic structure. Mainly Br and CrBr forests grow here (with the total frequency of occurrence of nearly 60%). However, if the former belong to Manchurian flora entirely, the latter belong there only by 49%. In cedar forests, the representatives of Manchurian–Okhotian and Manchurian–Angaridian floras take a particular place (27% and 14%, respectively). At the same time, up to 27% of the area of flat interfluves is occupied by valley transitional spruce/broad-leaved forests belonging to Okhotian flora by 45%. Thus, the diversity of floras and coenotic structures of the forests of valley flat interfluves as local representatives of background (zonal-regional) bioclimatic norm quite definitely characterize the transitional (buffer) character of the Priamurye boreal ecotone.

The accumulative piedmont-plain microlandscapes are in sharp contrast to eluvial (flat-interfluve) microlandscapes with respect to flora. Here there is the almost exclusive predominance of larch forests of Angaridian taiga phratries and, in part (at the TA sites of foothills), the forests of the buffer MA flora.

The flat crests and ridge-top sunny slopes are only 22% of eluvial-transeluvial mountain landscapes (E–TEm) in the upper zone of low mountain country (height intervals of 310–680 m). They are occupied almost exclusively by representatives of the Mc flora comprising mountain cedar/broad-leaved forests. The major territory of these microlandscapes, with the shady and neutral solar exposure of middle-steep slopes (up to 15–25°), is covered by dark-coniferous Ajan forests of the Okhotian flora. In the middle and lower zones of low mountains (transit microlandscapes with the maximum slope steepness – up to 20–55° and more) are predominant: the Ajan fir-spruce forests on shady slopes ($T^m{}_{shad}$) and the buffer Manchurian–Okhotian communities on sunny and neutral slopes ($T^m{}_{sun}$). The transition to the foot of the ridge (TA) and then to inter-mountain plains and river valleys (accumulative microlandscapes Av) is accompanied by the substitution of East-Siberian floras for Manchurian and Okhotian floras. The cedar/broad-leaved and fir/spruce forests are replaced by larch forests (pure or with an admixture of broad-leaved species) and larch mari on closed highly overwetted superaqual (Saq) sites. The ecological dominants of the above forest communities are at subhorizontal sites.

The maximum variegation in spatial distribution of the groups of forest types is typical of sunny and neutral transit slopes, as well as foothills. In the system of geomorphological differentiation of forest types, it is a kind of "diffusion" altitudinal zone of forest cover with the maximum uncertainty of relationship between phytocoenological units and the characteristics of mountain-valley relief (see Figure 10.9,*b*). There are no ecological dominants of forest types in this zone. The neighbors are mostly cedar/broad-leaved and fir/spruce forests. They are supplemented with the buffer spruce/broad-leaved communities on sunny slopes and larch forests, mainly of the Manchurian–Angaridian floristic phratries, in the foothills.

The Angaridian flora is represented by only larch mari and transitional taiga communities on vast supra-floodplain terraces of the Amur River. The main group of larch forests belongs to the buffer Manchurian–Angaridian floristic phratrie, where broad-leaved species are mixed with larch. Simultaneously, boreal and nemoral species are mixed in the soil cover. These forests are widespread mainly in the lower zone of low mountains (at the absolute heights of 100–200 m) at eluvial sites of small uplands and, more rarely, at the TA–A sites of piedmont trails.

10.8. THERMO- AND HYDROEDAPHIC CONDITIONS OF FOREST GROWTH

10.8.1. Soil Temperature

Soil temperature and moisture content are the most important local factors determining the floristic and phytocoenotic structure of forests, as well as their productivity. Information models were used for detecting binary relations of floristic phratries and groups of forest types with hydrothermal soil parameters (Tables 10.4 and 10.6), taking into account the adverse effect of forest canopy on soil temperature through the lighting of soil cover as an indicator of the screening effect of this canopy on penetrating total solar radiation. In general, there is an inverse relationship between soil temperature and soil cover lighting.

The summer temperature field of model territory soils in the Lower Priamurye is characterized by highly marked contrasts, quite clearly correlating with forest types of different floristic phratries. The maximum amplitude of temperature fluctuations at the same soil depths between these categories of forest communities of the Priamurye Sub-Pacific is up to 5–7°, which demonstrates their high sensitivity to the changes in heat and energy resources. At the same time, there is a rather tight relationship between forest types and summer moisture content in forest litter and different genetic horizons of soil (see below).

In the forest types of Manchurian flora (broad-leaved and cedar/broad-leaved), the summer soil temperature is much higher than in all other types of forest (Tables 10.4 and 10.5; Figure 10.10). This gap is 1.5–1.5° at the depths of 30–40 cm and up to 3.5–4.0° in the deeper soil layers. In the first place, it is due to the role of favorable solar exposure of the sites of these forests (see Figure 10.7). At the same time, the soil in xerophytic broad-leaved forests is slightly warmer than in mesophytic cedar forests due to much lower amounts of total skeletal phytomass and green mass and, partially, undergrowth and understory mass (see below Figures 14.2 *A, B* and 14.3 *A, B*).

As a result, the lighting of soil cover is up to 27–37000 Lx in the former case and no more than 9000 Lx in the latter case (Table 10.6). The temperature of long-frozen soils (with sporadic permafrost) under larch mari is the lowest. The exception is only the upper 30-cm layer, which is well heated in the daytime under the conditions of sparse tree layer and the absence of dense shrub. The soils in the Ajan fir/spruce forests of the Okhotian phratrie are slightly warmer, mainly because they occupy shady mountain slopes (see Figure 10.4). This factor prevails here over the positive role of enhanced lighting under forest canopy determined by the moderate values of live above-ground phytomasses.

Table 10.4. The areas of soil temperature dominance at different depths in different groups of forest types

Depth, cm	Groups of forest types					
	Br	CrBr	SBr	FS	Lar	FB
30	15.1–17.2	13.5–17.2	12.3–13.5	8.6–11.0	11.0–12.3	12,3–13.5
40	13.3–14.3	14.3–16.6	11.0–11.5	8.5–11.0	11.0–11.5	8.5–11.0
50	12.0–13.5	13.5–16.6	7.1–10.7	8.2–10.3	10.0–10.7	7.110.0
60	12.0–13.3	12.0–16.6	7.1–10.0	8.0–9.5	9.0–10.0	7.1–10.0
70	12.0–13.0	9.5–15.5	6.0–8.0	6.0–8.9	3.7–6.0	3.7–6.0

Table 10.5. The normalized partial coefficients of connection between floristic phratries and soil temperature at a depth of 50 cm

Temperature of soil, °C	Floristic fratries					
	Okh	An	MO	Mm	MA	Mx
7.7–10.0	**0.51**	**0.46**				0.20
10.0–10.7	0.38	0.23	0.24		0.24	0.10
10.7–12.0	0.11	0.31	**0.67**	0.27	0.22	
12.0–13.5			0.12	**0.73**	0.24	0.20
13.5–16.6					**0.30**	**0.50**

Footnote. The domination fields of ecological niches are showed by half-rich print.

Table 10.6. The normalized partial coefficients of connections between the lighting of soil cover and the groups of forest types

Lighting, Lx	Groups of forest types					
	CrBr	SBr	Lar	FS	Br	FB
1780–9000	**0.50**	0.23	0	*0.15*	*0.13*	0
9000–12800	0.29	*0.18*	0.26	*0.11*	*0.20*	0
12800–18200	*0.11*	**0.41**	**0.44**	0	0.22	0
18200–27000	0.10	0.18	*0.13*	**0.45**	*0.20*	0
27000–38100	0	0	*0.17*	o.29	0.25	**1.00**

Footnote. The domination fields of ecological niches are showed by half-rich print, the insignificant connections – by the italics.

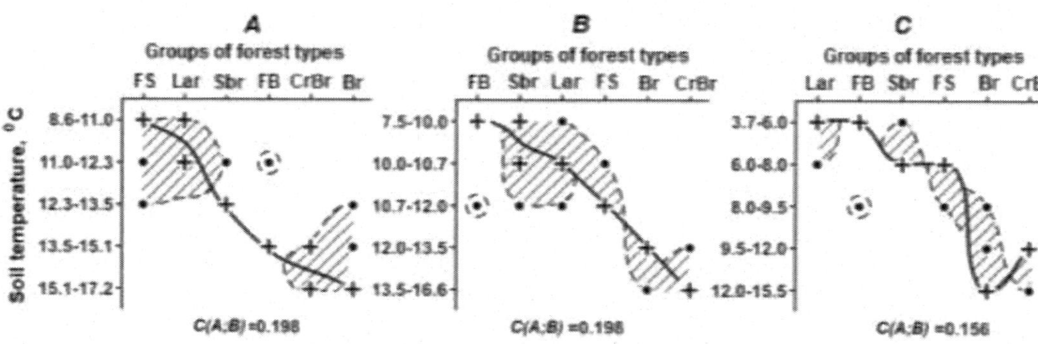

Figure 10.10. The ecological niches of groups of forest types in the space of soil temperatures at the depths of 30 cm (*A*), 50 cm (*B*), and 70 cm (*C*). Legends are as in Figure 10.3.

In spruce/broad-leaved forests of the buffer Manchurian–Okhotian phratrie, soils are warmer than in fir/spruce forests due to almost exclusive confinement of these forests to sunny mountain slopes (see Figure 10.4, *B*). Indeed, soil temperatures here must be slightly lower because of the higher mass of forest stand and, particularly, dense undergrowth reducing soil cover lighting.

In larch forests of the Angaridian and Manchurian–Angaridian floras, where forest canopy is not so dense, the upper half-meter soil layer is warmer than in dark-coniferous taiga, and soil temperature drastically decreases (to the temperature level in larch mari) only beginning from the depth of 70 cm affected by winter cold reserve. It is especially typical of the valley larch forests growing under conditions of winter temperature inversions.

The phenomenon of temperature inversion under conditions of mountain-valley relief is in good agreement with numerous facts of disturbance of altitudinal zonality: at the middle parts of slopes there are more thermophilic groups than in the valleys. There are especially high contrasts between sunny slopes and the bottoms of narrow valleys. Even northern slopes have the higher summer temperatures and the longer duration of frostless and vegetation periods compared to valley bottoms (Kolesnikov 1956). In addition to winter chilling, which results in the deep freezing of soil, the durable convection fogs typical of deep intermontane valleys and depressions also play a negative role here.

The revealed patterns of forest type ordination by soil temperature form a certain geophysical basis for making local landscape-ecological predictions by the methods that we have developed (Kolomyts 2008). Soil temperature variations can lead to quite significant transformations of Manchurian flora forest types into all other types of Okhotian–Kamchatian or East-Siberian floristic phratries, and vice versa. The equally distinct forward and reverse transitions could be expected between larch and spruce/broad-leaved forests with the respective temperature signal (Br ↔ FS, CrBr ↔ FS, Br ↔ Lar, CrBr ↔ SBr, etc.). At the same time, it will be difficult to reveal the Br ↔ CrBr and FS ↔ SBr mutual transitions, especially at a weak temperature trend, because both members in each pair of forest types have similar basic soil temperature conditions.

10.8.2. Soil Moisture Content

In contrast to temperature, soil moisture content was assessed only qualitatively and rather roughly – by the edaphic grid of the types of habitat by P.S. Pogrebnyak (1955). As it has already been mentioned, five categories of moisture content were distinguished: waterlogged (wet), moist, humid, fresh and dry soil.

With respect to moisture content in forest litter (hor. A_0) and humus horizon A_1, floristic phratries, as well as coenotic forest types, are differentiated quite distinctly (Table 10.7; Figure 10.11) and confine themselves to the regular series of their ecological dominants. The abnormally hydromorphic larch mari with the waterlogged soil category are followed by dark-coniferous Ajan forests with wet soils, then moderately moist spruce/broad-leaved, broad-leaved and larch forests and, finally, cedar forests with fresh and dry soils. At the same time, mountain fir/spruce forests growing on steep rocky slopes give an enclave of dry soils on the graphs of binary ordination. The type of broad-leaved forests is also characterized by distinct bifurcation by moisture content in horizons A_0 and A_1. The two of its areas of dominance distant from each other (moist and dry soils) belong to the two floristic phratries of nemoral forests – mesophytic and xerophytic.

The narrowness of hydroedaphic niches of horizons A_0 and A_1 consisting mainly of the dominant areas is indicative of rather high sensitivity of forest communities to the above-mentioned very gross changes in the moisture content of upper soil layers. Especially strong connections are observed for forest litter ($C(A;B) > 0.23$), which thereby has the highest hydroecological gradient and, consequently, can be used as one of the basic predictors in local landscape-ecological forecasts. At the same time, mutual phytocoenotic transitions with respect to moisture content in forest litter will manifest themselves most clearly between the forests of the Angaridian and Okhotian floras, on the one hand, and the communities of the Manchurian floras, on the other hand. The mutual transitions of forests of the following floras may also be significant: An ↔ MO, An ↔ MA, Okh ↔ MO, Okh ↔ MA, Okh ↔ Mm, and Okh ↔ Mx. The hydrogenic phytocoenotic transformation in binary connectives Mm ↔ Mx, Mm ↔ MA, MO ↔ MA, and MO ↔ Mm will become much more problematic.

Table 10.7. The soil moisture content in forests of different floristic phratries (partial coefficients of connection)

a) Llitter-fall index (*C(A;B)* = 0.219)

State of the soil	Floristic fratries					
	An	**Okh**	**MO**	**MA**	**Mm**	**Mx**
Wet	**0.52**					
Humid	0.34	**0.53**				
Moisture			**0.48**	**0.46**	**0.50**	
Fresh	0.14	0.21	0.19	**0.46**		0.38
Dry		0.26	0.32	0.08	**0.50**	**0.63**

b) Horizon B (*C(A;B)* = 0.192)

State of the soil	Floristic fratries					
	An	**Okh**	**MO**	**MA**	**Mm**	**Mx**
Wet	**0.54**	0.23				0.23
Humid		**0.46**	**0.51**			
Moist	0.17	0.19	0.11	**0.72**	**0.44**	
Fresh	0.14	0.12	0.38	0.28	0.26	0.11
Dry	0.15				0.30	**0.66**

Footnote. The domination fields of ecological niches are showed by half-rich print.

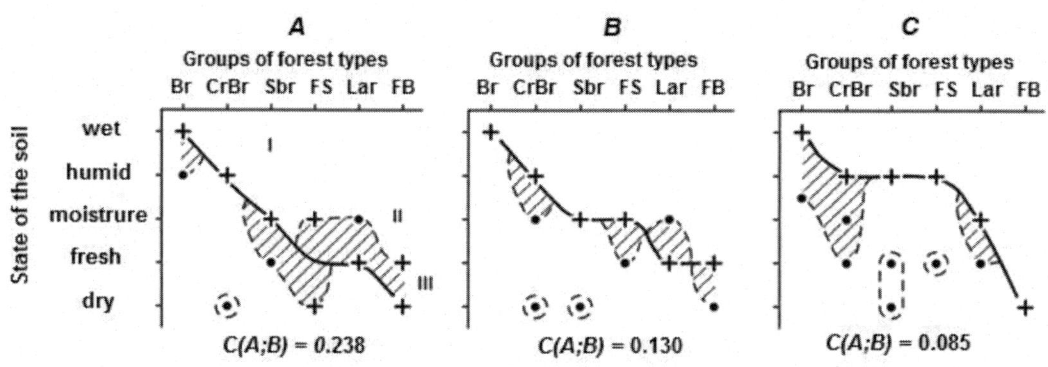

Figure 10.11. The ecological niches of groups of forest types in the space of moisture content in soil horizons A_0 (*A*), A_1 (*B*) and B (*C*). Legends: I–III, three levels of soil moisture content; other legends are as in Figure 10.3.

In contrast to the upper soil layers, the illuvial-metamorphic horizon B has less definite hydroedaphic niches for most forest types, with a broad range of moisture content and with enclaves – from the wet category to fresh and dry. Only overwetted larch mari and dry cedar forests form the two contrasting "poles" of soil moisture content. It seems that forest communities are rather tolerant to moisture variations in soil horizon B.

Generally, the "fuzzy" parts of hydroedaphic niches over the entire soil mass are directed to the lower moisture content relative to their dominants. This fact indicates that phytocoenological changes in forest cover can be most significant and stable under humidization of the climate, whereas forest communities of the forest type rank must demonstrate higher tolerance to the arid signal, with a shift of ecological dominant of the niche within its "fuzzy" part.

The combinations of mechanical composition (trophicity) and moisture content of soils determine the respective *types of habitat* (litho-hydroedaphotopes). This attribute is quite precisely distributed over microlandscapes and exerts the most significant effect on the groups of forest types (see Figure 10.1). The optimal forest growth conditions formed by the richest (middle- and heavy-loamy and clay) substrate with weak and moderate moistening are formed on sunny and neutral mountain slopes, as well as at accumulative valley sites. Under these conditions, not only broad-leaved forests are predominant but also, as it should be emphasized, spruce/broad-leaved forests (Figure 10.12). Thus, the mixing of the Manchurian and Okhotian floras, with formation of buffer forest communities, occurs under the most optimal conditions of habitat both on mountain slopes and in river valleys.

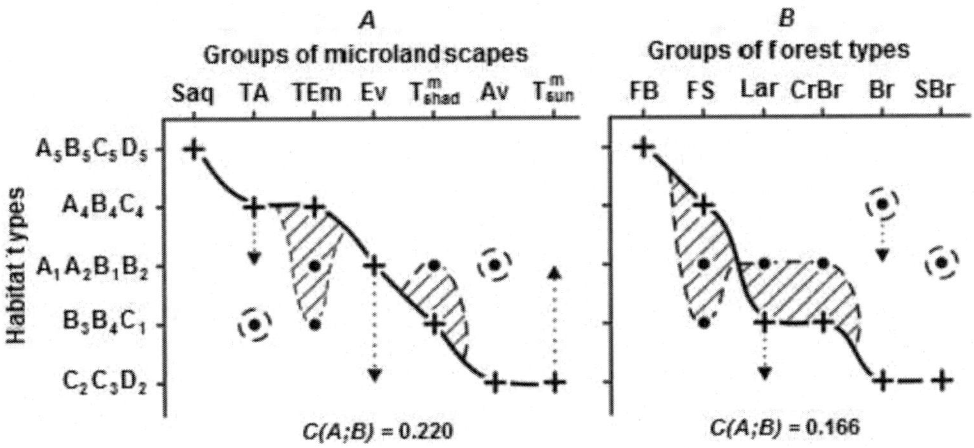

Figure 10.12. The binary ordination of microlandscapes and groups of forest types by the types of habitat. Gradations of the types of habitat (Pogrebnyak 1955): $A_5B_5C_5D_5$ – sands, sandy loams and clays (moist and wet); $A_4B_4C_4$ – sands, sandy loams and wet light loams; $A_1A_2B_1B_2$ – sands and light sandy loams (dry and fresh); $B_3B_4C_1$ – sandy loams and light loams (moist and wet); $C_2C_3D_2$ – medium and heavy loams and clays (moist and fresh). The predominant areas of ecological niches are in bold. Other legends are as in Figure 10.3.

Transeluvial mountain and transaccumulative piedmont microlandscapes, on the contrary, are generally characterized by less favorable litho-hydroedaphotopes (wet, more rarely fresh and dry sandy loams and sands), though with enclaves of a richer substrate (light loams). These habitats are almost solely occupied by the Ajan fir/spruce forests. Flat-interfluve microlandscapes (Ed sites) have a rather diverse set of habitat conditions – from dry and fresh sands and sandy loams to wet light loams and moist clays, which predetermined the proper wide range of forest types occupying valley flat interfluves (see Chapter 9).

It would be interesting to mention that the microlandscapes of shady mountain slopes with the least favorable heat and energy resources are characterized by the relatively satisfactory quality of habitat: here, moist and wet sandy loams and even dry and fresh loams are predominant. These litho-hydroedaphotopes provide a kind of compensation for the lack of heat to the cedar/broad-leaved and larch forests that grow here.

10.8.3. Evaluation of Edaphic Moistening of Forests

The assessments were performed for the ecological groups of soil cover. The binary ordination of floristic phratries and forest types with respect to the given factor yielded high normalized coefficients of contingency (Figure 10.13). Hence, it was possible to establish the following dominant categories of edaphic moistening for the forests of different phratries:

a) mesophilic and meso-xerophilic – for mesophytic cedar/broad-leaved forests;
b) xerophilic and xero-mesophilic – for xerophytic cedar/broad-leaved forests with oak;
c) meso-hygrophilic and hygrophilic – for Okhotian flora forests as a "core" of moderately moist ecological group of herbage; and
d) xero-, meso- and hygrophilic mixture, less often meso-hygrophilic – for larch forests and larch mari of the Angaridian phratrie, as well as mixed forests of the buffer Manchurian–Angaridian phratrie.

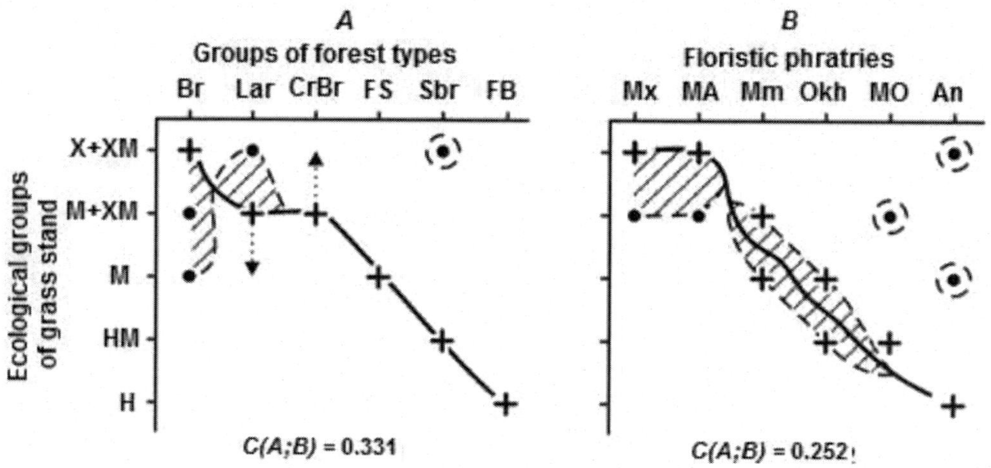

Figure 10.13. The binary ordination of groups of forest types and floristic phratries by the factor of edaphic moistening. Ecological groups of grass stand: X+XM – xeromorphic + xero-mesomorphic; M+XM – mesomorphic + xero-mesomorphic; M – mesomorphic; HM – hydro-mesomorphic; H – hydromorphic. Rest designations are the same as in Fig. 10.3.

In general, the mesophytic and xerophytic types of Manchurian flora, which we have distinguished on the basis of species composition of forest stands and, additionally, species composition of undergrowth, are in good agreement with the ecological groups of herbage. It also characteristic that in the forests of Manchurian–Angaridian and Angaridian floristic phratries (mainly in larch forests with the involvement of broad-leaved species) there is a widely represented mixture of meso- and xerophilic species of soil cover (in the form of evergreen dwarf shrubs), as well as hygrophytes (sphagnum, sedges, etc.).

Interestingly, there is a hydroedaphic dualism of spruce/broad-leaved forests (see Figure 10.13, *a*) determined by different soil-geomorphological conditions of the mixing of Manchurian and Okhotian floras. In the foothills (at the heights of 40–110 m), at the accumulative geotopes with loose and moist lake-alluvial deposits, the meso-hygrophilic variant of this Manchurian–Okhotian phratrie is formed (Figure 10.14, *a*). The herbage in these forests confined mainly to shady gentle slopes is more hygrophilic than in the initial hygro-mesomorphic fir-spruce forests of the basic Okhotian phratrie situated in the middle and upper zones of low mountains (at the heights from 200–300 to 550–650 m) on slopes of different exposure. The elements of Manchurian flora come here from below: from the neighboring superaqual broad-leaved forests with meso-hygrophilic herbage on fresh soils (Figure 10.14, *b*).

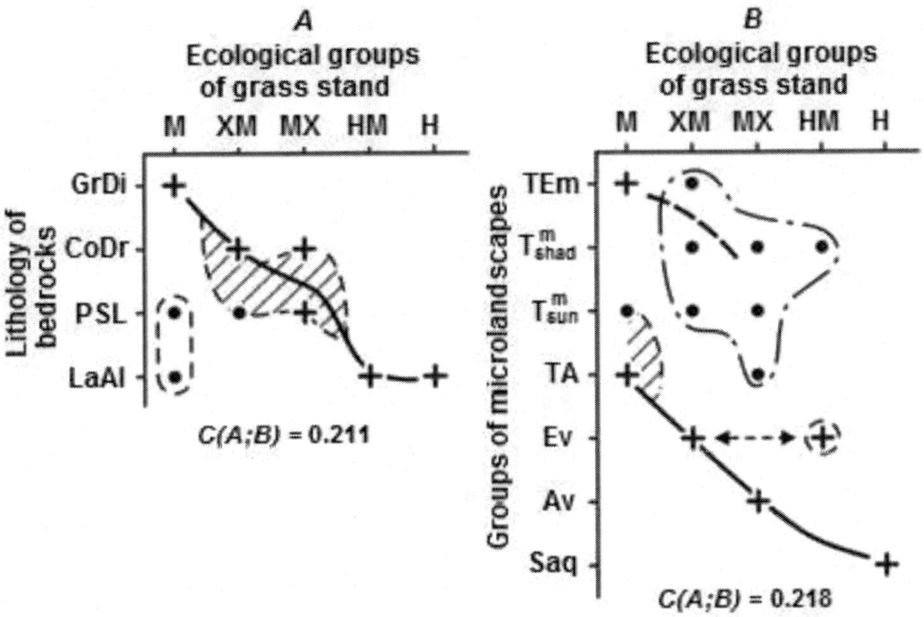

Figure 10.14. The binary ordination of edaphic moistening with respect to bedrock lithology and microlandscapes. Legends see in the Supplement to Table 10.2 and in Figure 10.3.

The second, xero-mesophilic variant of spruce/broad-leaved communities is formed in the lower zone of low mountains (the heights of 190–310 m), also with predominance of fresh soils on conglomerates and breccias. Here, the Manchurian xero-mesophilic and Okhotian floras are mixed at the transit and transaccumulative geotopes, also on shady and neutral but steeper slopes (up to 15–20° and more) providing good drainage and, accordingly, higher xeromorphism of this variant of the buffer Manchurian–Okhotian phratrie.

Table 10.8. The ecological niches of edaphic moistening in the space of states of moisture content in forest litter and soil humus layer

State of soil	Hor. A_0 of soil (*C(A;B)* = 0.134)					Hor. A_1 of soil (*C(A;B)* = 0.115)				
	Hy	**HM**	**M**	**XM**	**MX**	**Hy**	**HM**	**M**	**XM**	**MX**
Wet	**0.75**					**0.67**				
Humid	0.25	**0.51**				0.33	**0.36**			
Moist			**0.73**	**0.42**			0.14	**0.59**	**0.40**	
Fresh		0.20	0.15	0.34	0.44		0.26	0.27	0.31	0.32
Dry		0.29	0.12	0.24	**0.56**		0.24	0.14	0.29	**0.68**

Note. The ecological groups of soil cover: Hy – hygrophytes; HM – hygro-mesophytes; M – mesophytes; XM – mesophytes + xero-mesophytes; MX – xerophytes + xero-mesophytes. The dominant areas of ecological niches are in bold.

Then significant relationships between the ecological groups of herbage and the moisture content in forest litter and soil humus layer were obtained (Table 10.8). These relationships, first, confirmed the objectivity of distinguished (undoubtedly rough) categories of soil moisture content and, second, showed the possibility of assessing the degree of xeromorphism/hydromorphism of forest community as a whole by ecological groups of soil cover (only the differentiation between HM and MH categories of ecological groups with respect to soil moisture content is still uncertain). Both of these empirical conclusions seem to be very important for quantitative estimations of hydroedaphotopes as the main intermediate of influence of climatic system on forest plantations and types of forest in different floristic phratries of Priamurye Sub-Pacific. Thereby, it becomes possible to use under these conditions the developed (Kolomyts 2003, 2005, 2008) numerical methods of local and regional landscape-ecological prediction.

10.8.4. On Assessment of Possible Ecological Consequences of Climate Changes

The most destructive influence of climatic changes unfavorable for cedar and fir-spruce forests occurs at the stage of their overmaturity with spasmodic interruptions in the development of these forest stands. "A weak impetus from outside is enough for ... the beginning of diffusive die-away of trees in a mature or overmature forest stand ..." (Kolesnikov 1956 126), which can proceed vary rapidly – during 2–5 years. In the marginal-continental part of the Russian South of the Far East, the following three hydro-climatic signals occurred previously and can be observed in future: (1) the intensification of autumn and winter winds causing windfalls; (2) the nonperiodical strong droughts resulting in overmature coniferous forest stand drying up (with mass reproduction of bark beetles); and (3) the equally strong floods destroying supra-floodplain terraces with valley cedar forests and including them into the regime of floodplain with long-term inundations, which is highly unfavorable for cedar growth. Leaf tree stands are much more stable in this respect; therefore, under such climatic signals they rapidly substitute for the previously predominant cedar generation, especially if it has reached the die-away stage.

Based on the above, it can be supposed that, in case of climatic changes towards the intensification of summer droughts (global warming) and/or the activation of winter monsoon, it can lead to the appearance of vast "foci" of degradation and destruction of cedar and fir/spruce forests, mostly in the inner mountain-valley regions of the Lower Priamurye and Primorye in the former case and on the coast of the Sea of Japan in the latter case (along "windward slopes" and "windy valleys," as local residents call them).

At the stages of maturity and overmaturity of indigenous closed cedar and fir-spruce forest stands, eth lower forest layers are developed comparatively weakly; their spotted location by the "windows" of forest canopy is characteristic, with the predominance of small shade-tolerant plants. The absence of noticeable traces of erosion of well-matted soil indicates the potential development under these phytocoenotic conditions of complete-profile zonal soils to stage IV – the relatively equilibrium state (Polivanov 1984). At the same time, hor. A_2 demonstrates the development of the most marked signs of podzolization (Kolesnikov 1956).

During the catastrophic destruction of coniferous forest stands in the presence of negative climatic signals (see above), deciduous species strengthen their positions, rapidly penetrating into the upper forest layer and substituting for cedar, spruce and fir trees. The rapid development of lower layers starting simultaneously is accompanied by the increase in their floristic richness and transition to the dominance of light-demanding nitrophilic species forming dense tall herbage with ferns in the soil cover and almost impassable thickets entwined with lianas in the undergrowth. It is favored, first, by the inflow of large amounts of organic matter into soil from deadwood mass (Kolesnikov 1956) and, second, by the acute (due to stand thinning) erosion-denudation processes that "rejuvenate" soil and favor the inflow of fresh portions of mineral nutrients for plants from the deeper layers of weathering crust. The replacement of cedar forests by deciduous forest stands is also accompanied by the intensification of sod-forming processes – the development of humus hor. A_1 with weakening podzolization.

The described biogeocoenotic process implies interruption of the natural alternation of age-related stages of indigenous cedar and spruce-fir forests under their typical habitat conditions and transition to the demutation (restoration) dynamics of derivative (secondary) leaf communities – small- and broad-leaved (birch, aspen, oak-lime forests, etc.). It seems to be the general pattern of exogenous transformation of cedar and spruce-fir forests under the extreme climatic changes mentioned above.

Chapter 11

Functional Organization of Forest Ecosystems

Abstract

The functioning of forest ecosystems in the subtaiga ecoregion of Priamurye Sub-Pacific is characterized by discrete parameters of autotrophic biogenesis, namely: the reserves of different live phytomasses and productivity. We have succeeded in more integral representation of the regularities of formation of buffer forest communities typical of ecotone systems. The forests of the Priamurye boreal ecotone have comparatively similar values of organic matter production – the basic characteristic of their functioning – under the conditions of their rather marked floristic and phytocoenological contrasts. Here we can see one of the phenomenal ecological consequences of unique forest-forming processes described by B.P. Kolesnikov for the forests of the Primorye–Priamurye Sub-Pacific. The cedar and fir/broad-leaved and nemoral fir/spruce forests of sunny and neutral mountain slopes and flat crests of ridges have the maximum reserves of live phytomass. The reserves of phytomass decrease 1.5 times due to insufficiency of summer warmth on the shadowy slopes of lowlands occupied mainly by fir/spruce, spruce/broad-leaved and larch forests. The forests of eluvial plain sites (plakors) of the Priamurye boreal ecotone are characterized by both the highest total annual production and the maximum values of total live phytomass. They have the most developed mechanisms of adaptive functioning, which allow them to maintain the maximum possible total production of organic matter in the given hydrothermal and edaphic conditions. It has been established that the buffer dark-coniferous/broad-leaved forests of the Priamurye sub-Pacific are a phenomenon of evolutionary landscape genesis. They have more developed structure and more diverse functions compared to ecosystems of the basal floras. The forests of the Manchurian–Okhotian phratrie proved to be the most similar to the *evolutionary climax*, according to A.N. Krištofovič, as the last, the highest stage of vegetation development under the given zonal-regional climatic conditions.

11.1. Phytomass and Productivity

The functioning of forest ecosystems is characterized by the discrete parameters of autotrophic biogenesis (Glazovskaya 1988), the reserves of different live phytomasses and the primary biological productivity (Tables 11.1–11.4). These parameters are the output characteristics of the biogeochemical cycle (Perelyman 1975; Bazilevich and Titlyanova2008) expressing the efficiency of utilization of the abiotic resources by phytocoenoses (Utekhin 1977; Gerasimov 1985).

Table 11.1. The taxonomic base norms of phytomass (t/ha) for different groups of forest types

Phytomass (see in the Chapter 2)	Groups of forest Types					
	Br	CrBr	SBr	FS	Lar	FB
BS	98.52	255.96	128.62	120.35	111.94	37.00
BV	7.06	9.18	10.72	8.70	4.46	2.63
BC	139.20	361.60	204.09	140.63	142.31	52.40

Table 11.2. The taxonomic base norms of phytomass (t/ha) for forests of different floristic phratries

Phytomasses (see in the Chapter 2)	Floristic phratries					
	Mm	Mx	MO	Okh	Ma	An
BW	95.10	122.60	251.18	131.75	79.10	70.70
BS	110.13	138.35	279.21	139.98	92.10	67.61
BB	1.33	2.42	2.14	1.31	0.68	1.56
BG	0.66	0.58	2.29	2.34	0.47	1.51
BV	6.12	9.24	11.27	10.85	4.74	3.31
BL	121.06	153.02	284.78	136.88	91.00	81.96
BC	140.30	197.36	330.04	170.42	124.68	96.94

The parameters of autotrophic biogenesis have been mentioned in Chapter 2. The descendent branch of the biological cycle was assessed indirectly according to the theory of soil-forming processes (Duchaufour 1968; Glazovskaya 1981; Polivanov 1984) using soil characteristics: the thickness and ratios of horizons A_0 and A_1, as well as their acidity. This approach revealed, in the first approximation, the most common features of the detritus branch of metabolism: the tempos of decomposition of dead organic matter under different thermo- and hydroedaphic conditions, the ratio of mineralization and humification as the two antagonistic processes of detritogenesis.

The sufficiently distinct floristic and phytocoenotic contrasts in forests of the Priamurye boreal ecotone are also manifested in live phytomass reserves (see Tables 11.1 and 11.2); however, they have comparatively similar values of annual productivity. It seems to be one of the phenomenal ecological consequences of the unique forest-formation processes peculiar to sub-Pacific. In respect of ecological dominants and taxonomic norms, the maximum reserves of timber, as well as above-ground and total live phytomass, are typical for the forests of sunny and neutral mountain slopes and flat crests of ridges. The "fuzzy" parts of niches of different groups of forest types meet here: the cedar and fir/broad-leaved and nemoral fir/spruce forests belonging to the same buffer MO phratrie. The predominant reserves of *BS* (over 265 tons/ha) make up 73% of cases, while the probability of prevalent values of *BC* > 320 tons/ha is 63%. The undergrowth and understory layer demonstrates the maximum development on sunny slopes.

The reserves of timber and herbage mass decrease 1,5 times due to insufficiency of summer warmth on the shadowy slopes of lowlands occupied mainly by fir/spruce and spruce/broad-leaved forests (sporadically larch forests) of the Okhotian and, to a lesser extent, Manchurian–Angaridian flora (see Table 11.1). The undergrowth and understory green mass takes no more than 3–7% of total green mass of the forest. All these data suggest that *heat supply is a limiting factor of forest productivity in the Priamurye boreal ecotone.*

The low values of reserves of live phytomass are typical for piedmonts and small river valleys, where the valley cedar forests of the Mm flora, as well as the birch forests and larch mari of the Angaridian flora, occur together with the fir/spruce and larch forests. Timber reserves decrease especially dramatically. Here, the limiting factor is not summer heat resources (the valleys are well heated in summer) but soil-ground overmoisturing together with deep winter freezing.

The plakor natural complexes of piedmont plains, where the broad-leaved, cedar and spruce broad-leaved forests of the Manchurian and Manchurian–Angaridian floras are predominant, take an intermediate position between the above two "poles" of phytomass reserves (optimal and pessimal). The low reserves of timber and skeletal phytomass in general, with an about 35–40% probability of timber mass values $BW \leq 90$ tons/ha, are indicative of insufficiently favorable bioclimatic conditions for woody vegetation on upland-plain sites. These forest communities acquire a certain *adaptive structure* (Utekhin 1977), which replenishes the loss of phytomass reserves by forest stands. Such adaptation manifests itself in the enhanced (compared to pessimal habitats) ground cover productivity (up to 0.95–2.80 tons/ha and more), as well as the developed shrub layer with the respective increase in its productivity (see Table 11.2).

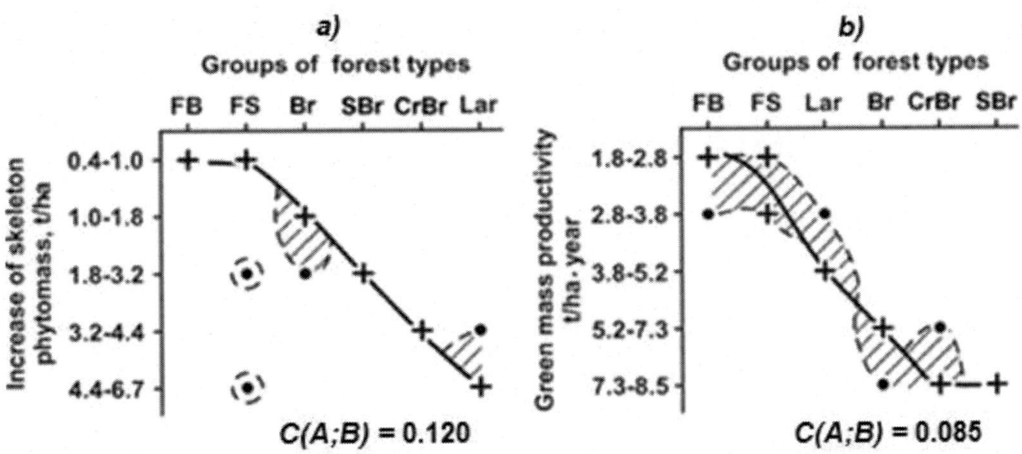

Figure 11.1. The ecological niches of groups of forest types in the space of various parameters of forest productivity. Legends are as in Figure 10.3.

Let us proceed to the assessment of annual forest productivity. Floristic phratries have a very great spread of productivity values; therefore, the *C(A;B)* parameter for the annual increase in skeletal mass (*PS*) remains low. The relationship between the increase in skeletal mass and the groups of forest types is much more marked (Figure 11.1, *b*; Table 11.3), and ecological niches here are represented mostly by their ecological dominants. It allows the arrangement of forest communities as a following series of productivity (in ascending order).

The mountain Ajan fir/spruce forests and the plain larch mari are characterized by minimal productivity (*PS* = 0.4–1.0 t/ha·yr), though the former have communities with an average and even high increase in skeletal phytomass. The comparatively low productivity is also typical of broad-leaved forests of the Manchurian xerophilic flora with oak (*PS* = 1.0–3.2 t/ha·yr). It is followed by spruce and cedar broad-leaved forests with the *PS* ecological dominant up to 3.2 and 4.4 t/ha·yr, respectively. Finally, the maximal annual increase in skeletal phytomass (up to 6.5 t/ha·yr and more) is typical for larch forests of the buffer Manchurian–Angaridian phratrie.

Table 11.3. The taxonomic norms of productivity (t/ha·yr) of the complex parameters of autotrophic biogenesis and the average age of forest-forming species in forests of different floristic phratries and groups of forest types

Parameters (see in the Chapter 2)	Floristic phratries						Groups of forest types				
	Mm	**Mx**	**MO**	**Okh**	**MA**	**An**	**Br**	**CrBr**	**SBr**	**FS**	**Lar**
PS	2.56	2.84	2.82	2.01	4.01	2.38	2.77	2.58	2.51	2.39	4.16
PG	–	–	–	–	–	–	0.71	0.49	3.37	0.92	1.28
PV	5.17	6.44	5.63	3.50	4.50	3.26	5.25	6.36	5.36	2.93	4.02
PC	7.68	9.45	9.08	5.61	9.34	5.62	8.07	9.52	8.10	6.02	8.57
IEE	0.050	0.051	0.031	0.034	0.092	0.069					
CR	0.036	0.040	0.022	0.021	0.039	0.084					
CA	0.39	0.43	0.39	0.31	0.47	0.33					
Forest age, years	101	83	133	116	39	64					

Table 11.4. The normalized partial association coefficient of total annual production with floristic phratries and groups of forest types

Common продукция, т/га · год	Floristis phratries (C(A;B) = 0.154)						Groups of forest types (C(A;B) = 0.105)					
	Mm	Mx	MO	Okh	Ma	An	Br	CrBr	SBr	FS	Lar	FB
2.4–5.0	0.14		0.14	0.51		0.44	0.11			0.50	0.14	1.00
5.0–6.5	0.26	0.14	0.12	0.15	0.26	0.26	0.30	0.10	0.55	0.11	0.13	
6.5–9.0	0.29	0.32	0.27	0.17		0.15	0.22	0.33		0.25	0.14	
9.0–11.0	0.14	0.16		0.17	0.58	0.15	0.11	0.33	0.21		0.43	
11.0–13.8	0.17	0.37	0.47		0.17		0.26	0.25	0.24	0.14	0.16	

Footote. The domination regions of ecological niches are in bold.

The information models of functional organization of forest cover in the Priamurye sub-Pacific (by the partial coefficients of connections) made it possible to arrange forest type groups into certain rows of productivity. The row of increase in the annual increment of skeletal mass is as follows: mountain spruce/fir forests and plain larch-tree maries → broad-leaved forests with oak of the Manchurian xerophilic flora → spruce and cedar/broad-leaved forests → larch-tree forests of the buffer Manchurian–Angaridian phratrie. With respect to the total annual production, forest communities are arranged as follows: spruce/fir forests → cedar/broadleaved forests → larch-tree forests → spruce/broad-leaved and broad-leaved with oak forests.

The forest communities of different floristic phratries are arranged in the following series by the ecological dominants of total annual productivity *PC* (Table 11.3): Okhotian fir/spruce forests (≤ 5.0 t/ha·yr) → Mm cedar/broad-leaved forests (6.5–9.0 t/ha·yr) → MA larch forests (9.0–11.0 t/ha·yr) → MO spruce/broad-leaved and Mc broad-leaved forest with oak (11.0–13.8 t/ha·yr).

Table 11.5. The taxonomic norms of productivity (t/ha·yr) and the complex parameters of autotrophic biogenesis in forests of different microlandscapes

Parameters (see in the text)	Microlandscapes						
	TEm	T^m_{shad}	T^m_{sun}	TA	Ev	Av	Saq
PS	2.80	2.01	2.64	4.49	2.36	2.60	0.68
PV	5.85	3.32	4.90	4.94	6.31	3.57	2.68
PC	8.39	5.98	7.29	9.54	9.82	6.14	3.45
CE	0.033	0.059	0.039	0.085	0.054	0.056	0.095
CR	0.027	0.028	0.024	0.039	0.042	0.025	0.099
CA	0.46	0.33	0.33	0.37	0.48	0.39	0.28

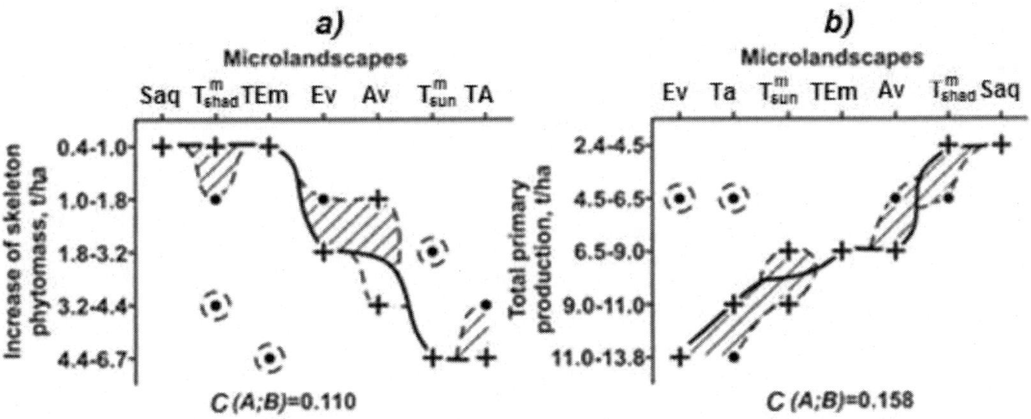

Figure 11.2. The ecological niches of microlandscapes in the space of parameters of forest productivity and vertical distribution of green mass. Legends are as in Figure 10.3.

The floristic phratries and forest types differentiate the parameters of the primary productivity by microlandscapes of the experimental testing ground. The TE apical landscapes and transit geocomplexes of sunny and neutral slopes are most productive as regards timber and skeletal phytomass (Table 11.5; Figure 11.2 *a*), the dominant *PS* = 3.2–6.7 t/ha·yr with a probability of 52–58% (the abnormally high *PS* and *PV* values on piedmont TA landscapes will be discussed below). In respect of the total productivity of phytobiota, they take the second place after the plakor valley landscapes. The latter, together with the accumulative natural complexes, have the average values of skeletal mass production (the weighted average normal value is 2.4–2.6 t/ha·yr). However, plakor landscapes take the first place in the total forest productivity: normal *PC* = 9.82 tons/ha·yr; the probability of the domination region of 11.0–13.8 t/ha·yr is about 48%. It is determined by the above-mentioned effect of structural and functional adaptation – abnormal development of the grass and shrub layers and considerable increase in the net production of green mass (normally, more than 6.3

t/ha·yr). The forests of shadowy slopes and especially waterlogged superaqual sites show the lowest productivity (see Table 11.6).

Table 11.6. The taxonomic base norms of productivity (t/ha·yr) of the complex parameters of autotrophic biogenesis and the average age of forest-forming species in forests of different floristic phratries

Parameters (see in the text)	Floristic phratries					
	Mm	Mx	Mo	Okh	MA	An
PS	2.56	2.84	2.82	2.01	4.01	2.38
PV	5.17	6.44	5.63	3.50	4.50	3.26
PC	7.68	9.45	9.08	5.61	9.34	5.62
IEE	0.050	0.051	0.031	0.034	0.092	0.069
CR	0.036	0.040	0.022	0.021	0.039	0.084
CA	0.39	0.43	0.39	0.31	0.47	0.33
Age, years	101	83	133	116	39	64

It should be noted that the reserves of different categories of phytomass far from always correspond to the annual production for these categories. The forests of TA microlandscapes with the maximum increase in skeletal wood-shrub mass (see above) eventually have comparatively low values of BS = 62–165 t/ha. On the contrary, the higher reserves of skeletal phytomass (165–265 t/ha but with an enclave of $BS \leq 62$ t/ha) prevail in the forests of TE mountain landscapes, with the dominant minimal values of their annual production. Only the E valley landscapes are characterized by the conformity of moderate values of the *PS* and *BS* parameters.

The absence of overall directly proportional relationship between phytomass reserves and annual productions is due to a significant "noise" effect of the most important index of the quality of locality of a forest: the age of basic forest-forming species.

11.2. Age of Basic Forest-Forming Species and Forest Productivity

The empirical selection of testing areas encompasses almost the entire range of forest age (from 25–50 to 200–240 years) for each category of microlandscapes, as well as floristic phratries (see Table 11.6) and groups of forest types. Forest communities are at different stages of succession: restorative after fires and cuttings and endoecogenetic during century-long forest successions. The youngest forest communities (up to 60 years old on average) are widespread in the foothills. The stages of progressive growth and, less frequently, overmaturity are typical for the most of valley plakor (eluvial) forests. Young forests are predominant also in transit mountain microlandscapes; on sunny and neutral slopes, they are slightly older than on shadowy slopes. The mature and overmature forests are typical of the bald peaks and ridges and the adjacent transeluvial slopes (Figure 11.3). Here, forest cover was apparently least disturbed and now is in the most spontaneous regime.

The multivariate correlations of live phytomasses (*BS* and *BC*) with the productivity (*PS* and *PC*, respectively) and the average age of basic forest-forming species (τ) were established using the data of our statistical selection for the Komsomolsky Nature Reserve. Significant connections are a kind of a linear multivariate correlation. For the skeletal phytomass of forest stand and undergrowth:

$$BS = 29.86 \cdot PS + 1.187 \cdot \tau - 46.05; R = 0.721; R^2 = 0.520. \quad (11.1)$$

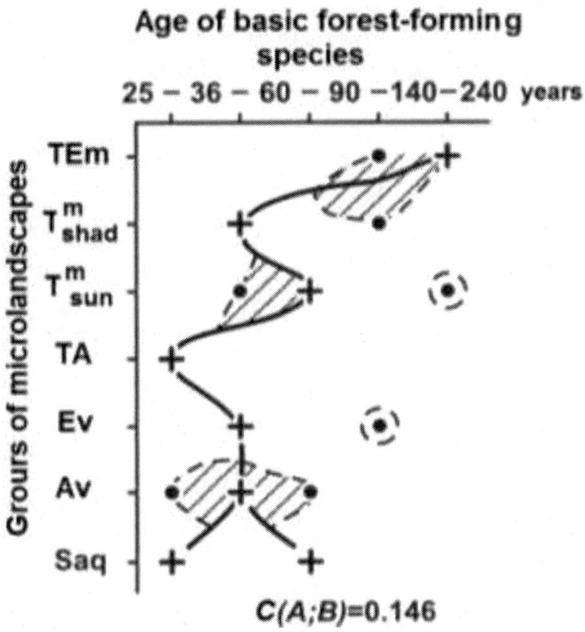

Figure 11.3. The ecological niches of average-aged basic forest-forming species in the space of microlandscape groups.

At the same time, the partial connections of *BS* with *PS* and τ are insignificant ($R \leq 0.44$–0.58). Only the joint effects of both factors seem to be significant. The total live phytomass of forest phytocoenosis has a closer multivariate correlation with these factors:

$$BC = 21.66 \cdot PC + 1.183\, \tau - 91.86; R = 0.827; R^2 = 0.683. \quad (11.2)$$

Here, the partial connection between *BC* and *PC* is more marked though rather weak:

$$BC = 2.522 \cdot PC + 1.061 \cdot PC^2 + 73.72; R = 0.631; R^2 = 0.398. \quad (11.3)$$

On the other hand, the connection between *BC* and τ is still insignificant ($R = 0.46$).

Different correlations between annual forest productivity and the age of forest-forming species give a highly varied pattern of reserves of skeletal tree and shrub phytomass and total live phytomass in different forest types and in their respective floristic phratries. The *forests of eluvial plain sites (plakors)* of the Priamurye boreal ecotone are characterized by both the highest total annual production and the maximal values of total live phytomass. Consequently, our previous empirical generalization (Kolomyts 2005, 2008) suggests that they *possess the most developed mechanisms of adaptive functioning*, which contribute to their survival, i.e., allow them to maintain the maximum possible total production of organic matter under the given hydrothermal and edaphic conditions.

One of the most important conditions of existence and development of natural ecosystems is a *balance between the processes of production and decomposition of vegetable mater* (Odum 19071; Glazovskaya 1981; Gerasimov 1985). This condition can be indirectly assessed by the combination of specific values of the *BC* and *PC* parameters, as well as the age of forest-forming species, for different groups of forest types. The relatively balanced biological cycle with the highest intensity is typical of mesophilic Manchurian cedar forests. Being mainly in the mature and ripening states (48% are more than 90 years old), they have the maximal taxonomic norms of reserves of total live organic matter and the maximal normal value of total net production (see below Table 11.9). The no less balanced but much slower cycle is typical for overmature fir/spruce forests of the Okhotian flora (the probability of the age of 140–240 years is 43%), with the minimal values of the *BC* and *PC* parameters. At the same time, rather young Manchurian xerophilic broad-leaved forests (41% of cases are up to 60 years old) are characterized by distinct unbalance of the biological cycle: the comparatively high values of total net production are combined here with the minimal reserves of live phytomass.

Thus, a developing *forest community passes along a certain logistic trajectory of biological cycle intensity and balance.* Such a trajectory describes, as is known (Armand 1967), the evolution of self-developing natural ecosystems. The exponential (ascendant) branch of this trajectory realized in the young and ripening forest stands describes a functional system with positive feedback. The turning point of the development curve observed after the forest has reached the state of overmaturity is accompanied by emergence of the asymptotic branch representing a system with negative feedback and characterizing the stabilization of metabolic balance achieved against the background of a certain decrease in metabolic rate.

11.3. The Descending Branch of the Biological Cycle

The revealed regularities of autotrophic biogenesis described only the ascendant, creative branch of biological cycle. In the meantime, it is known that the intensity of metabolism in forest ecosystems is determined first of all by its detritus but not creative branch; therefore, forest is considered to be a detritus ecosystem (Odum 1971). The detritus branch of biological cycle consists of a series of elementary soil processes creating a certain genetic profile of the soil (Glazovskaya 1981). Detritogenesis is based on the two antagonistic processes (Duchaufour 1968): mineralization and humification. Their balance relationships with the entry of dead organic matter (above-ground tree waste and drain, as well as root waste) and reciprocal relations determine the thickness of forest litter and humus layer. Such morphological approach makes it possible to reveal, in the first approximation, the most common features of the detritus branch of metabolism: the tempos of decomposition of dead organic matter under different thermo- and hydroedaphic conditions, the ratio of mineralization and humification processes.

The forests of Manchurian phratries growing on well-drained sunny slopes are characterized by the highest productivity of green mass (*PV*); so, in xerophilic forests, the taxonomic norm is *PV* = 6.44 t/ha·yr. The mainly leaf waste yields a soft (mull) litter containing a lot of water-soluble substances, with the prevalence of humic acids, as well as aggressive and complexing compounds (Duchaufour 1968). As a result of intensive intrasoil weathering of the primary minerals, a good deal of bases get into soil solutions (Liverovsky, 1969); therefore, horizons A_0 and A_1 have weakly acidic and close to neutral reactions (water pH = 5.0–6.2, Table 11.7) providing for comparatively rapid biodegradation of the compounds. Due to this fact, the processes of decomposition and mineralization of dead organic matter are most intensive, with the active utilization of mineral substances in plant nutrition directly from litter (Odum 1971). Hence, in spite of abundant tree waste, the thickness of forest litter remains insignificant (normally about 5–6 cm, often not more than 3 cm). Humus reserves are comparatively low as is confirmed by the low thickness of horizon A_1 (7–8 cm on average, less frequently 10 cm), as well as the minimal thickness of the humus profile (Tables 11.8 and 11.9).

Table 11.7. The distribution of soil acidity (water pH) among floristic phratries (the normalized partial coupling coefficients

C(A;B) = 0.149)

Water pH	Floristic phratries					
	Okh	**An**	**Mm**	**MA**	**Mx**	**MO**
4.2–4.7	**0.55**	0.20				0.31
4.7–5.0	0.27	0.31	0.17	0.16	0.18	0.12
5.0–5.3		**0.49**	**0.41**	0.25	0.14	
5.3–5.7			**0.41**	**0.38**	**0.43**	0.09
5.7–6.2	0.18			0.21	0.24	**0.47**

Footnote. The domination regions of ecological niches are in bold.

Table 11.8. The taxonomic base norms of morphological characteristics of soil in different groups of forest types

Parameters	Groups of forest types					
	Br	**CrBr**	**SBr**	**FS**	**Lar**	**FB**
Thickness of layer A_0	5	6	10	6	6	8
Thickness of layer A_1	7	8	15	12	15	14
Thickness of layer A_2	17	22	29	15	29	2
Thickness of layer B	39	43	35	17	34	40
Ratio A_0/ A_1	0.71	0.90	0.67	0.50	0.40	0.57
Ratio (A_0+ A_1)/ B	0.32	0.33	0.71	1.12	0.62	0.55

Table 11.9. The taxonomic base norms of humus profile thickness in soil section with floristic phratries and groups of forest types

Thickness of humus profile, cm	Floristic phratries (*C(A;B)* = 0,093)						Groups of forest types (*C(A;B)* = 0,075)					
	Mx	**Mm**	**Okh**	**Ma**	**An**	**Mo**	**Br**	**CrBr**	**FS**	**Lar**	**FB**	**SBr**
6–10	**0,45**	0,18	0,20	0,13	0,13	0,14	0,23	0,22	0,29	0,13		0,18
10–16	0,24	**0,57**	0,16		0,10	0,22	**0,37**	**0,35**	0,12	0,10		
16–22		0,26	**0,43**	0,29	0,09	0,20	0,17	0,08	**0,42**	0,09	0,39	0,26
22–31			0,22	**0,31**	**0,43**	0,16		0,13	0,17	**0,43**	**0,61**	0,20
31–58	0,30			0,27	0,27	**0,28**	0,23	0,22		0,25		**0,36**

Footnote. The domination regions of ecological niches are in bold.

The presented data demonstrate that the rates of the small biological cycle in the forests of Manchurian phratries are maximal for Priamurye Sub-Pacific. The brown forest illuvial-humus weakly podzolic soils formed here can be considered most mature in the series of forest formations, i.e., maximally close to the equilibrium (climax) state.

The Ajan fir/spruce forests of the Okhotian phratrie grow on brown-taiga illuvial-humus soils under the conditions of edaphic excessive moistening and the lowest summer temperatures. Therefore, these forest communities are least productive: *PS* = 0.4–1.0 and *PV* = 1.7–2.8 t/ha·yr, with the minimal taxonomic norms of these parameters (see Table 11.9). Coniferous waste yields a forest litter with an acidic low-active medium (pH = 4.2–5.0).

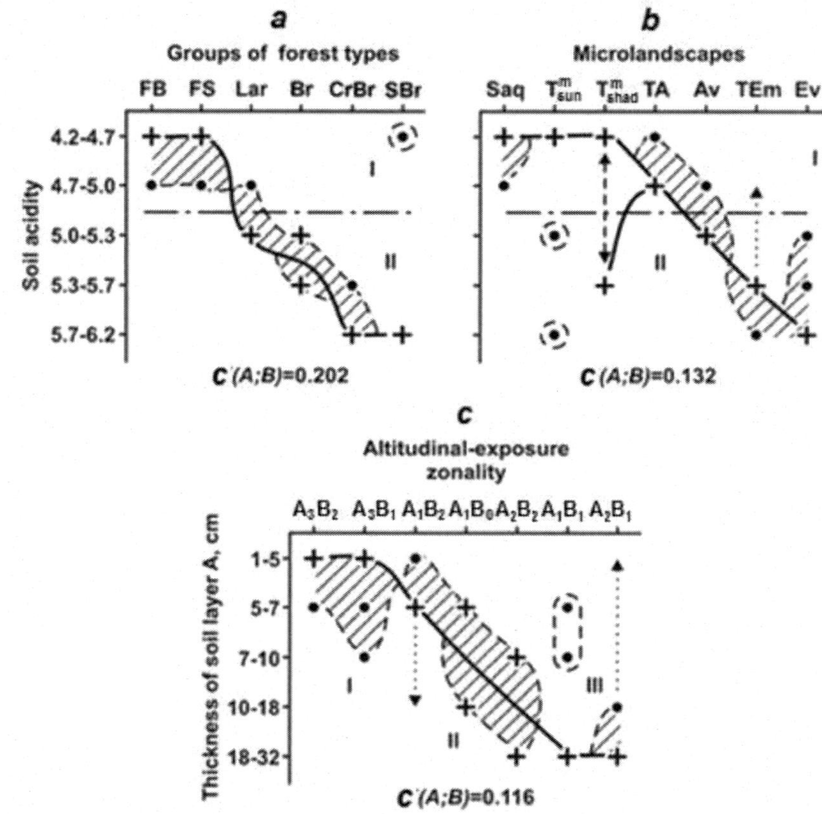

Figure 11.4. The binary ordination of groups of forest types and microlandscapes in the space of acidity of the humus layer of soil. The ordination of horizon A_1 thickness by altitudinal-exposure zonality. Legends are as in Figure 10.3.

Due to weak decomposition of dead organic matter ($A_0/A_1 \approx 0.9$–1.5), the litter in most cases is rather thick: 5.5–10 cm. Under such conditions, the evolution of infiltrating organic matter must be comparatively slow (Duchaufour 1968). The horizon A_1 is composed of coarse (mor- or moder-type) humus of rather high content: up to 8–11% (Liverovsky 1969), while the thickness of the humus layer varies within 7–32 cm. The humus profile of soil is also very deep (see Table 11.9).

The spruce/broad-leaved forests of the buffer MO phratrie are a peculiar phenomenon against this hard background of the detritus branch of metabolism. These forests produce quite a substantial amount of green mass: even greater than the Manchurian broad-leaved and cedar/broad-leaved forests: 7.3–8.5 t/ha·yr (see Table 11.9). Their soils are characterized by the optimal moisture and exposed to the lower summer temperatures compared to the soils of Manchurian forests. Here, the detritus branch of biological cycle functions in a soil medium with the same subneutral reaction as in the cedar/broad-leaved forests (see Figure 11.4). The horizon A_0 reaches the maximum thickness among all forest types (up to 10–12 cm). It indicates that the decomposition of forest litter is a slower process than the annual inflow of tree waste.

Then, the portion of mortmass utilized for mineralization and plant nutrition also decreases, which leads to the average and high values of the taxonomic norm (see Table 11.8), and the dominant of thickness of the humus horizon is 10–32 cm. Accordingly, the entire humus profile has maximum thickness values (see Table 11.9), penetrating even into the illuvial horizon B. In contrast to the Ajan forests, the decreased mineralization here is accompanied by increased moistening. Finally, the maximum thickness of horizon A_2 is observed in the broad-leaved forests. It demonstrates that podzol formation processes in brown-taiga soils are most developed.

In the larch forests of the An and MA phratries, the temperatures of upper soil horizons are as low as in the soils of Ajan fir/spruce forests; the soils are moist and fresh, with a weakly acidic and acidic reaction of soil solutions (see Figures 11.4 and 11,6, *a*). Under these conditions, the processes of dead organic matter decomposition are inhibited. However, due to insignificant green mass production (3.26–4.50 t/ha·yr), forest litter has an average thickness (3.5–8.0 cm). The processes of litter mineralization are also significantly weakened; therefore, the humus layer has maximum thickness.

If the described intensification or, on the contrary, weakening of mineralization and humification processes are designated by symbols «+» and «–», respectively, then the general pattern of the detritus branch of biological cycle can be represented as follows:

Groups of forest types	**Mineralization**	**Humification**
(a) broad-leaved and cedar broad-leaved	+	–
(b) fir/spruce	–	+
(c) spruce/broad-leaved	–	+
(d) larch tree	–	–

11.4. Degree of Soil Maturity

Soil maturity is one of the most representative indicators of the level of structural and functional development of the entire natural complex (Mamay 1994). The criterion of representation "… is the correspondence of diagnostic characters of a particular ecosystem and a particular soil profile to a particular classification taxon" (Kerzhentsev 2001 21). For the highly dynamic marginal-continental landscapes of the Lower Amur Sub-Pacific (see above), the characteristics of the degree of soil maturity seem to be especially important for ecological assessments of their condition.

Under climatic conditions of the south of the Far East, the following major trend of soil formation processes is formed (Ivanov 1976). Humid tropical weather in the summer months favors periodic overwetting of soil-grounds, active decomposition of dead organic matter and intensive weathering of primary minerals. The drastic transition to low temperatures in autumn, the subsequent severe soil freezing in winter and the dry cool spring result in conservation of the products of biogenic decomposition and weathering. Consequently, the soils are characterized by a shortened profile, general argillification as a result of metamorphism (Zonn 1969) and the high humus content in upper layers. The generally higher degree of soil hydromorphism against the background of rather high summer temperatures intensifies chemical weathering and migration of active chemical elements: oxygen, humic and mineral substances, and CO_2 (Duchaufour 1968). It inevitably activates the small biological cycle and must increase the productivity of natural ecosystems. Empirical data on the productivity of boreal forests of marginal-continental and insular land of the Asian fragment of the Pacific Ocean Ring (Bazilevich 1981) confirm this assumption.

V.M. Fridland (1970) has noted that in the south of the Far East, in contrast to the intracontinental regions of European Russia and Siberia, soil formation occurs on less mature (younger) weathering crusts and, therefore, the processes of intrasoil weathering here are more intensive. Rejuvenation of weathering crusts is associated with the greater soil-forming role of erosion processes under conditions of mountain-valley relief (Kolesnokov 1965). On the crests of ridges and in the upper parts of steep mountain slopes, the soils of cedar/broad-leaved forests are rejuvenated due to removal of fine-grained sediment and involvement of new layers of weathering rock inthesoil formation process. Denudation on the mountain slopes stabilizes soil formation process, retarding it at stages II and III of ascending development – the formation of zonal soils with simplified and complete profiles (Polivanov 1984). The process does not reach stage IV, i.e., the relatively equilibrium state of soil profile. On the middle, less steep parts of slopes there is a transit zone, with accumulation of mainly large detrital weathering products as landslide scree tongues. Skeletal shallow eluvial–deluvial soils develop. Finally, the foot of mountain slopes and the adjacent upper, usually relict, river terraces are overlapped with a continuous mantle of fine-grained deluvium (piedmont trails). Soil "rejuvenation" occurs in connection with the continuous increase in soil mass from above – due to accumulation of alluvia. Here there are the outflows of mineral ground and soil waters. As a result, the soils on piedmont mesosites are characterized by the highest fertility, which decreases in both directions – towards the tops of the bald mountains and towards river floodplains (Kolesnikov 1956).

Soil maturity is anyhow associated with the isolation and development of illuvial-metamorphic horizon B as a zone of intrasoil weathering (Glazovskaya 1981). The morphological characteristics of this horizon (thickness, clay content, etc.) are the indicators of soil profile maturity, the degree of closeness to the climax. This degree is determined both by the rate of soil formation and by the dynamics of development of the loose sheath on the slope – the ratio of weathering and matter removal processes (Polivanov 1984). The thickness of zone (A_2 + B) gives a more generalized idea of the degree of podzolization and intrasoil weathering (Tsyganenko 1972). Among the morphological properties of soils, the content of lithomasses (bedrock debris) in horizon B is of no less important diagnostic significance.

The "slope" elementary morpholithosystem has a three-floor structure: above the eluvial horizon of bedrocks (El), there are successively the slope (Sl) and cover horizons (Karpukhin et al. 1984). During morpholithosystem development, the eluvial horizon gradually penetrates into the bedrock, while the upper part of El passes into the slope horizon. In the first approximation, it can be supposed that the total content of lithomasses in this system will depend on the correlation of weathering processes (to the stage of claying) in the eluvial horizon and the slope horizon movement together with fragmentary material. The increase in lithomass content in the slope horizon must indicate that deluvial processes dominate over eluvial processes (Sl > El), which supply material for the formation of soil horizon B. On the contrary, in case of Sl < El ratio, the lithomass content in this horizon must be reduced.

Under forest cover, the most widespread deluvial process is defluxion (Karpukhin et al., 1984), or seasonal soil creep – a slow sliding of soil mass caused by seasonal fluctuations in its temperature and moisture content (Gerrard 1981). In this case, it is a deep winter freezing of soil and its periodical summer overwetting in the period of monsoon rains. The defluxion-slope process can develop only in a viscous-plastic consistency of soil (Voskresensky 1971), which suggests the presence here of sufficient amounts of finely dispersed (clayish and heavy-loamy) material and moderate saturation with lithomasses. Defluxion is clearly manifested in forest stands with mass upslope inclination of trunks, which gives the impression of "drunken forest."

Many properties of the soil depend on slope steepness, though this relationship is unambiguously manifested in different elements of the catena (Gerrard 1981). In low-mountain landscapes of the Lower Priamurye, there are three types of binary relations between the morphological characteristics of structural horizon B and slope steepness.

Type 1. It relates to subhorizontal and very gentle slopes (0–2° steepness, Figure 11.5, *a*; Table 11.10). Horizon B is characterized by the maximum argillization (by 60% and more), while the lithomass content in this horizon is no more than 35% (Table 11.11). The horizon reaches considerable thickness (40–50 cm, with a probability of 60%), though not maximum possible in this ecoregion (Table 11.12). This is a morphological result of stable El >> Sl (or El > Sl) ratio, i.e., the most completed and, apparently, maximally intensive processes of intrasoil weathering, as well as podzolization (the inflow of slimy particles susceptible to chemical degradation from above) and lessivage (the inflow of slimy particles without degradation). Following the proposition concerning the formation of more mature soils on more gentle slopes (Gerrard 1981), it can be concluded that the presented data exactly characterize the state of the most mature soils for oro-climatic conditions of the Lower Amur Sub-Pacific.

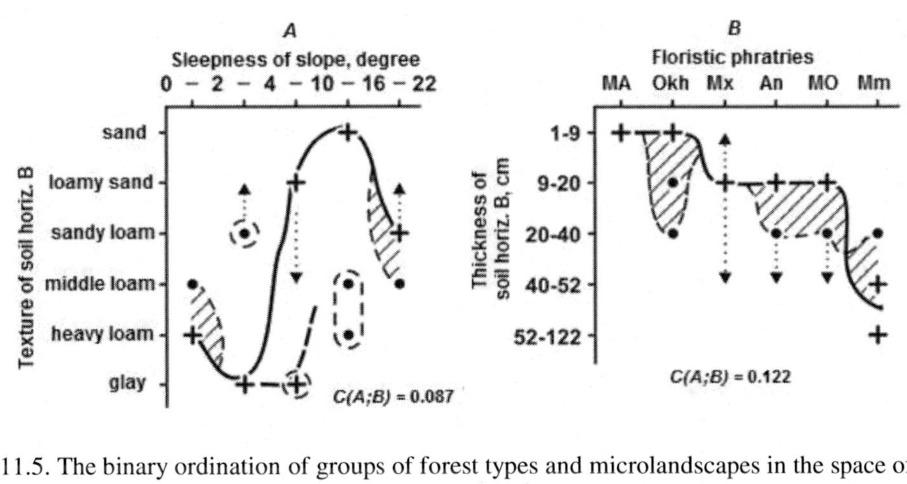

Figure 11.5. The binary ordination of groups of forest types and microlandscapes in the space of allometric coefficient and the age of forest-forming species. Designations are the same as in Figure 10.3.

Table 11.10. The content of lithomasses in hor. B on the slopes of different steepness (normalized partial coefficients of connection)

Content of lithomasses. %	Steepness of the slope				
	0–2	2–4	4–10	10–16	16–22
0–18	0.23	**0.31**	0.29	0.10	
18–36	**0.69**				
36–54	0.09	0.28	0.22	0.29	0.17
54–72		0.15		**0.46**	**0.54**
72–95		0.26	**0.49**	0.16	0.29

Footnote. The domination regions of ecological niches are in bold.

Table 11.11. The mechanical composition of hor. B on the slopes of different steepness (normalized partial coefficients of connections)

Mechanical composition of hor. B	Steepness of the slope				
	0–2	2–4	4–10	10–16	16–22
Sandy			0.10	**0.62**	
loamy sand	0.17	0.16	**0.43**	0.08	0.17
Sandy loam		**0.24**	0.12		**0.63**
Light loam	0.21	0.12	0.24	0.15	0.20
Heavy loam	**0.62**	0.12	0.11	0.15	

Footnote. The domination regions of ecological niches are in bold.

Table 11.12. The thickness of hor. B on the slopes of different steepness (normalized partial coefficients of connection)

Thickness of hor. B, cm	Steepness of the slope				
	0–2	2–4	4–10	10–16	16–22
1–9	0.18	0.15	0.17	0.17	**0.43**
9–20	0.22	0.15	0.21	0.22	0.27
20–40		0.22	0.12	**0.36**	0.30
40–52	**0.35**	0.15	**0.50**		
52–122	0.25	**0.33**		0.24	

Footnote. The domination regions of ecological niches are in bold.

Type 2. It includes the slopes of the greater steepness (2–10°). Argillization of hor. B considerably decreases (to 35–50%), with simultaneous appearance of gravelly loam sandy with light clay loam (up to 40%). At the same time, lithomass content in the horizon increases to 50% and more with a probability of about 40–50% (see Tables 11.10 and 11.11). All the above is indicative of the El < Sl ratio, i.e., that the soil profile has moved far away from the mature (climax) state due to incomplete processes of weathering and continuous "rejuvenation" of the slope horizon through deluvial, mainly defluxion, processes. However, hor. B reaches maximum thickness – up to 1 m and more, with a probability of 33%, suggesting the maximum possible involvement of the upper bedrock layers in weathering on these slopes.

Type 3. It relates to very steep slopes: 16–22° and more. Here, hor. B is characterized by almost complete absence of argillaceous and heavy loamy fractions; sand with sandy loam, or light loam, makes 70–80% of its mechanical composition. The horizon is actually clogged with bedrock debris: the probability of 60–80% lithomass content is more than 54%, with 75–95% content in almost 30% of the horizon at slope angles above 15°. Here there is a quite obvious El << Sl ratio, i.e., eluvial processes are suppressed by defluxion and surface-deluvial removal, which continuously "rejuvenates" soil and drastically throws back its maturity. At the same time, the predominant thickness of hor. B decreases in this steepness range from 20–40 cm to 5–10 cm and less.

Generally, the presented data (see Table 11.16) confirm "... the inverse dependence between slope angle and the depth of structural horizon B" (Gerrard 1981 93).

Let us proceed to the analysis of maturity of soils under forest communities of different types and different floristic phratries. Judging by the taxonomic norms and distribution of hor. B thickness, as well as its mechanical composition (see Tables 11.10 and 11.13), the most mature, i.e., maximally close to the climax state, are brown-taiga soils of spruce/broad-leaved and larch forests belonging to the buffer phratries: MO and MS, respectively. Consequently, the more equilibrium annual regimes of the biological cycle should be expected here.

Table 11.13. The normalized partial coefficients of connections of mechanical composition of soil horizon B with floristic phratries and groups of forest types

Mechanical composition of hor. B	Groups of forest types					
	FS	B	CrBr	SBr	Lar	FB
Sandy	**0.55**					
loamy sand	0.18	**0.33**	0.06	0.07	0/19	
Sandy loam		0.25	0.23	**0.28**	**0.24**	
Light loam		0.17	**0.47**	0.08	0.08	**1.00**
Heavy loam	0.14	0.25		**0.24**	**0.24**	
Clay	0.13		0.23	**0.24**	**0.24**	

Footnote. The domination regions of ecological niches are in bold.

The genesis of horizon B is not only metamorphic but also illuvial. There is, first, the intensive lessivage of clay particles (the content of heavy loam and clay is up to 50–60% and more, see Table 11.3) and, second, the deep penetration into soil of "mature" organic mass (with sesquioxides) causing intrasoil transformation of mineral mass (as is demonstrated by considerable thickness of the humus profile, see Table 11.9). This is the case when horizon B increases both from below and from above (Polivanov 1984) and thereby becomes thicker, generally bringing the paraeluvial weathering crust being formed here (Polynov 1934) closer to the syallitic-ferralitic type, with Fe and Al hydroxides (Rode 1975).

However, these forest communities occupy a rather broad range of slope steepness: from 2–5° to 16–22° (Figure 11.6, *b*); therefore, the lithomass content in hor. B is 55–75% and more (see Table 11.10). The EL > Sl (Type 1 of connections, see above) and El < Sl (Type 2) ratios are realized on flat and steep slopes, respectively. In the latter case, active defluxion processes substantially reduce the degree of maturity of these soils, as is also confirmed by the low and moderate thickness of horizon B (see Table 11.12; Figure 11.5, *b*).

The second place, with respect to the degree of maturity, is taken by the brown forest soils of Manchurian broad-leaved forests and cedar forests. There, argillization of horizon B is much weaker but its thickness is maximal: apparently due to intensive intrasoil weathering under conditions of high temperatures of the entire soil profile (see Table. 10.4). Here, horizon B is almost solely metamorphic.

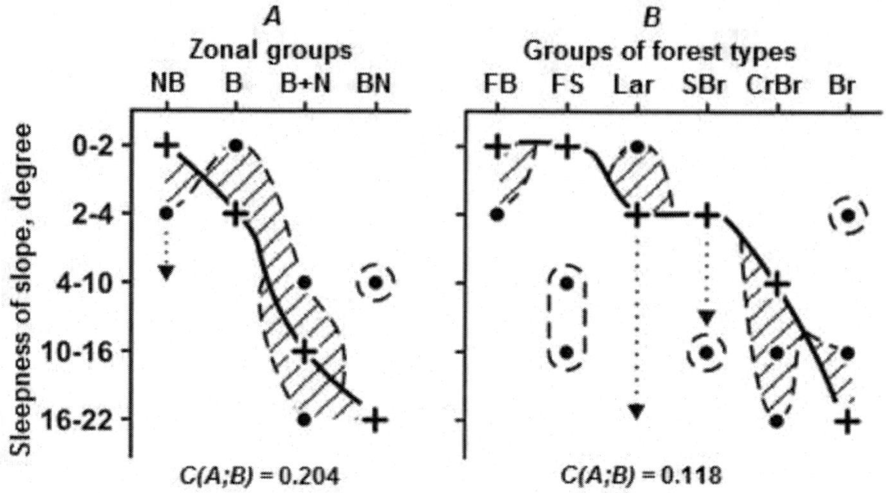

Figure 11.6. The ecological niches of zonal groups and groups of forest types in the space of slope steepness. Zonal groups: NB – nemoral–boreal; B – boreal; B+N – boreal + nemoral; BN – boreal–nemoral. Legends are as in Figure 10.3.

These forests occupy mainly the slopes of moderate and great steepness; however, the soils with lithomass content up to 18% and above 55% occupy more than 75% of forest area in the former and latter group, respectively. It seems to be due to the influence of bedrock lithology, which is one of the most important factors of weathering (Karpukhin et al. 1984), as well as the moisture content in hor. B. The soils of broad-leaved forests are well-developed on the softer and looser rocks and under conditions of higher summer edaphic moistening compared to the soils of cedar forests. As a result, we have El > Sl (closer to Type 1) in the former case and El < Sl (Type 2) in the latter case.

Apart there are the brown-taiga podzolized soils of the Ajan fir-spruce forests – representatives of the Okhotian flora, with small and, more rarely, moderate thickness of horizon B (see Table 11.12), with a 55–65% content of gravelly sand and very negligible argillization (see Table 11.13) and with the high percentage (80–85%) of moderate lithomass content (18–54%). These data demonstrate the much lesser maturity of these soils compared to the soils described above. Since the Ajan fir/spruce forests occupy the slopes of almost the entire range of steepness (see Figure 11.5), the least mature soils must occur on the steepest slopes, where horizons B almost entirely composed of bedrock debris make up to 10–20%.

Such a negative anomaly in the degree of soil maturity is probably associated with quite a number of causes – lithogenic, climatic, geomorphological, and phytocoenotic. The Ajan fir-spruce forests are generally widespread in the upper zone of low mountain ridges composed of massive-crystalline rocks – in contrast to the middle and lower zones with a not so dense sedimentary-metamorphic complex of rocks and with loose quaternary deposits, where the Manchurian, Manchurian–Okhotian and Manchurian–Angaridian forests grow (see above Figure 10.6). It seems to the major factor of slower evolution of the soil profile under the Ajan fir/spruce forests. Special studies performed in the middle reaches of the Amur River have shown (Liverovsky 1969) that horizon B on granites and basaltic andesites is only metamorphic and grows from below. It is formed mainly due to intrasoil weathering, with the minor lessivage and weak podzolization (the latter is confirmed by the minimal thickness of horizon A_2 in fir/spruce forests, see Table 11.9). Thereby, the syallitic orthoeluvial crust of weathering is formed (Polynov 1934).

Further, the Ajan fir/spruce forests are confined by 55–65% to shady slopes and rarely occur on sunny slopes (see above Figure 10.4, *b*). Here, at relatively low temperatures (see above), the biotic and, accordingly, soil processes substantially slow down, thereby contributing to manifestation of the first (lithogenic) factor. The undergrowth and herbage masses, as well as the production and annual turnover of green mass, are minimum. It drastically decreases the content of leaf litter which, as is known (Glazovskaya 1981), is decomposed much quicker than needle litter. Accordingly, the flows of "mature" organic mass in hor. B become weaker.

Thus, the results of the analysis can be summarized as follows. The establishment and functions of buffer forest communities at the Priamurye boreal ecotone resulted in the development of more mature soils compared to the soils of forests of the basic floras (Manchurian, Okhotian and Angaridian) providing the basis for these buffer formations. It was most clearly seen in the forest ecosystems with Manchurian–Okhotian flora, which thus proved to be most well-developed with respect not only to the structure and function of phytobiota (see above) but also to the degree of soil maturity. Accordingly, it can be supposed that the maximum structural and functional development in the low-mountain Lower Amur Sub-Pacific is typical of transit microlandscapes of sunny and neutral mountain slopes, as well as flat-interfluve landscapes of intermontane plains, where forest vegetation is formed by the buffer Manchurian–Okhotian flora (see Figure 10.3, *b*). Generally, the following series of microlandscapes of the Komsomolsky Nature Reserve is formed in the order of decreasing degree of soil maturity and the respective level of structural and functional development of landscapes:

Transit (sunny and neutral slopes) and flat-interfluve plain landscapes with dark-coniferous/broad-leaved forests of the MO phratrie

↓

Transaccumulative piedmont landscapes with larch forests of the MA phratrie

↓

Accumulative and flat-interfluve valley landscapes with broad-leaved forests of the Mm phratrie

↓

Transeluvial mountain, transaccumulative and flat-interfluve plain microlandscapes with cedar/broad-leaved forests of the Mc phratrie

↓

Transit (mainly shadowy slopes) and transaccumulative landscapes, with the Ajan fir/spruce forests and broad-leaved/spruce forests of the Okh phratrie

↓

Superaqual and accumulative valley landscapes with larch forests and larch mari of the An phratrie.

11.5. Polysystemic Structural-Functional Models of Forest Ecosystems

This structure was analyzed using the fuzzy set theory-based models of similarity relations (N_k), the structural schemes of inclusion relations (J_k, Figures 11.5–11.8), and the integral parameters of the states of different floristic phratries and groups of forest types ($\boldsymbol{G}_k$ and $\boldsymbol{S}^m{}_k$; Table 11.14).

The graphs of the binary relations of similarity actually characterized the isomorphism between the members of each pair of objects, the similarity of their structures or functions. In general, the structural characteristics of forests show the minimal similarity (see Figure 11.7,*a*). The isomorphism for phytomass parameters is much higher (see Figure 11.6), reaching the maximum values in productivity parameters (see Figures 11.9, *a*; 11.10, *a*). It means that the forests of the Priamurye boreal ecotone with their rather pronounced floristic and phytocoenotic contrasts have comparatively similar values of organic matter production – the principal characteristic of their functioning. Here there seems to be one of the phenomenal ecological consequences of unique forest-forming processes described by B.P. Kolesnikov (1956) for the forests of the Primorye–Priamurye sub-Pacific.

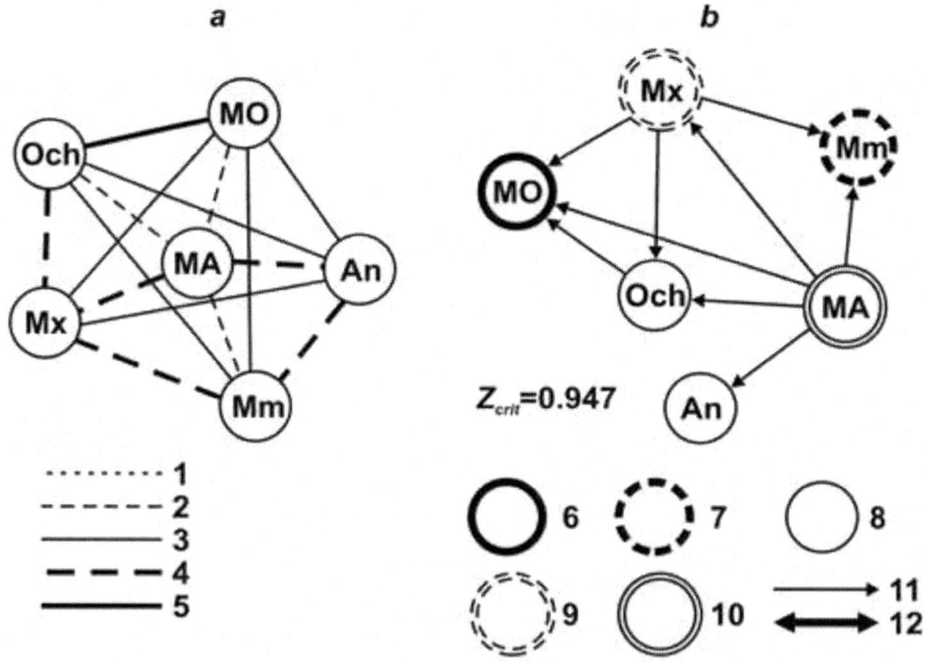

Figure 11.7. The schemes of nontransit relationships between floristic phratries of the Priamurye boreal ecotone by the structural parameters of forest communities: the graph of similarity relations (a); the orgraph of relations of domination (b). Similarity measures: 1, 0.70–0.75; 2, 0.76–0.80; 3, 0.81–0.85; 4, 0.86–0.90; 5, 0.90–0.96. The ecotypes of forest communities: 6, dominants; 7, subdominants; 8, temporal ecotypes; 9, subpessimal; 10, pessimal. 11, Direction of inclusion; 12, connections of tolerance.

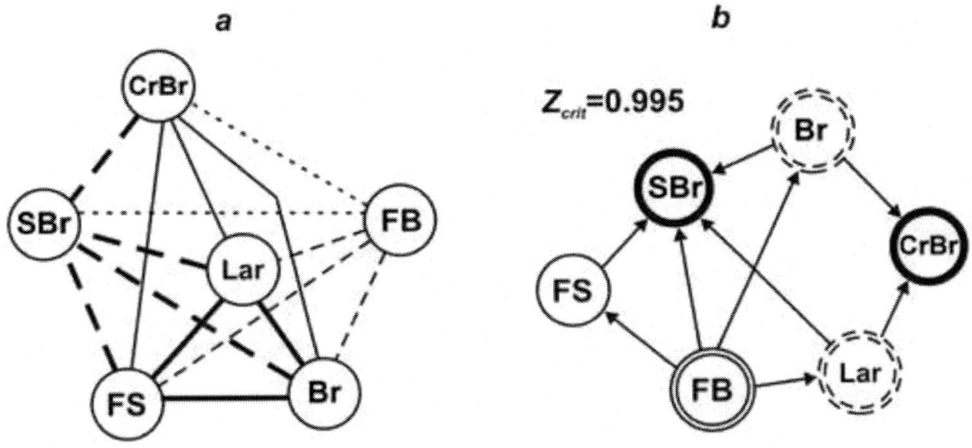

Figure 11.8. The schemes of nontransit relationships between the groups of forest types of the Priamurye boreal ecotone by the parameters of living phytomasses in forest communities: the graph of similarity relations (a); the orgraph of relations of domination (b). The symbols are the same as in Figure 11.7.

Table 11.14. The average measures of similarity (I_k) and homeostatisity (G_k) of floristic phratries and groups of forest types, as well as their structure-metabolic indices ($S^m{}_k$), in the respective ecological spaces

Parameters		Floristic phratries					
		Mm	**Mx**	**MO**	**Okh**	**MA**	**An**
Structural phytocoenotical	I_k	0.85	0.86	0.84	0.84	0.83	0.84
	G_k	**0.355**	*0.170*	**0.367**	0.271	*0.126*	**0.324**
	$S^m{}_k$	**1.220**	*0.770*	**1.340**	**1.110**	*0.450*	**1.110**
Living phytomasses	I_k	0.81	0.79	0.61	0.80	0.78	0.71
	G_k	*0.167*	**0.500**	**0.500**	**0.500**	*0.167*	*0.167*
	$S^m{}_k$	*0.750*	**1.125**	**1.500**	**1.500**	*1.250*	*0.750*
Productivity and biological cycle	I_k	0.88	0.85	0.88	0.81	0.85	0.82
	G_k	0.263	**0.400**	0.263	*0.141*	**0.398**	*0.158*
	$S^m{}_k$	1.000	**1.430**	1.000	*0.430*	**1.280**	*0.850*
Parameters		Groups of forest types					
		Br	CrBr	SBr	FS	Lar	FB
Structural phytocoenotical	I_k	0.84	0.82	0.80	0.80	0.81	0.76
	G_k	0.279	**0.427**	0.279	**0.426**	*0.139*	*0.158*
	$S^m{}_k$	1.000	**1.340**	1.000	**1.330**	*0.500*	*0.850*
Living phytomasses	I_k	0.76	0.62	0.73	0.78	0.78	0.45
	G_k	*0.211*	**0.355**	**0.365**	0.240	*0.209*	*0.115*
	$S^m{}_k$	*0.890*	**1.220**	**1.440**	1.000	*0.890*	*0.560*
Productivity and biological cycle	I_k	0.85	0.80	0.85	0.79	0.81	0.61
	G_k	0.238	**0.366**	0.240	*0.115*	**0.357**	*0.156*
	$S^m{}_k$	1.000	**1.330**	1.000	*0.560*	**1.230**	*0.890*

Note. Metabolic ecotypes of forest communities: macrobolites and submacrobolites a showed by half-rich print with underline and without one corresponding; microbolites and submicrobolites – by italics print with underline and without one corresponding; mesobolites (ordinary ecotypes) are not marked.

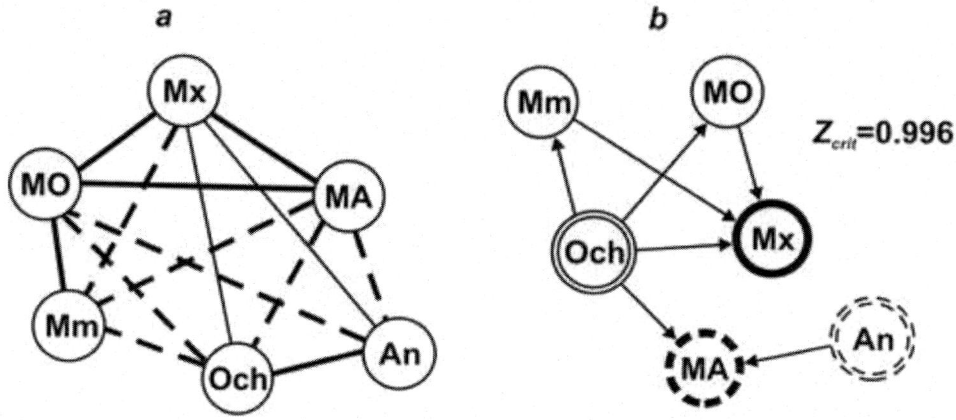

Figure 11.9. The scheme of nontransit relationships between floristic phratries of the Priamurye boreal ecotone by the parameters of productivity of forest communities: the graph of similarity relations (a); and orgraph of relations of domination (b).
Legends are as in Figure 11.7.

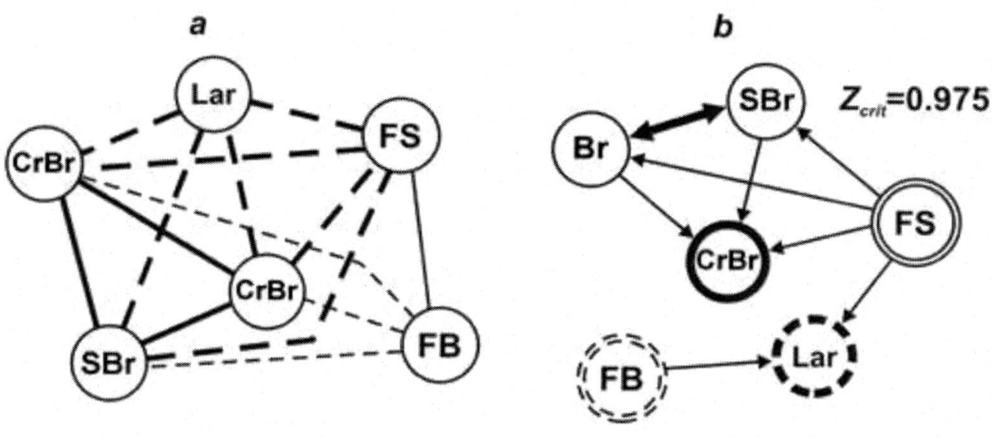

Figure 11.10. The schemes of nontransit relationships between the groups of forest types of the Priamurye boreal ecotone by the parameters of productivity of forest communities: the graph of similarity relations (a); and orgraph of relations of domination (b).
Legends are as in Figure 11.7.

Floristic phratries as the taxonomically largest definitions of plant cover are at a slightly higher level of structural isomorphism compared to the groups of forest types. It is due to the fact that the modern phratries have been formed by way of mixing different initial floras and, therefore, the same forest types belong to several phratries, thereby enhancing the degree of their structural similarity. The particularly distinct structural homogeneity is typical of the forests that form a bunch of floras: Mm – Mx – MA – An, with $N_k > 85\%$. As we can see, the meridional mixing of the Manchurian floras with the Angaridian flora in the geological epochs of cold snaps and aridization resulted in the formation of more uniform phytocoenological structure of forest cover. The epochs of cold snaps accompanied by high humidity were characterized by zonal transgression of the Okhotian flora mixing with

Manchurian floras. As a result, much more complex and diverse buffer phytocoenological structures were formed.

It is no coincidence that this maximum mutual structural similarity (the level of similarity over 90%) is typical of the forests of the basal Okhotian and buffer Manchurian–Okhotian floras, while the representative Ajan fir/spruce and spruce/broad-leaved forests have quite substantial differences ($N_k \leq 75\%$). However, having inherited many of their structural characteristics from the Ajan forests, the spruce/broad-leaved communities substantially differ from them in the higher reserves of live phytomasses (see Table 11.1) and productivity (see Table 11.2). Accordingly, the measures of functional similarity of the forests of Manchurian–Okhotian and Okhotian floras are in the range of 81–90% (see Figures 11.9 and 11.10). In general, the buffer forests of the MO phratrie are similar in the intensity of autotrophic biogenesis to the communities of another basal phratrie, mesophilic broad-leaved forests ($N_k = 91–96\%$), as they have nearly the same productivity (about 8.1 t/ha · yr) with much higher reserves of total live phytomass (204.1 t/ha · yr vs. 139.2 t/ha · yr).

Let us proceed to the ecotypes of forest cover, their structural and functional development and potential stability. On the orgraphs of inclusion relationships (see Figures 11.7–11.10), there are almost no one-to-one relationships (relations of tolerance), as well as critical couplings, determining the integrity of the entire coupling network. The orgraphs are represented by exclusively weak connections indicating, under the conditions of high critical values of indistinguishability thresholds ($Z_{crit} = 0.90–1.00$), the relative structural autonomy of phytoecosystems under study, i.e., the enhanced fractionation of their ecological space on the Priamurye boreal ecotone. It should predetermine rather variegated mosaicity of local ecosystem responses to the same climatic background. A similar situation is observed in the boreal ecotone of the Volga River basin (Kolomyts 2008), so that it can be considered the common polysystemic regularity for this transcontinental transitional zone.

First of all, let us turn to the most ancient and exotic basal group of forest types of the Primorye/Priamurye sub-Pacific (Sochava 1946; Kolesnikov 1956; Kurentsova 1973): cedar/broad-leaved forests, which make 33% of the Manchurian mesophilic phratrie and 22% of the buffer Manchurian–Okhotian phratrie (see Table 10.1). These forests are the dominant ecotypes both in the degree of development of its phytocoenotic structure and in the level of functioning (particularly autotrophic biogenesis). They have rather high values of the structural metabolic index (S^{m}_{k} = 1.22–1.34) and the maximum potential stability (homeostatisity reserve $G_k = 36–43\%$). It can be assumed that the cedar forests of Priamurye simultaneously realize the two alternative evolutionary strategies: the strategy of maximal structural development (species and coenotic diversity, the substrate parameters of phytobiota) and the strategy of producing maximal biomass. In other words, the forests of this ecological group possess, to the proportionate extent, the properties of patience and violence (Ramensky 1971).

The analogous high indices of structure and function are characteristic of forest communities of the buffer Manchurian–Okhotian phratrie: mainly nemoral fir/spruce forests and, to a lesser extent, spruce/broad-leaved forests (see Figures 11.7, *b* and 11.8, *b*; Table 11.14). They are the dominant ecotypes by the level of structural development and also the leaders by the reserves of live phytomass (see Figures 11.9, *b* and 11.10, *b*). Possessing the respectively maximal homeostaticity reserves (G_k), the buffer communities of the MO flora are potentially most stable phytocoenological formations. At the same time, as regards current productivity, these forest ecosystems descend from the dominant positions of macrobolites to the temporal (medium) level, and nemoral fir/spruce forests descend even to the pessimal (microbolite) level characterized by the minimal values of autotrophic biogenesis.

The much more complex pattern of florogenetic and phytocoenological ecotypes is formed by the broad-leaved forests that make a basis of Manchurian phratries: both xerophilic (by 67%) and mesophilic (by 71%). The broad-leaved forests with oak of the Mx phratrie have a less developed phytocoenotic structure than the cedar forests: temporal and even pessimal (S^{m}_{k} = 0.77), with low (or moderate) homeostaticity reserves (G_k = 17–28%, see Table 11.14). However, as regards phytomasses and productivity parameters, they get into the category of macrobolites or, at least, submacrobolites (S^{m}_{k} = 1.25–1.43), with much more stable function ($G_k \approx 50\%$).

Mesophilic broad-leaved forests, on the contrary, are characterized by more highly developed structure (subdominant, S^{m}_{k} = 1.22) but comparatively low indices of functioning: in respect to phytomass reserves and productivity, they are referred to mesobolites or even submicrobolites (S^{m}_{k} = 1.00–0.75) and have a weak potential stability (G_k = 17–26%).

Thus, xerophilic broad-leaved forests (with oak) were formed through the violent evolutionary strategy – the tendency to gaining the maximum levels of productivity and biomass. Such immanently "programmed strategy of biotic community" (Sochava, 1974 24) could be ensured only by the stable favorable conditions of the abiotic environment. This upper-mountain enclave of forests corresponds to the flat tops and crests of ridges and near-peak sunny slopes, with the types of rich (loamy and clay) moderately moistened habitats. At the same time, mesophilic broad-leaved forests growing on subhorizontal piedmont plains and shadowy slopes of the lower band of lowlands were formed under less favorable conditions: excessive moistening of barren sandy-sabulous soils with low summer temperatures. Hence, their development was directed along the patience strategy, with the keen interspecies competition, which reduced the total autotrophic biogenesis but resulted in the formation of more diverse phytocoenotic structures of broad-leaved forest.

The ecotypes of dark-coniferous forests of the Okhotian flora are much less distinct, because in our statistical sample the Ajan fir/spruce forests make up 56% of this phratrie, while the mixed forests make up other 44%. It can be altogether assumed that the development of these forest formations occurred by analogy with the mesophilic broad-leaved forests: through the patience strategy (the achievement of structural diversity to the prejudice of biomass). The larch forests making up 45% of the buffer Manchurian–Angaridian phratrie, on the contrary, developed though the violence strategy, with simplification of its phytocoenotic structure but the higher level of functioning.

11.6. Buffer Dark-Coniferous/Broad-Leaved Forests of Priamurye Sub-Pacific – The Phenomenon of Evolutionary Forest-Forming Process

The performed analysis of the structural and functional parameters of forest ecosystems of the Lower Amur sub-Pacific leads to a number of conclusions that emphasize the *evolutionary nature* of its *forest-forming process* based on the floristic richness of forest phytocoenoses and including the diversity of their successions. What is meant here is the formation of *buffer forest communities* typical of ecotone systems, the striking example of which is the forests of the Manchurian–Okhotian phratrie (see Figures 1.3 and.1.4). As it has already been mentioned, the latter was formed on the basis of two basal phratries: Manchurian mesophilic and Okhotian. According to the data of spore and pollen analysis (Akhmetyeva 1977; Korotky et al. 1999), the polydominant broad-leaved forests represented by the modern Manchurian flora were widespread in the southern Far East already in the warm climatic epochs of the Early Eopleistocene (within 1.6–1.2 million years ago). The penetration of Okhotian dark-coniferous taiga into the Lower Priamurye dates from the climatic minimum of the Early Pleistocene (about 0.9 million years ago), while that of the boreal Angaridian forests dates from the cold epochs of the Middle Pleistocene (not later than 300 thousand years ago).

The buffer Manchurian–Okhotian forest formations differ from the forests of not only their basal but also other phratries, first of all, in the thicker tree layer, which is manifested in drastic increase of the average height of forest stand and the quality of locality, as well as in preservation of the maximum floristic diversity of the tree layer of Manchurian forests (see above Table 9.1). They are among the leaders also as regards the development of the lower layers (undergrowth and soil cover), yielding only to the forests of the Manchurian mesophilic phratrie.

The Manchurian–Okhotian spruce/broad-leaved and nemoral fir/spruce forests are characterized also by the highest indices of accumulation of live organic matter: the maximal skeletal, green and total phytomass (both above-ground and summarized, see Tables 11.1 and 11.2. Then, they are characterized by the maximal or close to maximal values of partial and total productivity (see Tables 11.3 and 11.6). Accordingly, the maximum homeostaticity reserves potentially make the buffer communities of the MO flora the most stable phytocoenotic formations.

There forests have generally the maximum age (the taxonomic norm is 133 years); however, it is hardly possible to explain the extreme values of organic deposition only by the above facts. The forest phytocoenoses of both basal phratries (Mx and Okh) are much younger (101–116 years old, according to the norm); however, they show much lower partial and total results of autotrophic biogenesis at nearly the same productivity.

Finally, the forests of the MO flora have peculiar features of the detritus branch of metabolism, distinguishing them from other forest types, which are represented in the morphological and physicochemical properties of soil. Here, the rate of mortmass decomposition is less, which results in the thickest forest litter; however, the weakening of mineralization processes is accompanied by intensification of the synthesis of humic substances; therefore, the humus soil layer here has high values of thickness, and the humus profile spreads most deeply into the soil, penetrating even into horizon B. Simultaneously, the soils of these forests have the characteristics of the most intensive processes of illuviation, lessivage and intrasoil weathering, so that these soils can be considered most mature in the series of floristic and phytocoenological formations under study, i.e., maximally similar to the equilibrium (climax) state. It seems that the microlandscapes of transit (sunny and neutral) slopes, as well as the plakor plain landscapes with dark-coniferous/broad-leaved forests of the MO phratrie reached the maximal level of structural and functional development.

Thus, there was not merely trans-zonal mixing of the Manchurian mesophilic and Okhotian floras. The buffer forest ecosystems were formed, with the more developed structure and more diverse functions compared to the ecosystems of the basal floras. The spruce/broad-leaved and nemoral fir/spruce forests of the MO phratrie proved to be the most similar to the *evolutionary climax* (Krištofovič 1946) as the last, the highest stage of vegetation development under the given zonal-regional climatic conditions. As is known (Vasilyev 1946), climax is formed by *migration processes* in the plant cover, i.e., the driving forces for its evolution that create buffer forest communities. There is a convincing evidence for the previously proposed (Yurtsev 1974; Kolomyts 1987; Panfilov 2005) conceptual statements on the Pacific Ocean mega-ecotone of Northern Eurasia as a focus of evolutionary processes in the continental biosphere.

It can also be assumed that the main biosphere function of the Manchurian–Okhotian mixed forests on the Pacific Ocean mega ecotone is the high indices of biogenic accumulation of substances, which prevents the leaching of chemical components. According to A.I. Perelman (1975), it is nothing else but the most developed *mechanism of negative reverse bio-inert connection* under the given zonal-regional conditions, which contributes to stabilization of the soil-biotic "core" of natural complexes. The phenomenon of Manchurian–Okhotian phratrie – a novel, more highly organized system of forest formation – is considered as one of the results of evolutionary phytocoenological processes that occurred in the recent geological past on the boreal ecotone of the Primorye/Priamurye sub-Pacific. There is convincing evidence of the concepts proposed in (Kolomyts 1987; Panfilov 2005) about the Pacific Ocean mega ecotone of Northern Eurasia as a focus of evolutionary processes in the continental biosphere. A little earlier, analogous ideas had been suggested by B.A. Yurtsev with respect to florogenesis of the northern wing of the mega ecotone – Beringia, which he referred to as "… a peculiar florogenetic laboratory of the northern zones of the Earth" (1974 144).

The emergence of another buffer phratrie, Manchurian–Angaridian, based on the Manchurian xerophytic and East-Siberian phratries, had no such striking effect (see Tables 11.2, 11.3 and 11.6). By all of the mentioned parameters, these buffer communities take an intermediate position between the forests of the basal phratries and are similar to Manchurian xerophilic forest only in the measures of floristic diversity of forest stand and undergrowth.

There is one more important conclusion logically following from the above data: the latitudinal-zonal (north ↔ south) transgression of the tree and shrub species and phytocoenoses in general in the marginal-continental sector of the Pacific Ocean mega ecotone of Eurasia was much more marked and had an incomparably greater structural and functional effect in the forest cover of the region than the longitudinal-sectoral (east ↔ west) transgression. Such an obvious prevalence of the forest-forming process of the first group over the process of the second group could only be a result of long-term sustainable preservation of oceanic features in the regional bioclimatic system of the Primorye/Priamurye sub-Pacific. The modern geological epoch in the southern Far East is characterized by the combination of marine transgression with the hydrothermal conditions similar to the Middle Quaternary (Riss–Würm) climatic optimum (Nikolyskaya, Timofeev 1966; Korotky et al. 1999), when the latitudinal migration of flora elements was predominant and its longitudinal (west-east) contrasts were more intense (Yurtsev 1974). This is a manifestation of longitudinal bioclimatic sectorality, which is so marked on the Pacific Ocean mega ecotone of Northern Eurasia and which substantially refracts and complicates the general background of latitudinal zonality.

11.7. INITIAL TENDENCIES OF MODERN FOREST FORMATION IN MICROLANDSCAPES

The polysystemic phyto-ecological space of microlandscapes will be assessed by the simultaneous analysis of their similarity relations (Figures 11.11, I *a*; 11.11, II *a*; 11.11, III *a*) and inclusion relationships (Figures 11.11, I *b*; 11.11, II *b*; 11.11, III *b*), as well as by the integral parameters of spatial organization (Table 11.15). Here, it is important to pay attention to the critical thresholds of indistinguishability Z_{crit} of inclusion relationships between the objects. This parameter indicates the minimal values of their J(ai; bi), when the continuous chain of overlapping structural or metabolic niches of phytobiota is still maintained. It ensures the appropriate integrity of regional landscape system, as well as the continuity of reciprocal transformations of its elements (microlandscapes) into each other under certain external influences.

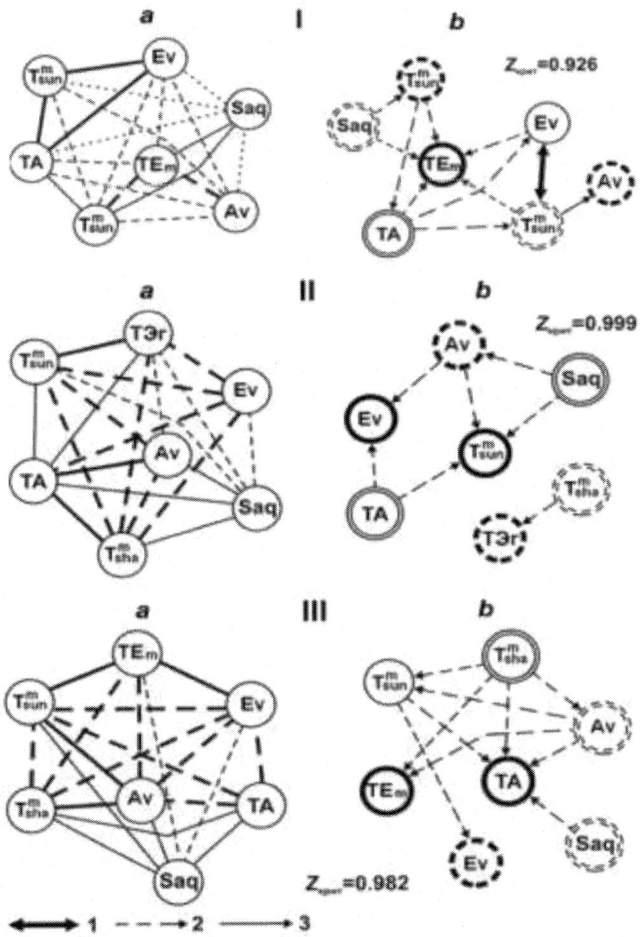

Figure 11.11. The schemes of nontransit relationships between plain and low-mountain microlandscapes of the Priamurye boreal ecotone. The parameters of microlandscape states with respect to: I – structure; II – live phytomasses; III – productivity. The structural schemes of: *a* – similarity relations; *b* – relations of inclusion. The connections of inclusion relations: 1 – tolerant (mutual inclusion); 2 – weak; 3 – ordinary. Legends are as in Table 9.2 and Figure 11.5.

Table 11.15. The measures of homeostatisity (G_k) and structural-metabolic indices ($S^m{}_k$) in the phytocoenotic block of different microlandscapes

Parameters		Groups of microlandscapes						
		TEm	**$T^m{}_{shad}$**	**$T^m{}_{sun}$**	**TA**	**Ev**	**Av**	**Saq**
Structure	G_k	**0.333**	0.201	**0.249**	*0.103*	0.218	**0.282**	*0.126*
	S_k	**1.526**	*0.923*	**1.143**	*0.471*	1.000	**1.292**	*0.577*
Living phytomasses	G_k	**0.367**	*0.158*	**0.400**	*0.127*	**0.399**	**0.232**	*0.127*
	S_k	**1.14**	*0.86*	**1.43**	*0.71*	**1.29**	**1.14**	*0.71*
Productivity	G_k	**0.339**	*0.109*	0.224	**0.343**	**0.294**	*0.180*	*0.154*
	S_k	**1.27**	*0.64*	1.00	**1.37**	**1.09**	*0.82*	*0.91*

Footnote. Conventional meanings see in Note to Table 11.14.

The plain/lowland landscape space on the boreal ecotone of the Priamurye sub-Pacific is characterized by evident fragmentation and territorial contrasts corresponding to the same discreteness of floristic phratries and groups of forest types (see above). The predominant closeness of structural phytocoenotic connections between microlandscapes is in the range of 76–85%. At the same time, as regards phytomass reserves and productivity, these connections are mostly above the 85% threshold and often approach 92–98%.

The connectivity of the orgraph of inclusion relationships of microlandscapes by their structural phytocoenotic characteristics (see Figure 11.11, I *b*) is provided by relatively low values of the banality threshold Z_{crit} with not only weak or critical connections but also with the connections of tolerance: one-to-one correspondence (Andreev, 1980) forming the *typological core of regional ecological space.* It means that the polysystem space of the Priamurye boreal ecotone is also characterized by the relatively high degree of intercomplex (subhorizontal) interactions, which is represented by the mutual penetration of above described floras and phytocoenotic structures and has resulted in the formation of buffer natural complexes – a phenomenal event in the moderate belt of the Asian marginal-continental sub-Pacific.

In respect of phytocoenotic structure, the closest connections between the measures of similarity (N_k = 93–94%) are formed by the three groups of microlandscapes: transit shadowy mountain slopes ($T^m{}_{shad}$), piedmont transaccumulative complexes (TA) and the adjacent plain/valley plakors (Ev). This triad is characterized by the minimal reserves of homeostaticity (see Table 11.15), i.e., has the lowest resistance to external perturbations and forms, as it has already been mentioned, the pessimal/subpessimal pleiad of forest types by phytomass reserves and productivity. At the same time, the typological core of regional polysystem space is a tandem of elivual valley (plakor) and transit shadowy slope microlandscapes (see Figure 11.11, I *b*). Consequently, it would be proper to consider these very two groups of natural complexes (Ev and $T^m{}_{shad}$) as subregional representatives of zonal bioclimatic background on the Priamurye boreal ecotone. Accordingly, it can be assumed that the *regional bioclimatic norm of the ecotone must have relatively low values of the biological cycle and the enhanced susceptibility of forest communities to climatic fluctuations.*

The second, less evident bunch of the measures of similarity in phytocoenotic structure includes the microlandscapes of low-mountain watersheds, transit sunny slopes, and accumulative valley sites. Here, the positive extrazonal hydrothermal conditions are created and the optimal "pole" of productivity and phytomass formation appears. As regards the level of structural development, the TEm group of microlandscapes with relatively high stability is a dominant ecotype ($G_k > 33\%$), while $T^m{}_{shad}$ and Av, the less stable landscape groups, are subdominants. Thus, a single "unipolar" system of spatial connections between microlandscapes by phytocoenotic structure is formed, with the flat tops and crests of low ridges and near-peak (mainly sunny) slopes in its center.

On transition from structural phytocoenotic to functional parameters, the total degree of similarity between microlandscapes considerably increases. However, the orgraphs of inclusion relationships are substantially complicated against this background. They are represented by exclusively weak connections of both biomass and productivity at the high values of Z_{crit} (see Figure 11.11, II *b* and III *b*). The regional functional space proves to be "multipolar," fractionated into separate structural elements, with the emergence of several macrobolites (Ev, $T^m{}_{sun}$, TEm, TA) and submacrobolites (TEm, TA, Av) not connected with each other, each of them possessing its own subordinate microbolites or submicrobolites. Hence, there appears considerable variegation of the regional functional space, more marked than the structural and phytocoenotic variegation.

Thus, the above-described forest-forming processes in the marginal-continental sector of the Pacific Ocean mega ecotone have created an ordered regional system of structural phytocoenotic connections between microlandscapes. At the same time, the functional (metabolic) space proved to be "multipolar," i.e., fragmented into separate units with their own vectors of inclusion relationships. There is obvious unconformity (mismatch) between the two characteristics of polysystemic organization of regional geospace: structural and functional. There is no reason to consider it manifestation of the properties of patience and violence of different species of forest phytobiota relative to microlandscapes, since most of the groups of forest types are distributed over the entire spectrum of altitudinal-exposure zonality and occur in all groups of landscape types.

It seems that the cause of the revealed unconformity should be sought in the new trends of changes in the forest cover of the given ecoregion. As it is known (Zlotin 1987; Kolomyts 2008), the first cause and driving force of the functional and structural rearrangements in geo(eco)systems is the aggravation of lateral (intercomplex) natural relationships and interactions in the periods of disturbance of the previously achieved (to some extent) stability of these systems. Disturbances may take place under the influence of both external (climatic) signals and internal (endoecogenetic) processes. At the same time, the first and foremost shifts occur not in the structural characteristics of geo(eco)systems but in the parameters of their functioning, including the small biological cycle (phytomass production and decomposition). These shifts take the first several years in the taiga zone and occur during a year in the subzone of broad-leaved forests (Osipov 2004; Kolomyts 2005). This period is quite adequate, as regards its duration, to the total shift of landscape-ecological conditions; therefore, the territorial mosaic of functional relaxation of geo(eco)systems must reveal the foci of their forthcoming evolutionary transformations. The structural rearrangement of forest phytobiota needs 2–3 generations of the basic forest-forming species: up to 250–300 years (Puzachenko 1996).

Thus, the "multipolar" fragmentation of the functional phytocoenological space on the Priamurye boreal ecotone, with the forest communities of individual microlandscapes passing from one ecotype to another and with the entire system of intercomplex relationships and connections (compared to the structural space) being complicated, can be considered as certain mechanisms of origination of new phytocoenological structures. Since phytobiota from the geosystem-stabilizing factor (Sochava 1978) turns into the dynamics-transforming factor, the respective changes must take place also in the structure of microlandscapes.

Let us give a few examples of such evolutionary foci on the territory of the Komsomolsky Nature Reserve. Fir/spruce forests and, to a lesser extent, cedar and larch forests of piedmont TA microlandscapes fall into the category of pessimums (the least developed communities, see Figure 11.11, I *b* and II *b*) by their structure and phytomass reserves. However, their current productivity drastically increases, and they become the dominant macrobolites, on a par with forests of the TEm microlandscapes, to which they will tend in their structural transformations (see Figure 11.11, III *b*). Certain shifts towards the activation of autotrophic biogenesis also take shape in the forest communities of eluvial valley plakors (Ev), which will lose the relations of tolerance with the forests of shadowy mountain slopes and, apparently, approach the piedmont TA communities.

The group of broad-leaved and mixed forests of accumulative valley (Av) microlandscapes, on the contrary, shows a tendency to reducing their productivity. Remaining the suboptimal structural ecotype, they pass into the category of submicrobolites by the level of functioning. The structural prototype of their future states may be fir/spruce forests and spruce/broad-leaved forests of adjacent landscapes on shadowy mountain slopes (T^{m}_{shad}). The less marked negative functional dynamics is demonstrated by the large group of forest types (cedar and spruce/broad-leaved and fir/spruce) in the landscapes of sunny and neutral mountain slopes.

On the whole, the maximum variability of phytobiota must be typical for the microlandscapes of lowland plains, river valleys and piedmonts (TA, Ev, Saq) having the minimum reserves of homeostaticity by their phytocoenotic structure (see Table 11.14). The moderate variability is typical for the forest cover of shadowy mountain slopes. Finally, the forests of low-mountain watersheds, as well as the adjacent sunny and neutral mountain slopes, are the least variable. They possess the maximally high values of the structural-metabolic index and reserves of not only structural but also functional stability.

These are some tendencies of the forthcoming functional and structural changes in forest cover which, according to our set-theoretical model, take shape in the low-mountain/valley microlandscapes of the Lower Amur sub-Pacific. This model undoubtedly has certain *prediction properties*, though in itself it does not disclose the first causes of evolutionary processes. It reveals the foci and directions of the most probable structural transformations in forest ecosystems, the prototypes of which are their already existing functional states.

11.8. Some Inferences

The experience of empirical-statistical modeling of the structure and function of forest ecosystems of the Lower Amur sub-Pacific leads to a number of conclusions.

1) The general information-statistical model of monosystemic organization of plain-lowland subregional and local geo(eco)systems of the Lower Amur sub-Pacific has been used to characterize the most common regularities of formation of phytocoenological and landscape structures under the conditions of epigeosynclinal mountainous country of Mesozoic age and boreal bioclimatic ecotone. The causal mechanisms of appearance of the known floristic richness and phytocoenotic diversity of forest cover typical for marginal-

continental lands of the southern Far East have been established. The more detailed mechanisms of these biospheric processes have been revealed using discrete partial models of the binary ordination of geocomponent characters. These models emphasize the *evolutionary nature* of the *forest-forming process* based on the floristic richness of forest phytocoenoses and including the diversity of their successions. What is meant here is the formation of *buffer forest communities* typical for ecotone systems; the striking example is the forests of the Manchurian–Okhotian phratrie formed on the basis of two basal phratries: Manchurian mesophilic and Okhotian.

2) The Manchurian–Okhotian spruce/broad-leaved and nemoral fir/spruce forests differ from the forests of not only their basal but also other phratries, first of all, in the thicker tree layer and the highest class of bonitet, with the maximum floristic diversity of forest stand, undergrowth and soil cover. These buffer forest formations are also characterized by the highest indices of accumulation of live organic matter – the maximal skeletal, green and total phytomass (both above-ground and cumulative). Forest litter reaches maximum thickness due to slower mortmass decomposition; however, the weakening of mineralization processes is accompanied by intensification of the synthesis of humic substances; therefore, the humus layer of the soil also becomes very thick. Finally, the processes of illuviation, lessivage and intrasoil weathering are most intensive here and, hence, the soils of these forests are closest to the evolutionary climax.

3) Thus, there was not merely transzonal mixing of the Manchurian mesophilic and Okhotian floras. The buffer forest ecosystems formed had more developed structure and more diverse functions compared to ecosystems of the basal floras. The forests of the Manchurian–Okhotian phratrie proved to be the most similar to the *evolutionary climax* (Krištofovič 1946) as the last, the highest stage of vegetation development under the given zonal-regional climatic conditions. The climax is formed by *migration processes* in the plant cover (Vasilyev 1946), i.e., the driving forces for its evolution that create buffer forest communities. There is convincing evidence of the previously proposed concepts (Yurtsev, 1974; Kolomyts, 1987; Panfilov, 2005) about the Pacific Ocean mega ecotone of Northern Eurasia as a focus of evolutionary processes in the continental biospher

4) It can also be assumed that the main biosphere function of the Manchurian–Okhotian mixed forests on the Pacific Ocean mega ecotone is the high indices of biogenic accumulation of substances, which prevents the leaching of chemical components. As it has been shown (Perelyman 1975), it is nothing else but the most developed *mechanism of negative reverse bio-inert connection* under the given zonal-regional conditions, which contributes to stabilization of the soil-biotic "core" of natural complexes. The phenomenon of Manchurian–Okhotian phratrie – a novel, more highly organized system of forest formation – is considered as one of the results of evolutionary phytocoenological processes that occurred in the recent geological past on the boreal ecotone of the Primorye/Priamurye sub-Pacific.

5) The appearance of another buffer phratrie (Manchurian–Angaridian) had no such structural/functional effect. Its communities take an intermediate position between the forests of the basal phratries by all of the above parameters and are similar to Manchurian xerophilic forests only in the measures of floristic diversity of forest stand and undergrowth.

6) It follows from the above that he latitudinal-zonal (north ↔ south) transgression of the tree and shrub species and phytocoenoses in general in the marginal-continental sector of the Pacific Ocean mega ecotone of Eurasia was much more marked and had an incomparably greater structural and functional effects in the forest cover of the region than the longitudinal-

sectoral (east ↔ west) transgression. Such an obvious prevalence of the forest-forming process of the first group over the process of the second group could only be a result of long-term sustainable preservation of oceanic features in the regional bioclimatic system of the Primorye/Priamurye sub-Pacific.

PART IV: CONTINENTAL-ISLANDFADING EVOLUTION LANDSCAPE

Chapter 12

PUTYATIN ISLAND AS A MODELING OBJECT

ABSTRACT

The structural and functional analysis of continental-insular landscape was performed by the example of Putyatin Island (Peter the Great Bay, South Primorye). Such landscape characterizes the mega state of this part of continental biosphere on the regressive branch of its geological development. The island geosystem is at the stage of damped evolution: the "senile" erosion-denudation relief, the uniform anthropogenically modified forest cover (secondary prariefied oak forests), and the active manifestation of exogenous morpholithogenesis. The oak forests of Putyatin Island, as a subarid phytocoenological structure, do not correspond to the modern zonal-climatic conditions of southern coastal-island regions of Primorsky Krai. Here, such *zonal representative is the soil cover* which has yet preserved many morphological features since the time of dominance of coniferous/broad-leaved forests. Under the conditions of anthropogenically altered and, to a considerable extent, degraded forest vegetation of the continental-island geosystem, the main object of landscape structural-morphological analysis is the soil and the related processes of exogenous morpholithogenesis.

12.1. ZONAL AFFILIATION OF THE ISLAND

The extreme south of the Russian Far East, including the coast and the continental islands of the Peter the Great Gulf, belong to the zone of coniferous/broad-leaved forests of the Manchurian Province of the Far East (Kolesnikov 1961). The Correlation Ecological-Phytocoenotic Map of Asian Russia (Sochava and Bayborodin 1977) shows that this region is a part of subzonal ecological-phytocoenotic complex of nemoral and subnemoral broad-leaved and coniferous/broad-leaved, mainly tall-grass, forests and their restorative series, maximally productive (more than 10 tons/ha), very warm and moderately humid habitats. According to the data obtained at the Vladivostok (Port) weather station, the region is characterized by the following climatic parameters: (1) the mean August temperature (the warmest month) is about 21.0; (2) the mean January temperature (the coldest month) is –13.5; (3) the sum of biologically active temperatures is 2648; (4) the annual evaporation is 744 mm; (5) the annual precipitation is 813 mm; (6) the annual Vysotsky–Ivanov atmospheric humidity factor (F_{hum}) is 1.09.

According to the scheme of natural regions of the Earth (Volobuyev 1973), with regard to the humidity factor, the southern coastal and continental-insular regions of Primorye must be referred to the transition belt between the zone of mixed forests (F_{hum} = 1.20–1.24) and the subzone of broad-leaved forests–northern forest steppe (F_{hum} = 0.99–1.03). If we compare this ecoregion with the western, most humid sector of the East European Plain (Isachenko 1985; Kolomyts 2003), the atmospheric humidity of the Putyatin continental-island landscape corresponds to the southern limit of broad-leaved forests and their transition to meadow steppes of the subzone of typical forest steppe. As is known (Bazilevich et al. 1986), the climaareals of meadow steppes and oak forests (together with oak/pine and hornbeam/oak forests) have a considerable reciprocal overlap, which underlies their genetic unity forming the subzone of northern forest steppe. Thus, Putyatin Island is part of the *regional bioclimatic forest-meadow-steppe ecotone of South Primorye as a trigger system with hysteresis properties* (Armand and Vegyushkin 1989; Vedyushkin 1989). Such ecotone is characterized by competitive relationships (Armand 1989) between forest and meadow communities, their unstable equilibrium states, high dynamicity, and rather rapid mutual transitions under the conditions of even minor climatic fluctuations or anthropogenic impacts (Kolomyts, 2005). Indeed, the recent history of vegetation on the islands of the Peter the Great Bay, including Putyatin Island, confirms this conclusion.

While staying under the conditions of long-term tectonic downwarping, Putyatin Island has acquired the features of strongly leveled, smoothed hill country with absolute heights up to 250–350 m (Figure 12.1). The outlines of the test ground relief are gentle and flattened, with minor differences in height between watersheds and stream bottoms. The stream valleys are slightly cut in. The flat bottoms of wide and short valleys are often waterlogged. It is favored by the predominant sandy clay loam soil-forming substrate, as well as by the combination of rather high values of summer temperatures and atmospheric precipitation.

Figure 12.1. The general view of Putyatin Island.

In Upper Holocene, Putyatin Island underwent several stages of plant succession associated with the region-wide climate changes. The spore-pollen analyses of soil samples from the islands in the Peter the Great Gulf, including Putyatin Island, has shown (Rodnikova et al. 2012; Lyashchevskaya et al. 2010, 2014) the prevalence of polydominant broad-leaved forests with oak, elm, walnut, hornbeam and maple in the beginning of the Upper Holocene (before 2.0–2.5 thousand years ago), when climate was similar to the modern one. This time interval characterizes the age of the contemporary morphological and geochemical structure of soil horizon B. In the subsequent period (no later than 1200–1400 years ago), due to cooling these forests were replaced by coniferous/broad-leaved forests, with disappearance of some broad-leaved species and spread of lime woods. At the third stage, mixed forests were replaced by the secondary broad-leaved forest vegetation (mainly oak forests) as a result of anthropogenic influence, with the general tendency towards aridization and meadow formation (Figure 12.2). This stage began 720 ± 150 years ago, with the maximum development in the past 100 years. It was the stage of formation of the morphological and geochemical properties of soil horizon A_1.

Figure 12.2. The general types of relief and plant cover of Putyatin Island. An open scrubby oak forest and a mixed-grass-cereal meadow are in the forefront.

Thus, the upper humus-accumulative paedohorizon can be considered (at least tentatively) to be a "soil-moment" (Targulyan, Sokolov 1978), which represents the latest, anthropogenic stage of supercentury phytocoenological successions that have led the forest cover of the island to the contemporary state of predominant xerophilic oak forests. The lower illuvial soil horizon is a "memory soil" preserving the properties of geosystem invariant (Mamay 1994). The structure of this horizon exposes the biogeocoenotic processes of the previously existing more humid coniferous/broad-leaved forests. However, it is not a soil relict but rather reflects the contemporary zonal bioclimatic conditions in the south of Primorye, with the same subnemoral type of soil formation as in the beginning of the Upper Holocene.

Hence, under the conditions of anthropogenically altered and largely degraded forest vegetation of continental-island geo(eco)system, *soil and the related processes of exogenous morpholithogenesis become the major objects of landscape structure-morphological analysis.* For the forest plant cover, there still has to be the analysis of productivity and some complex parameters of the biological cycle.

According to the landscape scheme (Lyashchevskaya et al. 2014), the lithogenic basis of landscapes in the central part of Putyatin Island is composed by granitoids and granites superposed with a mantle of sabulous-argillaceous sediments. The territory of the test ground bordering on the eastern coast of the island (Figure 12.3) is represented by a "senile" erosion-denudation hill country, with local watersheds in the northwest and south and with predominance of gentle and moderate slopes.

12.2. Plant Cover

There are two types of plant formations widespread on the test ground: (1) broad-leaved forests and (2) mixed-grass-cereal meadows in combination with shrub-herbaceous communities (see Figures 12.2 and 12.3). Both formations are secondary, anthropogenic, originating from the third paleobotanic stage mentioned above.

The Putyatin oak forests are usually park-like. They have low, tapering trunks and branched "umbrella" crowns attached to the treetop. The main types of oak forests and the related soil differences are rather clearly distributed along the slopes of different steepness (Lyashchevskaya et al. 2014).

The open broad-leaved forests with typical and weakly matted chernozems are predominant on gentle slopes. The dominant species is Mongolian oak (*Quercus mongolica*). The admixtures are the Amur lime (*Tilia amureensis*), the Siberian (Manchurian) alder *(Alnus hirsuta*), painted, pseudo and micro sieboldianum maples (*Acer mono, A. pseudosie-boldianum, A. microsieboldianum*). The undergrowth is formed by the alder Maksimovič (*Duschekia maximowiczii*), the Amur barberry (*Bereberis amurensis*), heterophyllous hazel (*Corylus heterophylla*), false spirea (*Sorbaria sorbifolia*), and other shrubs. The soil cover includes goat's beard (*Aruncus dioicus*), the Dahurian (Asiatic) moonseed (*Menispermum dauricum*), *Cacalia praetermissa*, bottle brush (field horsetail) (*Equisetum arvense*), the Korean meadowsweet (*Filipendula glaberrima*), oriental geranium (*Geranium orientale*).

The medium-steep slopes are occupied mainly by close oak forests on typical brown soils, as well as open broad-leaved forests on typical weakly matted brown soils. In addition to oak, linden and maples, the species growing here are: the Japanese elm (*Ulmus japonica*), rhynofolious ash (*Fruxinus rhynchophylla*), heartleaf hornbeam (*Carpinus cordata*), the Amur velvet (*Phellodendron amurensis*), the Japanese angelica tree (*Aralia elata*), the Dahurian and flat-leaved birch (*Betula davurica, B. platyphylla*), and dimorphant (*Kalopanax septemlobus*). The herbage is composed by Wilford's cranesbill (*Geranium wilfordii*), Solomon's seal (*Polygonum odoratum*), *Arisaema amurense*, *Thalictrum tunberiferum*, *Adiatum pedatum*, etc.

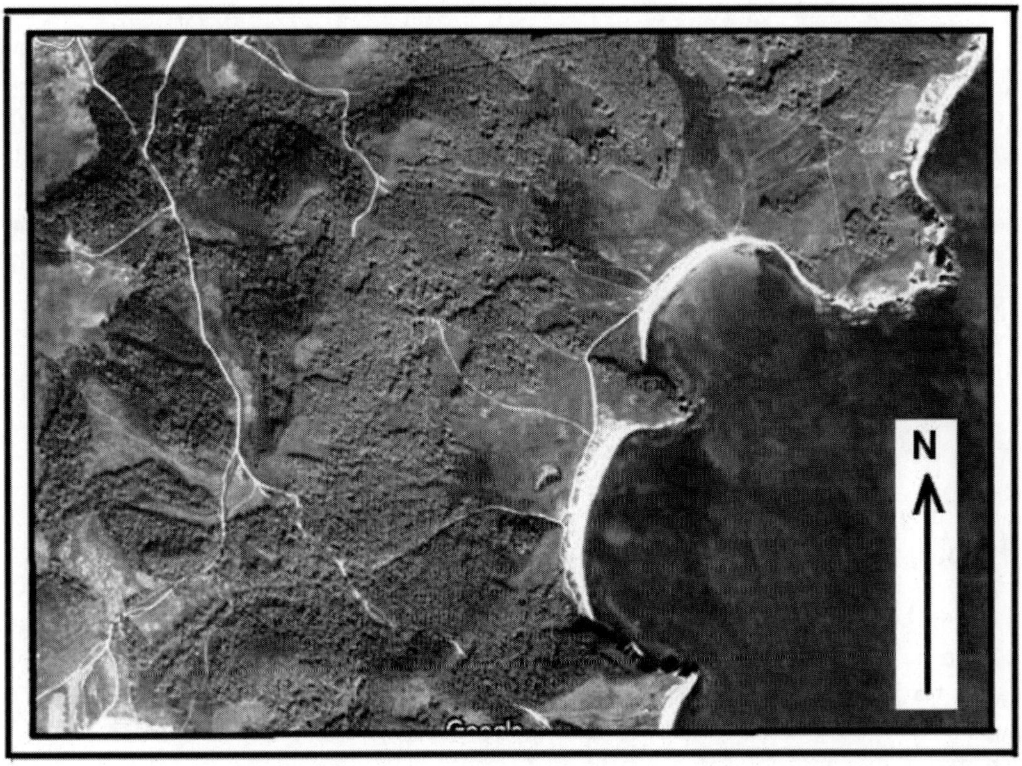

Figure 12.3. The space photograph of a plot in the test ground on Putyatin Island. There are the clearly marked areals of forest and meadow associations. In the upper left part of the photograph there is a local watershed: the hill-mountain ridge with an absolute height of 80–120 m.

The oak forests growing on dark brown soils with the thicker dark-humus layer are represented by the wild rose Maksimovič (*Rosa maximowiczii*), Japanese silver grass (*Miscanthus sinensis*), hagi (*Lespedeza bicolor*), broad-leaved beadruby (*Maianthemum dilatatum*), large brown clover and bustard lupine (*Trifolium spadiceum, T. lupinaster*), etc.

Finally, light oak forests with lime, maples, *Micromeles alnifolia*, Ruprecht's honeysuckle (*Lonicera ruprechtiana*), *Carex siderosticta*, touch-me-not balsam (*Impatiens noli-tangere*) are widespread on steep (including riverine) slopes with incompletely developed skeletal brown soils and fragmented primitive soils. Thin undergrowth is formed by cuspidate rhododendron (*Rododendron mucronulatum*), Amur barberry (*Berberis amurensis*), and Manchurian hazel (*Corylus mandshurica*). Under the same soil-geomorphological conditions, there are shrub-herbaceous communities with rhododendron, barberry, hazel, bush-clover, Russian wormwood (*Artemisia gmelinii*), whorled ladybell (*Adenifora vertisillata*), broad bean (*Vicia uniqua*), and yellow bedstraw (*Galium verum*), etc.

One can clearly trace two phytoindicators characterizing the alternative conditions of lithomorphism/hydromorphism. Korean rhododendron (*Rhododendron mucronulatum*) prefers the driest gravelly edaphotopes, whereas *Cacalia hastata* is widespread on hydromorphic loamy habitats.

In addition to meadow formations, our statistical sample includes wet grassy meadows on meadow gley soils. They are composed by Japanese buttercup (*Ranunculus japonicus*), silverweed cinquefoil (*Potentilla anserina*), narrow-leaved bluejoint (*Calamagrostis angustifolia*), slough grass (*Beckmania syzigachne*), Haupt's alkali-grass (*Puccinella hauptiana*), Mongolian pilea (*Pilea mongolica*), *Sanguisorba parviflora*, Japanese inula (*Inula japonica*), etc. The shrubby-grassy meadows on meadow sod soils occur much less frequently. The predominant species of these communities are goat willow (*Salix caprea*), hawthorn raspberry (*Rubus crataegifolius*), Nippon strawberry (*Fragaria nipponica*), *Euphorbia komaroviana*, *Lamium barbatum*, stocky Solomon's seal (*Polygonum humile*), and the Siberian geranium (*Geranium sibiricum*).

12.3. THE GROUPING OF FOREST BIOGEOCOENOSES

Thus, the oak forests of Putyatin Island, as a subarid phytocoenological structure, do not correspond to contemporary zonal-climatic conditions of the southern coastal-insular regions of Primorsky Krai. Such *zonal representative here is soil cover,* which undoubtedly retains many of the morphological features since the time of predominance of coniferous/broad-leaved forests. This is why we consider the soil-morphological characteristics as priority indicators for biogeocoenotic processes and spatial differentiation of the selected groups of biogeocoenoses in Putyatin Island.

The prevalent uniformity of forest-forming species, with predominance of oak stands, the weak development of undergrowth, and the general prariefication of soil cover eliminated the possibility of using the traditional phytocoenotic criteria for selecting *the groups of biogeocoenoses* (*BCG*) on the territory of the test ground. These groups were selected by the two local abiotic characteristics: geomorphological and hydroedaphic. The first characteristic encompasses the catenary system of *types of sites* (TS) (Glazovskaya 1964, 1988), or *geotopes* (Nef 1968; Krauklis 1979): from eluvial (E) and transeluvial (TE) via transit (T) to transaccumulative (TA) and superaqual (Saq) sites. The second diagnostic characteristic was the ecological scales of soil plant cover (Ramensky, 1971) characterizing the extent of surface hydration of the edaphotope. According to L.G. Ramensky and A. Cajander (cited by: Sukachev 1972), soil cover is an indicator of *biologically equivalent habitats.* Based on the floristic composition of grass cover, we have distinguished the following gradations of *edaphic moistening:* (a) meadow steppe (xerophytes, x); (b) dry meadow (xero-mesophytes, xm); (c) fresh meadow (mesophytes, m); (d) moist meadow (meso-hygrophytes, mg); and (e) wet meadow (hygrophytes, h).

Table 12.1. The distribution of gradations of edaphic moistening by the types of sites (the matrix of the normalized partial coefficients of connection)

Types of sites	Edaphic moisture				
	Xeromorphic (x)	**Xero-meso-morphic (xm)**	**Mesomor-phic (m)**	**Hydro-meso-morphic (hm)**	**Hydromorphic (h)**
Eluvial (E)	0	**0.48**	0.21	0.27	0
Transeluvial (TE)	**0.84**	0.09	0	0	0
Transit (T)	0.16	0.27	0.17	**0.45**	0
Transaccumulative (TA)	0	0.16	**0.62**	0.27	0
Superaqual (Saq)	0	0	0	0	**1.00**

Note. The dominant areas of ecological niches are in bold.

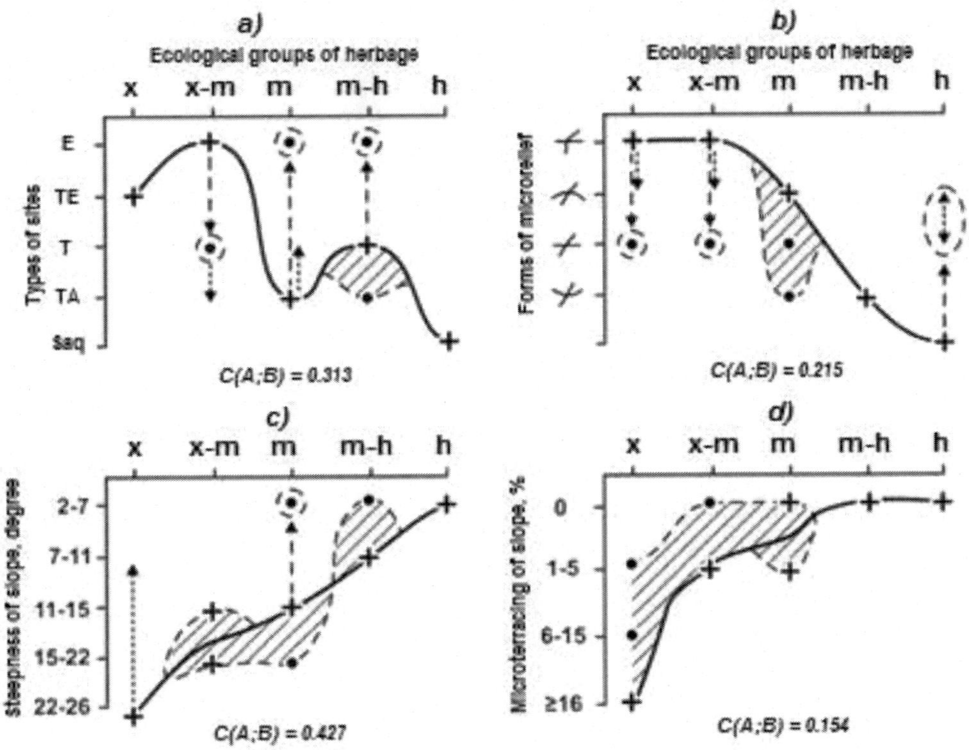

Figure 12.4. Putyatin Island. The abiotic factors of formation of ecological groups of grass canopy. Legends are as in Figure 6.3.

Binary ordination (Table 12.1) has shown a rather high conformity of edaphic moistening categories to the types of sites and micro-relief forms (Figure 12.4, *a*, *b*). There is a clear geomorphological "polarization" of two alternative ecotypes: xerophilic (based on TE types of sites) and hydrophilic (based on Saq types). Flat interfluves (the E types of sites) are occupied by the variety of ecotypes: from the dominant (with a 50% probability) xeromesophilic to enclave-type 25% mesophilic and mesohydrophilic. The T type of sites is characterized by the increased moisture content: up to the moist-meadow level with the 45%

probability of this dominant. TA geotopes located further down the slope show more moderate moisture content with the 62% probability of the mesophytic dominant.

In general, the upper elements of catena (the E and TE types of sites) are characterized by the maximum spread of the ecological groups of grass canopy. With transition to the lower elements (TA and Saq), their group diversity drastically decreases, and the dominant moist-meadow, and then wet-meadow ecotypes become more and more marked.

We have selected *five groups of forest biogenocoenosis* (*BGC*) based on the combination of the types of sites and edaphic moistening of the Putyatin landscape:

Ekm (); TEk (); Thm (); TAm (); SAh ().

The characteristics of their phytocoenotic structure are briefly presented in Table 12.2.

Table 12.2. The biogenocoenotic groups (landscape facies) of the Putyatin regional ecosystem and their designations

Designation and brief description	Symbol
1. **Ekm.** Oak forest, sometimes with lime and maple, eluvial, xero-mesomor-phic (with deviations towards hydromorphism), sparse, motley grass-cereal, with Pale Indian Plantain (Cacalia), on sabulous, light and medium loamy typical brown soils.	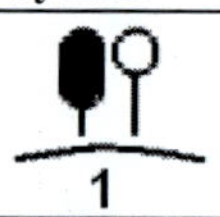
2. **TEk.** Oak forest, transeluvial, xeromorhpic, open, rhododendron, motley grass-cereal, on sabulous, incompletely developed, strongly skeletal brown soils and fragmentary primitive soils.	
3. **Thm.** Oak forest, sometimes with lime, birch and Amur cork tree, transit (less often eluvial and transaccumulative), hydro-mesomorphic (with deviations towards xeromorphism), sparse and open, motley grass, with Cacalia, on sabulous and light loamy typical and weakly matted brown soils.	
4. **TAm.** Oak forest, sometimes with lime and birch, transaccumulative and transit, mesomorphic (with the signs of hydromorphism), from open to high closed, Cacalia-motley grass, on sabulous-loamy typical and polygenetic brown soils.	
5. **SAh.** Oak forest, hydromorphic, superaqual (stream bottoms on the val-ley pediment), sparse, motley grass-Cacalia, on gley and dark-brown soils.	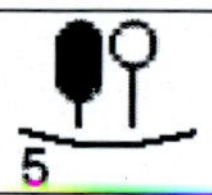

In addition, the following three groups of meadow biogeocoenoses were selected: 6, eluvial and transeluvial; 7, transit and transaccumulative; 8, superaqual. The distribution of biogeocoenotic (*BGC*) groups is presented in Figure 12.5.

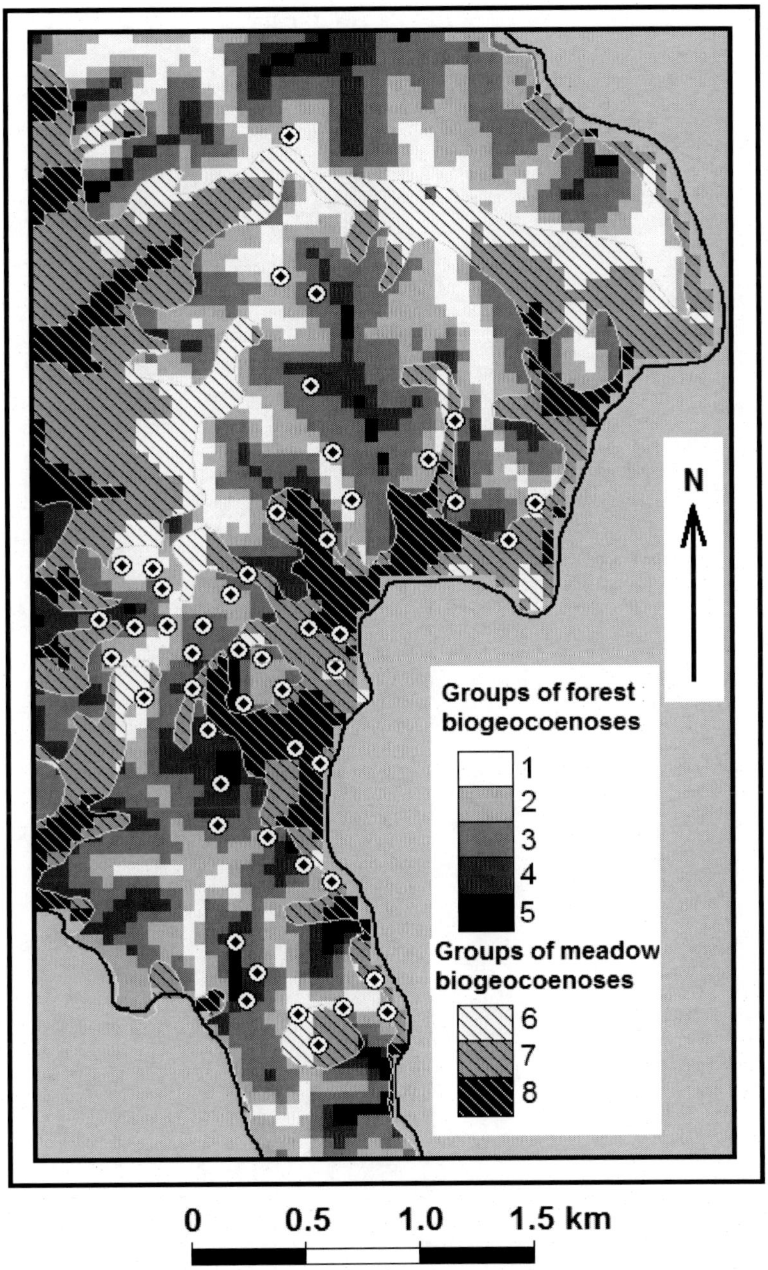

Figure 12.5. The raster map of forest and meadow biogeocoenotic groups on the test area of Putyatin Island. Calculated and plotted by L.S. Sharaya. The test sites are marked with punches. Other legends see in the text.

Chapter 13

Monosystem Organization of Continental-Island Landscape

Abstract

The information-statistical models of monosystem spatial organization of Putyatin landscape have demonstrated the highly developed system of chain reactions, the distinct echelonment of intercomponent relationships creating the hierarchic ordering of monosystem organization of forest biogeocoenotic cover of continental-insular land, and the relative homogeneity of intercomponent relationship, which provide a certain stability of monosystem structure. The role of slope steepness as the most powerful abiotic factor determining the entire system of local intercomponent relationships has been revealed. It is a *phenomenon* of morpholitho- and biogenesis in a regressive continental-island subboreal landscape, which shows that the main factor of edaphic humidification under the conditions of "senile" relief is the degree of drainage of the territory but not its radiation-thermal regime. The hydrothermal structure of the geotopological space of continental-insular sub-Pacific was analyzed in comparison with other regions. The local hydrothermal diversity creates a specific monsoon-continental type of catenary spectra of forest biogeocoenoses, which is clearly represented in the landscape of Putyatin Island. The main features of soil formation under the conditions of monsoon-continental climate of subboreal South Primorye have been revealed. Under the conditions of monsoon-continental climate of marginal-continental Sub-Pacific, the cross-correlation between soil temperature and moisture is very weak, being evidence of their certain mutual independence, *autonomy* and, consequently, their separate contributions to the system of intercomponent relationships. The analysis of soil morphology made it possible to elucidate the main features of soil formation under the conditions of monsoon-continental climate of subboreal South Primorye. The hydromorphic factor is more powerful here than the thermal factor.

13.1. Generic Statistical Information Model

The model (Figure 13.1) shows the most essential features of Putyatin continental-island landscape. It characterizes the mega state of this part of continental biosphere on the regressive branch of its geological evolution (Kolomyts 1987, 2015 b). The model provides a general idea of the topological mechanisms of landscape formation in the coastal-insular Primorye Sub-Pacific. The following features of these mechanisms can be distinguished.

1. Rather high density of significant inter-component relationships, where nearly all features play the role of retransmitter factors by our terminology (Kolomyts 2003). It indicates a well-developed system of chain reactions that can be traced from the initial abiotic edificatory factors (geomorphological and hydrothermal) to the structural edificatory-biotic features and further to the parameters of productivity of forest communities and, finally, to the complex characteristics of the biological cycle.

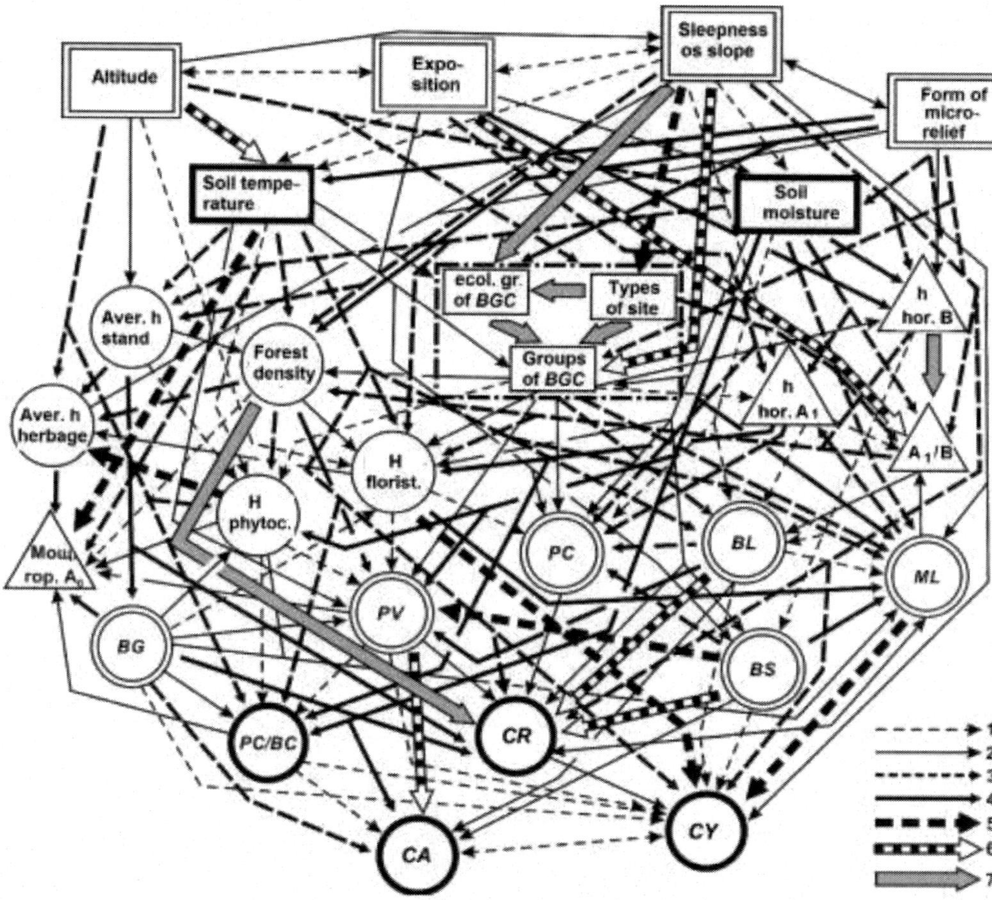

Figure 13.1. The generic statistical information model of monosystem organization of Putyatin continental-island landscape. The standardized contingency coefficients: 1, 0.101–0.130; 2, 0.131–0.160; 3, 0.061–0.190; 4, 0.191–0.220; 5, 0.221–0.250; 6, 0.251–0.300; 7, 0.301–0.450. TS, the types of sites; EG BGC, the ecological groups of biogeocoenoses; H, the informational measures of diversity. Phytomasses: *BL*, total above-ground; *BS*, skeletal; *BG*, grass canopy; *ML*, forest litter. Productivity: *PC*, total; *PV*, green mass. *CR*, the coefficient of the annual cycle of above-ground phytomass; *CY*, the litter/tree waste coefficient; *PC/BC*, the productivity to phytomass ratio.

2. Distinct echeloning of intercomponent interconnections. It manifests itself in the serial transfer of the strongest information signals between retransmitter features of different structural levels (geomorphological, hydrothermal, edaphic, structural and functional

phytocoenotic). Thereby, the hierarchic order of monosystem organization of forest biogeocoenotic canopy of the continental-insular land is created.

3. Relative homogeneity of inter-component relationships, with predominance of the higher normalized coefficients of contingency *(C(A;B)* ≈ 0.160–0.220), which provide certain stability of the acquired monosystem landscape organization.

Thus, the multiplicity, hierarchic order and stability of component interconnections – these are the three main features of Sub-Pacific landscape structure acquired as a result of long-term subaerial development on the Pacific Ocean mega ecotone. With these attributes, the continental islands of Primorye Sub-Pacific enter their regressive state. The presented model is apparently related to the initial stage of its fading evolution in terms of disintegration of continental margin and senile erosion-denudation relief.

Generally, absolute height under the conditions of erosion-denudation relief of a hill country (Shchukin 1980) is not the primary factor of phyto- and biocoenotic differentiation (see Figure 13.1). Here, the wind-solar exposure and especially slope steepness play a much more significant role in landscape formation. Their influence is especially obvious under the conditions of rather uniform mechanic composition of soil-forming rocks (from sandy loam to medium-textured loam), with the insignificant differentiating role of lithomass: the content of gravely-rocky components in soil. Let us trace the channels of chain reactions from these factors to the geomorphological and hydrothermal conditions forming the system of hill-country forest biogeocoenoses.

13.2. The Main Channels of Component Interconnections

The groups of forest biogeocoenoses are rather clearly distributed by the steps of altitudes of the Putyatin hill country and form a generalized catenary system (Table 13.1). The upper element of landscape couplings is presented by the Tek topogeosystems, while the lower element of composed of the Tam and SAh complexes. The flat-interfluve (eluvial) BGCs are spread almost over the entire altitudinal interval of the hill country. The most part of them (about 60%) occupy the geotopes of valley pediments bordering the foot of hills and ridges, with absolute heights of 15–40 m. However, about the quarter of them can be found on flat hilly watersheds and thus are part of the upper element of the catena. The general erosion-denudation land form of flat interfluves predetermines the xeromesomorphic type of their hydroedaphotopes. Even the more "fuzzy" elevation range is typical of the meso-hydromorphic biogeocoenoses of T sites, with the equally low (≤ 25-30%) probability of their ecological dominants confined both to watersheds and to the feet of the slopes. As the result, a generally complex hydro-mesomorphic range of edaphic moistening is formed here. That is why the forest communities of the Tgm group are characterized by the maximum spatial structural and functional diversity, as it will be shown below.

Rather tight coupling between the types of sites and slope steepness should be noted (Table 13.2), which predetermines the respective distribution of biogeocoenotic groups (Table 13.3). Rather strong, almost functional relationship between the ecological groups of grass canopy and slope steepness (see Figures 12.4 *c* and 13.1; Table 13.2), with their much lower expositional dependence (*C(A;B)* = 0.098), shows that it is the *degree of drainage of the territory but not the radiation-thermal regime* is the main factor of edaphic moistening.

Xerophytes are confined to the steepest slopes (≥ 20–25°), while hydrophytes are almost solely confined to the gentle (≤ 2-5°) and subhorizontal slopes. Slope microterraces caused by defluxion (Voskresensky 1971; Gerrard 1981), as well as steepness, contribute to xerophytization of soil cover (see Figure 12.4 *d*). With the development of terracettes, dehydration of the upper soil layers increases.

Table 13.1. The distribution of types of sites by slopes of different steepness (the matrix of normalized partial coefficients of connection)

Altitude, m	Groups of biogeocoenoses				
	Ekm	TEk	Thm	TAm	SAh
15–25	**0.39**	0.15	0.13	0	**0.55**
25–42	0.20	0	0.25	**0.56**	0
42–60	0.16	0.13	0.21	0.23	0.45
60–80	0	**0.54**	0.09	0.20	0
80–105	0.25	0.19	**0.32**	0	0

Note. The dominant areas of ecological niches are in bold.

Table 13.2. Distribution of site types by the slopes of different steepness (the matrix of normalized partial coefficients of connection)

Sleepness of slope, degree	Types of sites				
	E	TE	T	TA	Saq
2–7	0	0	0	0	1.00
7–11	0.28	0.17	0.21	0.27	0
11–15	0.49	0	0.12	0.46	0
15–22	0.23	0.27	0.34	0	0
22–26	0	0.57	0.12	0	0

Forest biocoenotic groups are accordingly distributed by slope steepness (Figure 13.2 *a*), and this information signal can be traced further in their structural and functional characteristics (see Figure 13.1). Our empirical-statistical model confirms and deepens the above-described relationships between the differentiation of species and variety composition of the broad-leaved forest communities of Putyatin Island with slope steepness.

The transeluvial xerophillic group of biogeocoenoses is confined mostly to the steepest ridge-top lopes, while the Tam and especially SAh groups are confined to the graded foot of slopes and river valleys. In the series of T→ TA→ SA geotopes, slope steepness decreases from 15–25° to 2–7°(see Table 13.2). The percentage of bedrock outcrops decreases in the same direction: from 20% to 0% (Figure 13.3 *b*), which contributes to drainage weakening and increased hydromorphism of edaphotopes. Let us note considerable slopes of day surface in the eluvial and transit groups of *BGC* having the areas of sporadic spreading towards the high steepness of slopes and, on the contrary, towards lesser steepness, respectively. It is characteristic that the *BGC* Tgm covers the broadest range of slope steepness, with the lower (40%) probability of its ecological dominant falling on the moderately steep slopes (15–20°). This very fact must determine the maximum structural and functional diversity of transit forest phytocoenoses.

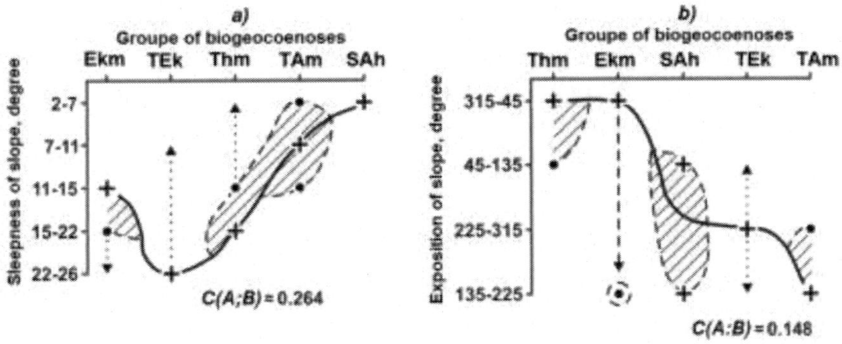

Figure 13.2. Putyatin Island. The distribution of the groups of forest biogeocoenosis by the steepness and solar exposure of the slopes. Legends are as in Figure 6.3.

Thus, *the role of slope steepness as the most powerful abiotic factor* determining the entire system of local inter-component relationships is a *phenomenal event of morpholitho- and biogenesis in regressive continental-island subboreal landscape.* Such information channel is actually ineffective in the forest belt of young volcanic island landscape and is substantially weakened in the low-mountain boreal forests of continental margin. The landscape-forming role of slope steepness increases only in the intracontinental regions of the boreal ecotone of the Volga River basin (Kolomyts 2008), which are similar in their evolutionary mega-state (at the geological time scale) to the offshore continental islands of Northwestern Sub-Pacific.

Let us further consider the contributions of the geotopic and hydroedaphic factors to the qualitative definiteness of the groups of forest biogeocoenoses (Tables 13.3 and 13.4). Both factors proved to be highly significant; however, the force of the organizing effect of surface hydration of the edaphotope is 1.5-fold stronger than the force of ecotope affiliation of a forest biogeocoenosis. It is indicative of the fact that the spatial differentiation of forest communities of the Putyatin island landscape is almost functionally related to the distribution of ecological groups of grass canopy, according to L.G. Ramensky's scales.

Table 13.3. Distribution of the groups of forest biogeocoenoses by the types of sites (the matrix of normalized partial coefficients of connection)

Types of sites	Groups of biogeocoenoses				
	Ekm	TEk	Thm	TAm	SAh
Eluvial	**0.68**	0	0.28	0	0
Transeluvial	0.13	**0.82**	0.08	0	0
Transit	0.19	0.18	0.23	0.45	0
Transaccumulative	0	0	**0.42**	**0.55**	0
Superaqual	0	0	**0**	0	**1.00**

Note. The dominant areas of ecological niches are in bold.

The system of geomorphological ordination is characterized by the comparatively high spread of Tgm groups of *BGC* by the types of sites, with the exception of the superaqual one, with a probability of 25–30% to 42%. This group includes, accordingly, almost the entire range of micro-relief forms (see Figure 13.3 *a*): from ridge crests and terrace cusps to *saddle-like depressions.* Obviously, the biogeocoenotic group of smooth slopes is most widely represented in the system of hill-mountain relief of Putyatin Island and must demonstrate the maximum diversity of local representations of the zonal-regional bioclimatic background (as it turned out in reality).

Table 13.4. Distribution of the groups of forest biogeocoenoses by edaphic moistening (the matrix of normalized partial coefficients of connection)

Ecologicasl groups of herbage	Groups of biogeocoenoses				
	Ekm	TEk	Thm	TAm	SAh
Xeromorphic	0.16	**1.00**	0	0	0
Xero-mesomorphic	0.84	0	0.29	0	0
Mesomorphic	0	0	**0.54**	0.21	0
Meso-hydromorphic	0	0	0.17	**0.79**	0
Hydromorphic	0	0	0	0	**1.00**

Note. The dominant areas of ecological niches are in bold.

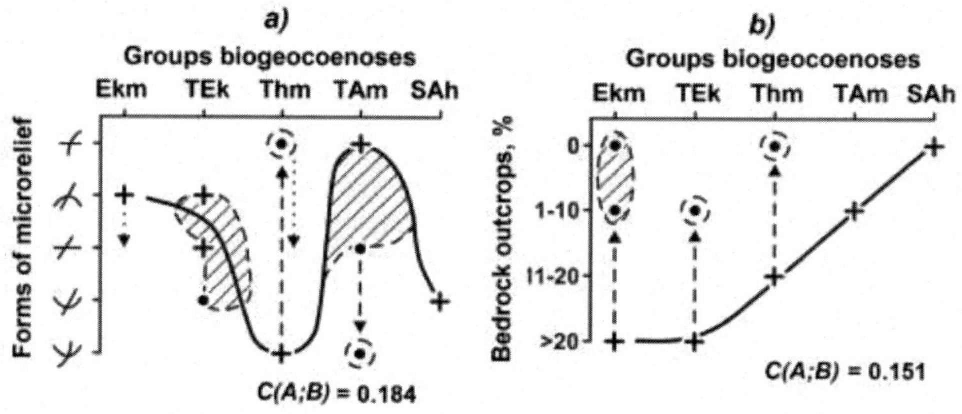

Figure 13.3. Putyatin Island. The distribution of forest biogeocoenotic groups by the forms of microrelief depending on bedrock outcrops. Legends are as in Figure 6.3.

The same can be said of edaphic moistening. The Tgm group of *BGC* is only half distributed over the moderately moistened slopes. Its ecological optimum with a low probability falls on the hydro-mesomorphic foot of the slopes; therefore, this biogeocoenotic group is put into the mixed hydro-mesomorphic category of edaphic moistening.

The much lower "fuzziness" of geotropic and hydroedaphic niches is typical of the flat-interfluve Ekm group of *BGC*. A significant (nearly 70%) probability of its ecological dominant falls on flat interfluves which, by all the rules of geotopology, must be characterized by moderate moistening optimal for these zonal-regional conditions (Vysotsky 1960; Ramensky 1971). However, in this case, the predominant flat interfluves are valley pediments with the greater steepness of slopes (see above) and frequent bedrock outcrops (see Figure 13.3 *b*), which imparts certain features of xerophytization to the eluvial group of *BGC*.

It should also be noted that the transit type of sites in the Putyatin island landscape is characterized by the higher edaphic moistening compared to the transaccumulative type, which is due to the influence of different dominant affiliations of these geotopes to the slopes of solar/wind exposure, as it will be discussed below.

13.3. Hydrothermal Structure of Topological Space

The hydrothermal structure of a natural complex is considered as the ordered spatial distribution of different characteristics of heat and moisture exchange between natural components over its territory (Kolomyts 1977). The hydrothermal structure (Krauklis 1974) characterizes the transit matter-and-energy part of natural-territorial organization. At the local level, this organization is based on the structural and functional, including hydrothermal, characteristics of edaphotopes (Vysotsky 1960; Ramensky 1971; etc). This fact was represented in the monosystemic model of Putyatin continental-island landscape. The summer temperatures and soil moistures are the two regime retransmitter factors formed as a result of transformation of background hydroclimatic signals by geomorphological ediphicator factors. Retransmitters substantially amplify these signals and exert a great influence on the entire vast network of communication channels leading to the soil-biotic features of forest topogeosystems (see Figure 13.1).

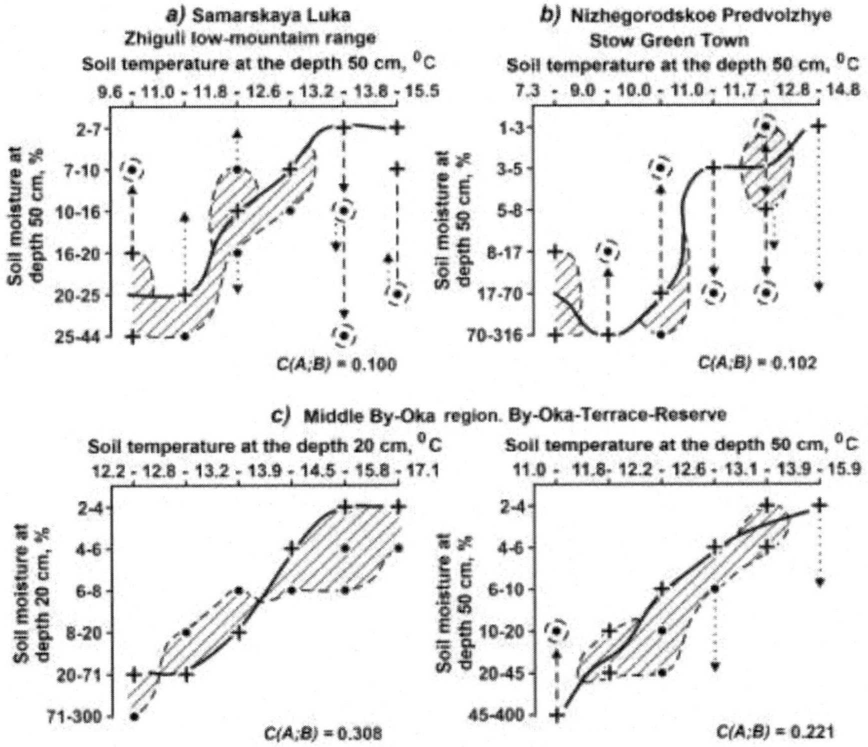

Figure 13.4. The binary ordination of soil temperature and moisture content at different depths in the intracontinental regions of the Middle Povolzhye. Legends are as in Figure 6.3.

The refractory hydrothermal role of local factors is substantially determined by the macrocirculation contrast ratio in the background climatic system per se. In the middle part of the Russian Plain, with the westerlies prevailing throughout the year and creating a series of moderately continental climates (from suboceanic to subcontinental), there is in general the inversely proportional relationship between soil moisture content and soil summer temperature (Figure 13.4). The increase in summer soil temperature is accompanied by the general tendency of decrease in its moisture content. This fact is in agreement with the general macroclimatic field of the boreal belt, where the increase in the mean July temperature is accompanied by the decrease in both annual and warm-season precipitation (Kolomyts 2005; Kolomyts and Surova 2014). In the forest biogeocoenoses of the Middle Povolzhye, the probabilities of the dominant areas of thermo- and hydrotopes vary from 45–55% to 75–80%, which is indicative of a certain spatial relationship between soil temperature and water regimes.

Against this predominant background, the frequency of occurrence of hydrothermal enclaves is not very high (about 20–25% in the dominants, rarely up to 40%). At the lowest soil temperatures, these abnormalities usually shift towards lower humidity (arid-cold sporadic edaphotopes are formed); in the area of higher temperatures, they shift towards the higher moisture content (thermal-humid edaphotopes are formed). Both of them occur in the suboceanic regions of the Middle Oka River basin and in the vicinity of Nizhny Novgorod, as well as in the subcontinental Zhiguli low mountains (more markedly in the latter case).

In the island volcanic landscape of the Neo-Pacific sector of the Pacific Ocean mega ecotone of Northern Eurasia, the above-described "regular" trend of relationship between soil temperature and moisture content is broken by the geothermal factor. In the area of the Mendeleev Volcano on Kunashir Island, the height-related decrease in soil (and air) temperature is accompanied by the increase in soil-ground moisture content only on the sites with insignificant effects of geothermal fluxes on the hydrothermal regime of soils (Figure 13.5 *a*, the left branches of the graphs of relations). The oceanic environment, similar to the continental one, creates its own background unity of thermo- and hydroedaphotopes. In the upper soil layers of the sites with sufficiently effective geothermal fluxes there is a substantial (by 30–40%) decrease in soil moisture content at the same temperature, while in the middle and lower layers these hydrothermal relations are absolutely unbalanced. As a result, the generalized curves of ecological dominants undergo drastic oscillations (the straight branches of the graphs, Figure 13.5 *a*).

Under conditions of monsoon-influenced continental climate of the marginal-continental and coastal-insular Sub-Pacifica, even the background unambiguous hydrothermal trend of thermo-hydroedaphotope is broken (Figure 13.5 *b*; Table 13.5). Here, *the mutual correlation between soil temperature and moisture content is very weak, which demonstrates their certain mutual independence* and, consequently, the independent contribution of each of them to the system of inter-component relationships. The probabilities of ecological dominants usually do not exceed 30–40%. The general tendency towards a decrease in soil moisture content along with an increase in its temperature (that can be judged by the trajectories of the dominants) is complicated by numerous enclaves giving rise to significant deviations from this tendency: towards both cold-arid and thermal-humid states.

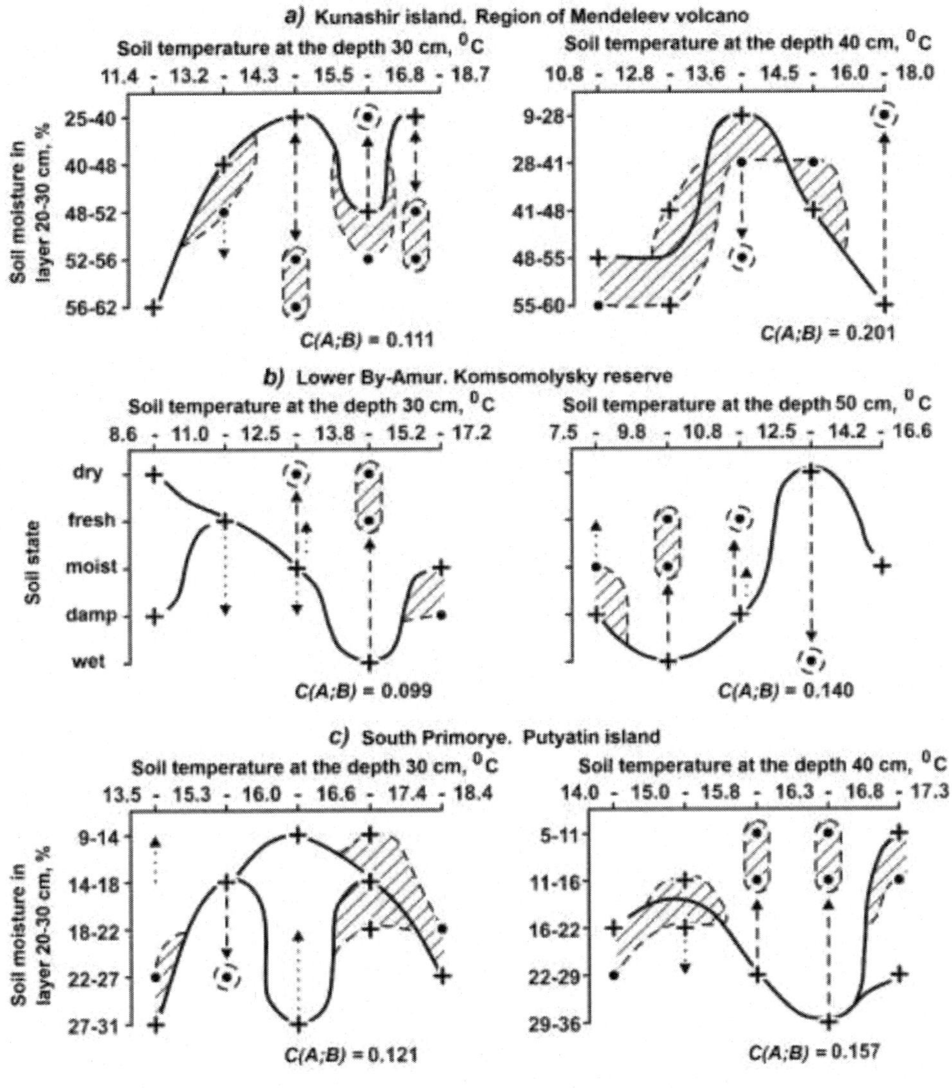

Figure 13.5. Binary ordination of soil temperature and moisture content at different depths in different regions of the Pacific Ocean mega ecotone of Northern Eurasia. Designations are the same as in Figure 10.3.

Table 13.5. Putyatin Island. The normalized coefficients of relations between soil temperature and moisture content in the upper layers

CA;B) = 0.099

Soil moisture in layer 10–20 cm, %	Soil temperature at the depth 30 cm, °C				
	13.5–15.3	15.3–16.0	16.0–16.6	16.6–17.4	17.4–18.4
5 – 15	**0.26**	0.09	0.13	**0.42**	0.17
15–20	**0.26**	0.19	**0.32**	0	0.26
20–25	0	0.09	0.06	0.33	**0.57**
25–30	0.23	0.25	0.23	0.25	0
30–35	**0.26**	**0.38**	0.26	0	0

Note. The dominant areas of ecological niches are in bold.

As a result, the abnormal edaphic regimes, i.e., xero-microthermal (dry and cold) and hydro-macrothermal (moist and warm), occur en masse along a common trajectory of hydrothermal soil regimes of continental-island landscape from xero-macrothermal to hydro-microthermal. Strictly speaking, here they play the role of not anomalies but regular components of the hydrothermal field of edaphotopes. In contrast to the Middle Belt of the Russian Plain, the first abnormal edaphic regime is especially pronounced here. It is due to the seasonal rhythms of monsoon circulation on the oceanic margin of the continent, with pronounced ecological consequences of the winter continental monsoon. The second hydrothermal anomaly is obviously associated with the summer oceanic monsoon. It is important to emphasize that such geophysical anomalies at the topological level are clearly localized by the forms of plain-hilly relief, which results in the formation of a *highly diverse and contrasting pattern of thermo- and hydroedaphic regimes of subboreal forest biogeocoenoses in continental-island landscape.*

Under conditions of the monsoon-continental climate of Sub-Pacific Primorye (Gartsman 1971) and the uniform sandy clay loam soil-forming substrate of Putyatin Island, the hydrothermal parameters of local natural-territorial organization are largely determined by the *solar-wind exposure of slopes* providing manifestation of both radiation-thermal and circulation climatic factors. It combines the sunny–shady exposure of the slopes (for the warm season), on the one hand, and their windward–leeward exposure (for the cold season, with the respective processes of snow removal and accumulation in winter), on the other hand.

The duality of exposure factor predetermines various combinations of thermal and hydrological regimes of edaphotopes, which are reflected in the respective structural and functional characteristics of forest biogeocoenoses. Such local hydrothermal diversity creates a specific *monsoon-continental type of catenary spectra of forest biogeocoenoses*, which is clearly represented in the regional landscape of Putyatin Island. This catenary type substantially differs from the intercontinental types of upland-plain spectra (Kolomyts 1979, 2015 c): suboceanic (e.g., the Middle Povolzhye) and subcontinental (the Middle Ural), and even more drastically differs from the oceanic type of catenary spectra of island-arc volcanogenic low mountains (the Mendeleev landscape on Kunashir Island). The catenary hydrothermal spectra of the analogous monsoon-continental type are also manifested in the marginal-continental sector of the Pacific Ocean mega ecotone, in particular, over the territory of the Lower Amur Sub-Pacifica. Here, the characteristic feature of this type is the complicated local *bioclimatic factor of altitudinal-exposure zonality* that we have distinguished.

The catenary biogeocoenotic organization of monsoon-continental type cannot be presented as a single system of local landscape couplings, because their dominant elements are scattered over different solar-wind exposures. It is characterized by the similar combinations of hydrothermal conditions for the soils of two slopes of alternative exposure: northern and southern. In both cases, there are sufficiently broad ranges of both summer soil temperatures and soil moisture content (Figure 13.6). Though in respect of ecological dominants, the soils of northern slopes are colder and furthermore wetter compared to the southern ones.

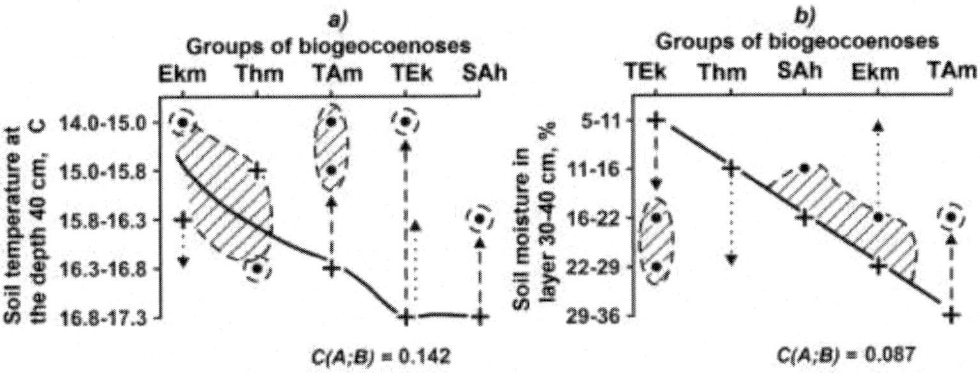

Figure 13.6. Putyatin Island. The hydrothermal niches of the groups of forest biogeocoenoses. Legends are as in Figure 6.3.

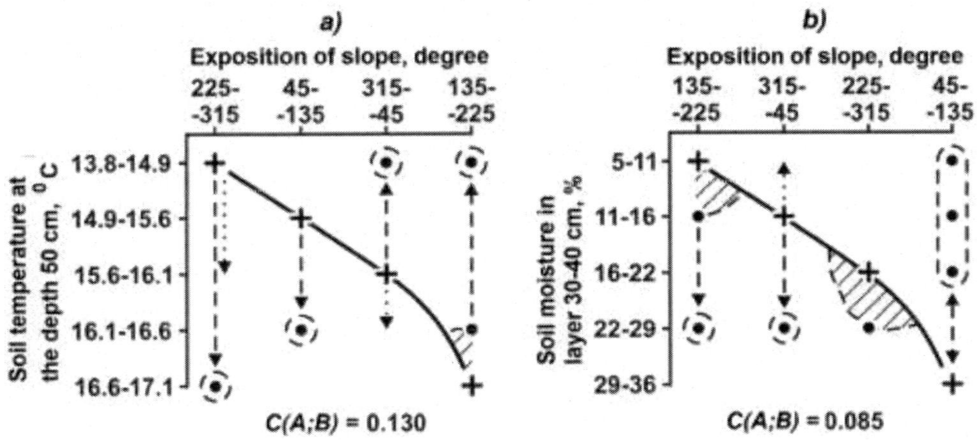

Figure 13.7. Putyatin Island. Soil temperature and moisture content distribution over the slopes of solar exposure. Legends are as in Figure 6.3.

The groups of forest biogeocoenoses were selected in accordance with the multiple-valued refractive effect of solar-wind exposure of slopes on the background catenary hydrothermal structure of edaphotopes. The following categories of soil summer temperature and moisture content ratios are distinguished.

1. Tgm group of BGC – *mesothermal arid-humid.* This biogeocoenotic group most widespread over the catena (see above) occurs almost solely (up to 90%) on the slopes of northern and, to a lesser extent, eastern exposure (Figure 136 *b*), with moderately-cold and moderately-warm soils within a wide range of their moisture content (Figure 13.7 *b*). The highest moisture content (up to 30–35%) is observed on the parts of eastern slopes, with the accumulation of snow windblown from the opposite slopes by continental monsoon in winter and the soil moistening by oceanic monsoon in summer.
2. The dominant of the flat-interfluve Ekm group– *micro(meso)thermal humid.* It is confined mainly to northern moderately cold slopes with higher soil moisture content

(*W(30–40)* ≈ 25–35%). However, more than 33% cases of this group occur on southern slopes – the warmest (t_{40} ≥ 16.5–17.0°) and comparatively dry (see Figure 13.7 *b*).

3. TEk group of biogeocoenoses, which is widespread mainly on western slopes, is for the most part *macrothermal arid*. In general, soils of the western slopes of island hill mountain country facing the continent are characterized by mainly low summer temperatures (see Figure 13-7 *a*), with strong aridization of upper soil layers (*W(10–20)* ≤ 10–15%). Though, soil moisture content increases with depth. In winter, these slopes are often exposed to the direct cryoarid influence of continental monsoon, with snow blowing off and deep freezing of soil. The after-effect of this monsoon is maintained here also in the summer season. However, in our statistical sample, the TEk biogeocoenoses are confined by nearly 65% to the high-temperature enclave of near-top western slopes (see Figure 13-6 *a*); therefore, forest communities were formed here under conditions of high summer temperatures and low soil moisture content.
4. TAgm and SAg groups of the lower element of catena – *macrothermal moderately and highly humid.* They are characterized by the confinement of their ecological dominants to the slopes of the southern quarter of the horizon (with a 45–55% probability), where the highest summer soil temperatures are predominant ($t_{40–50}$ = 16.5–17.5°). At the same time, soil moisture content is higher at the foot of slopes, while the better drained bottoms of river valleys are characterized by moderate humidity (Figure 13.6 *b*).

13.4. SOILS AND SOIL-FORMING PROCESSES

Organic matter formation in the soil of forest ecosystem begins with the litter, which is a good integral indicator of its state, inter alia, the small biological cycle intensity. As is mentioned by G.F. Morozov (1949), the transformation of soil-forming processes under external influence begins with the changes in forest litter mass (thickness) and chemical properties.

Forest litter is formed of the tree waste of above-ground phytomass of trees, shrubs and herbs. In the Putyatin oak forests, the thickness (hA_0) and mass (*ML*) of forest litter are determined first of all by the average height of timber stand and the total reserves of green mass *BG* (Figure 13.8). With the increase in the average height of timber stands from 6–8 m to 11–12m, tree waste becomes more abundant and forest litter thickness increases from 0.1–0.3 cm to 3.5–5.0 cm (*C(A;B)* = 0.102).

The correlation between hA_0 and *ML* proved to be rather distinct, with a more than 50% probability of ecological dominants in most cases (Figure 13.9 *a*). However, the closest relationship between these parameters, similar to the functional relationship, is observed only at their maximal values (≥ 3.5 cm and ≥ 4.5 tons/ha, respectively). As the thickness of and the reserves of dead organic matter in the forest litter decrease, this relationship becomes "fuzzy." At hA_0 = 3–7 cm, the dominant value *ML* = 1.5–2.2 tons/ha has a probability of only 32%.

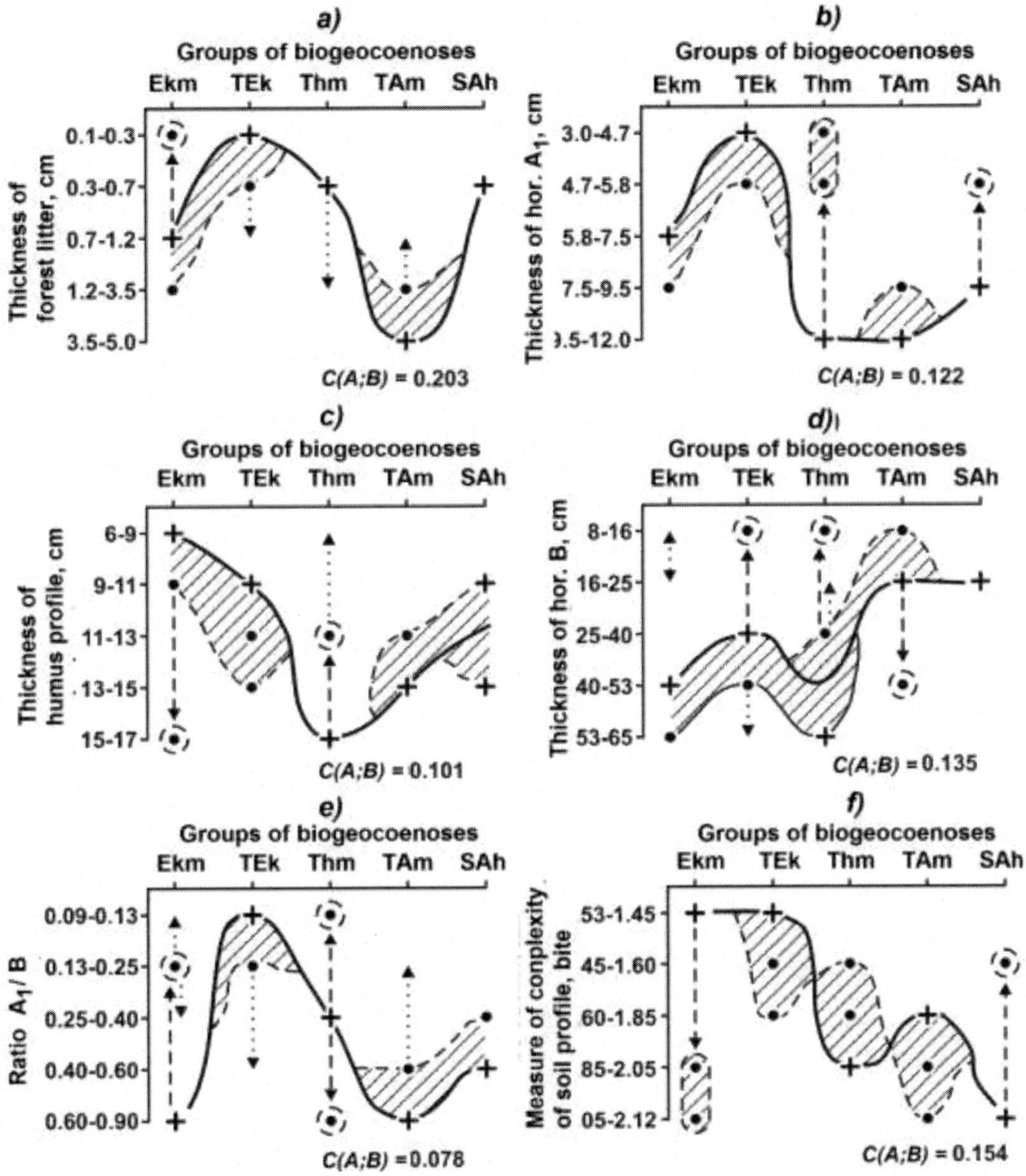

Figure 13.8. Putyatin Island. The distribution of different soil morphological characteristics over the groups of forest biogeocoenoses. Legends are as in Figure 10.3.

Preservation of forest litter considerably depends on mortmass decomposition processes, the intensity of which is determined by hydrothermal conditions of edaphotope. Partial connections here are ambiguous and rather "fuzzy" (Figure 13.9 *b*; Tables 13.6 and 13.7). At the moderate soil temperatures ($t_{30} \approx$ 15–16°) and the maximum moisture content in the upper soil layer ($W_{10\text{–}20} \geq$ 30–35%), forest litter mass is decomposed very slowly; therefore, the hA_0 and *ML* parameters reach their maximum values (≥ 3.5–5.0 cm and ≥ 4.5, up to 8.5 tons/ha, respectively). In the maximum temperature range ($t_{30} \approx$ 16.5–18.5°) with retained maximum soil moisture content, the forest litter thickness and mass slightly decrease (to 1.5–

2.5 cm and to 1.5–3.0 tons/ha, respectively), indicating certain intensification of mortmass decomposition processes. These processes become most intensive when humidity decreases ($W_{10–20} \leq 10–20\%$) under conditions of high temperature. In this case, forest litter thickness and mass can decrease to ≤ 0.5 cm and 1.0–1.5 tons/ha, respectively.

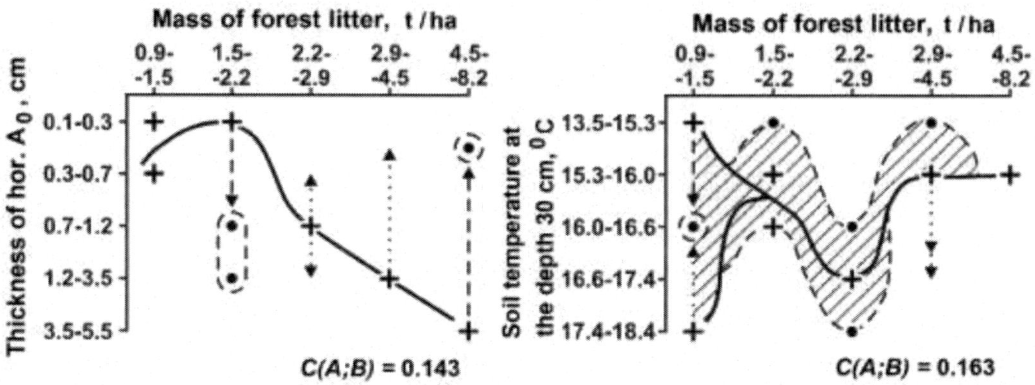

Figure 13.9. Putyatin Island. The ecological niches of forest litter mass in the space of hor. A_0 thickness and soil temperature. Legends are as in Figure 6.3.

Table 13.6. The distribution of forest litter thickness (soil hor. A_0) in the space of upper soil layer temperature (the matrix of normalized partial connection coefficients)

C(A;B) = 0.180

Soil temperature at the depth 30 cm, ^{0}C	Thickness of forest litter, cm				
	0.1–0.3	**0.3–0.7**	**0.7–1.2**	**1.2–3.5**	**3.5–5.0**
13.5–15.3	**0.53**	0	0.27	0	0
15.3–16.0	0.12	0.24	0.18	0	**1.00**
16.0–16.6	0	**0.27**	**0.36**	0.25	0
16.6–17.4	0	0.24	0.18	**0.75**	0
17.4–18.4	0.35	0.24	0	0	0

Note. The dominant areas of ecological niches are in bold.

Table 13.7. The distribution of forest litter thickness (soil hor. A_0) in the space of upper soil layer moisture content (the matrix of normalized partial connection coefficients)

C(A;B) = 0.115

Soil moisture in the layer 10–20 cm, %	Thickness of forest litter, cm				
	0.1–0.3	**0.3–0.7**	**0.7–1.2**	**1.2–3.5**	**3.5–5.0**
5 – 15	0	0.22	**0.33**	0.29	0
15–20	**0.31**	0.19	0.28	0	0
20–25	0.21	**0.27**	0	**0.35**	0
25–30	0.27	0.22	0	0	**1.00**
30–35	0/21	0.09	**0.39**	0.35	0

Note. The dominant areas of ecological niches are in bold.

Thus, under the conditions of monsoon-continental climate of subboreal South Primorye, forest litter (the reserve of organic matter in a forest ecosystem) is preserved due to high hydromorphism of edaphotopes in combination with their moderate and low summer temperatures. At the same time, the hydromorphic factor proved to be more powerful that the thermal factor. Intensive decomposition of forest litter leading to a drastic decrease in its thickness and mass can also occur at high soil temperatures but certainly under conditions of low soil moisture content (see Table 13.6 and Figure 13.9 *b*).

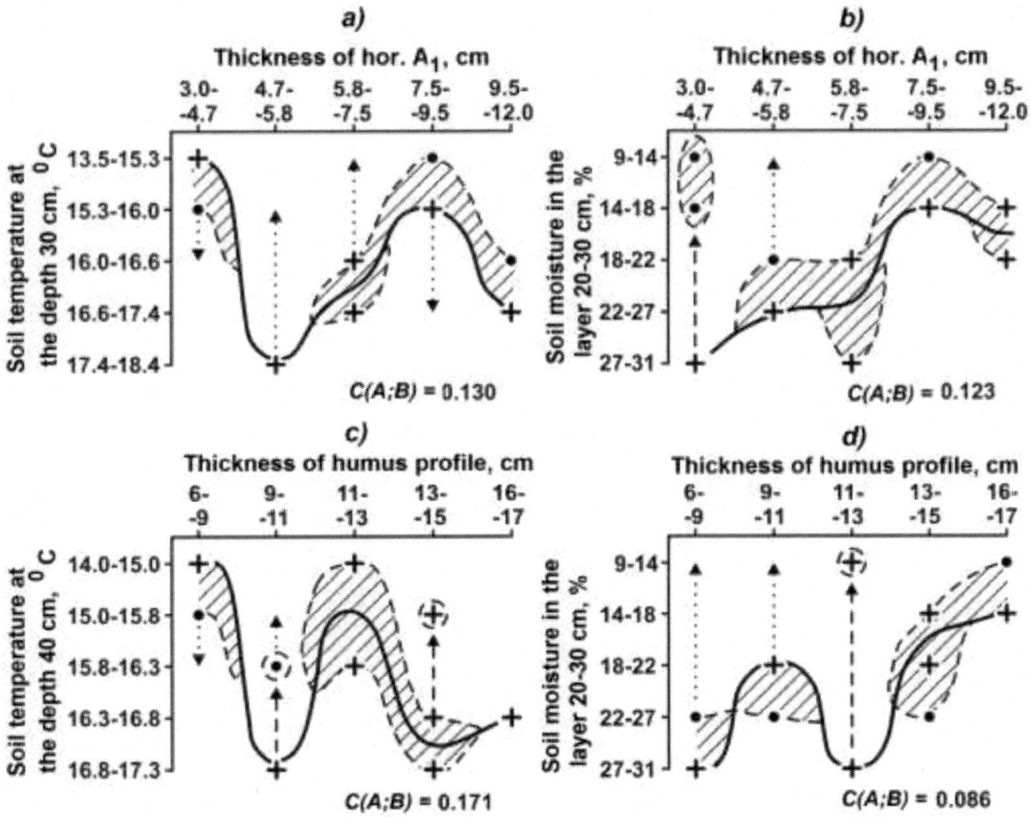

Figure 13.10. Putyatin Island. The hydrothermal niches of different morphological characteristics of soil. Legends are as in Figure 6.3.

The degree of development of humification processes assessed by the thickness of hor. A_1 and the entire humus profile (HP) is also determined by certain combinations of the summer temperature and moisture content of the upper soil layer (Figure 13.10). The weakest humidification is observed at lower soil temperatures (t_{30} ≤ 15.5–16.0°) and high soil moisture content ($W_{10–30}$ ≥ 25–30%). The ecological optima of thickness of hor. A1 and HP are ≤ 4.5 cm and ≤ 5–10 cm, respectively, with a probability of 65–80% and more. At the higher, but not maximum, temperatures (t_{30} ≤ 16.5–17.7°) and moderate humidity ($W_{20–30}$ ≥ 15–20%), humification processes show maximum development and the thickness of hor. A1 reaches 10–12 cm, while the depth of penetration of humic substances is 15–20 cm. Contrariwise, extremely high soil temperatures (t_{30} ≥ 18.0–18.5°) at moderate or high humidity drastically decrease the intensity of humus formation, which causes the reduction of hor. A1 and HP (with the optima up to 5.5–4.5 cm and 5–10 cm and a probability of 55–70%). Finally, the maximum soil moisture content and rather high temperature cause the moderate development of humification, so that the thickness of hor. A1 and HP is 6.0–7.5 cm and 13–15 cm, respectively.

Generally, we can speak about the predominant positive effect of soil temperature on humus formation processes, though its manifestation is not always quite clear. The combination of both hydrothermal factors mentioned above is of key importance. The partial increase in soil moisture content is accompanied by the general weakening of humus formation and the decreased penetration of humic substances into soil mass (see Figure 13.10 *c, d*). It seems to be determined largely by defluxion slope processes which, as shown below, become aggravated exactly with the increase in soil moisture content.

The inversely proportional relationship between hor. A_1 thickness and forest stand density (*C(A;B)* = 0.099) demonstrate the negative screening effect of forest canopy on the processes of the destructive branch of metabolism. With the increase in forest stand density (from ≤ 0.5–0.7 to 1.5–2.0, as calculated by phytomass), the inflow of solar radiation to soil cover decreases, the temperature drops and, hence, the processes of litter decomposition and humification become weaker, which results in the reduction of thickness of humus-accumulative soil horizon (from 9–11 cm to 3 cm and less).

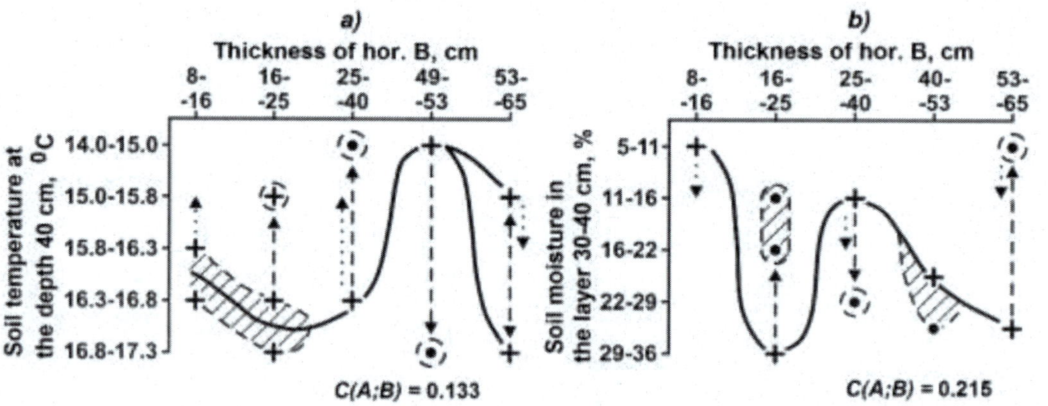

Figure 13.11. Putyatin Island. The hydrothermal niches of illuvial-metamorphic soil horizon. Legends are as in Figure 10.3.

The degree of soil maturity can be assessed, as is known (Duchaufour 1968; Polivanov 1982), by the thickness of its illuvial-metamorphic horizon. In contrast to humus formation processes, soil moisture content is more significant than soil temperature for the development of illuvial processes in continental-island landscape (Figure 13.11). It slightly resembles the situation with forest litter. Under conditions of dry soil (W_{30-40} ≤ 10–15%) and high temperatures (t_{50} ≈ 16–17°and higher), the processes of intrasoil weathering, migration of sesquioxides and lessivage slow down, and hor. B has the maximum thickness (in the ecological optimum of ≤ 10–15 cm). The maximum values of horizon thickness correspond to the combination of higher soil moisture content (W_{30-40} ≈ 20–30% and more) and rather low soil temperature (with the ecological dominant of t_{50} ≤ 15°). The probability of such hydrothermal situations is 40–55%. With moderate soil moisture content, the range of thickness of illuvial horizon is rather broad: from 55–60 to 70 cm and more. However, further increase in moisture content leads to a decrease in the degree of its development. Thus, both high aridization and excessive soil moistening have a negative effect on the development of illuvial horizon. The most favorable conditions for the soil to reach the mature (climax) state are its moderate moistening and lower temperatures.

The distribution of soil-morphological characteristics by the groups of forest biogeocoenoses in the Putyatin continental-island landscape is determined by: (1) the above-described characteristics of productivity of forest communities; (2) the revealed regularities of hydrothermal ordination of both *BGC* per se and the structural parameters of soil profiles; (3) the degree of pasture digression, which generally weakens down the catena. The thickness of forest litter, organomineral hor. A_1 and the entire humus profile (HP) is uniformly differentiated by *BGC* groups (see Figure 13.8*a–c*). The analogous distribution is observed for the A_0/A_1 ratio, which reflects the rate of destruction of dead organic matter – the initial stage of the destructive branch of the biological cycle.

The highest rate of destruction with $A_0/A_1 \leq 0.05–0.10$ is typical of the two opposite "poles" of the catena: TE and Saq. These sites are drastically different in their hydroedaphotopes (TE are xeromorphic and SA are hydromorphic); however, they are similar to each other as catenary elements that are the warmest in summer. On the comparatively cool and moist E and TA types of sites, litter mass decomposition is much weaker, as is shown by the maximum values of A_0/A_1 (up to 0.30–0.50 and more). Thus, under conditions of monsoon-continental climate of the boreal-nemoral Sub-Pacifica, *the major factor that regulates the initial stage of detritogenesis is the thermal factor* – the heat but not moisture supply of soil. Since forest litter is known (Morozov 1949; Dylis 1978) to be the main channel of the organic matter cycle in forest ecosystem, it can be supposed that *the spatial-temporal dynamics of thermoedaphotopes must underlie the predictive calculations and estimates of functional shifts in forest ecosystems under different scenarios of climate changes.*

Humification processes show an equally distinct catenary differentiation. The middle and lower elements of landscape couplings, consisting of the Tmg and Tam groups of *BGC*, respectively, are characterized by the maximum development of these processes, giving hor. $A_1 \geq 8–10$ cm, as well as the maximum (up to 15–20 cm) penetration of humic substances into soil. At the foot of the slopes, oak forests produce, as mentioned above, the minimum amounts of green mass. Nevertheless, the thickest forest litter is formed on these sites. It can be accounted for by the inflow of dead organic matter with the surface runoff from above-lying slopes and by the weak pasture digression. In the Tmg group of *BGC* there is no such inflow, and forest litter is formed by the abundant tree waste of oak forests maximally productive on these geotopes (see below Figure 14.4; Table 14.1). Here, the favorable mesothermal arid/humid soil regime (see above) contributes to rapid decomposition of dead organic matter and intensive humus formation. The deluvial process intensively developed on transit geotopes (see below) facilitates deep penetration of humic acids into soil. As a result, hor. A_1 and the entire humus profile of soil reach their maximum values at insignificant thickness of forest litter. The latter can also be due to the weak and moderate pasture digression of soil cover.

On the SA bottoms of river valleys, with the low productivity of oak forests, transit water regime and macrothermal humid edaphotopes (see above), the thickness of horizons A_0 and A_1, as well as the humus profile of soil, is less. It is also favored by the slightly higher degree of pasture digression compared to the adjacent slopes.

Still weaker development of organogenic soil horizons is observed in the *BGC* groups of the upper catenary element. On the near-top TE sites, where the thickness of hor. A_1 and HP decreases by ≤ 3–5 cm and down to 9–13 cm, respectively, it is associated primarily with the minimum production of green mass (see above). Here, it seems to be affected also by the prevalence of eluvial processes. The deluvial mixing of fine grained soil is yet low and involves only the upper soil-ground layers, as is indicated by the high microterraceness of near-top slopes (see below). The eluvial *BGC* of flat watersheds maximally exposed to pasture digression are characterized by an equally minimal thickness of HP, and only in 30% of cases the decrease in digression here results in the complete development of soil profile (see Ekm enclave in Figure 13.8, *c*).

One of the most representative indicators of the level of structural and functional development of the entire natural complex is the *degree of soil maturity* (Mamay 1994) – the phenomenon associated with the isolation and development of illuvial-metamorphic horizon B (Glazovskaya, 1981), as well as with other signs of "memory soil" (Targulyan and Sokolov 1978). The thickness of soil hor. B in the forest biogeocoenoses of Putyatin Island suggests that the maximum maturity here is typical of the soils of transit hydro-mesomorphic and in a less degree eluvial xero-mesomorphic oak forests (see Figure 13.8, *d*). On transit geotopes, the processes of intrasoil weathering, illuviation and lessivage are most developed. The minimal soil maturity is typical of forest biogeocoenoses of the lower element of the catena – the foot of slopes and the river valleys, where the constant accumulation of alluviums results in certain "rejuvenation" of soils (Kolesnikov 1965). Accordingly, the TA and Saq geotopes are characterized by high values of the measure of complexity of the soil profile (see Figure 13.8, *f*).

The derived conclusion is also confirmed by the soil-morphological parameter A_1/B characterizing, as is known (Polivanov 1984), the ratio of the two basic soil-forming processes: humification and intrasoil weathering. In continental-island landscapes, the maximum values of this parameter are typical of forest biogeocoenoses of the TA and Saq sites. These values drastically decrease upwards the hill slopes of the T and TE geotopes (see Figure 13.8, *e*). Consequently, the accumulation and transformation of organic matter are predominant in soil formation in the lower elements of the catena, while intrasoil transformation of mineral matter is predominant in its middle and upper elements, predetermining the higher maturity of soils in the latter case. The eluvial geotopes, with their dual micro(meso)thermal humid hydrothermal regime (see above), are characterized by a considerable spread of values of the A_1/B parameter. This spread indicates the contrasts between flat-interfluve soils by degree of maturity.

Thus, with regard to the three morphological characteristics of "memory soil," i.e., the thickness of hor. B, the A_1/B parameter, and the information measure of soil profile complexity, the soils of Thm forest biogeocoenoses can be considered as most mature, maximally close to the climax state. Being currently under the secondary, anthropogenic oak forests, these soils have maximally preserved the zonal properties acquired in the preceding period of their development (2500–1000 years ago) under the primary forest formation – first polydominant broad-leaved and then coniferous/broad-leaved forests.

Chapter 14

Functional Organization of Fading Evolution Landscape

Abstract

The subboreal forests and the light forests of continental-island geosystems of the regressive series are generally characterized by low productivity. With respect to phytomasses, productivity and biological cycle, the Putyatin secondary oak forests are at a much lower level of development compared to the much more northern primary and quasi-primary broad-leaved and mixed forests of the Lower Priamurye and Kunashir. By these parameters, the forest communities of the landscape of Putyatin Island form a single "unipolar" spatially ordered functional system with a transit group of biogeocoenoses in its center. The entire described complex of structural and functional phytocoenotic and soil characteristics of forest geosystems of Putyatin Island allow us to recognize that the transit hydro-mesomorphic group of biogeocoenoses as a certain zonal-regional model of continental-island landscapes of the boreal-subboreal Sub-Pacific is their topological "core." The flat-interfluve (eluvial) topogeosystems in this case have the lower bioproductivity and much lower reserves of homeostaticity. The Putyatin oak forests have considerable territorial contrasts by the integral parameters of polysystem organization, characterizing their total level of structural development and reflecting their homeostatic reserve, i.e., resistant stability. The mechanisms that close the biological cycle in the oak forests of Putyatin Island are inadequate to the processes of production and accumulation of live organic matter. These are substantially different characteristics of the functioning of forest communities in a continental-island landscape. The analysis of soil morphology made it possible to elucidate the main features of soil formation under the conditions of monsoon-continental climate of subboreal South Primorye. The hydromorphic factor is more powerful here than the thermal factor.

14.1. Phytomasses and Productivity

The Putyatin secondary oak forests generally have much lower productivity (Figure 14.1; Table 14.1) compared to the much more northern primary and quasi-primary broad-leaved and mixed forests of the Lower Priamurye and Kunashir (see Kolomyts, 2015 b, 2016 a).

Thus, the maximum weighted average values of living above-ground phytomass (*BL*) on Putyatin Island are no more than 94 tons/ha, while in the above regions they are about 200 and more than 350 tons/ha, respectively. There is also an equally drastic contrast between the minimum *BL* values: 41–43 tons/ha vs. 95–100 tons/ha. It seems that subboreal and light forests of continental-island geosystems of the regressive series (Kolomyts, 1987, 2015 a) are generally characterized by low productivity. According to the data of our landscape-ecological survey at the Vityaz Bay testing ground (in birch/lime/oak and pine herbage/cereal and light forests of coastal territories of Gamov Peninsula, Peter the Great Gulf), the reserves of living above-ground phytomass were 50–70 to 150–170 tons/ha, only rarely up to 250–400 tons/ha.

This conclusion is confirmed by the set-theoretical models of production space (see below Figure 14.4 *A, B*). The models show even higher Z_{crit} values than in the structural phytocoenotic space, against the background of equally weak relationships. The ascending (creative) branch of biological cycle in Putyatin secondary oak forests is characterized by considerable territorial fragmentation implying the generally low level of their autotrophic biogenesis.

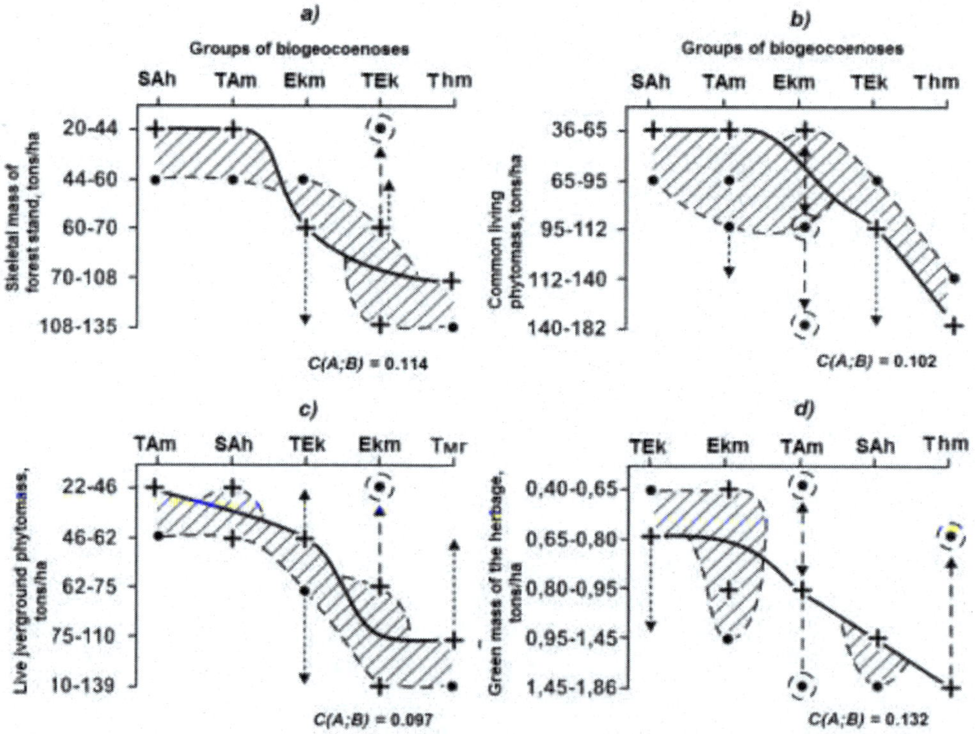

Figure 14.1. Putyatin Island. Distribution of different parameters of phytomasses by the groups of forest biogeocoenoses.

Among geomorphological factors, the solar/wind exposure of slopes exerts a substantial effect on the living above-ground phytomasses and productivity (see above Figure 13.1), and the situation here is very paradoxical (Figure 14.2 *a, b*). The southern and western slopes of hill country, being the warmest in summer, are not characterized by high phytomass reserves:

the dominant values on the southern and western slopes are *BL* ≤ 75 tons/ha and 22–46 tons/ha, respectively. Accordingly, productivity here has the minimum values: the dominant is 2.9–4.5 tons/ha yr for total above-ground production *PC* and 1.7–2.5 tons/ha yr for green mass production *PV*.

Table 14.1. The taxonomic (weighted average) norms of different phytomasses (tons/ha), productivity (tons/ha · yr), and discrete parameters of the biological cycle in the groups of biogeocoenoses of Putyatin island landscape

Functional parameters (see Figure 13-1)	Groups of biogeocoenoses				
	Ekm	TEk	Thm	TAm	SAh
BS	70.84 (±)	70.55 (±)	92.75 (+)	42.99 (–)	41.11 (–)
BL	81.89 (±)	73.06 (±)	93.90 (+)	45.52 (–)	44.05 (–)
PC	5.01 (±)	4.53 (–)	5.52 (+)	4.24 (–)	4.41 (–)
PV	3.26 (±)	3.10 (–)	3.82 (+)	2.51 (–)	2.80 (–)
PG	1.01 (±)	0.75 (–)	1.19 (±)	0.95 (–)	1.38 (+)
PC/BC	0.054 (–)	0.047 (–)	0.048 (–)	0.065 (±)	0.076 (+)
KR	0.057 (–)	0.046 (–)	0.043 (–)	0.071 (±)	0.086 (+)
KA	0.282 (±)	0.268 (–)	0.290 (±)	0.250 (–)	0.334 (+)
ML	1.92 (–)	1.63 (–)	4.01 (±)	4.67 (+)	2.57 (±)
KY	0.69 (–)	0.53 (–)	0.81 (±)	1.66 (+)	0.83 (±)

Note. The generalized ideas of the value levels of metabolic parameters (with comments). The value levels: (+) – high (intensive autotrophic biogenesis, high coefficient of forest phytocoenosis, low ability to handle of the descending branch of metabolism); (±) – moderate (moderate indices of productivity, coefficient of forest phytocoenosis and ability to handle of the detritus branch); (–) – low (weak autotrophic biogenesis, low coefficient of efficiency of forest phytocoenosis, high throughput ability to handle of the descending branch of the biological cycle). Other designations see in the text and in Figure 13-1.

The maximum phytomass reserves are confined to the coldest slopes of eastern (dominant *BL* = 75–110 tons/ha) and especially northern exposure (the dominant *BL* = 110–139 tons/ha). The area of dominance of the total production *PC* = 5.8–7.1 here has a probability of more than 60%. In our sample, Tmg and Ekm of the BGC group are widespread on these slopes. The dominant of the highest green mass production (*PV* = 3.8–5.4 tons/ha · yr) is also observed here, though such values can occur on the slopes of all exposures. In our sample, Tmg and Ekm of the BGC group are widespread on the northern and eastern slopes. Thus, the forest communities of the Putyatin subboreal landscape can function under the conditions of excessive summer warmth, which has a withering effect on edaphotopes and a negative effect on autotrophic biogenesis.

The above conclusion is confirmed by the general pattern of distribution of forest, shrub and meadow phytocoenoses over the hill slopes of different solar exposure. The densest forest communities are widespread mainly on cool NW slopes. On even colder northern slopes, there often occur the suppressed associations of rhododendron. At the same time, the SE and southern slopes are the area of predominance of meadow-steppe plants, in particular, grasslands, with prevalence of the subtropical cereal, Chinese silver grass (*Miscanthus sinensis* Anderss).

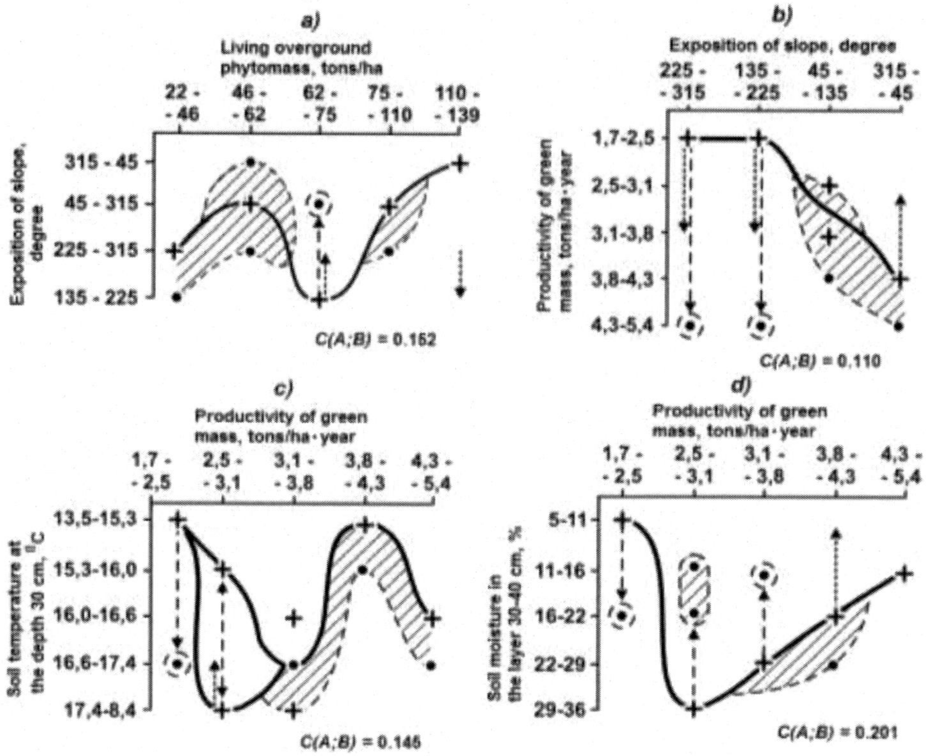

Figure 14.2. Putyatin Island. Ecological niches of phytomasses and productivity in the exposure and hydrothermal spaces. Legends are as in Figure 10.3.

The binary ordination of green mass productivity by soil temperature and moisture content gives analogous results (Figure 14.2 *c, d*). First of all, there is no unambiguous relationship between the *PV* parameter and soil temperature: there is a wide scatter of productivity at both low and high temperatures. One can only note that in the range of the highest temperatures (t_{30} = 17.5–18.5) green mass production decreases to 2.5–3.5 tons/ha yr, with a probability of 50–60%. However, under conditions of the coldest soil (t_{30} = 13.5–15.3), productivity is even lower (*PV* = 1.7–2.5 tons/ha yr) in up to 35% of cases. The maximum productivity (*PV* ≥ 4.5 tons/ha yr) is within the range of moderate temperatures (t_{30} = 16.0–16.5, with a probability of about 60%).

The hydroedaphic ordination (see Figure 14.2 *d*) demonstrates the following two trends of productivity. The first, basic trend covers the range of high and moderate soil moisture content and describes a rather clear inverse relationship between productivity and this factor: as $W_{30\text{–}40}$ decreases from 4.5–5.5% to 2.5–3.0%, productivity increases to the maximum (*PV* = 11–16 → 29–35 tons/ha yr). However, further soil drying (up to 5–10%) causes a drastic decrease in green mass production to 1.5–2.0 tons/ha · yr and less. The maximum productivity requires a combination of rather high, but not maximum, soil temperatures (t_{30} = 16.0–17.5) and a moderate soil moisture content ($W_{30\text{–}40}$ ≈ 10–20%).

14.2. Polysystem Structure of Soil-Phytocoenotic Space

The structural development of oak forests of the Putyatin landscape was assessed by the average height of timber stand, forest density, crown density, and soil cover characteristics (Table 14.2). The set of these characteristics proved to be sufficient to find out, using the set-theoretical models (Table 14.3; Figure 14.3), the topological regularities of "structural metabolism" (Odum 1975). The measures of similarity relationships demonstrate rather high structural homogeneity of forest biogeosystems. The secondary, anthropogenic forest formation resulted in the spatial alignment of morphological characteristics of forest communities (the measure of phytocoenotic similarity between the most of biogeocoenoses is more than 90%).

Table 14.2. The characteristics used for calculating the parameters of polysystem structural and functional organization of forest biogeocoenoses of the Putyatin continental-island landscape

Structural-functional units	Geocomponent characteristics
I. Phytocoenotic structure	Average height, project coverage and floristic diversity of grass stand; forest density; crown density; average height of timber stand
II. Living phytomasses	Skeletal mass of trees; total living phytomass; living above-ground phytomass; leaf mass and green mass of herbage
III. Productivity	Production of the green mass of trees, shrubs and herbage; total productivity of forest phytocoenosis
IV. Biological cycle	Productivity to phytomass ratio. Coefficient of the annual turnover of above-ground phytomass, allometric coefficient, litter-fall index *)
V. Soil morphology	Thickness of hor. Ao, A1 and B; thickness of the humus profile. The Ao/A1*) and A1/B ratios; the measure of complexity of soil profile

*) For calculations, the values of these parameters were taken with the opposite sign.

Table 14.3. The measures of homeostaticity (Gk) and structural-metabolic indices (Smk) of different soil-phytocoenotic characteristics in the groups of forest biogeocoenoses of the Putyatin continental-island landscape

Characteristics		Groups of forest types				
		Ekm	TEk	Thm	TAm	SAh
Soil morphology	G_k	**0.294**	0.232	**0.249**	*0.155*	*0.158*
	S^m_k	**1.15**	*0.85*	**1.29**	*0.85*	*0.86*
Structure of phytocoenoses	G_k	**0.460**	*0.142*	**0.460**	**0.435**	*0.168*
	S^m_k	**1.40**	*0.60*	**1.40**	**1.20**	*0.40*
Live phytomass	G_k	**0.294**	0.231	**0.384**	*0.125*	*0.158*
	S^m_k	**1.15**	*0.85*	**1.57**	*0.57*	*0.86*
Productivity	G_k	**0.305**	0.132	**0.418**	0.139	**0.400**
	S^m_k	**1.16**	*0.67*	**1.50**	*0.50*	**1.17**
Biological cycle	G_k	**0.311**	**0.312**	*0.125*	*0.096*	**0.311**
	S^m_k	**1.33**	**1.34**	*0.33*	*0.67*	**1.33**

Footnote. Conventional meanings see in Note to Table 11.10.

Nevertheless, the set-theoretical criteria of selection of phytocoenotic ecotypes (see Figure 14.3, *A*, *b*) have revealed a rather high threshold of indiscernibility of objects (Z_{crit} = 0.991) and the presence of exceptionally weak spatial connections between them. It is indicative of structural space fragmentation and weak inter-complex interactions during the formation of forest communities.

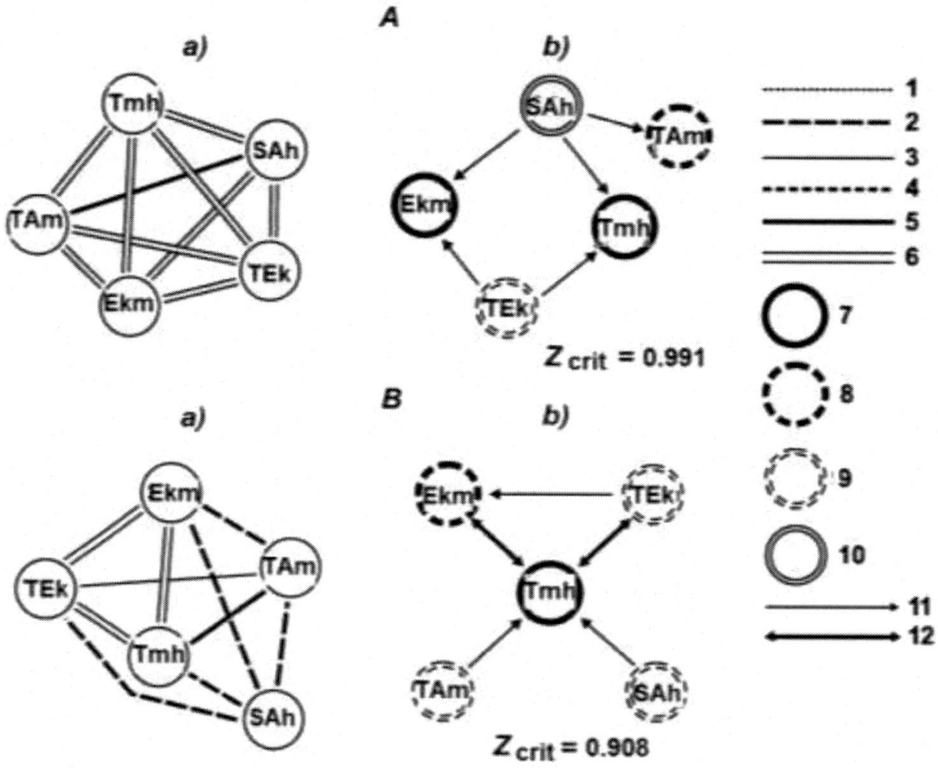

Figure 14.3. The schemes of non-transitive relations of similarity (*a*) and inclusion (*b*) between the groups of biogeocoenoses of the Putyatin landscape by phytostructure (*A*) and soil morphology (*B*). The measures of similarity: 1, ≤ 0.70; 2, 0.71–0.75; 3, 0.76–0.80; 4, 0.86–0.90; 5, 0.91–0.95. Ecotypes of forest communities: 6, macrobolites; 7, submacrobolites; 8, mesobolites; 9, submicrobolites; 10, microbolites. 11, direction of inclusion. 12, connections of tolerance. Z_{crit}, the critical threshold of indiscernibility of objects.

The Putyatin oak forests have considerable territorial contrasts by the integral parameters of polysystem organization: $S^m{}_k$ characterizing their total level of structural development and G_k reflecting their homeostatic reserve (Armand 1989), i.e., resistant stability. The spread of both parameters is up to 3–4-fold values. The forest communities of transit meso-hydromorphic *BGC* widespread on the middle, rather gentle slopes of the hill country are maximally adapted to the geomorphological and hydrological conditions of Putyatin ecoregion, as well as most developed and stable (see Table 14.3; Figure 13.8 *a*).

The analogous structural properties are typical of the oak forests of flat-interfluve (eluvial) geotopes. These are structural macrobolites. The bases of slopes and stream valleys (TA and Saq geotopes), on the contrary, are occupied by submicrobolites – insufficiently developed, weakly adapted forest communities with the minimum homeostatic reserves. Submicrobolites also include the oak forests of transeluvial sites (near-top convex slopes); however, their stability is rather high.

The soils of transit hydrogeotopes take a leading position in respect of both measures of similarity and measures of inclusion in the set-theoretical models of nontransitive relations between the soils of different biogeocoenotic groups (see Figure 14.3 *B*). In the series of metabolic ecotypes (Bykov 1973), they are macrobolites (see Table 14.3) and establish relations of tolerance with soils of the upper catena elements (E submacrobolite and TE submicrobolite) on the structural scheme of dominance. Together with the soils of eluvial geotopes, they maximally display the diversity of states of the entire complex of the abiotic factors (absolute height, exposure, slope steepness, microclimatic conditions, and hydrological regime), are maximally adapted to these factors and, naturally, possess the maximum reserves of homeostaticity.

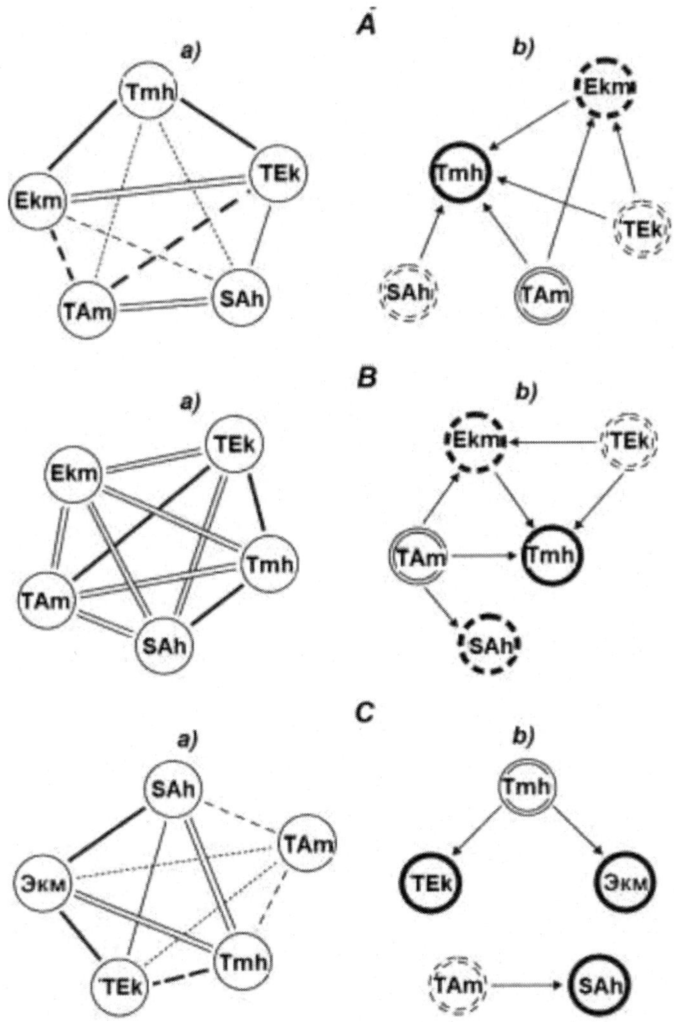

Figure 14.4. The schemes of non-transitive relations between the groups of biogeocoenoses of the Putyatin landscape with regard to live phytomass reserves (*A*), productivity (*B*), and discrete parameters of the biological cycle (*C*).Legends are as in Figure 14.3.

With regard to the main parameters of live phytomasses (*BS*, *BL* and *BC*), all biogeocoenoses of the Putyatin island landscape are combined into two contrasting pleiads (see Figure 14-1). Pleiad I characterized by the highest phytomass reserves includes the Tmg and Ekm groups of *BGC*; pleiad II with the lowest reserves includes the TAm and CAg groups. The transeluvial group takes an intermediate position.

The transit meso-hydromorphic group of biogeocoenoses has the maximum reserves of timber, as well as the reserves of total living and above-ground phytomass and the green mass of soil cover. There are the sufficiently distinct dominant areas: $BL \approx$ 75–140 tons/ha and $BC \approx$ 110–180 tons/ha, with a probability of 70–85% and more. They are also leaders with regard to the parameters of productivity of both total and green mass: the dominant $PV \approx$ 3.8–5.4; $PG \approx$ 1.4–1.9; $PC \approx$ 5.8–7.1 tons/ha · yr. On the graph of non-transitive relations of similarity in living phytomasses (Figure 14.4, *A*), the Tmg phytocoenoses are somewhat apart from the two pairs of the most similar phytocoenotic groups: Ekm–TEk and TAm–SAh. On the structural schemes of dominance of phytomass reserves and productivity, the transit forest communities, as by their structure as (see above), join the category of macrobolites, with a high level of functional development and adaptation to the abiotic environmental conditions ($S^m{}_k$ = 1.57), as well as with the maximum resistant stability (G_k = 0.384). Thus, by the parameters of autotrophic biogenesis, the forest communities of the Putyatin island landscape form a single, "unipolar" spatially ordered functional system with the transit meso-hydromorphic group of biogeocoenoses in the center.

According to the Grigoryev–Budyko concept of geographical zonality (Grigoryev 1966; Budyko 1977), primary bioproductivity is one of the main criteria for distinguishing zonal types of landscapes, and each of these types is characterized by the maximum values of productivity under the given background hydrothermal conditions. In light of this concept, the *forest communities of the transit meso-hydromorphic group of biogeocoenoses of Putyatin Island* with the maximum parameters of autotrophic biogenesis, being the leaders in the system of functional dominance relations, *can be rightfully considered as the priority representatives of this zonal-regional geosystem.* They are the intra-landscape "core" of physical-geographical background (Sochava 1974). Flat-interfluve (eluvial) topogeosystems, which are traditionally considered to be such a "core" (Vysotsky 1960; Ramensky 1971; Krauklis 1974), in this case have a lower bioproductivity and much lesser reserves of homeostaticity. As a result, they are shifted into the lower functional category of submacrobolites. This is one of the key landscape-ecological shifts in the boreal-subboreal topogeosystems of continental-island Sub-Pacific under conditions of monsoon-continental climate and long-term anthropogenic influence.

At the two opposite "poles" of the catena, TEk and TAmh–SAh forest communities are characterized by the lower level of autotrophic biogenesis. At the bases of slopes and in stream valleys, there is a drastic decrease in the reserves of living phytomasses (the dominants $BL \approx$ 22–62; $BC \approx$ 35–95 tons/ha, with a probability of 55% to 100%), as well as a decrease in productivity (PC = 2.9–5.1 and PV = 1.7–3.1 tons/ha · yr), in spite of the higher summer temperatures and moderate moisture of edaphotopes. In the oak forests of near-top convex slopes, with their macrothermal arid edaphotopes (see above), the reserves of skeletal, above-ground and total living phytomasses are rather substantial (see Figure 14.1; Table 14.1) and, in respect of these parameters, these communities fall into the category of submacrobolites. However, very low productivity (the dominants PC = 2.9–4.5 and PV = 1.7–3.1 tons/ha · yr, with a probability of 50–60%) shifts these phytocoenoses into the category of microbolites, demonstrating their extremely weak functional development and minimal resistant stability.

14.3. BIOLOGICAL CYCLE

The complex estimation of ascending and descending branches of the biological cycle makes it possible to disclose the main driving forces of functional organization of forest natural complexes. This analysis involves four complex discrete parameters of forest ecosystem metabolism (Kolomyts 2008, 2015 a): (1) the *PC*/*BC* ratio which, according to Yu. Odum (1986), is an ecological efficiency indicator (EEI) of autotrophic biogenesis and also considered as an efficiency factor (EF) of forest phytocoenosis; (2) the coefficient of above-ground phytomass annual turnover, $CR = PV/BL$, which is a measure of involvement of living organic matter in its annual turnover; (3) the allometric coefficient $CA = \log PV/\log BL$ characterizing the functional degree of similarity of the given local geo(eco)system to its respective zonal type (Perelyman 1975); and (4) the coefficient of above-ground phytomass annual destruction, or litter-fall index (Bazilevich and Rodin 1969): $CY = ML/PV$, representing the ratio (balance) of the rates of annual inflow of organic matter to forest litter and litter decomposition.

In contrast to the autotrophic biogenesis represented by living phytomass reserves and annual productivity, the biological cycle generally has an essentially different spatial structure (Figure 14.4, *C*). First, in spite of the fact that all objects remain weakly connected, the value of the critical threshold of their indiscernibility Z_{crit} decreases, indicating an increase in the total level of lateral paragenetic interactions by the turnover of organic matter, on the metabolic interrelationships between different elements of local landscape coupling – the higher compared to their connectedness by phytomass reserves and productivity.

The metabolic paragenesis produces a marked functional "polarization" between the upper and middle floors of the catena (Ekm, TEk and Tmh groups of biogeocoenoses), on the one hand, and the lower floor (TAm and SAh groups), on the other hand, manifesting itself in the bipolar system of domination on the or-graph of inclusion relations. Such "polarization" manifests itself first of all in the ecological efficiency of production process. The upper floor of the catena is characterized by the lower CE of forest phytocoenosis (Table 14.4). Down the hill country slopes, the EEI parameter increases 3–4-fold (from 0.03–0.04 to 0.08–0.10), reaching the maximum at slope bases and in river valleys. However, the "fuzziness" of ecological niches of biogeocoenoses increases in the same direction.

Second, the affiliation of *BGC* groups to some phytocoenotic ecotypes or other (Bykov 1973; Odum 1983) drastically changes in the biological cycle. The transit meso-hydromorphic BGC group which, as it has already been shown, is a leader by its phytocoenotic structure, living organic matter reserves and productivity, has proved to be a microbolite with regard to the biological cycle (with the minimum stability of the latter). At the same time, the TEk group from the submicrobolite state, where it falls to by phytomass and productivity, is shifted into the leading category of "turnover" macrobolites. Thus, *the mechanisms closing up the biological cycle, in the oak forests of Putyatin Island, are inadequate to the processes of production and accumulation of living organic matter. These are substantially different characteristics of the functioning of forest communities of the continental-island landscape.*

Table 14.4. Distribution of the *PC/BC* ratio among the groups of forest biogeocoenoses (the normalized partial coefficients of connection)

C(A;B) = 0.120

Ratio *PC / BC*	Groups of biogeocoenoses				
	Ekm	TEk	Thm	Tam	SAh
0.029–0.039	**0.44**	**0.35**	0	0	0
0.039–0.045	0.11	0.18	**0.43**	0.16	0
0.045–0.055	0	0.24	**0.43**	0.21	0
0.055–0.070	0.15	0.24	0.14	0.21	0.33
0.070–0.095	0.30	0	0	**0.42**	**0.67**

Note. The dominant areas of ecological niches are in bold.

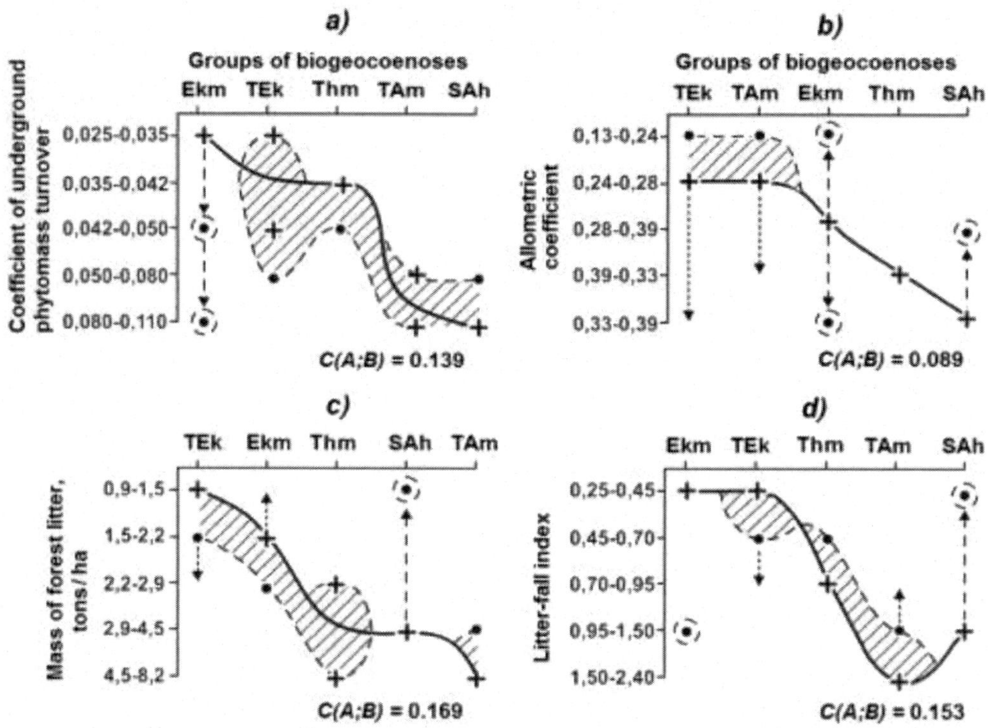

Figure 14.5. Putyatin Island. The complex parameters of the biological cycle in different groups of biogeocoenoses. Legends are as in Figure 10.3.

Let us consider the patterns of the biological cycle in more detail. Judging by the values of the coefficient of annual destruction *CY* (see Figure 14.5 *d*), i.e., litter–fall index, which is a generalized indicator of the work of metabolic mechanisms (Bazilevich and Rodin 1969; Odum 1975; Dylis 1978), it can be assumed that subboreal forests of the continental-island landscape are generally characterized by rather high rates of the small biological cycle and fall into the categories of *intensive and moderate turnover* (Kovda 1971). It exceeds the zonal standards for not only mixed but also broad-leaved forests and makes anthropogenic Putyatin oak forests similar to the moist monsoon "prairie-like forest steppe" (Liverovsky 1974), including the "Amur prairies" (Isachenko 1985), the neighboring Khanka–Ussuri Plain, and the coasts of the Peter the Great Gulf with a secondary origin (Sochava 1980). Such is the *functional shift* (substantially zonal) in the forest cover of continental-island Sub-Pacific, which is determined by the long-term exposure of this cover to anthropogenic impacts.

The degree of balance between the ascending and descending branches of the biological cycle is crucially important for the functional stability of forest biogeocoenosis (Odum 1971; Gerasimov1985; Glazovskaya 1992). What is the situation with continental-island landscape?

In the metabolically "polarized" catenary system of the Putyatin hill country there are two different types of unconformity between the parameters of the biological cycle, two types of their unbalance. The former type is typical of forest communities of the upper floor of the catena: TE and partially E geotopes. The moderate and low green mass production (the dominants *PV* = 3.1–4.3 tons/ha per year with a probability of 75%) are combined here with the high rate of forest litter decomposition as is demonstrated by the minimum *ML* values (≤ 1.5–2.2 tons/ha, with a 70–80% probability of the dominants) and *CY* (≤ 0.45–0.70; see Table 14.1; Figure 14.5 *c, d*). The near-top convex slopes in our sample are mostly shadowy, with the moderate values of soil temperature and moisture content. Here, this metabolic unconformity is more marked than on flat interfluves, the one third of which with southern exposure, which causes an insufficiency of soil moisture and prevents litter decomposition. The high carrying capacity of the detritus branch of metabolism on the E and TE geotopes are insufficiently realized due to the deficiency of dead organic matter becoming tree waste (and drain), which leads to a decrease in the entire annual turnover of above-ground phytomass (see Figure 14.5 *a, b*). The unbalance of metabolism is intensified due to the low EF of forest phytocoenoses (the *PC*/*BC* ratio).

The second type of metabolic unconformity is widespread in the lower floor of the catena – on the TA and Saq geotopes occupying the slopes of southern and western exposure with a hydrothermal regime unfavorable for forest phytobiota. Here, green mass productivity is even lower than on the upper floor (the dominants *PV* = 1.7–3.1 tons/ha · yr, with a ≥80%probability of the dominants). However, due to the minimal reserves of above-ground phytomass *BL* (see Table 14.1; Figure 14.1 *c*), there is the highest annual turnover of organic matter (*CR* ≥ 0.050–0.080) and EF of forest phytocoenosis reaches its maximum. It suggests that the mechanisms of autotrophic biogenesis work intensively and produce the maximally possible annual inflow of dead organic matter to forest litter. At the same time, the high summer temperatures and low moisture content in the upper soil layer drastically slow down litter decomposition; therefore, the *ML* and *KY* parameters reach their maximums (up to 4.5–8.2 tons/ha and 1.5–2.4, respectively). The descending branch of the biological cycle is inhibited, thereby producing metabolic unconformity in biogeocoenoses of the lower floor of the catena.

Against this background, the maximum balance of the biological cycle and, consequently, the highest functional stability were typical of the transit meso-hydromorphic oak forests. The exceptionally high reserves of living phytomass in these forests (see Table 14.1) provide their relatively high productivity (*PV* ≥ 3.8–4.3 tons/ha· yr with a probability of 75%). However, the mean values of forest litter mass and the coefficient of annual destruction are indicative of moderate carrying capacity of the descending branch of metabolism, which restricts the possibilities of autotrophic biogenesis (the *KR* parameter decreases to 0.035–0.050) and provides its certain balance with detritogenesis.

– « » –

Thus, the entire described complex of structural-functional phytocoenotic and soil characteristics of forest geosystems of Putyatin Island allow us to accept *the transit hydro-mesomorphic group of biogeocoenoses* as a certain *zonal-regional model of continental-island landscapes of the boreal-subboreal Sub-Pacific as their topological "core."* The phytocoenoses of these intra-landscape formations are predominant macrobolite ecotypes, with the most optimal adaptation to regional geomorphological and bioclimatic conditions, the maximum diversity of their structural-functional states and the highest reserves of homeostaticity. In respect of soil morphology and phytocoenotic structure, they are similar to flat-interfluve biogeocoenoses; however, as regards the parameters of forest productivity, these geosystems decrease to the subordinate level of submacrobolites. It can be assumed that, in contrast to the accepted canons of landscape science (Vysotsky 1960; Isachenko 1965; Sochava 1978; etc), in the continental-island landscape exposed to the conditions of monsoon-continental climate and long-term anthropogenic influence, the biogeocoenoses of the flat-interfluve xero-mesomorphic series substantially lose the properties of local representatives of the zonal-regional bioclimatic background. These properties are transferred to topogeosystems of the transit meso-hydromorphic series.

14.4. Pasture Digression

For a long time, beginning from the last decade of the XIX century, the forest cover of Putyatin Island was exposed to intensive anthropogenic influence: deforestation as a result of construction and work of brick and porcelain factories and ironworks, raising orchards and vineyards, setting up a stud farm and a fur farm and, finally, systematic cattle grazing and deer habitat in the forests. All the above reduced the "quality" of forest ecosystems (Schwarts 1975), as is demonstrated by very low values of the allometric coefficient *CA* (Figure 14.6 *d*). The boreal south-taiga and subboreal rates of C*A* are 0.54–0.55 and 0.58–0.60, respectively (Perelyman 1975), i.e., 1.5–2 times higher than even the maximal coefficients in the Putyatin oak forests (see Figure 14.5. *b*). Such coefficients characterize the zonal autotrophic biogenesis of only subarctic biogeocoenoses. At the same time, intensification of pasture digression of the soil cover in oak forests from the null category to enhanced and high categories causes a decrease in the *CA* parameter from 0.33–0.39 to 0.13–0.24. The forests of the two "polar" geotopes (the near-top slopes in the upper floor of the catena and at slope bases in its lower floor) are maximally exposed to digression changes (see Figure 14.5 *b*). The middle parts of even slopes and stream valleys are minimally exposed to pasture digression.

We have assessed pasture digression by the retrogressive stages of changes in herbage as a result of its grazing and trampling by cattle and deer (Kolomyts and Tsepkova 2009). The damage to herbage and upper soil layer occurs not only on the vast meadows of Putyatin Island but also in forest stands. The herbage with the dominant and codominant species typical of the given ecoregion and with almost complete absence of weeds is considered as undisturbed. The state of a community with predominance of rather valuable but grazing-resistant plants, with poorly grazed motley grass as subdominants, is considered to be a stable working state of grazing ecosystem at the stage of weak digression.

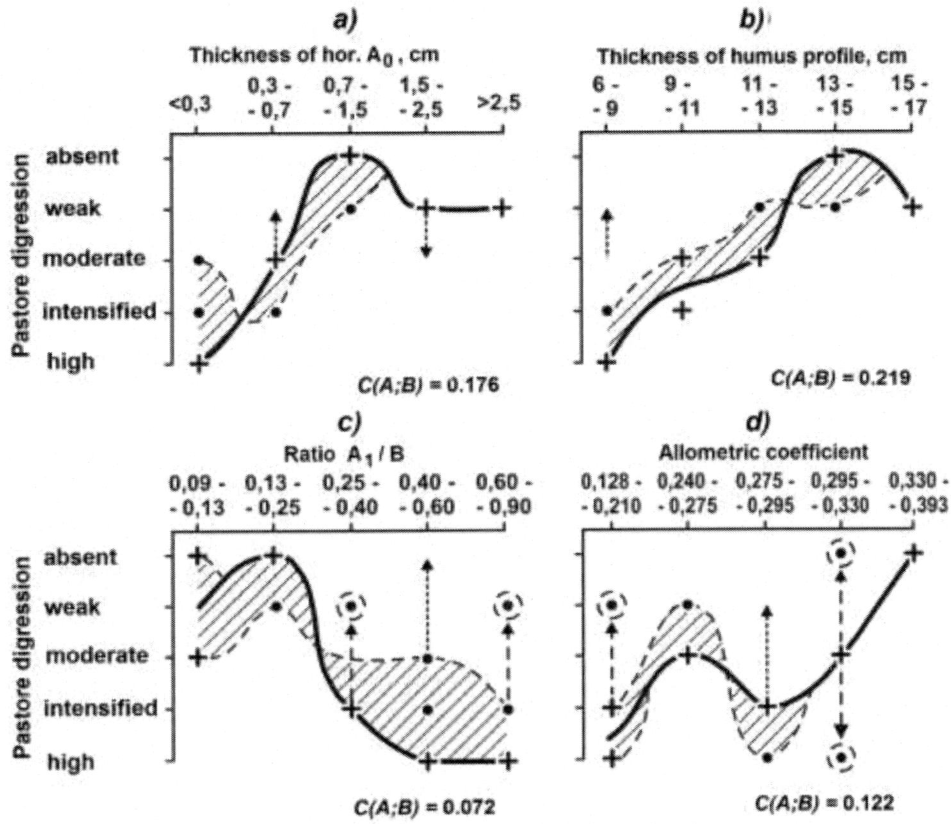

Figure 14.6. Putyatin Island. The effect of pasture digression on the morphological properties of soil and the allometric coefficient. Legends are as in Figure 10.3.

Under moderate grazing loads, the predominant motley grass species are those of the low feed value but resistant to trampling. The enhanced stage of digression begins when firm bunchgrass plants start to dominate in the herbage. Finally, the grass phytocoenoses consisting of the species of low feed value or absolutely uneatable species, as well as the communities of ruderal, wild motley grasses, are attributed to the high digression stage.

The grazing and trampling of soil cover cause the degradation first and foremost of forest litter (or dead grass on the meadow) and the humus horizon of soil (see Figure 14.6). In the Putyatin oak forests exposed to weak grazing influence, horizons A_0 and A_1, as well as the entire humus profile, are yet in the state close to natural, with a thickness of ≥ 2.6–11, and 13–17 cm, respectively. At the same time, the stable working state of herbage (see above) has a certain positive effect, since it contributes to the development of humification as is evidenced by the dynamics of humus profile thickness within an interval of 13↔17 cm (see Figure 14.6 *b*). The thickness of humus horizon per se can increase from 5.5–7.5 cm to 9.5–11 cm.

The drastic changes in soil occur at the enhanced and high degrees of herbage digression. The thickness of hor. A_1 decreases from 8–10 cm to 3–4 cm, while the forest litter and humus profile decrease 2.5–4 times. In spite of degradation of the humus horizon, there is a substantial increase in the A_1/B ratio. Though this tendency is expressed as the dominants with a probability of only 35–55% (see Figure 14.6 *c*), it indicates that grazing often intensifies the decomposition of dead organic matter and humification.

Table 14.5. Putyatin Island. Distribution of pasture digression by the types of sites of forest and meadow biogeocoenoses (the matrix of normalized partial coefficients of connection)

$C(A;B) = 0.124$

Pastore digression	Forest			Meadow		
	E, TE	T–TA	A–Saq	E, TE	T–TA	A–Saq
Absent	0.08	0	0	0.14	**0.62**	**0.69**
Weak	0.07	**0.71**	0.31	0.09	0.28	0.31
Moderate	0.16	0.29	**0.69**	0.28	0.10	0
Intensified	0.20	0	0	**0.50**	0	0
High	**0.49**	0	0	0	0	0

Note. The dominant areas of ecological niches are in bold.

There are quite obvious contrasts between forest and meadow in the grazing disturbance of herbage. Under the same geotope conditions, forest soil cover is more exposed to pasture digression compared to meadow communities (Table 14.5). Apparently, meadow ecosystems have more developed mechanisms of resistance to grazing influence compared to subsystems of the grass layer in the forest. The meadow herbage on T–TA and A–Saq geotopes is hardly exposed to grazing influence, while the herbage in the forest is at the weak digression stage by 70%. The equally obvious contrasts are observed in the upper elements of the catena (on E–TE geotopes). Here in the forest, herbage is disturbed to the maximum extent; on the meadow, it stays within the limits of enhanced digression.

Chapter 15

EXOGENOUS MORPHOLITHOGENESIS

ABSTRACT

The Putyatin continental-island landscape is a scene of extensive development of the processes of exogenous morpholithogenesis, which perform the total planation of the relief to a residual-denudation, "senile" state, with the development of pediments and pediplains. Morpholithodynamic system (MLDS) has been considered as a morphosculptural analog of the catenary system of the types of sites and groups of forest biogeocoenoses, which we have distinguished on Putyatin Island. Exogenous morpholithogenesis is characterized by the four types of structures of the morpholithodynamic system, which differently involve eluvial and defluction/solifluction processes, with the respective development of microterrace structure of slopes. Terracettes are an impressive indicator of defluction/solifluction processes in the upper soil layers. There is a relationship between the distribution of terracettes and the pasture degradation of ground cover. The complex model of a combined meadow-forest continental-island ecosystem was constructed (by calculating, first, the measures of similarity and inclusion relations between objects and, second, the information-statistical indices of diversity of these relations). Based on this model, the coupled mono/polysystem analysis of spatial organization of meadow-forest ecosystem at the experimental site was carried out. It was established that all topological couplings of the Putyatin island landscape have maintained an approximately integrated level of development of illuvial-metamorphic processes achieved in the previous historical periods of the Holocene.

15.1. THE CONCEPT OF MORPHOLITHODYNAMIC SYSTEM

The characteristic features of subaerial landscape genesis on its descending (reduced) branch in the period of morphostructural disintegration of land and formation of continental islands (Kolomyts 1987) are intensive and diverse manifestations of exogenous lithodynamic processes of general planation of the relief to a residual-denudation, "senile" state, with the development of pediments and pediplains, according to Davis–Penck–King's concepts (Rice 1978). Such exogenous lithomorphogenesis involves the Putyatin continental-island landscape.

As it has already been mentioned (see Chapter 11, §11.4), the "slope" elementary morpholithosystem includes the eluvial horizon of bedrocks (El), with the slope (Slp) and cover horizons lying sequentially above it. In addition, it is generally accepted that the enhanced content of lithomass in this system indicates the predominance of eluvial processes (Slp > El), while its decrease shows the predominance of deluvial processes (Slp > El).

Slopes as elements of erosive-accumulative morphosculpture are formed under the influence of the descending lithodynamic flow (Florensov 1978); therefore, slope is not only a morpholithosystem but also a lithodynamic system. Hereinafter, we will refer to slope as a *morpholithodynamic system* (MLDS) and to slope layer per se as (more simplified) a *deluvial layer* (DL), though it contains the roof of the eluvial layer in its lower part. In vertical section, this system encompasses all three horizons (Karpukhin et al. 1984): the cover (soil) horizon, the slope (deluvial) horizon, and the eluvial horizon. We distinguish *deluvial layer thickness* (h_{DL}) as an important independent characteristic of the state of MLDS.

According to the Carson–Kirkby's classification, there are three main types of movements of clastic rock masses: sliding, flowing, and heaving. They are realized in particular slope processes – the seasonal creep of soil (defluxion), landslips, solifluction, earth and mud flows (Rice 1978). The lateral 2D model of morpholithodynamic system consists of three parts (Voskresensky 1971): (1) the upper – decerption slope; (2) the middle – defluxion slope; and (3) the lower –slow solifluction slope. The above parts are in conformity to the basic elements of landscape couplings, or catenas (Polynov 1956; Glezovskaya 1964): eluvial-transeluvial (E–TE), transit-transaccumulative (T–TA), and accumulative-superaqual (A–Saq). As is known, "... slopes and slope sediments ... make an absolutely definite paragenetic composition" (Simonov 1972 144) and "... a particular shape of slope corresponds to a particular sequence of soil differences" (Gerrard 1981 70). This fact allows us to consider MLDS *as a morphoscuptural analog of the distinguished catenary system of the types of sites and the groups of forest biogeocoenoses.* The change of the types of lithodynamic flow down the catena is generally accounted for by: (1) decreasing steepness of the slope; (2) general comminution of mechanical composition of loose sediments; and (3) increase in soil-ground moisture content with the respective changes in consistency from semisolid (dry mass) in the upper element of the catena via viscous-plastic in its middle part to viscous-flow in the lower element (Voskresensky 1971).

15.2. Types of Morpholithodynamic Structures

The vertical composition and lateral structure of slope morpholithosystem are determined by competitive relationships between the two processes of exogenous morpholithogenesis: eluvial (weathering, including intrasoil) and deluvial (defluxion/solifluction drift). Sheet erosion under conditions of forest cover is of subordinate significance. We have assessed these relationships in the first approximation, according to the published data (Voskresensky 1971; Simonov 1972; Karpukhin et al. 1984; Kopotky et al. 2011), by the state of bedrock fragments (crushed stone and boulders): the extent of their weathering or, on the contrary, remaining monolithic; by the predominance of acute-angled or smoothed forms; and by the density of saturation of soil horizons A_1, B and BC with lithomass (fragments). In the presence of weathered fragments, up to the state of rotten stone, and at a low content of lithomass in soil, we could assume the predominance of eluvial process. At the same time, the high content of crushed stone and boulders, with their clear smoothness, was indicative of the well-developed deluvial process. At our experimental ground, the mechanical composition of fine grained soil in soil hor. B proved to be rather monotonous and, therefore, could not be used as a diagnostic character of slope MLDS structure. With regard to all four categories of this structure, sandy loam was 42–46% of total composition of the fractions. It contained almost equal admixtures of light and medium sandy clays enveloping lithomass and imparting the plastic-dynamic properties to the entire lithodynamic layer.

The four types of structures of the "slope" morpholithodynamic system were selected on the basis of the above criteria: two extreme alternative types and two intermediate types (Table 15.1). Each type corresponds to a particular functional state of slope MLDS determined by predominance of each of the two alternative exogenous processes – weathering and mass movement of the material down the slope, or their combination in different proportions.

Table 15.1. The types and diagnostic characteristics of the structures of the "slope" morpholithodynamic system

Slope MLDS structure	Morphological characteristics of bedrock debris	Lithomass content in soil, %
1. Eluvial (El)	Strongly weathered crushed stone and boulders with rabble	≤ 10–20
2. Deluvial-eluvial (De–El)	Acute-angled weakly weathered crushed stone and boulders	From 10–20 to 40–50
3. Eluvial-deluvial (El–De)	Nonweathered acute-angled crushed stone and boulders	From 40–50 to 60–70
4. Deluvial (De)	Smoothed monolithic crushed-stone/boulder fragments	≥ 60–70

The two types of deluvial processes are widespread under forest cover. The first, most frequent type is defluxion, or seasonal soil creep, i.e., slow downhill creep of moistened clayey soil caused by the seasonal fluctuations of its temperature and moisture content (Voskresensky 1971; Simonov 1972; Rice 1978; Gerrard 1981). Under conditions of subboreal Sub-Pacific, it is the deep winter freezing of soil and its periodical summer waterlogging in the period of monsoon rains, which causes the alternating volume changes in soil-ground mass and the additional pressure setting it into motion. The defluxion slope process can manifest itself only at a viscous-plastic consistency of the ground (Voskresensky 1971), which implies the necessary gravimetric moisture of ≥ 20% (Karpukhin et al. 1984) and the presence in MLTS of a sufficient amount of fine-dispersed (clayey) material, with its moderate saturation with lithomass.

The increase in soil moisture content (up to 30–40% and more) leads to the viscous-flow consistency of soil-grounds, and defluxion changes into slow solifluction (Voskresensky 1971) termed as "monsoon solifluction" under conditions of the south of the Russian Far East (Korotky et al. 2011), which places emphasis on manifestation of this process in the period of summer monsoon rains.

Defluxion and slow solifluction clearly manifest themselves in forest stands with massive incline of trunks up the slope, giving the impression of "drunken forest." The long-term erosion-denudation planation of the relief of Putyatin Island resulted in the formation of uniform morpholithogenic substrate represented by gruss-loamy, with crushed stone, eluvodeluvial mantle, which makes it a good aquiclude. In creek valleys, the ground water level is often at a depth of 50–60 cm; in micro depressions, water rises up to the surface. Along the steep scarp slopes of ravines, the entire strata of soil-grounds with tree trunks and roots slide down and crumble from the depth of 2.5–4.0 m. Below the root system there is a bare wet viscous sandy-loam mass of horizon C, which acts as an aquiclude, along which the sliding occurs.

15.3. Geotopological Space of Denudation Morpholithosystem

The "slope" morpholithodynamic system, with respect to its structure and the thickness of lithodynamic horizon (H_{LDH}), is rather difficult to be differentiated between geotopes and, accordingly, the groups of biogeocoenoses (Figure 15.1 *a*; Table 15.2). The types of MLD structures also have their particular dominants of thickness of the lithodynamic horizon (Table 15.3). On flat-interfluve sites (coastal plains), weathering processes are predominant with a probability up to 60% and the El structure develops with an almost complete absence of deluvial drift. Here, accordingly, the H_{LDH} parameter reaches high values (up to 50–75 cm). However, on the sites with steepness of more than 15–20°, the activation of slope processes decreases LDH thickness to 40–20 cm.

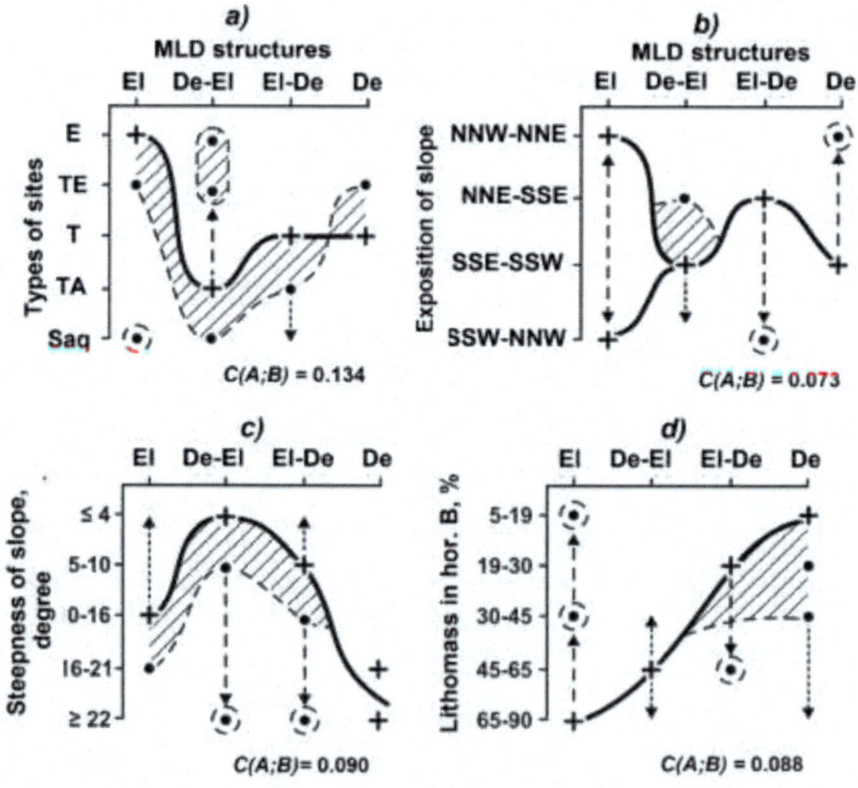

Figure 15.1. Putyatin Island. The binary ordination of morpholithodynamic structures by the characteristics of relief and sediments.

Table 15.2. The coupling of lithodynamic horizon thickness with the groups of forest biogeocoenoses (the matrix of normalized partial coefficients of connection)

C(A;B) = 0.101

Thickness of lithodynamic horizon, cm	Groups of biogeocoenoses				
	Thm	TEk	TAm	SAh	Ekm
13–25	**0.37**	0.25	0	0	0
25–40	0.15	**0.31**	0.43	0	0
40–50	0.07	0.14	**0.57**	0.40	0.25
50–65	0.20	0.10	0	**0.60**	**0.38**
65–85	0.20	0.21	0	0	**0.38**

Note. The dominant areas of ecological niches are in bold.

Deluvial processes begin to appear (as decerption) on TE steep near-watershed slopes but develop to the maximum extent, already as defluxion, on the even slopes of the middle (transit) element of the catena, with a more than 50% probability of the dominant. Indeed, the mixed El-De structure of MLDS also occurs here in nearly equal proportion. On the tops and crests of the hills, under conditions of the mixed El–De and De–El structures, the H_{LDH} parameter drastically decreases (to 15–20 cm).

In the biogeocoenoses of transit sites with the optimal development of defluxion, the LDH thickness is maximal (up to 60–85 cm and more). It remains the same also in BGC of the underlying TA and Saq geotopes, where the total defluxion/solifluction flow slows down, followed by accumulation of material and intensification of weathering processes, which give an almost 40% manifestation of El morpholithogenic structure. At slope bases and in stream valleys, the El and De processes are combined, providing the mixed structure of the morpholithogenic system with a probability of 40–48%.

Generally, the strip of depths of 40–50 cm can be considered as a boundary that separates the upper part of soil-ground with predominant eluvial processes from its lower part with predominant deluvial processes.

The exposure differentiation of morpholithodynamic structures is rather weakly marked (see Figure 15.1 *b*). One can only note that their maximum diversity is typical of slopes of the SE half of the horizon, which face the Sea of Japan and thus are exposed to the effective influence of moist summer monsoon. Under these conditions, both biochemical weathering and defluxion/solifluction processes are actively manifested. At the same time, on the "pluvially shadowy" slopes of NW bearings, which in winter are directly influenced by the dry continental monsoon, predominance is shifted to the processes of physical weathering, with the slowed motions of soil-ground material. Here, the less developed El structure of MLDS is predominant.

Slope steepness, which determines many soil properties, is also a certain factor of diversity of exogenous morpholithogenesis, though this relationship is ambiguous in different elements of the catena (Gerrard 1981). Judging by the ecological dominants (see Figure 15.1 *c*), the El and De-El lithodynamic structures are widespread on relatively gentle slopes (the steepness of ≤ 5–10°) with a dominance of 60%. Deluvial processes are drastically intensified and become predominant on steep slopes (>15–20°). Nevertheless, the merely eluvial structures occur within a rather broad range of slope steepness. The intermediate structures (De-El and El-De) are confined to gentle slopes and have the dominants with a probability of 50–60%. However, they also have 20% enclaves confined to very steep slopes, which altogether reduces this partial connection.

Morphological structures quite clearly differ in the content of lithomass in soil hor. B (see Figure 15.1 *d*). The eluvial process decreases the percentage of crushed stone and boulders in soil-ground, since there is their intensive destruction and dispersion into rotten stone. Therefore, eluvial processes altogether decrease the percentage of bedrock exposure at the day surface, and the relatively small thickness of deluvial layer (20–40 cm) corresponds to the low content of lithomass in soil hor. B (≤ 20–40%, Table 15.3). With intensification of defluxion/solifluction processes, the bedrock exposure, soil-ground skeletality and the percentage of angular, unweathered rock fragments in these soil-grounds increase (in hor. B, up to 70–90%).

Table 15.3. The distribution of morpholithodynamic structures in the space of lithodynamic layer thickness (the matrix of normalized partial coefficients of connection)

C(*A*;*B*) = 0.112

Thickness of lithodynamic horizon, cm	Types of MLDS structures			
	De	El–De	De–El	El
13–20	0	**0.30**	**0.50**	0
20–32	0.05	0.11	0.19	**0.53**
32–48	0.14	0.25	0.11	0.33
48–60	**0.51**	0.13	0.08	0
60–85	0.30	0.21	0.12	0.14

Note. The dominant areas of ecological niches are in bold.

Against this background there are quite marked abnormalities (enclaves) determined by catenary affiliation of soils. At the TA and SA sites, deluvial process (slow solifluction) often occurs with the minimum skeletality of soil-grounds, because the material at this stage of displacement consists almost exclusively (by 80–90%) of fine grained soil. At the TE and T sites, where defluxion begins and develops, loose material has a high content (up to 70–90%) of bedrock fragments (see Figure 15.1 *d*). The skeletality of soil hor. B drastically decreases (up to 45–65%), especially with the minimum thickness of deluvial layer (see Table 15.3). However, on the flat tops and crests of the hills with the most active eluvial processes, the skeletality of soil-grounds is minimal (≤ 20%). At the same time, in stream valleys and at the foot of the hills there is a gain of not only fine grained soil but also weakly weathered fragments; therefore, soil skeletality here can increase to 30–45% and more.

15.4. Terracettes

The powerful indicator of defluxion/solifluction processes occurring in the upper soil layers is the microterracing of slope, i.e., the development of terracettes (Gerrard 1981). The emergence of terracettes is associated with the *process of the flow of waterlogged masses of fine material,* which is most marked just in the near-surface soil layer and gradually subsides with depth (Rice 1978). In general, the slopes of the Putyatin hill country are weakly terraced

– over no more than 15–20% of its area. The terracettes can vary from 3–7 cm to 10–15 cm in height and from 20–25 cm to 40–50 cm in width. Numerous bare spots with small gravel are formed as a result of sod breaks appearing on the slope. Microterraced slopes have a negative effect on the floristic diversity of herbage both in the forest and on the meadow, though this relationship is altogether rather low (*C(A;B)* = 0,096). The dominants of the information measure of floristic diversity decrease from 4.70–4.93 bit in the absence of terracettes to 3.71–4.15 bit under conditions of their maximum distribution. The above dominants characterize the maximum values of these measures. Numerous "blurs" of ecological niches are determined by sporadic deviations from the dominants, which usually tend to the lesser floristic diversity.

Under conditions of rather homogeneous sandy-loamy substrate and its ability to become viscous-flow during the summer monsoon rains, the microterraced pattern of slopes is determined first and foremost by their steepness (*C(A;B)* = 0.208). With steepness increasing from 2–10° to 20–25°, the extent of slope coverage by terracettes unambiguously increases from 0–5% to 15% and more. The terracetted pattern reaches its maximum in the middle and lower elements of the catena: on the T and TA geotopes with predominance of mixed De–El and El–De MLD structures, while the values of deluvial layer thickness are medium or low: up to 40–50 cm (Figure 15.2, *a, b*). The solifluction processes leading to the emergence of microterracettes on the slopes seem to be predominant under these conditions (Simonov 1972). However, the low probability of the dominants (≤ 35–40%) suggests that terracettes here have a limited area of distribution. With the maximum thickness of deluvial layer determined by defluxion, they are rare or absent at all. A. G. Gerrard (1981) shows the critical soil thickness of about 70 cm, when there may be no emergence of terracettes due to more massive sliding of the entire soil mass but not only its upper layers. On flat interfluves and near-top slopes of the hills (E and TE geotopes), where El and the related De–El morpholithostructures are predominant and the dominants of the deluvial layer thickness are no more than 20–30 cm, the microterracing decreases to the minimum and the coverage of slopes with terracettes is no more than 5–15%.

The relationship between the microterracetted organization of slopes and pasture digression is quite important. On the middle parts of slopes and at their bases with the most developed defluxion/solifluction processes (see Figure 15.1,*a*) and the deluvial layer reaching its maximum thickness (see Table 15.3), pasture digression is exhibited to the minimum extent or even absent. This fact is demonstrated by the very weak microterracing of slopes (see Figure 15.1, *c, d*). The deluvial MLD structure of the hill country provides thereby the maximum resistance of herbage to grazing influence, and one can suggest its negative feedback with the given digression factor.

The opposite situation is observed on the TE crests of the hills and near-top slopes with the dominance of mixed El–De and De–El structures, with the minimum thickness of deluvial layer and wide distribution of terracettes. Here, pasture digression of soil cover is most marked. The feedback is found to be positive: excessive grazing causes further drying and consolidation of soil on these already macrothermal arid geotopes (see above) and, hence, the defluxion and solifluction processes are inhibited, being confined to the upper soil layers, and the resistance of soil cover to grazing influence drastically decreases.

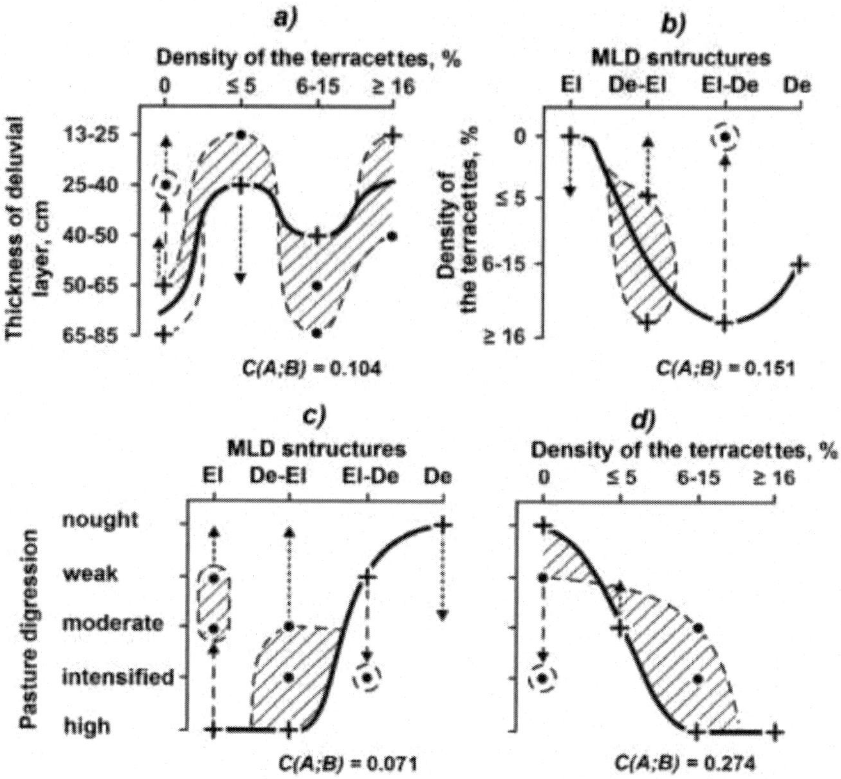

Figure 15.2. Putyatin Island. The development of terracettes in different morpholithodynamic structures and under conditions of pasture digression. Legends are as in Figure 10.3.

On flat interfluves with the dominance of El structure of MLDS, with a slightly thicker deluvial layer, pasture digression manifests itself to an equally high extent. Accordingly, the resistance of herbage to grazing influence is minimal here. The weak development and even absence of terracettes demonstrates that pasture digression is a secondary factor of microterracing of slopes. The primary factors are natural defluxion/solifluction processes determined by the periodical viscous-flow state of sandy-loamy soil-grounds and considerable steepness of the slopes.

Table 15.4. The coupling of microterracing of slopes with the groups of forest biogeocoenoses (the matrix of normalized partial coefficients of connection)

C(A;B) = 0.164

Microterracing of slope, %	Groups of biogeocoenoses				
	TEk	Thm	Ekm	TAm	SAh
0	0.06	0.18	**0.65**	**1.00**	**1.00**
≤ 5	0.24	**0.44**	0	0	0
6–15	0.21	0.38	0.35	0	0
≥ 16	**0.48**	0	0	0	0

Note. The dominant areas of ecological niches are in bold.

The groups of forest biogeocoenoses are clearly differentiated by the extent of terracing of slopes (Table 15.4). The maximally developed terracettes (with a slope coverage up to 15–20% and more) are observed at the upper decerption floor of the catena – on the steep tops and crests of the hills and on the near-top slopes (in the TEk group of BGC). Here, the height of terracettes is up to 15–20 cm (width of 0.3–0.5 m), with the bare sports of fine grained soil and with the plots of soil washed out down to the humus horizon. There are also the typical "inundations" of fine soil material on the butts of tree trunks. Down the catena, with the development of defluxion processes involving still thicker MLDS mass, the frequency of occurrence of terracettes drastically decreases and becomes less than 5% already in the Tmg group of BGC. The microterraced organization almost completely disappears over the entire lower element of landscape coupling (on the TA–SA geotopes). It is nearly absent also in the flat-interfluve Ekm groups of BGC. Instead, there is a massive defluxion/solifluction movement of the entire soil-ground mass, which is fixed by the morphophysiological characteristics of oak forest stands (the trunks begin to bend and become thick-set, with heavy branching; differently directed bends of the trunks give the impression of "drunken forest").

15.5. The Hydrothermal Regime of Morpholithogenesis

In the Putyatin continental-island landscape, exogenous morpholithogenesis actually does not depend on the fluctuations of thermal conditions of the warm period. There are no significant correlations between MLD structure/deluvial layer thickness and soil temperatures ($C(A;B) \leq 0.65$). The relationship between these parameters and soil moisture content is much more explicit (Figure 15.3, *a, b*), once again emphasizing the *significance of hydroedaphotope as a critical ecological factor under conditions of monsoon-continental climate.* At the most arid geotopes ($W(20\text{–}40) \leq 15\text{–}25\%$), the eluvial and similar De–El MLD structures are predominant with a probability up to 80% and more, while the deluvial layer thickness is minimal (≤ 40–50 cm). Defluxion begins to manifest itself already at a soil moisture content above 15% and is drastically aggravated at $W(20\text{–}40) > 25\text{–}30\%$, followed by the 60–80% dominance of De and El–De structures. At the same border of soil moisture content there is an abrupt, discontinuous increase in the thickness of deluvial layer (up to 65–85 cm). Thus, *as soil moisture content increases, the thicker and thicker mass of the morpholithodynamic system is involved in slope defluxion/solifluction processes.*

The microterracing of slopes is not unambiguously dependent on summer soil temperature (Figure 15.3 *c*). Terracettes are absent in 45% of cases at high temperatures ($t_{30} > 16.5°$); however, they can be absent within the entire range of low temperatures. On the other hand, the maximum development of terracettes falls on the two extreme values of soil temperature with nearly equal probability (about 50%).

Terracettes can be absent at any soil moisture content – from semisolid consistency (dry loose mass) to viscous-flow (solifluction). The mass emergence of terracettes (with more than 15% coverage of slopes) falls on a particular average moisture content in the upper soil layer (15–25%), which corresponds to the viscous-plastic consistency of fine grained soil, just when defluxion develops. When the soil acquires viscous-flow consistency ($W(10\text{–}20) > 25\text{–}30\%$), defluxion is replaced by slow solifluction and the formation of microterraces is stopped.

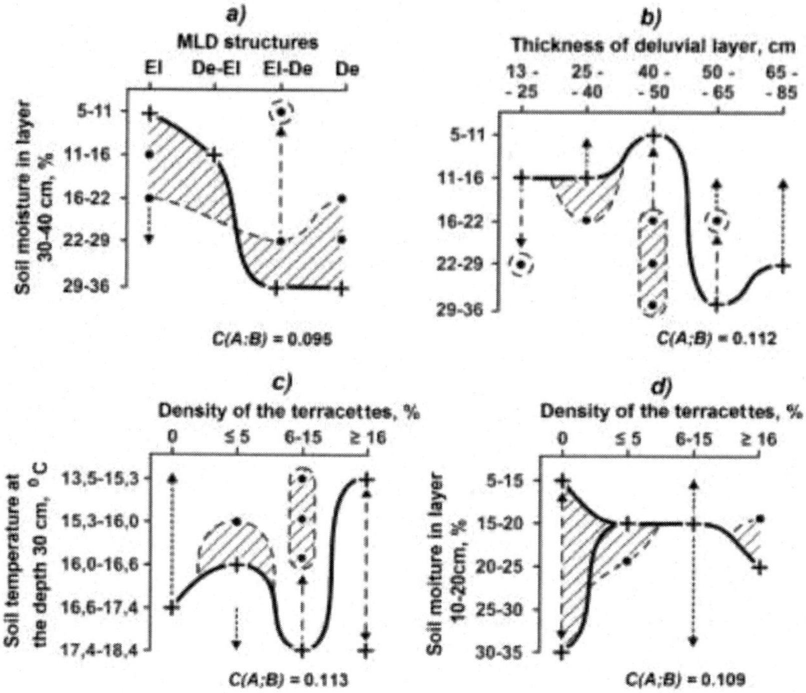

Figure 15.3. Putyatin Island. The characteristics of morpholithogenesis in the space of hydrothermal conditions.

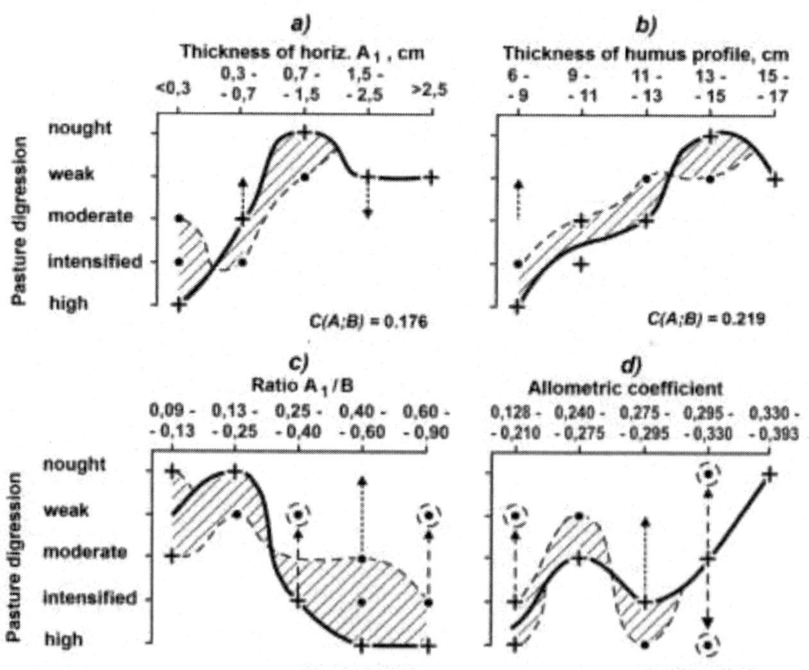

Figure 15.4. Putyatin Island. The ecological niches of morphological characteristics of soil and the allometric coefficient in the space of pasture digression.

Exogenous morpholithogenesis exerts diversified effects on soil formation processes in the Putytatin continental-island landscape. The ordination of the A_I/B soil horizon thickness by the MLD structure of the "slope" system (Figure 15.4 *a*) clearly demonstrates the "polar" spatial differentiation of two basic soil-forming processes (Glazovskaya 1981; Polivanov 1984): (1) the accumulation of organic matter and humus formation and (2) the intra-soil transformation of mineral mass. The indicators of dominance of the former process are the highest values of parameter $A_I/B \approx 0.60–0.90$, which by all 100% belong to El and the related De–El structures of morpholithogenesis. The low values of parameter A_I/B ($\leq 0.10–0.25$) showing the apparent dominance of the latter process are confined by 90% to De and El–De morpholithogenic structures.

The relationship between the thickness of hor. B and morpholithogenic structures is also characteristic (Figure 15.4 *b*). Though this relationship is rather "fuzzy," the main contrasts are quite distinct. In the upper (E and TE) elements of the catena, with the dominance of weathering processes and El and De–El structures, horizon B has the minimum and medium thickness ($\leq$ 25–40 cm), indicating that illuviation and lessivage processes are slightly inhibited here.

In the middle and lower catena elements (at the T, TA and SA geotopes), the dominance of deluvial drift over eluvial processes forms El–De and De morpholithogenic structures and contributes to the drastic increase in hor. B thickness (up to 55–65 cm and more). It has been shown (Simonov 1972) that soil displacement as a result of defluxion over a 100-year period is from 10 to 25 m at an average slope steepness of 15–30°. According to other estimates (Rice 1978), in moderately humid regions on a mid-steep slope, the upper soil layers usually slide down at a velocity of 1–2 mm per year. The vertical component of this vector is undoubtedly one of the causes of increased thickness of soil hor. B in the De and El–De structures.

During the defluxion/solifluction movement of material during its periodical waterlogging–drying and freezing–thawing, as well as under the influence of tensile and compressive stresses under gravity conditions (Voskresensky 1971; Gerrard 1984), there is the abrasion and smoothing of detrital rock, the grinding of particles to sandy-loamy fractions. Simultaneously, along the fractures there is the lessivage of loam particles enriching hor. B with Fe and Al sesquioxides, as is evidenced by "rust" stains on crushed stone. All these processes of deluvial mixing, or "paedoturbation," prevent consolidation of the soil-ground mass and lead to its certain *homogenization* (Boul et al. 1973), which imparts the respective characteristic features of their soil profiles to De and El–De morpholithogenic structures.

Thus, *eluvial process acts as the initial mechanism of humification,* which is known (Glazovskaya 1981) to involve the active processes of physicochemical weathering, with the formation of organomineral complexes making the basis of the soil humus profile. At the same time, *deluvial slope processes actively contribute to intra-soil transformation of mineral mass* and the respective increase in the thickness of soil horizon B, which can be considered with good reason as not only illuvial but also metamorphic.

15.6. The Complex Model of Island Meadow-Forest Ecosystem

Under bioclimatic conditions of the forest-steppe zone, phytobiota is represented by the two main types of edification (environment-forming) structures – forest and herbaceous (Utekhin 1977); therefore, the knowledge of natural-territorial structure of the Putyatin island landscape cannot do without its complex (mono- and polysystem) analysis. This analysis, in addition to the information and set-theoretical characteristics, also uses the *integral parameters* of structural-functional states of biogeocoenoses at different biotopes, which we have introduced: (1) the metabolic index of biogeocoenosis $S^m{}_k$ characterizing the total level of its structural-functional development and (2) the index of significance of biogeocoenosis in the given altitudinal-zonal geospace G_k reflecting its homeostaticity reserve (see Chapter 2).

The comparative analysis of forest and meadow biogeocoenoses could be performed only with the "through" geocomponent characteristics, which excluded the possibility of using the data on the tree-shrub layers of forest communities. The analysis was performed by using the 4 distinguished geocomponental units: (1) hydrothermal-edaphic; (2) morpholithodynamic; (3) soil morphological; and (4) phytocoenotic structural.

Table 15.5. The list of characteristics of geocomponental units

Geocomponental units	Geocomponental characteristics
Hydrothermal-edaphic	Soil temperatures at the depths of 30 and 40 cm; soil moisture content in the layers of 10–20 and 30–40 cm
Soil morphological	Thickness of hor. A_0 and A_1 and the humus profile; A_1/B ratio
Morpholithodynamic	Thickness of deluvial layer; lithomass in soil hor. B; density of terracettes; MLD structure of the system
Phytocoenotic structural	Average height, projective coverage, floristic diversity and ecological groups of herbage

The list of geocomponental characteristics for each unit is given in Table 15.5. Geocomponental characteristics were divided into points in the ascending order of unit organization. For soil morphological unit, it represented the increase in accumulation of dead organic matter on soil surface and the enhanced efficiency of humification processes. For morpholithodynamic unit, the ranking order corresponded to the trend of the increasing role of deluvial processes in MLD structure formation and the decreasing role of eluvial processes. Phytocoenotic structural unit took into account the level of development of herbal soil cover and the degree of its hydromorphism.

The hydrothermal-edaphic field of the Putyatin hill country in the vegetation period is highly homogenous, as is evidenced by the high total degree of hydrothermal similarity between the types of sites (≥ 85%) irrespective of whether it is forest or meadow (Figure 15.5 *A*). Nevertheless, certain topological contrasts can be traced. For all geotopes, with the exception of slope bases and river valleys, soil temperatures on the meadow are higher than in the forest. On the E–TE sites, this difference is 2–2.5°. At the same time, on watersheds and near-top slopes, the minimum and maximum temperatures were recorded in the forest ($t_{40} \leq$ 14.0–15.5°) and on the meadows ($t_{40} \geq$ 17.5–19.0°), respectively. It is probably due to the strong shadowing effect of forest cover. At the same time, the meadow eluvial and transeluvial geotopes are wetter compared to the analogous forest geotopes: *W(20–40)* ≈ 19–20% vs. 15–18%.

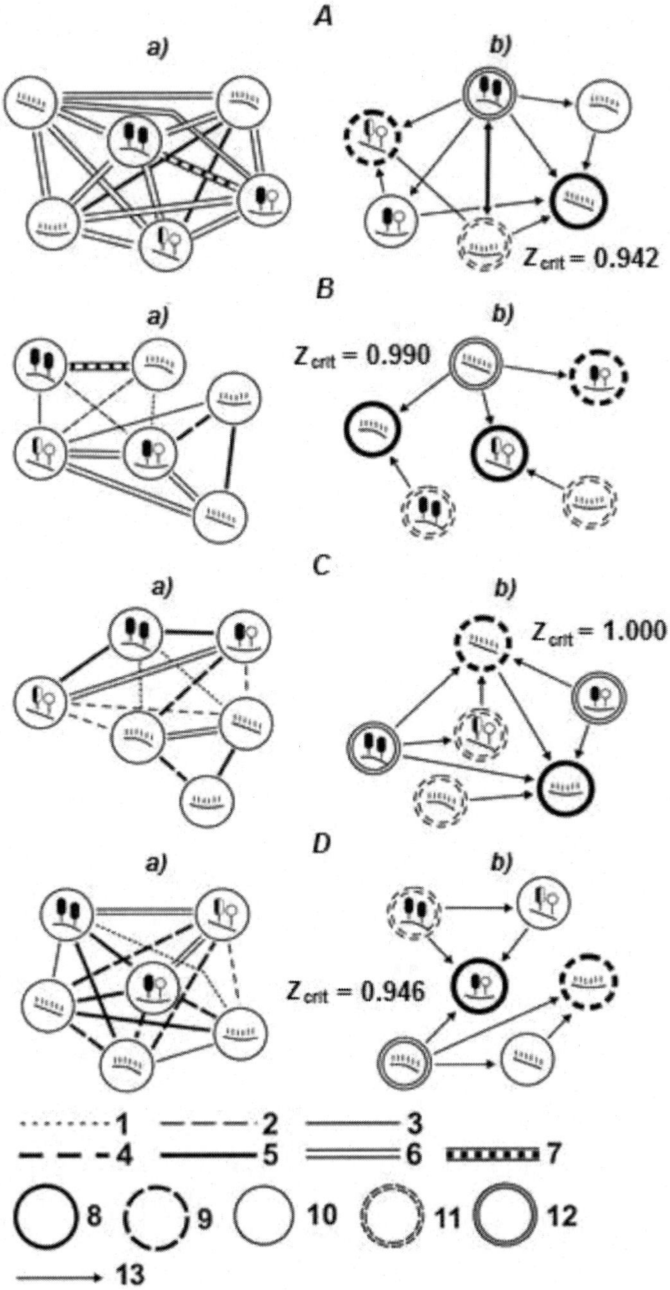

Figure 15.5. The schemes of nontransitive relations of similarity (*a*) and inclusion (*b*) between the components of complex meadow-forest ecosystem of Putyatin Island. Geocomponental units (see Table 19.6): *A* – hydro-thermoedaphic; *B* – morpholithodynamic; *C* – soil morphological; *D* – phytocoenotic structural. Other legends are as in Figure 7.5 and Table 12.2.

In the middle elements of the catena (at the T–TA geotopes), the above thermoedaphic contrasts between meadow and forest are maintained within a very broad range of soil moisture content in both cases: from 10–15% to 30–35%. In general, these types of sites were characterized by the highest heat and energy levels and therefore classified as macrobolites and submacrobolites (see Figure 15.5 *a, b*). At the same time, they have the maximum reserves of homeostaticity (Table 15.6). In the lower elements (at the A–Saq geotopes), the situation is the opposite: in the forest, soil is by 1–1.5° warmer and slightly drier than on the meadow, which can be accounted for by the enhanced drainage of forest soil. It is confirmed by the ecological groups of herbage at the A–Saq geotopes: in the forest, these are mesophytes and meso-hygrophytes; on the meadow, these are almost exceptionally hygrophytes.

The meadow soils of slope bases and river valleys have the lowest temperatures similar to the forest soils of watersheds, all of them falling into the categories of microbolites and submicrobolites, with minimal homeostaticity. They have the highest (>95%) degree of mutual similarity with regard to hydrothermal conditions and the highest ratio of tolerance (high isomorphism) with regard to inclusion measures.

The general hydrothermal homogeneity of the field of edaphotopes of the forest-meadow system of Putyatin Island seems to be a direct consequence of long-term and intensive anthropogenic transformation of the forests (see Chapter 12). The comparative analysis of woodless and much less transformed forest biogeocoenoses of the Lower Priamurye (Golubchikov and Mandych 1979) has shown that the difference between them in the hydrothermal characteristics of soil-grounds under conditions of low mountain country is comparable with the contrasts between these indices determined by the change of not only elements but also forms of mesorelief.

With reference to the development of the *morpholithodynamic system,* the situation is rather complicated (see Figure 15.6 *a*; Table 15.6). On hilly watersheds and near-top slopes with predominant weathering processes (see above), the El and De–El structures dominate in the forest (with a probability of more than 55%). Here, the displacement of loose material down the slope is prevented by two factors. First, it is the comparatively low soil moisture content (the range of $W_{20\text{–}30}$ = 5–20% with a more than 50% probability). Second, it is the characteristic high forest density with the dominants of 1.5–2.3 and 0.75–1.05 units by the mass of forest stands at the TE and E geotopes, respectively. As is known, a forest with a thriving root system usually slows down the defluxion/solifluction processes (Rice 1978; Gerrard 1981).

On the meadows, the entire set of morpholithogenic structures is sporadically distributed (with a probability of 15 to 35%) along the upper floor of the catena. Here, slope processes are activated by high moisture content (the significant $W_{20\text{–}30}$ = 20–35% with a 45% dominance). Accordingly, the "slope" system is submicrobolitic and least resistant ($G_k \leq 0.16$ units) in the former case and reaches the state of submacrobolites with the maximum resistance ($G_k = 0.46$) in the latter case. In the forest, the thickness of deluvial layer is somewhat less (h_{DL}, see Table 15.7) and the content of lithomass in soil hor. B and the occurrence of terracettes are minimal (Table 15.8).

Table 15.6. Structural-metabolic indices ($S^m{}_k$) and measures of homeostaticity (G_k) by the structural and functional characteristics of forest and meadow communities in different types of sites of the Putyatin island landscape

Characteristics		Forest			Meadow		
		E–TE	T–TA	A–Saq	E–TE	T–TA	A–Saq
Hydrothermal	$S^m{}_k$	*0.66*	**1.25**	1.00	1.00	**1.33**	*0.76*
	G_k	*0.158*	**0.327**	0.218	0.218	**0.333**	*0.171*
Soil	$S^m{}_k$	*0.67*	*0.89*	*0.76*	*0.89*	**1.21**	**1.56**
	G_k	*0.113*	0.209	*0.125*	*0.156*	**0.278**	**0.352**
Morpholitho-dynamic	$S^m{}_k$	*0.80*	**1.40**	**1.20**	**1.40**	*0.40*	0.80
	G_k	*0.161*	**0.460**	**0.441**	**0.460**	*0.168*	*0.161*
Phytocoenotic	$S^m{}_k$	*0.71*	1.00	**1.43**	*0.57*	1.00	**1.29**
	G_k	*0.127*	0.263	**0.400**	*0.125*	0.263	**0.399**

Footnote. Conventional meanings see in Note to Table 11.10.

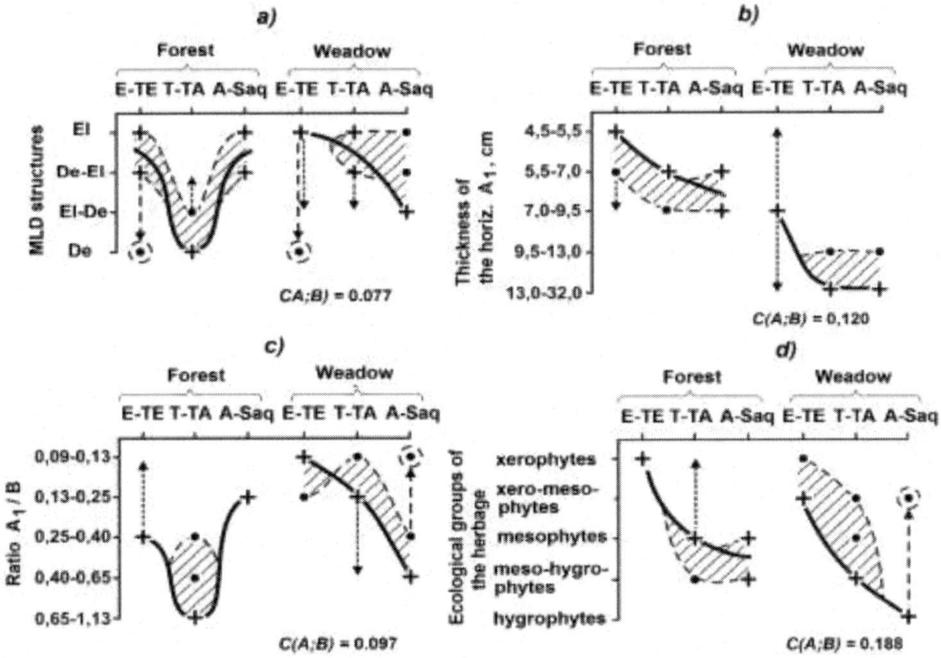

Figure 15.6. Putyatin Island. The distribution of structural characteristics of forest and meadow biogeocoenoses in the catenary system of the types of geotopes.

Table 15.7. The distribution of deluvial layer thickness by the types of sites in forest and meadow biogeocoenoses (the matrix of normalized partial coefficients of connection)

C(A;B) = 0.090

Thickness of deluvial layer, cm	Forest			Meadow		
	E–TE	T–TA	A–Saq	E–TE	T–TA	A–Saq
	0.30	0	0	**0.50**	0	0
20–32	0.17	0.15	0	0.19	**0.42**	0
32–48	0.15		0.42	0.22	0.07	0
48–60	0.07	**0.34**	**0.58**	0.08	0.20	**0.56**
60–85	**0.32**	0.13	0	0	0.31	0.44

Note. The dominant areas of ecological niches are in bold.

Table 15.8. The distribution of microterracing of slopes by the types of sites in forest and meadow biogeocoenoses (the matrix of normalized partial coefficients of connection)

C(A;B) = 0.125

Frequency of the terracettes, %	Forest			Meadow		
	E–TE	T–TA	A–Saq	E–TE	T–TA	A–Saq
0	0.14	0.20	**1.00**	0.14	**1.00**	**1.00**
1–5	**0.32**	0.38	0	0.16	0	0
6–15	0,27	**0.42**	0	0.18	0	0
> 15	0.27	0	0	**0.53**	0	0

Note. The dominant areas of ecological niches are in bold.

With regard to the middle T–TA element of the catena, the situation is absolutely different. On the meadows, the morpholithogenic dominant of more than 75% is shifted towards E + De–El structures, with deluvial layer thickness no more than 30–45 cm. Here, slope processes weaken due to dramatic aridization of soil-grounds (the dominant $W_{20\text{–}30}$ decreases to 10–15%) and probably their compaction. In the forest, defluxion and then solifluction processes are maximally developed due to low forest density (≤ 0.60–0.75, with a probability of more than 60%) within a broad range of soil moisture content (from 5–10 to 25–35%). It results in the dominance of De and E–De structures (with probability of the dominants more than 80%), as well as the increase in h_{DL} to 60–85 cm, which brings the morpholithosystem of transit-transaccumulative *BGC* to the rank of highly resistant macrobolites ($G_K > 0.44$).

Thus, *the stability of the "slope" system is provided by the dynamicity of slope morpholithogenesis* – so far as the deluvial and related El–De structures are predominant there, with the respective thickness of deluvial layer.

With regard to the extent of development of the ***soil-morphological unit,*** forest biogeocoenoses entirely "concede" to meadow ones by all catena elements, staying within the categories of microbolites and submicrobolites. The meadow biogeocoenoses, with the exception of eluvial and transeluvial ones, act as macrobolites in stream valleys and as submacrobolites on hill slopes and at their bases (see Figure 15.5 *C, b*).

Meadow soils critically differ from forest soils in more intensive humification, and this difference increases down the catena as edaphic humidification increases. The thickness of hor. A_1 on the meadows increases by 3–5 cm for the E–TE geotopes and reaches 10–15 cm and more at the lower A–SA geotopes (see Figure 15.6 *b*). Accordingly, the entire humus profile of the soil is more developed; its predominant thickness is about 6–15 cm in the forest and 20–30 cm on the meadow (probability of the dominants is 65–75%). Such soil-morphological contrasts can be explained by the fact that meadow ecosystems more efficiently use the heat and moisture reserves for organic matter production (Gerasimov, Grin 1976; Utekhin 1977). Thereby, the conditions for more intensive weathering and soil formation are provided (Kovda 1973). In addition, illuvial-metamorphic processes in the soil of forest bioegocoenoses at the E–TE geotopes drastically weaken, yielding a decrease in the A1/B ratio, in spite of low thickness of the humus horizon (see Figure 15-6,*c*). Humification processes reach their maximum at the transit and TA geotopes in the forest and in the lowest element of the catena on the meadows.

There is no significant relationship between the thickness of soil hor. B and the types of sites in the forest-meadow-steppe ecosystem. It means that the same level of development of illuvial-metamorphic processes as that reached in the past historical periods of the Holocene is maintained at all topological couplings of the Putyatin island landscape. The subsequent anthropogenic changes in plant cover (not only the change of forest formations from subtaiga to nemoral but also the emergence and wide distribution of meadow-steppe formations) has no substantial effect on the state of hor. B, i.e., on the degree of soil maturity in general.

In the overall morphological development of the soil unit, the maximum similarity (≥ 91–95%) was shown for two associations: T–TA ↔ A–Saq for the meadow and E–TE ↔ T–TA for the forest. The forest transit biogeocoenoses, being most representative in the zonal-regional respect (see above), tend to eluvial biogeocoenoses (*BGC*), while transit meadow biogeocoenoses tend to superaqual *BGC*.

Finally, the herbaceous soil cover comprising the ***phytocoenotic unit*** of our operation system is most developed (both in the forest and on the meadow) in the lower elements of landscape coupling: at the TA–SA geotopes, where the phytocoenotic structures under consideration are in the macrobolite and submacrobolite states with the maximum stability (see Figure 15.6 *d*; Table 15.6). Here, mesophytes and meso-hygrophytes are predominant in the forest and almost exceptionally hygrophytes are predominant on the meadow (see Figure 15.6, *d*). The height and projective cover of herbage, as well as the development of its floristic diversity, are favored by the maximum thickness of hor. A_1 and of the entire humus profile and by the moderate soil moisture content (the dominants $W_{20–30} = 10–20\%$). The upper floors of forest and meadow catenas, with the E–TE geotopes, are characterized by the dramatically lower degree of development of herbaceous cover – both in mass and in floristic diversity, rating the former among submicrobolites and the latter among microbolites, with the minimum stability. At the same time, the watershed and slope geotopes in forest biogeocoenoses show high similarity in the structure of soil cover (see Figure 15.5, *D, a*). Generally, forest and meadow geotopes form two rather autonomous associations of phytocoenotic structures.

The experience of mono- and polysystem analysis of the forest-meadow-steppe ecosystem of Putyatin Island may be of certain scientific and methodological value with respect to the development of a *complex geoecological model of continental-island landscapes of Sub-Pacific in the zone of transition from the boreal to subboreal belt.* The construction of such complex models will apparently provide a set of critical parameters of the states of geo(eco)systems necessary for the efficient – both bioecological and geosystem – regional monitoring.

PART V: CLIMATOGENIC MECHANISMS OF EVOLUTIONARY PROCESSES

Chapter 16

THEORY AND METHODS OF CLIMATOGENIC SIMULATION OF EVOLUTIONARY PROCESSES

ABSTRACT

The strategy of landscape-ecological prognosis was as follows: first to identify the selected ecosystem objects (local natural complexes) to certain values of contemporary climatic conditions and then to estimate the most probable transformation of the revealed ecological niches of the given objects according to the expected climatic changes for a given prognostic date. Predictions of ecosystem transformations include two stages of analysis: (1) evaluation of the probabilities of changes in the functional states of ecosystems and (2) calculation of the rates of ecosystem transformations. Quantitative predictive analysis was carried out by means of operations with the hydrothermal niches of regional or local ecosystems. The algorithms of predictive calculations have been described in detail for both stages of analysis. The results of local prognostic analysis are presented in both graphic-analytical models.

16.1. PECULIARITIES OF LANDSCAPE-ECOLOGICAL PROGNOSIS

We have described the *climatogenic mechanisms of evolutionary landscape-ecological processes* at the Pacific Ocean mega ecotone of North Eurasia via *simulation of these processes according to landscape-ecological predictions.* The regional climate scenarios of global prediction models E GISS and HadCM3 (version A2) for the forthcoming 100–200 years for island-arc Neo-Pacific and marginal-continental Sub-Pacifica were taken as a basis, in comparison with the analogous predictions for intracontinental regions of East Europe. The simulation tool was the previously developed method of regional and local landscape-ecological prediction (Kolomyts 2003, 2005, 2008). The mechanisms of evolutionary processes were revealed by the trajectories of functional-structural transformations of biogeosystems and landscapes in light of the given hydrothermal trends for a particular period of climate prediction, with predetermined temperature and precipitation deviations from the base period (for approximately a 100-year period till 1985). The probabilistic character of landscape-ecological prediction made it possible to reveal the local and regional diversity of responses of geo(eco)systems to the same background climatic signal, their stability, as well as the trajectories of their mutual and extrazonal transitions, being a multivalued pattern of climatogenic mechanisms of evolutionary processes.

As is known, "... all the diversity of biota is determined within a particular range of environmental factors" (Kerzhentsev, 2001 21). Landscape evolution is determined by global climate changes, on the one hand, and by shifts in the structure of the water balance, which is a matter-and-energy basis of the functioning of natural complexes, on the other hand (Dyakonov 2002). The super-century, thousand-year and million-year climatic fluctuations have been of paramount importance for biosphere evolution (Budyko 1984, 1990), creating conditions (over the entire geological epochs) for the natural selection of more perfect species, populations and biotic communities, as well as for the development of new ecological balances in nature, with higher bio- and geodiversity.

The important trend of landscape ecological analysis is the development of the problem of *chronoorganization* (Solntsev 1981 b) of the natural complexes of the Pacific Ocean mega ecotone, which is associated with external (in this case, climatic) regulation of their functional states. In this section we consider the climatogenic mechanisms of evolutionary transformations of forest ecosystems by using empirical simulation of this dynamics on the basis of landscape ecological predictions. The main goal was to establish how the local geo(eco)systems belonging to different structural subdivisions of the mega ecotone respond to century-long large-scale climatic changes – the leading exogenous landscape-forming factor, i.e., what are the regularities of interaction between natural-territorial structures of different age generations and the climatic system. The scientific search in this direction based on the property of temporary self-similarity of the operation system was aimed at determining the basic parameters of structural-functional organization and stability of local-level ecotone systems.

The known simulation models of forest responses to climatic impacts (Smith et al. 1992; Holten et al. 1992; Komarov et al. 2003; Tarko 2005; etc.) are in agreement with the narrow preset framework of habitat conditions, therefore the results of such modeling are insufficient for prediction of the state of forest community as a whole and do not cover the spatial diversity of succession changes under the same background influence. The local landscape-ecological prediction, which we have developed, is based on construction of *discrete empirical-statistical models* of natural geo(eco)systems (Krumbein, Graybill 1965; Rozenberg 1984; Kolomyts 2003, 2008). They are used to obtain probabilistic prognostic estimates of the behavior of regional landscapes and biogeocoenoses under various geomorphological, zonal-climatic and edaphic conditions of specific ecoregions. These models describe the category of self-organizing systems, which can adequately describe stabilizing selection as a response of the biota to climatic perturbations exceeding the adaptation threshold. In these models, the results of field observations are used as an empirical basis for modeling itself, rather than as reference data for testing results of calculations. This, first, minimizes the effect of the subjective factor in developing the model; second, provides a considerably higher spatial resolution than, e.g., simulation modeling; and, third, gives empirical grounds for wider geographic generalizations.

All predictive ecological models proceed somehow or other from the principle of actualism; however, they are unambiguously determined in some cases and probabilistic in other cases. The determined models do not take into account the diversity of states of subregional and local geo(eco)systems under the same background climatic conditions, which is especially typical of mountain territories. Ecological-geographical prediction is always more reliable and practically significant if it is characterized by certain stochasticity, diversity i.e., if it is probabilistic.

The previous generalization by the authors of field and office studies made it possible to develop a substantiated and internally consistent *ideology of local and regional landscape ecological predictions*. This ideology, also relying on authorities in its basic statements, can be summarized as follows:

1. The stable functioning of geo(eco)systems is provided first of all by the soil-biotic mechanisms of their monosystem organization and self-regulation (Sochava 1974, 1978). These mechanisms are based on *intercomponent interactions* – the vertical landscape connections realizing the stabilizing selection, which maintains the modal phenotype of this system (Pianka 1981), and providing it with certain *geotropic areal stability* (Kolomyts et al. 2015).
2. The impetus to geo(eco)system transformation at a new climatic signal is given by aggravation of *intercomplex interactions* (Zlotin 1987; Tishkov 1988), i.e., horizontal (lateral) connections and relations *in the system of landscape couplings* (Glazovskaya 1988), or catenas. The multivalued transformation of this signal by local geomorphological and hydroedaphic factors creates and aggravates a heterogeneous environment. This aggravation is determined by the multivalued transformation of a background signal in the elements of micro- and mesocatenas under the influence of local geomorphological and hydro-edaphic factors (Vysotsky 1960).
3. The developing disruptive selection promotes the formation of two and more modal phenotypes (Pianka 1981). There is inevitable *spatial diversity of local and regional* functional and structural *rearrangements* of ecosystems at the same background signal (Kolomytz 2003, 2008). This system-forming process is aimed at a new *evolutionary climax* of phytobiota (Krištofovič 1946) under the given zonal-regional conditions.
4. In case of disturbed catenary organization of ecosystems, *phytobiota* as the most dynamic geocomponent *plays the key transforming role, determines the "fate" of the entire ecoregion* (Semenov-Tyan-Shansky 1928; Simonov 1982). The main driving force of ecosystem rearrangements is *competitive relationship between the ecological niches of plant communities* (Svirezhev, Logofet 1978) in the space of new hydrothermal fields. Due to competition, there is ecological diversification, i.e., the division of niches that favors the development of more complex biotic formations (Pianka 1981). Evolution is implemented via migration processes in the plant cover (Vasilyev 1946) and, therefore, ecosystems change into the *new type of areal stability – migration*, which is accompanied by the enhanced efficiency of utilization of matter and energy resources, always more or less limited.

It has already been attempted to simultaneously apply the "micro-" and "macro-substrate" approaches to landscape analysis, which is a merely landscape-ecological problem. As is known, landscape ecology applies to interrelationship between the "ecological processes" underlying landscape heterogeneity and the landscape "pattern" formed at much larger scales than these processes (Turner and Gardner 1990).

The strategy of landscape-ecological prognosis can be put in the following algorithm: first to identify the selected ecosystem objects (either landscape or facie groups) to certain values of contemporary climatic conditions and then to estimate the most probable transformation of the revealed ecological (hydrothermal) niches of the given objects according of the expected climatic changes for a given prognostic date. For all that, it is suggested a plural of the transformation of ecosystems at a fixed value of the hydro-thermal trend, when a new state can share a number of trains of not one but several states which are known at present instants of time.

The general principle of the prognosis is expressed as follows: the magnitude of climate-induced transformation of one ecosystem into another increases with the decreasing degree of overlap of their climatic niches in the initial states, i.e., with the increasing contemporary contrasts of their functional states, as well as the overlap of niches after the exposure of ecosystems to the given climatic factor. For example, the transformation of object*A* into object*B*is greater if their positions in the multidimensional ecological space are more distant and they become closer to each other after the shift of object*A* in the coordinates of this space (given that the climatic niche of object*B* is unchanged). In this case, object*A* is *a reducer* and object*B* is *an absorber.* In turn, object*B* is transformed into object*C,* with the natural boundaries shifting accordingly; etc. As a result, the general pattern of ecosystem transformations in the region was obtained.

The maximum values of transformation were used to estimate the most probable directions and degree of ecosystem conversion. At the same time, the unambiguous character of transformation of regional ecosystems is proposed at a fixed value of geophysical trend, when the new state may have features of not one but several states existing at a given moment.

Note that *functional relaxation as the primary response of ecosystems to an external factor is a priority object of landscape ecological prediction.* The main prediction based on ecological estimation of the forthcoming global climate warming is expected to be functional rather than structural-morphological (Kolomyts 2003, 2008, 2012).

The prognostic landscape-ecological analysis was performed using a series of empirical models: (a) informational used as a basis of regional and local banks of ecological niches of geo(eco-)systems characterizing their parametric space and (b) fuzzy set theory-based describing the polysystem unit of nature-territorial organization by operations with ecological niches as descriptive vectors. Complex interconnections were revealed and the territorial ecological space of regional and local ecosystems was described on the basis of information-statistical and fuzzy set theory-based models.

16.2. EVALUATION OF PROBABILITIES OF CHANGES IN THE FUNCTIONAL STATES OF ECOSYSTEMS

The mechanism of estimation of probability of ecosystem transitions (for the predictions) is illustrated by Venn's diagram (Figure 16.1). Here, A_0 and A_1 are respectively an initial (contemporary) and final (future) values of the row vector of the ecological niche of object *A* (either landscape group or facie group), which must be absorbed by another object *B* if its niches intersect each other in the final state more than in the initial state. Landscape-ecological transition proceeds in the direction $A \rightarrow B$ and absorption proceeds in the opposite direction $B \rightarrow A$. The shaded area $a + b + c$ on the diagrams shows the total value of transformation of object *A*, which can be expressed in terms of the following measure of inclusion (Andreev 1980):

$$a + в = A_0 \cap A_1; \qquad в = A_0 \cap B_0; \qquad в + c = A_1 \cap B_0, \tag{16.1}$$

where "$\cap$" is the sign of intersection (logical product) of two sets. The elements of the intersection "zone" belong to two sets simultaneously.

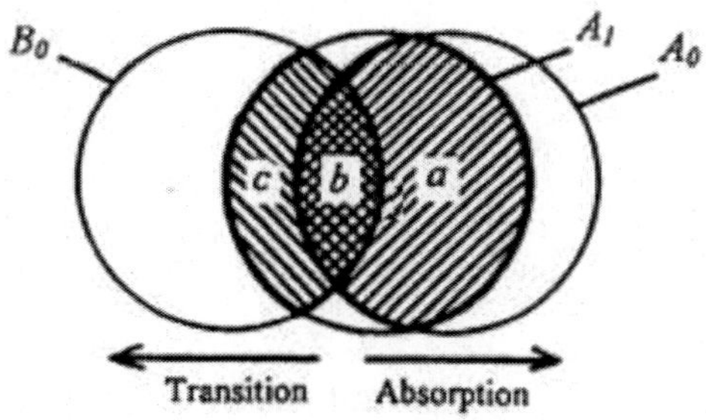

Figure 16.1. Venn's diagram illustrating the mechanism of probability estimation for landscape-ecological transitions (explanations see in the text).

It is supposed that any object *A* must be absorbed by another object *B*, if its niches intersect each other in the final state more than in the initial state. The initial (contemporary) A_0 and final (future) A_1values of the row vector of the climatic niche of object *A* are considered. In each studied pair of objects *A* and *B*, the two transformation parameters are calculated: the probability measure P_{ii} of stabilization of object *A*: $K(A_0 \rightarrow A_1)$; and the measure P_{ij} of its absorption by object *B*: $K(A_0 \rightarrow B_0)$. These parameters are expressed in the following measures of inclusion (Kolomyts, 2008):

$$K(A_0 \rightarrow A_1) = \frac{A_0 \cap A_1}{A_0}; \tag{16.2}$$

$$K(A_0 \rightarrow B_0) = \frac{A_1 \cap B_0 - A_0 \cap B_0}{A_0}. \tag{16.3}$$

Since the operations are performed with descriptive fuzzy sets, the operating formulas are in the form of:

$$K(A_0 \rightarrow A_1) = \frac{\sum \min\left[L_i(A_0), L_j(A_1)\right]}{\sum L_i(A_0)}; \tag{16.4}$$

$$K(A_0 \rightarrow B_0) = \frac{\sum \min\left[L_j(A_1), L_k(B_0)\right] - \sum \min\left[L_i(A_0), L_k(B_0)\right]}{\sum L_i(A_0)}. \tag{16.5}$$

Here, *i, j* and *k* are the order numbers of partial association coefficients in the row vectors describing the ecological niches A_0, A_1 and B_0, respectively.

By analogy, we calculated the values of transitions $K(A_0 \to C_0)$, $K(A_0 \to D_0)$, etc. For the following state, we have $K(B_0 \to B_1)$, $K(B_0 \to C_0)$, $K(B_0 \to D_0)$, etc. For the next stage of prediction, its own measures of transition were calculated: $K(A_0 \to A_2)$, $K(A_0 \to B_0)$, $K(A_0 \to D_0)$, …; $K(B_0 \to B_2)$, $K(B_0 \to C_0)$, $K(B_0 \to D_0)$, … .

The formulas for calculations at this stage are usually in the form of:

$$K(A_0 \to A_2) = \frac{A_0 \cap A_2}{A_0}; \tag{16.7}$$

$$K(A_0 \to B_0) = \frac{A_2 \cap B_0 - A_0 \cap B_0}{A_0}. \tag{16.8}$$

Thus, a specific quadratic matrix of probabilities of stabilization of each object (diagonal elements of the matrix, P_{ii}) and its transitions into other objects (P_{ij}) is obtained for each prediction period. Here, the probabilities may have zero and negative values. The former indicate the absence of transitions and the latter show the intensification of contrasts between the functional states of objects (at $A_1 \cap B_0 < A_0 \cap B_0$). Further operations are executed with transition probability matrices. All negative transitions are provisionally replaced by zeros.

If objects *A* and *B* are first-order neighbors, then the value of displacement of the boundary between them in the $A \to B$ direction is taken as proportional to the calculated measure of transformation of object *A*. In turn, object *B* can be transformed into the following objects (*D, C,* etc.). The evolution of ecosystems is usually multivariate; hence, all possible transitions have to be considered for each of them with revealing the maximum measures of transformation, which will indicate the most probable directions of transformation of landscape or phytocoenological structures as a whole. For first-order landscape neighborhoods, this process will be rather distinct, if $P_{cn}(B) > P_{cn}(A)$, i.e., given that absorbing object *B* has a higher power of climatic niche P_{cn} compared to absorbed object *A*.

16.3. Algorithm of Predictive Calculations for Regional Ecosystems

The first stage of our own prognostic analysis of regional landscape structure included the following procedures (Kolomyts 2003, 2008):

1. The information analysis of connections of each object (landscape group) with a particular climatic characteristic was used to obtain two matrices of partial association coefficients ($C(x_i/y_j)$): matrix T_0 for the base period and matrix T_1 for the given prognostic dates. The graduations of a given factor are on the columns of each matrix and the gradations of a given object are on the lines. Since the regional climatic prognosis was differentiated with respect to temperature and precipitation, we formed 4 pairs of matrixes: the mean January and July temperatures ($T_0(1)$ and

$T_1(1)$; $T_0(2)$ and $T_1(2)$ and the amount of atmospheric precipitation in the cold and warm periods ($T_0(3)$ and $T_1(3)$; $T_0(4)$ and $T_1(4)$.

2. The above formulas were used to calculate four square matrixes T(1), T(2), T(3) and T(4) of the measures of stabilization (resistant stability) P_{ii} of each *i*-th object (diagonal matrix elements) and the measures of its transformation P_{ij} into other *j*-th objects. The zero values of matrix elements show the absence of transition in the given pair of objects, and the negative values demonstrate the intensification of contrasts between them ($A_1 \cap B_0 < A_0 \cap B_0$).
   The P_{ii} and P_{ij} parameters represent the two alternative categories of *geo(eco)system stability: areal* (geotopic) and *migration* (transgressive), which characterize the state of stable equilibrium of the system and its forced behavior in response to external effects, respectively.
   For simplification of predictive calculations, the above condition $P_{cn}(B) > P_{cn}(A)$ is not taken into consideration at the given stage of the described algorithm development, which certainly reduces the reliability of results.
3. Simultaneous changes in all four (or three) cited climatic factors must be taken into account; then the first factor-weighted average matrix T_{wa} of object transformation is calculated on the basis of matrices T(1), T(2), T(3) and T(4). The "weight" coefficients for its calculation were the values of the $C(X;Y)$ parameter normalized once again to obtain a linear polynomial of dependencies between the distributions of objects and the geophysical factors under consideration. For the groups of regional landscapes in the given region of Greater Caucasus, the prognostic calculation matrix $T_{wa}(1)$ was calculated by the linear polynomial with the following coefficients:

$$T_{wa}(1) = 0.241 \cdot T(1) + 0.315 \cdot T(2) + 0.126 \cdot T(3) + 0.318 \cdot T(4). \quad (16.9)$$

4. At this stage of operations with the matrices, it is necessary to take into account the probable presence of "residual" transitions of a particular object into other prototype objects absent in our operational system (sampling), i.e., to extra-sampling objects. For some facial groups, these transitions seem to be dominant and even unique. Such transitional "residue" $P_{ij}(x)$ for each object (i.e., for each line of the weight-average transition matrix) can be easily found from the ratio:

$$P_{ij}(x) = 1 - P_{ii} - \sum P_{ij}. \quad (16.10)$$

The $P_{jj}(x)$ values are introduced as an additional column into matrix T_{wa}. If the "residue" is negative, it means that all predicted transitions of the given object are within the limits of the considered set of prototype objects and there are no other transitions. Negative "residues" of transitions change into zero.

5. Comparable values of the probabilities of transition of a particular object into different prototype objects can be obtained, if the sum of these probabilities in each line of the matrix is equal to 1. After normalizing each of the five T_{wa} matrices (with "residual" columns) by lines, we will obtain the second weight-average matrices $T_{wa}(2)$. The lines of the matrix with positive transition "residues" prove to be normalized already at the preceding (4th) stage of calculations.

6. In conclusion, the correcting procedure is performed. Matrix $T_{wa}(2)$ contains quite a lot of elements with very low (below 1–2%) probabilities going beyond the limits of measurement and calculation accuracy. It is necessary to calculate the guaranteed minimum of probabilities of landscape-ecological transitions: M_{min}. The simplest method for finding the guaranteed minimum of elements is based on application of one-sided criterion of their significance, $t = \sigma$, which provides the 5% level of significance (Mitropolysky, 1971). All values of M transitions or deviations (P_{ii} or P_{ij}) below the difference of $M - \sigma$ must be eliminated. For this purpose, we calculate $M_{min} = M - \sigma$and drop all vector elements below this difference.

Thus, each object acquires a much shorter transition vector. Residual vector elements are normalized again and, thus, we obtain the third matrix of transition probabilities average weighted by the given geophysical factor: the $\overline{T}_{wa}(3)$ matrix. It gives a rather clear notion of the potentials of transformation of particular objects or, on the contrary, their ability to resist external influence. The orientation graphs of probabilities of functional transitions between the objects are plotted on the basis of $T_{wa}(3)$ matrices (Figure 16-2). These graph-analytical models are the main tool of landscape-ecological prediction. They give the most general notion of the exogenous succession dynamics of ecosystems as whole natural-territorial complexes.

16.4. Calculations of the Rates of Ecosystem Transformations

The second stage of the analysis was as follows: the potential rates of transitions and the average times of full or partial absorption of the hydrothermal niche of one examined object (either landscape or facie groups) by the niches of other objects were calculated. For this purpose, we used the mathematical apparatus of Markov's chains theories. Markov's models are used in geology (Harbaugh & Bonham-Carter 1970), physical geography (Viktorov 86), ecology (Jeffers 1978), etc. They describe nature processes as random by determining the influence of each preceding state on each subsequent state, which is in agreement with the peculiarities of dynamics of geo(eco-)systems. Previously, the known methods have been used to ascertain that the studied chains of geo(eco)system transitions have Markov's properties.

The rates of landscape transitions were calculated by the model of W. Krumbein (Harbaugh & Bonham-Carter 1970). The rates h_{ij} (and, consequently, times) of transitions are defined by way of non-complicated algebraic transformation of the initial normalized matrix of transition probabilities, where both the relative transition probabilities (P_{ij}) and the a priori probabilities (P_i) (see below Table 17.2) are to be presented. The total rates of transition $M_{i\_}$ from the given state in one step, i.e., within a time interval $d = 1$, is calculated first of all. In general,

$$M_i = -\ln P_{ii} / d, \text{ and by } d = 1 \; M_i = -\ln P_{ii}. \quad (16.11)$$

The M_i values are put into the next additional column of the matrix.

Here, d is the number of steps. In this case, we have three steps: 1985–2050, 2050–2100, and 2100 – 2200. As is known (Albritton et al. 2003), the modern global warming began in the middle/end of the 80s of the XX century.

The expected total time of residence of the system in state P_{ii}, i.e., up to the moment when it will give way to all other states, like in:

$$\tau(P_{ii}) = 1 / M_i. \quad (16.12)$$

Here, $\tau(P_{ii})$ is expressed in conventional units – numbers of steps. For example, it may be expressed in the degree of plant transformation or in the percentage of area of a given object per one step of reduction of this area. At $\tau(P_{ii}) \leq 1$, it is evident that the given area would be entirely absorbed already at the first step. Multiplying the $\tau(P_{ii})$ value by the time equivalent to the first step (20–100 years in our case), we obtain the expected absorption time $T(P_{ii})$ (in years), which conceptually makes the final time of full transformation of the given object.

Then two square matrices are created for the relative and absolute rates of transitions of objects into each other. In case of $M_i = 1$, the relative partial rates of transitions m_{ij} from the i-th to j-th state are found by the correlation:

$$m_{ij}= P_{ij} / P_i. \quad (16.13)$$

Next, it is essential to calculate h_{ij} – the absolute partial rate of transition of the i-th object to the j-th object (as a conditional probability). This rate is equal to:

$$h_{ij} = m_{ij} \cdot M_i. \quad (16.14)$$

The h_{ij} values are calculated on all lines of matrix P_{ij} (except for its diagonal elements). The relation h_{ij}/M_i will show us the part of the total one-step reduction of the area of the i-th object (or the part of any other attribute of its transformation) that accounts for transgression of the j-th object. Since the Markov's process is stationary, this part remains on whole transition steps. Hence, it is easy to calculate what percentage of the area of the given object will be absorbed by the areas of other objects at time intervals (steps) of 20, 40, 60 years, etc.

The matrices of probabilities of mutual transitions P_{ij} of the objects were constructed for all prognostic intervals. However, in further estimations of transition rates on Markov's chains the duration of intervals itself, that is the number of steps (d), was taken into consideration. In particular, in h_{ij} calculations it is conventional that $d = 1$ for the first prognostic interval, $d = 2$ for the second prognostic interval, $d =$ for the third prognostic interval, etc.

The instability of Markov's chain has been considered by varying parameter h_{ij} at each next step of geo(eco)system transitions. For the first step, the h_{ij} value was calculated on matrix P_{ij} for the first prognostic interval (e.g., up to 2050), in the second step on matrix P_{ij} for 2100, etc. As a result, the chain of landscape transitions with an increasing total has been lined up.

16.5. Construction of Predictive Models

The matrices of relations of inclusion of an object with other objects of the given series (floristic phratries, groups of forest types, or microlandscapes) are used to calculate the *transit index* of the k object $\boldsymbol{S_k}$. For each object, the number of input connections (n_i) and the number of output connections (n_j) were calculated. In addition, the total number of connections in the systems (N) was calculated too. On this basis, the transit index $\boldsymbol{S_k}$ of an object was determined as

$$S_k = P_i - P_j + 1, \tag{16.15}$$

where $P_i = n_i/N$, $P_j = n_j/N$. Obviously, $0 \leq S_k \leq 2$.

With respect to the $\boldsymbol{S_k}$ parameter, the groups of forest phytocoenoses were subdivided into certain *transit ecotypes* (Odum 1983). At $S_k >> 1$, an object is included in the group of absorbers. It is a suboptimal category – a *subdominant,* which is referred to as *submacrobolic ecotype* in the functional aspect. If, in this case, $n_j = 0$, the object is included in the optimal category, becoming a *dominant – macrobolite.* With a decrease in its transit index, the object becomes first a *rank-and-file (temporal) member* of the given geospace, i.e., *mesobolic ecotype.* Finally, at $S_k << 1$, it proves to be mostly reducer and, therefore, passes over to the subpessimal and pessimal categories of *reduced ecotypes – submicrobolites* and *microbolites.* In the latter case, it has no input connections in the structural scheme of dominance ($n_i = 0$).

The predictive ecological analysis was performed for two regions: (1) island-arc – the territory and surroundings of the active Mendeleev Volcano on Kunashir Island of the South-Kuril Ridge; and (2) marginal-continental – the Komsomolsk Nature Reserve in the Lower Priamurye. Let us note the two main peculiarities of the performed predictive calculations. First, the following three climatic parameters were used as predictors: (1) the mean temperature of the warmest month ($t_{\text{VII–VIII}}$), (2) annual precipitation (r_{yr}) and (3) Vysotsky-Ivanov's atmospheric humidity factor (F_{hum}) as a ratio of annual precipitation to annual potential evaporation. Second, we took into consideration the probability of "residual" transitions of an object into other prototype objects absent in the given operation system (sample), i.e., into *extrazonal* (Sprygin 1986), or *extra-sample,* objects. These objects may be associated with different zonal-regional conditions, often very distant from the conditions of the given territory.

For the territory of the Lower Priamurye, according to the data from weather stations, the following linear connections of temperature and precipitation with the absolute height (H_{abs}, in the framework of low-mountain/valley relief) were obtained, with high correlation coefficients R:

$$t_{\text{VII}} = 21.6 - 0.009 \cdot H_{\text{abs}};\ R = 0.918; \tag{16.16}$$

$$r_{\text{yr}} = 1.187.3 + 0.489 \cdot H_{\text{abs}};\ R = 0.892. \tag{16.17}$$

Table 16.1. Komsomolsk Nature Reserve. The matrix of normalized partial coefficients of connection of floristic phratries and groups of forest types with absolute height

The steps of absolute elevation, m.	Floristic phratries (*C(A;B)* = 0.126)						Groups of forest types (*C(A;B)* = 0.103)					
	An	**Mm**	**Mx**	**MO**	**MA**	**Okh**	**FB**	**CrBr**	**Br**	**SBr**	**Lar**	**Fs**
20–40	**0.47**	**0.36**	0	0.11	0.21	0	**1.00**	**0.26**	0.18	0.15	0.09	0.06
40–110	0.12	0.24	**0.39**	0.21	0.21	0	0	0.17	**0.37**	**0.30**	0.09	0.06
110–190	0.17	0	0.18	**0.30**	0.15	0.17	0	0.17	0.18	**0.30**	0.36	0
190–310	0.24	0	0	0.21	**0.42**	0.12	0	0.14	0	0.25	**0.45**	0.09
310–450	0	**0.40**	**0.43**	0.18	0	0.10	0	**0.25**	0.26	0	0	0.24
450–680	0.16	0	**0**	0	0	**0.61**	0	0	0	0	0	**0.56**

Footnote. The domination fields of ecological niches are in bold.

On the other hand, there are quite definite regularities of basic altitudinal-zonal distribution of phytocoenological units over the territory of the Komsomolsk Nature Reserve (Table 16.1). Thereby, there is the *predetermined rather high sensitivity of these units to the changes in the background bioclimatic system, which must substantially enhance the representativeness of predictive landscape-ecological models.*

The third index is a function of the former two, mostly of the mean July temperature. Thus, the following linear connection was obtained for the territory of the Lower Priamurye according to weather station data:

$$F_{\text{hum}} = 4.37 - 0.138 \cdot t_{\text{VII}}; \qquad R = 0.850. \qquad (16.18)$$

Nevertheless, the atmospheric humidity factor plays an independent system-forming role, as is emphasized by the values of its "weight" coefficients (Table 16.2).

In the South Kunashir island-arc ecoregion, the sensitivity of phytobiota and microlandscapes to climatic changes must be much more marked than in the Lower-Amur marginal-continental region. It has already been mentioned (see Chapter 6) that the Mendeleev Volcano is also characterized by quite clear and full manifestation of altitudinal-zonal bioclimatic zonality, in spite of its relatively low height. The bioclimatic altitudinal zonality on the slopes of the Mendeleev Volcano is directly subordinated to the vertical hydrothermal gradients, as is demonstrated by very high values of the normalized coefficient of interrelation *C(A;B)*.

Table 16.2 shows that the distribution of phytocoenological and landscape units in the island-arc volcanic landscape of Neo-Pacific is much more tightly related to background hydrothermal conditions compared to the Sub-Pacific marginal-continental landscape. Accordingly, in the former case one should expect much more cardinal functional changes and then structural transformations of plant communities and landscape systems in the presence of climatic signals of similar intensity. The results of predictive calculations have confirmed these expectations.

Table 16.2. The normalized partial coefficients of interrelation *C(A;B)* between different natural-territorial units of experimental plots and climatic characteristics of the basal period, as well as their values as "weight" coefficients (in parenthesis)

***a*) Mendeleev insular volcanic landscape (Kunashir Island)**			
Natural-territorial units	**Climatic characteristics**		
	Mean August temperature	**Annual precipitation**	**Atmospheric humidity factor**
Floristic phratries	0.405 (0.318)	0.543 (0.426)	0.327 (0.256)
Groups of forest types	0.308 (0.321)	0.381 (0.398)	0.296 (0.281)
Groups of microlandscapes	0.313 (0.294)	0.457 (0.428)	0.295 (0.278)
b) Komsomolsk Nature Reserve (Lower Priamurye)			
Natural-territorial units	Climatic characteristics		
	Mean July temperature	Annual precipitation	Atmospheric humidity factor
Floristic phratries	0.088 (0.317)	0.106 (0.381)	0.084 (0.302)
Groups of forest types	0.118 (0.358)	0.115 (0.350)	0.096 (0.292)
Groups of microlandscapes	0.223 (0.350)	0.190 (0.299)	0.223 (0.351)

The landscape-ecological prediction presented below should be premised by the following important note. The supposed functional-structural shifts in geo(eco)systems that can be determined by climate changes (in this case, anthropogenic changes), with a period of fluctuations of 50 years and more, characterize not the future natural complexes per se but rather the landscape-ecological conditions and, accordingly, the limit of ecological equilibrium, which real biogeosystems will tend to in their variations. The time of reaching this equilibrium calculated in the number of steps τ(s) or in the number of years τ(y) relates only to the functional relaxation but not to the full period of structure transformation. The latter will be determined not only by the intensity of external effect but also by the characteristic times of different natural attributes (Armand and Targulyan 1974). For forest landscape, as is known, the prediction step not exceeding the lifetime of one generation of forest stands makes it possible to determine potential forest growth conditions.

Chapter 17

CLIMATOGENIC DYNAMICS AND SUSTAINABILITY OF FOREST ECOSYSTEMS

ABSTRACT

The quantitative analysis of responses of forest ecosystems in different sectors of the Pacific Ocean mega ecotone of Northern Eurasia to the forecasted climate changes by the global prediction models E GISS and HadCM3 for the forthcoming 50–200 years has been presented. The resulting prognostic ecological scenarios were compared with analogous predictions for the intracontinental regions of Eastern Europe. Two predictive climatic trends have been examined: cold-arid and thermoarid. The Mendeleev island-arc ecoregion (Kunashir Island) is characterized by the exceptionally high rate of functional absorption of the areals of almost all basal plant communities by extrazonal objects characterizing already the regional scales of climatogenic transformation of the biogeocoenological structure of island-arc landscape. In the marginal-continental low-mountain Lower Priamurye, against the background of analogous but much weaker phytocoenological transformations, buffer forest communities will be most developed.

The problem of ecosystem stability in forest ecology is briefly reviewed, and the applicability of well-known stability mathematic models for landscape-ecological systems is evaluated. The ideology and methodology of quantitative assessment of stability of forest ecosystems are presented. The parameters of resistant and elastic-plastic sustainability for forest geo(eco)-systems of Mendeleev island-arc landscape and the Lower Priamurye marginal-continental region were calculated. The chronological-organizing role of sustainability of forest communities in their climatogenic transformation was revealed.

17.1. CLIMATE PREDICTION MODELS

The hydrothermal trends for the period up to 2150were taken from two global coupled atmosphere ocean general circulation models (from the family of AOGSMs models): (1) GISS Model E (Goddard Institute of Space Search, USA) – more moderate (Hansen et al. 2007); and (2) HadCM3, Version A2 (Hadley Centre of Climate Research, UK) – more extreme (Gordon et al. 2000; Pope et al. 2000).The first model simulated quite well the real climate changes in the period of 1880–2003, especially in the past 30–40 years. It gives an increase in global temperature by 2.7–2.9°C with the doubled CO_2 in the atmosphere, which corresponds to the empirical climate sensitivity range determined on the basis of paleogeographical data. The model predicts the relatively temperate global warming. It is one of the most preferable general circulation models, according to the results of testing of statistical significance of data obtained and simulation of the state of contemporary climate, including its seasonal characteristics (Santer 1985). It has been successfully used in a number of prognostic-ecological researches (Izrael 1992; Holten et al. 1993; Smith & Shugart 1993; etc.).

The second model is a combined model of the atmosphere-ocean general circulation adequately reflecting weak climate changes. The A2 scenario was developed for the conditions of more rapid population growth but relatively slow technical progress, when the technogenic influx of greenhouse gases into the atmosphere decreases insignificantly (Albritton et al. 2003). This feature differentiates it from the B2 scenario and thus yields the stronger anthropogenic warming. The HadCM3 model is one of the most extreme models predicting the maximal warming for the recent 10,000 years. The results of numerical simulation (Albritton et al. 2003) envisage a much greater increase in the mean annual temperature compared to the mean global increase by 5–6°in version A2 for the entire East Europe (including the entire Volga River basin), West and East Siberia in the period of 1971–2100 compared to the period of 1961–1990. Ecological prognosis by the second model makes it possible to estimate the response of forest ecosystem to global climatic changes that may exceed the level of tree endurance and cause disintegration of both nemoral and boreal forests over vast territories.

Table 17.1. The comparison of predicted changes in the climatic parameters of intracontinental and marginal-continental regions of Northern Eurasia, as well as island-arc Neo-Pacifica

Region	Experimental plot (predictive period, yr)	Deviations of climatic parameters from basal values		
		$\Delta t_{VII-VIII}$, ^{0}C	Δr_{yr}, mm [Δr_{yr}]	ΔF_{hum}
(a) according to global model E GISS				
Middle Povolzhye	Kerzhensky Nature Reserve (2050)	– 0.8	– 67	– 0.13
	Prioksko-Terrasny Nature Reserve (2050)	– 1.9	– 67	– 0.02
Central Caucasus, Northern Prielbrusye	Zhilasu stow, the alpine belt (2100)	+1.1	– 24	– 0.21
Lower Priamurye	Komsomolsky Nature Reserve (2100)	– 0.2	– 53	+0.16
South Kunashir	Mendeleev Volcano (2100)	– 0.6	– 160	– 0.33
(b) according to global model HadCM3				
Middle Povolzhye	Nizhegorodskoye Predvolzhye (2100)	+ 7.1	– 100	– 0.83
Central Kunashir	Mendeleev Volcano (2050)	– 4.9	– 314	+ 1.2

The scenarios of phytocoenological transformations and the respective changes in landscape structures on the Pacific Ocean mega ecotone of Northern Eurasia predicted on the basis of global climatic trends describe a rather specific and complex system of landscape-ecological transitions of two hierarchic levels – regional and local. For assessing this system, we will first touch upon the known propositions of geoecology with regard to hierarchic regularities of geo(eco)system dynamics, with the involvement of our empirical materials for other ecoregions of Northern Eurasia. Table 17.1 presents the basic characteristics of climatic predictions for ecoregions under consideration by the two global models: moderate EGISS and extreme HadCM3, version A2.

17.2. Intracontinental Regions of East Europe

It has been shown by the example of the Central Caucasus (Zalihanov et al. 2010) that regional intracontinental landscapes undergo generally much weaker transitions into each other compared to their internal modifications resulting from the mutual transformations of constituent geo(eco)systems of topological dimension. As is known (Theurillat et al. 1998), the regional effects of global warming in mountains are traced back mainly by the general shift of the boundaries of vegetation zones, while the local effects are traced back by the dynamics of intrazonal structure of phytobiota. Substantial differences in the rates of response of regional and local ecosystems to the same climate changes are determined by their different times of relaxation: "… the higher is the rank of subdivision of natural environment, the greater is its age" (Sochava 1978 196) and, in continuation, the longer it retains the previous system of landscape connections at the same external impact, i.e., the slower it changes. If we consider geo(eco)system stability "… as a ratio of variability in the environment to variability in the system itself" (Ricklefs 1976 378), it is obvious that regional natural complexes must be more stable than local ones. Their internal organization is less dependent on the state of external environment, which decreases the rate of their response to the same external impacts. These are the common features of the hierarchic system of climatogenic dynamics of intracontinental geo(eco)systems that have had a long history of subaerial development.

Table 17.2. The generalized characteristics of probabilities and rates of functional and structural transitions for landscape groups of the northern macroslope of Greater Caucasus for the prognostic period of 2100, according to the thermo-arid climatic trend by E GISS (Zakihanov et al., 2010)

Altitudinal belt	Statistical parameter				
	P_{ii}	P_i	P_{ij}(mean)	τ (steps)	τ (years)
Piedmont	0.33 – 0.92	0.08 – 0.45	0.13 – 0.33	3.0 – 25.3	327 – 2830
Lower- and middle-mountain	0.39 – 0.75	0.25 – 0.61	0.11 – 0.14	3.3 – 8.0	359 – 882
Upper-mountain	0.40 – 0.46	0.54 – 0.60	0.12 – 0.13	3.3 – 3.7	364 – 404
High mountain	0.55 – 0.91	0.09 – 0.45	0.07 – 0.14	4.4 – 22.7	488 - 2498

Note. P_{ii}, the probability of stabilization of object i; P_{ij}, the probability of its transitions into another object j; P_i, the total probability of transitions of object i into all other objects; τ (steps) and τ (years), the total time of object staying in condition P_{ii} expressed in the number of steps and in years.

Let us demonstrate this fundamental geoecological regularity by the example of multiscale natural complexes of Greater Caucasus by comparing the climatogenic dynamics of regional landscapes and the constituent topological-level geosystems. In general, the regional landscape systems of this mountain country have a comparatively high resistance to climatogenic changes under study (the dominant values $P_{ii} \geq$ 55–75%, see Table 17.2), while the relative rates of their mutual transitions are very low (P_{ij}(mean) $\leq$ 15–30%). Obviously, the time of relaxation of the Caucasus regional landscapes is much greater than the half-period of these climatic fluctuations. According to the EGISS model, even by the end of the prediction period (2100) the internal rearrangement of high-mountain landscapes will prove to be yet insufficient for them to anyhow substantially lose their appearance and start to be entirely transformed into upper- and middle-mountain landscapes.

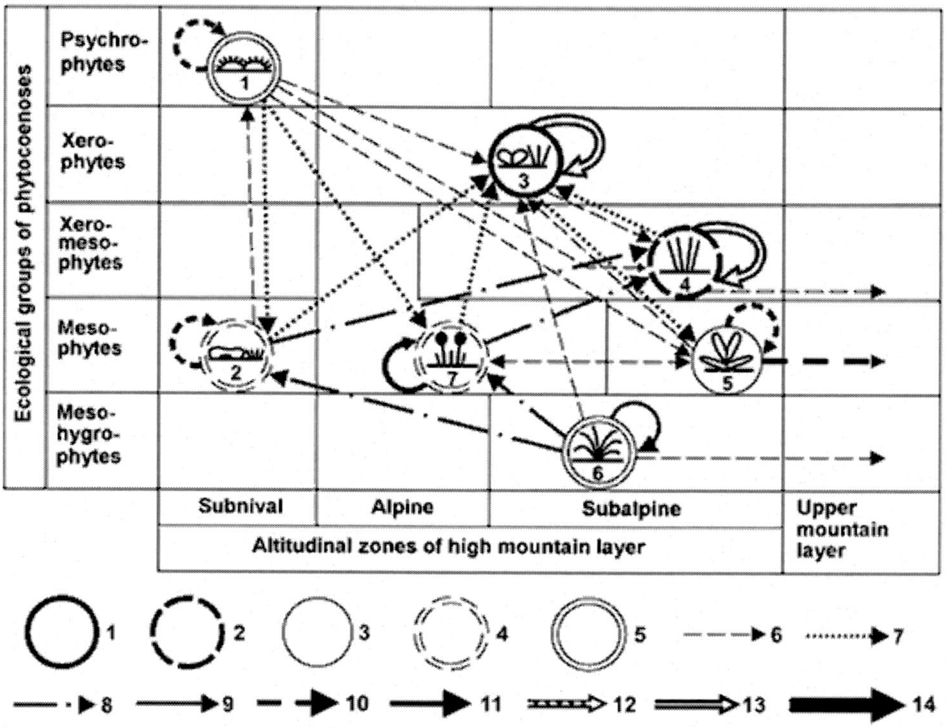

Figure 17.1. The northern vicinities of Mount Elbrus. The Zhilasu testing area. The oriented graphs of structural-functional transitions between the coenotic groups of alpine meadows (see in the text) located in different altitudinal layers and under different edaphic moistening conditions for the prognostic year 2100 according to E GISS climatic model. *(1–5).* Transitional ecotypes of meadow communities: (1) absorbers, (2) subabsorbers, (3) mesotrophs, *(4)* subreducents, (5) reducents *(6–14).* The probabilities of stabilization *(P_{ii})* and mutual transitions *(P_{ij}): (6)* 0.01–0.10; (7) 0.11–0.20; (8) 0.21–0.30; (9) 0.31–0.40; *(10)* 0.41–0.50; (11) 0.51–0.60; (12) 0.61–0.70; (13) 0.71–0.80; *(14)* 0.81–0.92.

Quite a different situation is typical of local-level high-mountain ecosystems, the brilliant example of which is the Zhilasu test site in Northern Prielbrusye (Figure 17.1). Even at a relatively weak climatic signal given by the EGISS model, the measure of stabilization P_{ii} of the most of high-mountain/meadow biogeocoenoses (BGC) decreases to 35–50%, and the probabilities of their mutual transitions prove to be several times higher than in the regional high-mountain landscapes in general (P_{ij}(mean) varies from 5–10% to 20–25%). The time of relaxation of topoecosystems is much more conformal to the half-period of climate changes under consideration and, therefore, their sensitivity to hydrothermal signals must be comparatively high. Thus, the alpine carpets *(Kobresia/*lichen, lichen/*Alchemilla/Kobresia)* *and* barren meadows (*Kobresia/Alchemilla/Sibbaldia*) of the subnival belt (biogeocoenosis group 1) will be transformed not only into the adjacent alpine carpets (group 2) but also into relatively distant subalpine grass communities. By the middle of the XXI century, alpine carpet per se will have been replaced by 20–28% with more xerophytic communities analogous to forb-cereal and fescue–sedge subalpine meadows (groups 3 and 4, *Alchemilla, Kobresia/Alchemilla*). Only weed communities and subalpine sedge meadows (groups BGC 5 and 6, *Veratrum/Urtica, Trifolium/Veratrum, Veratrum/Alchemilla*) can be transformed by 45–65% and 12–28%, respectively, into their extrazonal upper- and mid-mountain analogs.

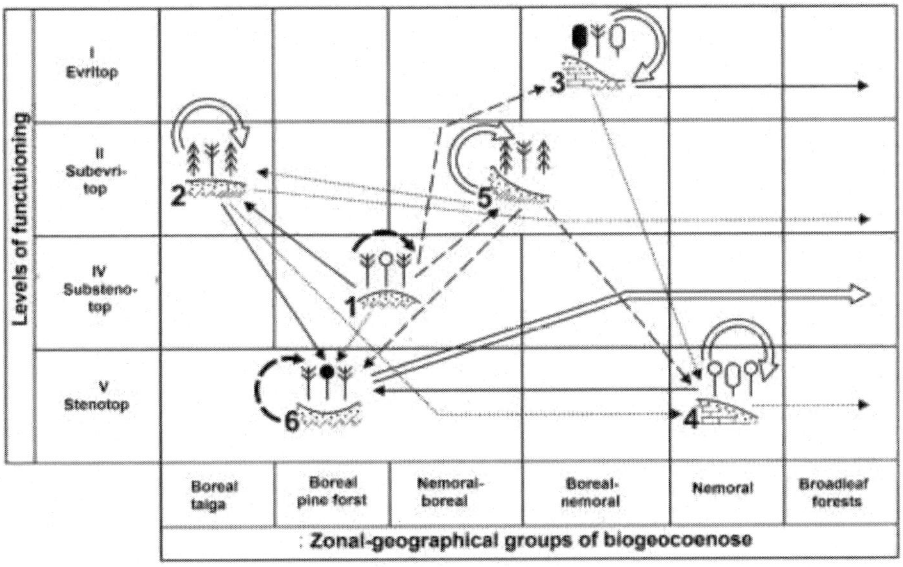

Figure 17.2. Middle Povolzhye. The E GISS climate prediction model for 2050. The ecoregion of the Prioksko-Terrasny Biosphere Reserve. The cold humid climatic trend. The orgraphs of functional landscape ecological transitions between the groups of facies (biogeocoenoses) belonging to different zones and being at different levels of basal functioning.

Here, let us give an example of predicted landscape-ecological consequences of the cold-arid climatic trend (according to E GISS) in the south of the subtaiga zone of the Middle Povolzhye, in the Prioksko-Terrasny Biosphere Reserve (Figure 17.2). Here, by 2050, against the background of rather high total stability of nearly all biogeocoenoses (the measure of stabilization is more than 60%), there will be a certain tendency (up to 25%) to transformation of xeromorphic transeluvial sandy green-moss reedgrass pine bors (biogeocoenosis group 2) into the neighboring eluvial meso-hydromorphic morain shamrock/motley grass spruce forests and spruce-pine forests (groups 5 and 2). The latter, in turn, will begin to be transformed with still higher intensity (about 30%) into hydromorphic eluvial-accumulative coniferous/small-leaved dwarf-shrub/sphagnum communities of interfluvial depressions with sandy loam/loamy soils (group 6). The more well-drained mesomorphic transeluvial and transit oak/lime/birch motley grass and broad-grass forests on limestone-loamy substrate (group 4) will change in the same direction (by 27%).

There will also appear the distinct signs of transgression of extra-sample south taiga communities into this ecoregion. These communities will penetrate mostly along river valleys, substituting wetland forests by more than 65%. The second, less intensive wave of transgression (up to 30%) will encompass the moderately drained watersheds and slopes with mesomorphic pine-lime-oak broad-grass forests on gleyey sandy-loam soils (BGC group 3).

Thus, even the relatively low cold-arid climatic signal, with the annual humidity factor changing from 1.52 to 1.78–1.79, can cause a distinct though rather localized process of distribution of forest and forest-swamp associations of south taiga over nearly the entire territory of the subtaiga zone. Nevertheless, the general landscape structure of this intracontinental ecoregion remains almost unchanged. Its climatogenic dynamics is confined mainly to the mutual transitions between its local elements, and anyhow significant regional shift is not observed.

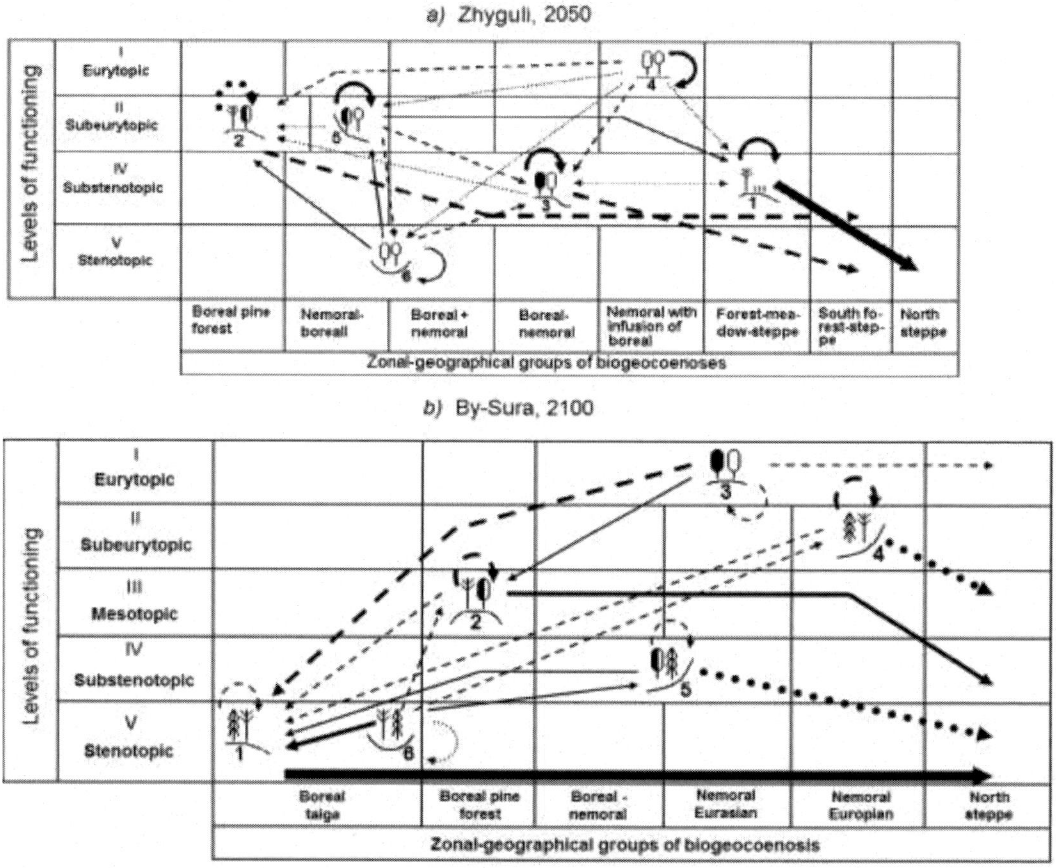

Figure 17.3. Vector graphs of functional transitions between facies groups: *a* – in the zonal forest and steppe ecotone (Zhiguli low-mountain range); *b* –in sub-taiga/forest-steppe ecotone (high right-bank By-Sura), according to GISS model. Transitions are designated as in Fig. 17.1. Levels of biogeocoenosis functioning was called, according to classification of (Odum 1971). Other designations see in the text.

In intracontinental regions, only the exceptionally strong climatic signals are able to transform local landscape-ecological transitions into regional ones. Such transformations may be associated, e.g., with the known extreme climate model HadCM3 giving a 5–7° increase in the mean January and July temperatures in the Middle Povolzhye by 2100, with a comparatively minor increase in precipitation (see Kolomyts 2008 41). Ecological prognosis by this model makes it possible to assess the forest ecosystem response to such global climatic changes, which can exceed the level of tree endurance and cause disintegration of both nemoral and boreal forests over large territories.

According to our estimates, there will be a century-long decrease in F_{hum} from 0.95–1.07 to 0.36–0.45 in the forest-steppe regions of Zhiguli and By-Sura and from 1.37 to 0.65 in the Trans Volga–Kerzhenets subtaiga region, indicating a sharply marked thermo-arid climatic trend. The progressive summer drying of soil will begin – the event expected in the south of the temperate belt under global warming by nearly all climate prediction models (Houghton et al. 1996). In this context, we expect a rather drastic starting "leap" of transformation of forest ecosystems for the entire middle zone of the Russian Plain (Figure 17.3). The functional stability of boreal and nemoral forests can be disturbed already by the middle of the XXI century.

Under the conditions of extreme global warming, the process of aridization of broad-leaved forests must progressively accelerate in the following exodynamic succession series: pure oak groves and spruce/oak forests → pine forests and spruce-pine subors → typical coniferous and small-leaved forest steppe → northern and middle steppe → southern (dry) steppe. Steppification of oak forests must certainly proceed via the intermediate stage of boreal, mostly pine forest, forest and light-forest associations.

The mixed forest zone of the Russian Plain differs from the nemoral-forest subzone in the higher resistance of forest communities to global warming. General climatic thermoaridization will cause local transitions directed mainly bottom-up along the local catenas – from superaqual and accumulative sites to transeluvial and eluvial sites. Under the conditions of lowland outwash plains, aridization of coniferous forest communities will cause their transitions to the states of typical and southern forest steppe. Xeromorphic and meso-xeromorphic spruce/pine forests of the upper elements of catenas will be exposed to steppification in the first place. According to the HadCM3 model, the communities of northern and middle steppes will develop here already in 2050.

17.3. Island-Arc Volcanic Landscape

At the initial island-arc stage of formation of the continental biosphere, the response of young volcanic and post-volcanic landscapes to fluctuations in the global climatic system was substantially different from the above-described response typical of intracontinental plain and mountain landscapes that have already had a long and very complicated history of subaerial development. These differences are determined by the two most important features of the island-arc Neo-Pacific bioclimatic system in the south of the boreal belt. It includes: (1) the simultaneous decrease and compression (compared to the neighboring continent) of the system of bioclimatic altitudinal zonality, with its almost entire spectrum (from forest belts to the subgoletz zone) falling within the limits of the low-mountain stratum of macrorelief; (2) the coverage by local structural and functional contrasts of the taxonomic standards of soil-biotic characteristics of quite a number of superior zonal-regional units; such geotopical contrasts on catenas quite definitely simulate the altitudinal-zonal and circulation-exposure differentiation of hydrothermal conditions. Thus, in the framework of low-mountain island-arc landscape, a particular *pseudoregional spectrum of altitudinal belts* is formed, which can be rightfully considered as a focus of origin of regional bioclimatic zonality typical of the mountain countries of continents.

Table 17.3. The changes in basic climatic parameters of the Mendeleev landscape (Yuzhno-Kurilsk station) predicted by EGISS and HadCM3 global models

Time period	Climatic parameters								
	Air temperature (t), ^{0}C			Sum of the biological active temperatures ($\sum t \geq 10^O$), ^{0}C	Annual potential evaporation (E_o), mm	Atmospheric precipitation, mm			Humidity factor (F_{hum})
	t_{yr}	t_{II}	t_{VIII}			r (year)	r (cold period)	r (warm period)	
Base period	4.8	– 6.0	15.8	1519	304	1319	410	909	4.33
Climate prediction model E GISS									
2100	4.3	– 6.6	15.2	1459	292	1159	278	887	3.97
2200	4.1	– 6.8	14.7	1411	282	1069	231	852	3.79
Climate prediction model HadCM3									
2050	3.4	–3.2	10.9	1048	209	1005	458	432	4.81
2100	6.3	–0.5	16.8	1615	323	1066	511	448	3.30

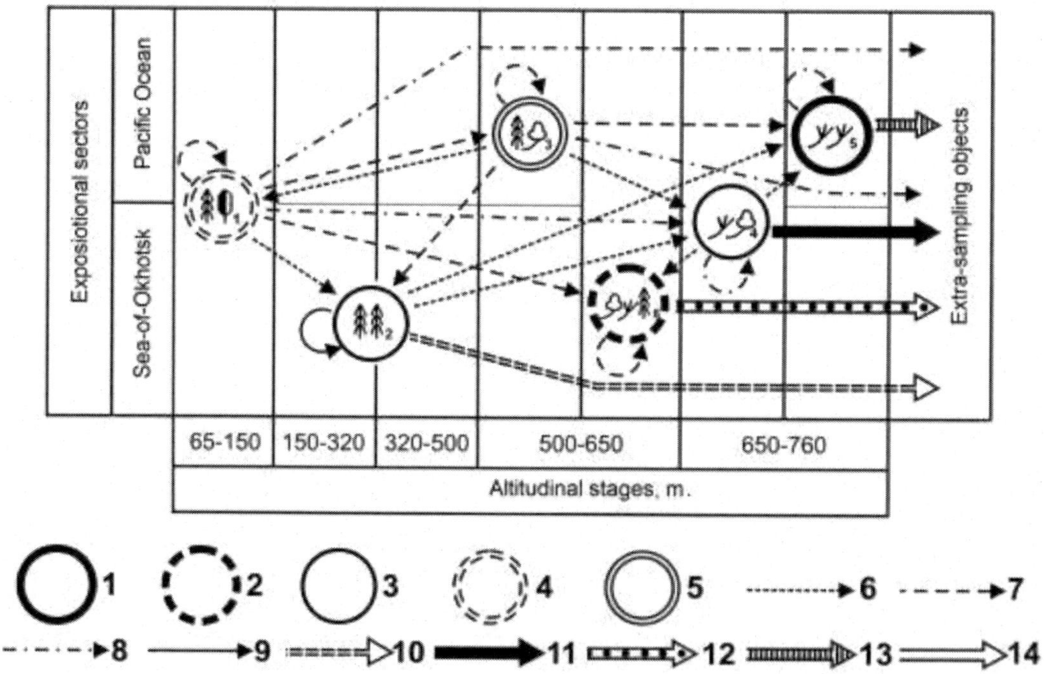

Figure 17.4. The Mendeleev Volcano landscape. The orgraphs of structural and functional transitions for the period of 2100 according to the cold arid trend of GISS climate prediction model among the groups of forest types located in different expositional sectors of Kunashir Island. The transit ecotypes of forest communities: 1 – "absorbents"; 2 – "sub-absorbents"; 3 – mesotropes; 4 – sub-reducers; 5 – reducers. The probabilities of stabilization *(P_{ii})* and mutual transitions *(P_{ij})*: *(6)* 0.01–0.10; *(7)* 0.11–0.20; *(8)* 0.21–0.30; *(9)* 0.31–0.40; *(10)* 0.41–0.50; *(11)* 0.51–0.60; *(12)* 0.61–0.70; *(13)* 0.71–0.80; *(14)* 0.81–0.90.

The response of island-arc Neo-Pacific local geo(eco)systems to fluctuations of the regional climatic system (Table 17.3) described below completely reflects the above-mentioned patterns of the primary subaerial landscape genesis and is one of the hydrothermal mechanisms of this process. We will refer to this system of reactions as the *Neo-Pacific type of climatogenic dynamics of plant cover* at the initial, island-arc stage of formation of continental biosphere. Above, we will examine the basic features of this type.

In the first place, in the Mendeleev landscape there are two major peculiarities of predicted climatogenic geo(eco)system transformations (Figure 17.4; Table 17.4):

1) The extremely low geotopic stability of nearly all floristic, phytocoenotic and landscape units under consideration, even under conditions of weak climatic signals, when the mean temperature and annual precipitation trends do not exceed ±0.7–1.1° and ±15–20%, respectively (see Table 17.4); it is represented by very low P_{ii} values (from 0 to 0.29 for floristic phratries and groups of microlandscapes and mainly in the range of 0.12–0.31 for the groups of forest types); and
2) Contrariwise, the exceptionally high rate of functional absorption of the areals of nearly all basic plant communities, as well as floristic and landscape complexes, by *extrazonal (extra-sample) objects* characterizing already the *regional scales of climatogenic transformation of the phytocoenological structure of boreal island-arc landscape* (P_{ij} (extrazon.): ≥ 0.25–0.40, often reaching 0.70–0.80 and 1.00 for the groups of microlandscapes.

Table 17.4. The generalized characteristics of probabilities and rates of functional and structural transitions for plant communities of the Mendeleev landscape for the prognostic period of 2100 according to the cold arid climatic trend of EGISS

a) Groups of forest types						
Group of forest types	Statistical parameters					Transit index S_k
	P_{ii}	P_i	P_{ij} (extrazonal)	τ (st)	τ (yr)	
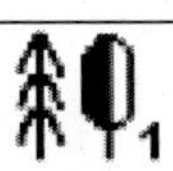	0.121	0.879	0.255	1.14	131	1.15 CA
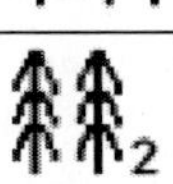	0.376	0.624	0.439	1.60	184	0.86 CP
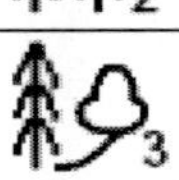	0.163	0.837	0.287	1.19	137	1.42 A
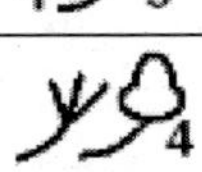	0.205	0.795	0.595	1.26	144	0.71 P
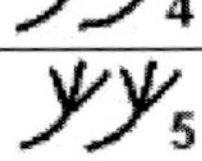	0.281	0.719	0.719	1.39	160	0.84 CP
	0.169	0.831	0.625	1.20	138	0.84 CP

Table 17.4. (Continued)

b) Floristic phratries						
Floristic phratries	Statistic parameters					Transit index S_k
	P_{ii}	P_i	P_{ij} (extrazonal)	τ (steps)	τ (years)	
MO	0.128	0.872	0.744	1.15	132	0.88 CP
Okh	0.288	0.717	0.457	1.39	160	0.62 P
BK	0.221	0.779	0.641	1.28	148	1.13 CA
Ber	0.053	0.949	0.949	1.05	121	1.37 A
BKO	0.154	0.846	0.602	1.18	136	1.00 M

Note: Transit ecotypes: A – absorbents; SA– sub-absorbents; M – mesotropes; SR – sub-reducers; R – reducers.

The most important feature of bioclimatic system at the stage of island-arc landscape genesis is the direct subordination of phytobiota to the vertical hydrothermal gradients. It forms a certain spectrum of altitudinal belts in the framework of low-mountain island-arc landscape. The tight relationship between phytobiota and climate in the Mendeleev volcanic landscape also predetermines the high intensity of climatogenic transformations of island plant communities – much higher than on the continent.

Thus, climatogenic transformations of natural-territorial structures of the Mendeleev volcanic landscape come up to the regional level. The displacements of altitudinal-zonal boundaries within a single low-mountain geomorphological stratum reach the level of major (regional-scale) changes. As we have already seen, in the intracontinental ecoregions under conditions of analogous climatic trends, these structural and functional changes remain mainly topological.

17.3.1. Regional Cooling

The EGISS model suggests an increasing cold arid climatic trend for the entire prognostic period – the decrease in temperatures (both in the cold and warm seasons) and in atmospheric precipitation. This trend is even more pronounced according to the extreme HadCM3 model for the closest period till 2050 (see Table 17.3). We will consider mainly the former, more moderate model. At the first prognostic step (up to 2100), the winter and summer temperature decreases are equal (about 0.6°); at the second step (by 2200), the summer cooling is greater by almost 0.5° than the winter one. Against this background, precipitation will show an evident seasonal disproportion.

Atmospheric precipitation of the warm season decreases only by 3–6%; in the cold season, it will decrease by 32% at the first step and by 25% at the second step. As a result, the humidity factor will decrease too. The reduced snow cover and more severe frosts in winter must cause deeper soil freezing, which will have a very unfavorable effect on the growth of thermophilic nemoral plants (especially subtropical relicts of the Manchurian flora) and contribute to their displacement by boreal species of the Okhotian and Kamchatian phratries.

According to the cold arid trend, the broad-leaved/dark-coniferous ecoregion of central Kunashir with the Mendeleev landscape (Yuzhno-Kurilsk station) on the diagram of climaareals of natural ecosystems of the Pacific Northwest (Figure 17.5) will shift in the northeastward direction, towards the neighboring Iturup Island (Kurilsk station). Here, the coastal-plain and low-mountain bioclimatic background is composed by stone birch forests and crooked forests, as well as larch forests and open woods with stone birch and cedar elfin wood; middle-mountain goltzy are occupied by heather tundras, heath lands and subalpine meadows (Kotlykov et al. 2009).

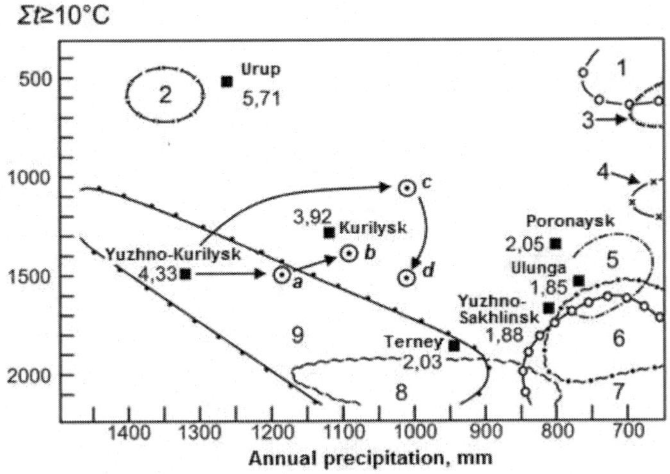

Figure 17.5. The climaareals of some natural ecosystems of the moderate belt of Eurasia (Bazilevich et al., 1986), with the locations of weather stations of the Pacific Ocean mega ecotone, as well as the predicted hydrothermal trends of the Mendeleev ecoregion in the southern part of Kunashir. Climaareals: 1 – plain tundras; 2 – the same, mountain; 3 – forest tundra; 4 - north taiga forests; 5 – the same, dark-coniferous forests; 6 – south taiga forests; 7 – subtaiga coniferous/broad-leaved forests; 8 – beech forests of the southern East Europe; 9 – mountain spruce forests of the southern East Europe. The predicted climatic trends (shown by arrows) according to global climate models E GISS (*a* – for 2100; *b* – for 2200) and HadCM3 (*c* – for 2050; *d* – for 2100).

What these extrazonal phytocoenotic objects can be? They seem to be highly diverse, and three generalized groups can be outlined here.

Group I. Judging from the fact that cedar elfin wood and crooked birch forests in the subgoltzy belt and at the upper boundary of the mountain-forest belt will be replaced by 60–70% and more by the associations of above-lying altitudinal belts, such objects are supposed to be mountain tundars, meadows and shrubs, the analogs of which are widespread, e.g., at the absolute heights of more than 1200–1300 m in the near-peak belt of the Ruruy volcanic massif (northeastern Kunashir). These communities are as follows (Kotlykov et al. 2009): heath lands (*Empetrum nigrum s.t., Rododendron aureum, Ledum macrophyllum, Vaccinium uliginosum*) in complex with thin prostrate woods (*Pinus pumila, Duschekia maximoviczii*) and fragments of subalpine meadows (*Solidago virgaurea, Geranium erianthum, Pedicularis resupinata, Nilingia ajanensis, Parageum calihifolium*), as well as with bamboo thickets (*Sasa spp.*).

Group II. The fir/spruce forests and birch/spruce communities of the mountain belt of the Mendeleev landscape will seem to be transformed (by 45–55%) into middle- and high-mountain alder prostrate woods (*Duchekia maximovizhii*) in complex with cedar elfin woods, meadows and heath lands. Their analogs are widespread at the heights of 800–1200 m in the Ruruy mountain ridge.

Group III. Finally, stone birch (*Betula ermanii*) forests with cedar elfin wood can transgress into piedmont coniferous/broad-leaved forests from the upper belt of the Mendeleev Volcano forest zone, being simultaneously transformed into crooked-forest herbaceous and elfin wood associations. The analogs of the latter can be alder elfin woods currently widespread in the zone of 600–900 m of the Tyatya Volcano, tall-grass (*Filipendula kamtschatica, Solidago spiraeifolia, Aruncus cliacus,* and *Cacalia hastata*), fern (*Dryopteris spp.*) and bluejoint (*Calamagrostis langsdorfii*).

On the coast, the following meadow communities will probably become widespread: sedge-cereal-herbaceous meadows with bamboo thickets and the species of dry habitats (*Artemisia*, *Cichorium*, Asteraceae, and Caryophyllaceae). The paleoanalogs of such associations were developed on the coastal plains of Kunashir Island during the cold snap in the first half of Late Holocene: 1700–1300 thousand years ago (Razzhigaeva et al. 2011).

The comparison of structural schemes of phytocoenotic transitions in 2100 and 2200 shows that the deepening cold arid climatic trend will be accompanied by the more and more extensive (total) transgression of new extrazonal objects (bluejoint/elfin-wood/tundra communities) from the upper low-mountain layer to the middle and lower layers. Their advance will be halted only in the piedmont layer of the Mendeleev landscape where, as we have already mentioned, the coniferous/broad-leaved forests of the Manchurian–Okhotian flora will be replaced mainly (by 40–50%) by birch-spruce forests and birch crooked forests (with cedar elfin wood in the undergrowth) largely belonging to the Beringian–Kamchatian, Okhotian and Beringian–Kamchatian–Okhotian phratries.

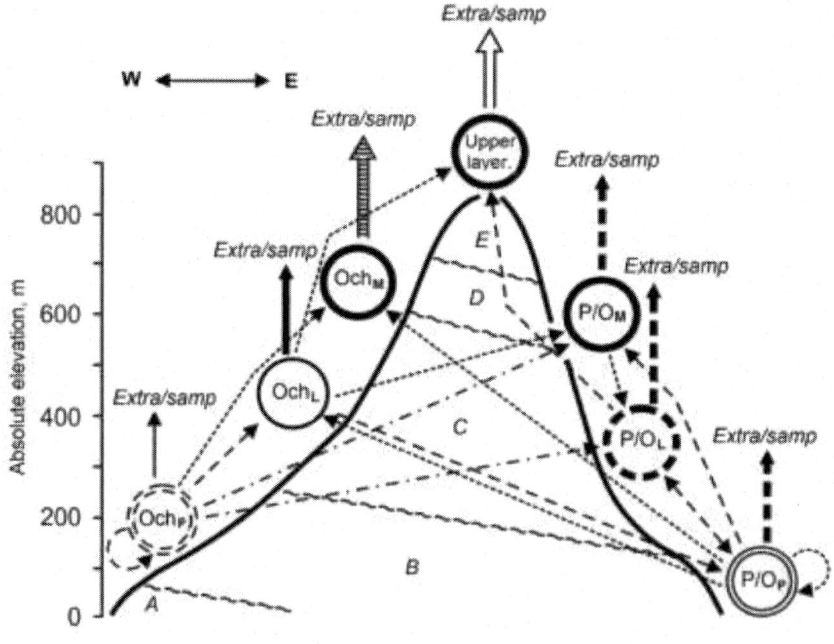

Figure 17.6. The altitudinal-sectoral scheme of functional and structural transitions between microlandscapes of the Mendeleev Volcano according of the cold arid trend of E GISS climate model for the period of 2100. *A – E,* the altitudinal bioclimatic belts. Sectors: Okh – Sea of Okhotsk; P/O – Pacific Ocean. Layers: P – piedmont; L – lower; M – middle. Extra/samp. – extra-sample objects. Upper layer. – upper layering. The symbols of transition probabilities and the transit ecotypes of forest communities are the same as in Figure 17.4.

The values of parameter $P_{ii} \approx 0.120$ show that coniferous/broad-leaved forests of the buffer Manchurian-Okhotian flora, as well as birch/spruce communities of the Beringian–Kamchatian–Okhotian phratrie under conditions of the cold arid climatic trend, will have minimum geotopic stability, being transformed into forest groups above the slope with a probability of 11% to 30% and simultaneously absorbed by heath lands emerging in the upper layer, as well as by alder and cedar elfin woods (see Table 17.4). The group of fir/spruce forests of the Okhotian flora will be much more stable in their areals (P_{ii} = 0.32–0.38). However, these communities will be flooded by 45–55% with transgression of new extrazonal goletz/tundra communities. These communities will also cover the areals of modern subgoletz cedar elfin woods and stone birch elfin woodland by 60–80% and more. Nevertheless, the latter will manifest a certain migration resistance, being distributed over the territory of modern coniferous/broad-leaved and birch/spruce forests, which gives them the properties of local absorbents and mesotropes.

According to the dynamics of plant communities, substantial changes will occur in the landscape structure of the Mendeleev Volcano per se (Figure 17.6). In the first place, we should note the general weakening of regional transgression of extrazonal (mountain/tundra) objects in the direction from the volcano peak to its foot. The most cardinal transformations will occur in the upper microlandscape layer with subgoletz vegetation. At the same time, relative to phytocoenoses of the lower layer, cedar elfin woods will act as an absorbent: their zero geotopic stability will be combined here with the maximum migration stability. The maximum geotopic stability (P_{ii} = 26–29%) will be manifested in stone birch and dark-coniferous microlandscapes of the midlle and lower layers of the Sea-of-Okhotsk sector of the volcano.

It is also obvious that regional transgression of the above-described mountain-tundra, meadow and heath land/elfin wood communities will cause a more drastic transformation of microlandscapes on the mountain slopes of the Sea-of-Okhotsk sector compared to the slopes of the Pacific Ocean sector which, being originally colder, will become less susceptible to the cold arid signal. It will also cause a 12–13% transformation of mixed-forest microlandscapes of the Sea-of-Okhotsk piedmont plains into the landscapes of fir/spruce and stone birch forests of lower and middle layers of the Pacific Ocean sector, respectively. The latter will imply certain alignment of the known bioclimatic contrasts between the Sea-of-Okhotsk and Pacific-Ocean sectors, with disappearance of many thermophilic forms and entire communities in the Sea-of-Okhotsk sector.

17.3.2. Regional Warming

The landscape ecological prediction for global warming has been characterized by the HadCM3 climate model for the period of 2100. According to this scenario, the temperature increase in the Neo-Pacific region under consideration will be accompanied by reduced precipitation (see Table 17.3), i.e., an obvious thermo-arid climatic trend is expected. In the island-arc volcanic landscape, this trend must cause its own specific phytocoenological rearrangements, highly different from those determined by the cold arid trend described above. First of all, there will be a drastic decrease (down to zero) in P_{ii} – the index of geotopic stability of local geo(eco)systems. In general, there will be less intensive intra-landscape phytocoenological transitions and incommensurably enhanced transgression of extrazonal objects – forest associations from more southern natural zones.

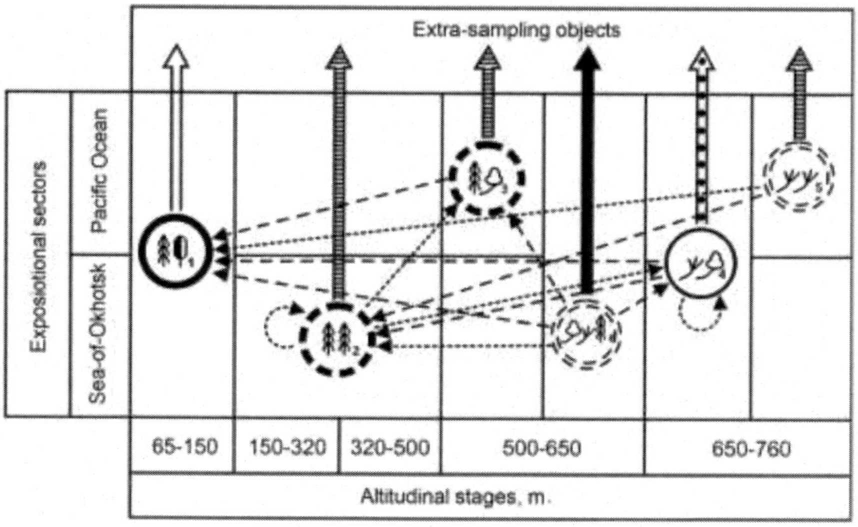

Figure 17.7. The Mendeleev Volcano landscape. The orgraphs of structural and functional transitions for the period of 2100, according to the thermo-arid trend of the HadCM3 climate prediction model, among the groups of forest types located in different expositional sectors of Kunashir Island. Legends are as in Figure 17.4.

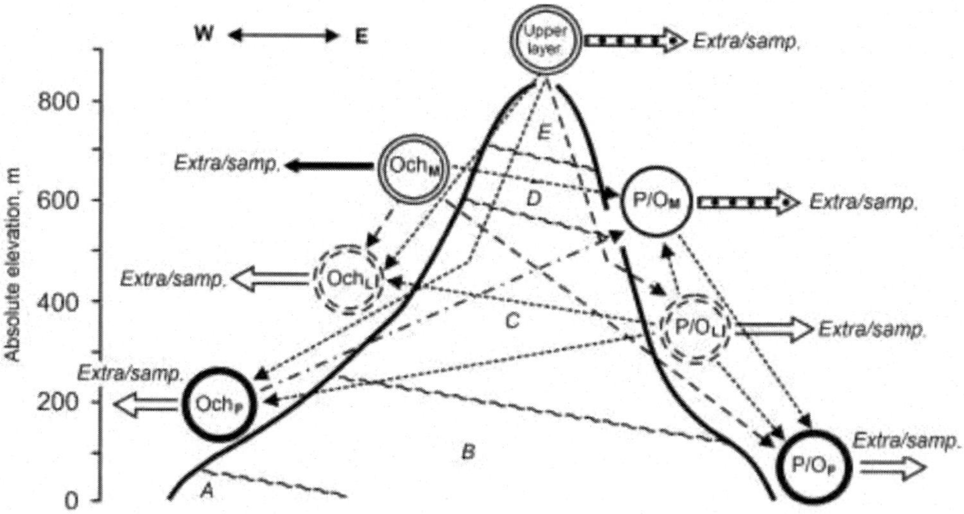

Figure 17.8. The altitudinal-sectoral scheme of functional-structural transitions between microlandscapes of the Mendeleev Volcano according to the HadCM3 thermo-arid trend for 2100. Legends are as in Figure 17.4.

In the thermo-arid trend, structural schemes of the dominance of floristic phratries, groups of forest types and groups of microlandscapes undergo dramatic changes of their climatogenic relationships (Figures 17.7 and 17.8; Table 17.5). The piedmont-plain coniferous/broad-leaved forests of the Manchurian–Okhotian flora, in contrast to their position in the cold arid trend, acquire the *maximum migration stability* and become *relative absorbents*, absorbing up to 12–17% of the area of mountain birch/spruce forests, stone birch crooked forests and ternary buffer communities, being spread even to the subgoletz zone of cedar elfin wood. The second local transgression is expected from the middle-layer fir/spruce sub-absorbents, which will move up the slopes and occupy 11–12% of the area of stone birch crooked forests and cedar elfin woods. Finally, the buffer elfin wood-birch-spruce associations of the Beringian–Kamchatian–Okhotian flora near the upper boundary of the mountain-forest zone will completely disappear.

However, all of the mentioned phytocoenological transitions will be negligible compared to the phenomenon of external transgression of more southern plant formations. In 115–125 years, one should expect the complete functional and structural transformation of the entire plant cover of the Mendeleev island landscape (see Table 17.5). The consistent pattern of such transformation can be as follows.

Table 17.5. The generalized characteristics of probabilities and rates of functional and structural transitions for plant communities of the Mendeleev landscape for the prognostic period of 2100, according to the HadCM3 thermo-arid climatic trend

Group of forest types	Statistic parameters					Transit index S_k
	P_{ii}	P_i	P_{ij} (extrazonal)	τ (steps)	τ (years)	
a) Groups of forest types						
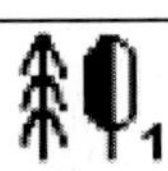	0	1.000	1.000	1.00	115	1.36 A
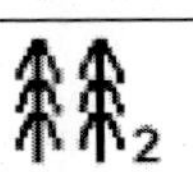	0.053	0.947	0.812	1.60	121	1.09 CA
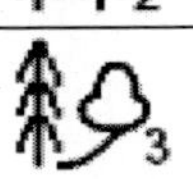	0	1.000	0.848	1.19	115	1.09 CA
	0.090	0.910	0.609	1.10	126	1.00
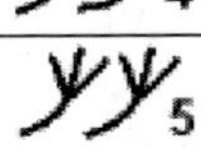	0	1.000	0.809	1.00	115	0.82 CP
	0	1.000	0.519	1.00	115	0.64 P
b) Floristic phratries						
MO	0	1.000	1.000	1.00	115	1.60 A
Okh	0.064	0.936	0.872	1.07	123	1.00 M
BK	0.082	0.918	0.836	1.10	125	1.00 M
Ber	0	1.000	1.000	1.00	115	1.00 M
BKO	0	1.000	0.772	1.00	115	0.40 P

Note: Symbols see in the Note to Table 17.4.

The bamboo fir/broad-leaved forests (*Abies sachalinensis, Quercus crispula, Acer mono var. monoratum, Sasa kurilenesis*) will begin to spread over piedmont plains from the neighboring Hokkaido Island, having much more continental climate compared to Kunashir.

However, these extrazonal absorbents will be little different from the modern coniferous/broad-leaved forests of South Kunashir coasts and foothills.

More important seems to be further advance of these associations up the volcano slopes to the upper layer, with absorption of the crooked birch forest belt and with emergence of cedar elfin wood here and there in the subgoletz belt. Palynological data show (Razzhigaeva et al. 2009) that dark coniferous/broad-leaved forests with the high percentage of oak were predominant on Kunashir plains in the warm epoch of Late Holocene (about 6100–2550 years ago).

In the context of further warming, it is quite acceptable that there will be a transgression of forest associations from more southern and still more continental regions of insular Neo-Pacific and marginal-continental Sub-Pacific to the lower and even middle low-mountain strata of the Mendeleev Volcano. These can be, e.g., the oak/beech communities of Hokkaido (*Fagus crenata, Quercus crispula*), the mixed and broad-leaved forests currently widespread in Manchuria, Northeastern China (*Pinus koraiensis, Quercus mongolica, Q. dentala, Fagus crenata, Ilex pedunculosas*), and the broad-leaved forests of Honshu. The penetration of subtropical coniferous forests of Honshu to the foothills of the Mendeleev Volcano (*Chamaecyparis obtusa*) cannot be ruled out either.

Finally, the ash/elm (*Fraxinus mandshurica, Ulmus propinqua*) communities from Sakhalin will possibly appear in river valleys and on ancient marine terraces and, with further thermoaridization, even xerophytic hazel oak forests (*Quercus mongolica*) analogous to the modern oak forests of South Primorye.

The described processes of thermo-arid functional-structural transformations of the plant cover will obviously be manifested, in the first place and most intensively, in the more continental Sea-of-Okhotsk sector of the Mendeleev landscape and, to a much lesser extent, in the Pacific Ocean sector.

The modern analogs of subboreal forests expected under thermoaridization in the lower and even middle strata of the Mendeleev landscape can be found in the vicinity of the neighboring Golovnin Caldera at the southwestern end of Kunashir. Here there are extrusive domes with active fumaroles in their periphery. The temperature of hot water springs is about 100° and the water temperature in Lake Kipyashchee is 35–65° to 90°, with the heating of vast surrounding areas including rather distant Lake Goryachee (Gorshkov 1967; Kozlov and Zharikov 2012). Intensive geothermal heating of soil-grounds determined the appearance in this region of two *areals-refugiums of extrazonal broad-leaved forest associations* against the background of predominant fir/spruce forests. The lesser areal is situated at the northeastern margin of the Golovnin Caldera adjoining Lake Kipyashchee and the larger one is situated near the southwestern foothills of external somma (Kotlykov et al. 2009). These are oak/stone-birch/maple communities (*Quercus crispula, Acer mayrii*) of the type of forest outliers enriched in broad-leaved species (*Kalopanax septemlobus, Phellodendron sachalinensa, Gerasus sachalinensis, C. maximoviczii, Ulmus laciniata, U. japonica*). The extrazonal (geothermal) genesis of these subboreal associations is emphasized by the fact that near the coast of Lake Kipyashchee they lie hypsometrically above the zone of boreal fir/spruce forests. Due to local geothermal energy, these broad-leaved forests simulate a forest cover specimen that can be formed in Mendeleev landscape under the thermo-arid climatic trend.

17.4. Marginal-Continental Low-Mountain Landscape

Both predictive landscape-ecological scenarios under consideration, according to the EGISS model for 2050 and 2200, describe aridization but on the background of warming and cooling in the former and latter case, respectively (see Table 17.3). At the same time, in 2050 precipitation is supposed to decrease more than twofold than in 2200 and, therefore, the thermo-arid trend will be much more pronounced than the cold arid trend. Accordingly, the processes of phytocoenological transformations in the former case must be more intensive as is shown by the orgraphs of landscape-ecological transitions.

17.4.1. Regional Warming

Under the conditions of thermo-arid trend (EGISS 2050), the bioclimatic situation developing in the Lower Priamurye will apparently be analogous to the situation that occurred here in the second climatic optimum of Holocene: in the Subboreal period (4.5–3.2 thousand years ago), according to the Blytt-Sernander scheme, when the mean July temperature was no less than 1° higher than the modern temperature, while precipitation decreased. That was the Upper Holocene phase 5g of cedar/broad-leaved forests in the Amur-Ussuri region at the latitude of 44–46° N (Korotky et al. 1999).

The major displacements of boundaries between microlandscapes will be directed bottom-up along mesocatenas: from TA, accumulative, eluvial and even superaqual elements to transit and transeluvial mountain elements. The most marked chain of landscape-ecological transitions may be outlined as follows: $TE^{m} \rightarrow T^{m}_{sun} \rightarrow Ev$, with a 17–22% probability and with the intensified process of consecutive absorption down the mesocatena.

Table 17.6. The Komsomolsky Nature Reserve. Distribution of habitat types (HT) by the groups of microlandscapes (the matrix of normalized partial coefficients of connection)

C(A;B) = 0.220

Type of habitat	Group of microlandscapes						
	Saq	**TA (pediment)**	**TEm (mountain)**	**Tv (valley)**	**Tshad (mounain/shadow)**	**Av (valley)**	**Tsun (mountain/sunny)**
$A_5B_5C_5D_5$	**1.00**	0	0	0	0	0	0
$A_4B_4C_4$	0	**0.63**	**0.40**	0	0	0	0
$A_1A_2B_1B_2$	0	0.14	0.30	**0.43**	0.38	0.26	0.18
$B_3B_4C_4$	0	0.23	0.30	0.27	**0.63**	0	0.15
$C_2C_3D_3$	0	0	0	0.31	0	**0.74**	**0.68**

Note. Gradations of the types of habitat, according to P.S. Pogrebnyak (1955): $A_5B_5C_5D_5$ – sands, sandy loams and clays (moist and wet); $A_4B_4C_4$ – sands, sandy loams and wet light loams; $A_1A_2B_1B_2$ – sands and light sandy loams (dry and fresh); $B_3B_4C_4$ – sandy loams and light loams (moist and wet); $C_2C_3D_3$ – medium and heavy loams and clays (moist and fresh). The predominant areas of ecological niches are in bold.

Groups of microlandscapes: TEm – mountain transeluvial; T^{m}_{shad} – mountain transit and transaccumulative of shadowy slopes; T^{m}_{sun} – mountain transit of sunny and neutral slopes; TAp – piedmont transaccumulative sunny and neutral; Ev – valley eluvial; Av – valley accumulative; Saq – valley superaqual. Rest meanings are the same as in Figure 17-4.

This lateral process will be accompanied by substantial changes in the types of habitats (HT) for transgressing forest communities. At present, HT types are rather clearly differentiated by the groups of microlandscapes (Table 17.6). The most favorable edaphic forest growth conditions (fresh and wet medium and heavy loams and clays) are formed on sunny and neutral mountain slopes, as well as on littoral accumulative plains of the Amur River, its tributaries (Gorin, Baturin, Ulimi), and in the estuary regions of small rivers (Khankuka, Bolshaya Talandinka, etc.). The forests of valley flat interfluves grow on much less fertile substrates (sands and sandy loams) in dry and fresh states. Even on shadowy mountain slopes, the habitats have a higher ecological status compared to flat interfluves.

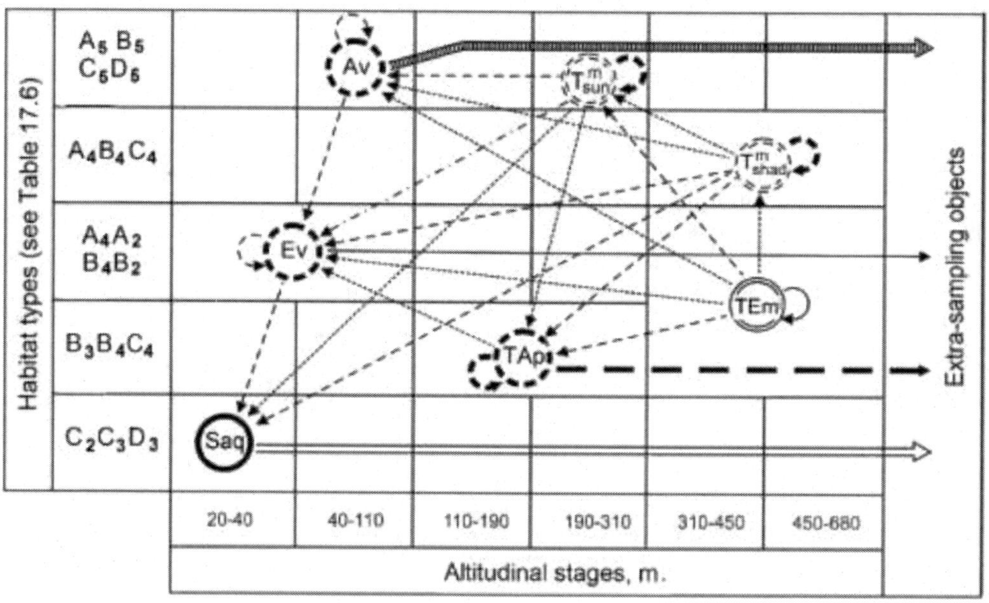

Figure 17.9. The Komsomolsky Nature Reserve. The orgraphs of structural-functional transitions between the groups of microlandscapes for 2050 according to the E GISS thermo-arid trend. Groups of microlandscapes: TEm – mountain transeluvial; T^m_{shad} – mountain transit and transaccumulative of shadowy slopes; T^m_{sun} – mountain transit of sunny and neutral slopes; TAp – piedmont transaccumulative sunny and neutral; Ev – valley eluvial; Av – valley accumulative; Saq – valley superaqual. Other legends are as in Figure 17.4.

In the series of the mentioned landscape-ecological transitions (Figure 17.9), the conditions of forest habitats must first improve and then worsen. In addition, the worsening of habitats will occur during the transformation of communities of shadowy transit microlandscapes into eluvial valley and superaqual ones, as well as during the transition of forests from the state of TE^m sites to TA piedmont states.

Against this background, intra-regional phytocoenotic rearrangements will be rather diversified (Figure 17.10, Table 17.7). The boreal fir/spruce forests of the Okhotian flora currently widespread in the middle and upper low-mountain strata (in the altitudinal range of 350–650 m) will be in the state of reducers, with the minimum stability (S_k = 0.77). They will be replaced by meso-xerophilic oak/lime forests (with a probability of about 0.21) and, to a lesser extent, by mountain cedar and larch-tree forests (by 18% and 14%, respectively). These two local subabsorbents (S_k = 1.04–1.08) will manifest a certain stability of both types: both geotopic (P_{ii} (broad-leaved) = 0.34; P_{ii} (larch-tree) = 0.56) and migration.

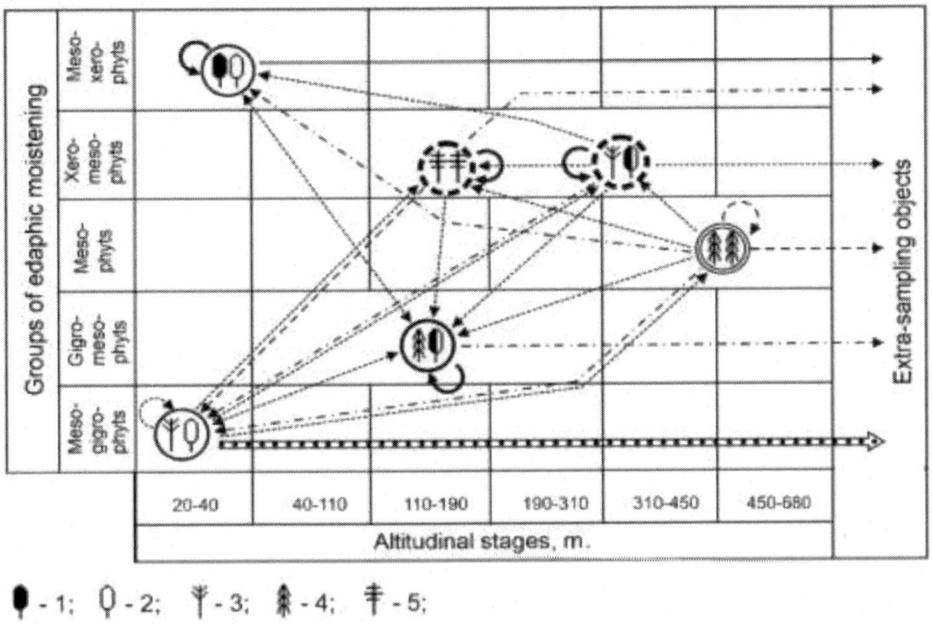

Figure 17.10. The Komsomolsky Nature Reserve. The orgraphs of structural-functional transitions between the groups of forest types for 2050 according to the thermoarid trend of the E GISS climate prediction model. Forest-forming species: 1, oak; 2, lime; 3, cedar; 4, spruce and fir; 5, larch. Other legends are as in Figure 17.4.

However, the internal functional-structural transformations of boreal forests of this ecoregion will be substantially "suppressed" by external phytocoenological expansion. The advance of extrazonal objects will be most efficient in the lower Ad and Saq elements of landscape couplings. Drastic transformation with a probability of 55–65% will occur in meso-hygrophilic cedar/lime forests on the lower terraces above the floodplain of the Amur River, as well as in the Gorin (with tributaries), Ulimi, Tuylia, and Baturin river valleys. The higher drainage density of these territories will contribute to the invasion here of the polydominant broad-leaved forests with needle fir (*Abies holophylla*) from the south, possibly comprising the moderately thermophilic species: Korean oak (*Quercus dentata*), Amur velvet (*Phellodendron amurensis*), Manchurian hazel (*Juglans manshurica*), and heartleaf hornbeam (*Carpinus cordata*). The latter two species now do not occur on the Lower Priamurye plains (Akhmetyeva 1977). The larch-tree sphagnum mari in vast depressions can completely disappear, being replaced by hygro-mesophytic forest/meadow communities. Such broad-leaved forests in the subboreal period were predominant in South Primorye (Rodnikova et al. 2012).

The transgression of polydominant broad-leaved forests to the slopes of the Lower Priamurye low mountains will be apparently limited by the height and very selective. These external absorbents can reach the absolute heights of no more than 200–300 m, with maximum penetration (by 29–33%) into the areals of Manchurian–Okhostian spruce/broad-leaved forests and Angaridian larch-tree forests. Their transgression will have a much weaker (by 16–17%) effect on the forests of the Manchurian flora. This transgression will not reach the above-lying belt of fir-spruce forests (300–680 m abs.).

Table 17.7. The generalized characteristics of probabilities and rates of functional-structural transitions for pant communities of the Komsomolsky Nature Reserve for 2050 according to the EGISS thermoarid climatic trend

The signs	Statistical parameters					Transit index S_k
	P_{ii}	P_i	P_{ij} (extrazonal)	τ (steps)	τ (years)	
a) Groups of forest types						
 1	0.574	0.426	0.321	2.35	99	1.00 M
 2	0.567	0.433	0.045	2.31	88	1.07CA
 3	0.064	0.936	0.703	1.07	41	1.00 M
 4	0.604	0.396	0.295	2.23	85	1.14 A
 5	0.184	0.816	0.151	1.23	47	0.71P
 6	0.554	0.456	0.259	2.19	839	1.08 CA
b) Floristic phratries						
Mm	0.282	0.718	0.483	1.39	53	1.06 CA
Mk	0.464	0.536	0.231	1.87	71	0.95 CP
MO	0.375	0.625	0.233	1.60	61	0.94 CP
Okh	0.367	0.633	0	1.58	60	0.94 CP
MA	0.373	0.627	0.434	1.59	60	0.94 CP
Ang	0.456	0.544	0.544	1.84	70	1.17 A

Note. Symbols see in the text, Chapter 9.

It should be noted that the buffer Manchurian–Okhotian forests are enriched in broad-leaved species under conditions of the thermo-arid climatic trend mainly due to *external transgression* (described above) but not internal transgression of the latter. It is confirmed by the known thesis (Krištofovič1946; Vasilyev 1946) that the *evolutionary climax* represented in this case by the buffer spruce/broad-leaved forests (see Chapter 11) is formed due to *regional migration processes* in the plant cover.

17.4.2. Regional Cooling

The paleogeographic analog of the predicted cold arid bioclimatic trend (EGISS for 2200) can be the 5h phase of Late Holocene cooling at the Subboreal–Subatlantic border, 2200–1500 years ago (Korotky et al. 1999). On the plains of the Amur–Ussuri region, it was the phase of coniferous/broad-leaved forests with widespread larch-tree and lime forests (Rodnikova et al. 2012). Summer temperatures were approximately 1° below the modern temperatures, and the regular marine regression decreased atmospheric moisture content. It generally corresponds to the climatic scenario according to E GISS for 2200.

The matrices and orgraphs of floristic and phytocoenotic transformations under conditions of the cold arid trend (Figure 17.11; Table 17.8) demonstrate the predominance of mutual (forward and reverse) arrows of transitions in the binary nontransitive relations of objects (with the similar values of their generally low probability: from 0.03–0.06 to 0.15–0.18). The multiplicity of such transitions is evidence of nearly *omnipresent processes of the mixing of floras and complication of forest phytocoenoses*, which is one of the mechanisms of formation of buffer ecosystems in the marginal-continental Sub-Pacific. The mixing of floras under conditions of the cold arid trend is favored by the lower (compared to thermoaridization) general pattern of phytocoenotic and microlandscape transformations, the more quiet climatogenic dynamics of geo(eco)systems, with the dominance of $P_{ij} \leq 0.10$.

In the low-mountain Lower Priamurye, the cold arid trend must cause two major flows of external phytocoenological transgression. The first flow will be the 35–40% absorption of the upper-stratum TE fir/spruce forests of the Okhotian flora, with the emerging and expanding areals of subgoletz cedar elfin woods of the Beringian flora. The center of origin of the second flow will be the enhanced winter cooling of river valleys and depressions, with the development of valley larch-tree forests, larch sphagnum mari and peat bogs with long-term seasonal permafrost. The transgression of valley larch-tree forests into oak/lime and cedar/spruce/broad-leaved forests of the neighboring flat interfluves (with a probability up to 25%) will imply that the winter inversion cooling will spread also to the Ed groups of microlandscapes, which will pass into the category of subreducents.

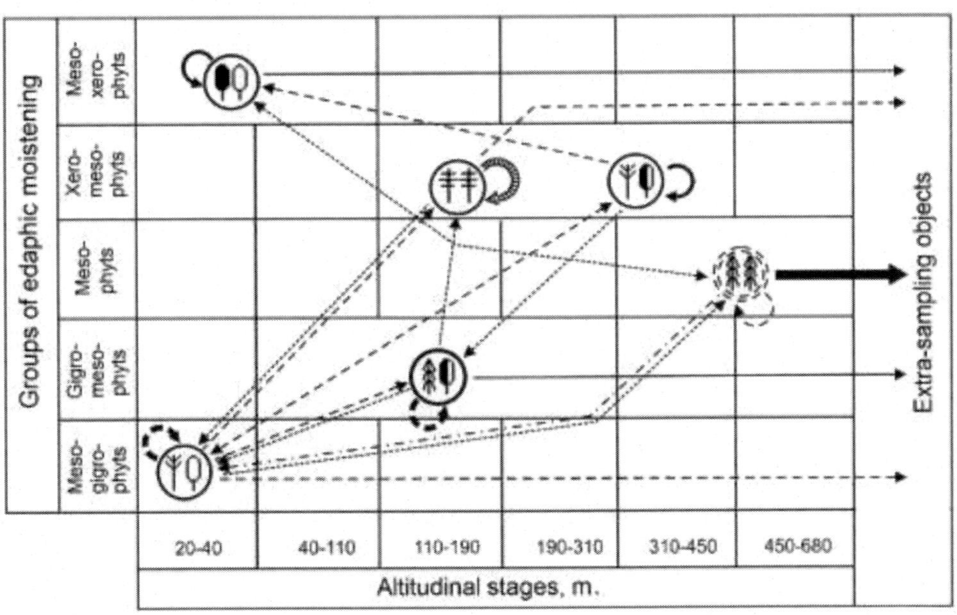

Figure 17.11. The Komsomolsky Nature Reserve. The orgraphs of structural-functional transitions between the groups of forest types for 2200 according to the cold arid trend of EGISS climate prediction model. Symbols are the same as in Figures 17.4 and 17.10.

Simultaneously, mountain larch forests will reinforce their positions by demonstrating high homeostatic stability (P_{ii} = 0.774) and begin to gradually capture the neighboring areals of fir/spruce, cedar and oak/lime forests (P_{ij} = 0.03–0.08). This process will spread mainly to mountain/transit shadowy microlandscapes; here, the transforming forests will be also characterized by high stability (P_{ii} = 0.632). At the same time, there will be a rather distinct (with a probability of 14–15%) tendency to replacement of mesophytic nemoral fir/spruce and hygro-mesophytic spruce/broad-leaved forests of the MO flora (subreducents) by meso-xerophytic and xerophytic lime/oak forests (subadsorbents), with penetration of larch tree.

In general, one can expect the general aridization of forest cover in the region, with reinforcement of the entire complex of the Angaridian flora passing into the category of subabsorbents (S_k = 1.05) with a stabilization measure of 0.54. Thus, the cold arid trend in the Lower Priamurye must intensify one of the processes of century-long successions of forest-forming timber stands (Kolesnikov 1956), namely, the durable steady replacement of the Manchurian boreal/nemoral flora by the Manchurian–Angaridian nemoral/boreal flora.

Table 17.8. The generalized characteristics of probabilities and rates of functional-structural transitions for plant communities of the Komsomolsky Nature Reserve for the prognostic period till 2200 according to the EGISS cold arid climatic trend

The signs	Statistical parameters					Transit index
	P_{ii}	P_i	P_{ij} (extrazonal)	τ (steps)	τ (years)	S_k
			a) Groups of forest types			
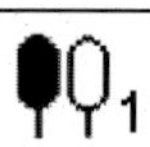	0.594	0.406	0.375	4.93	927	1.17 A
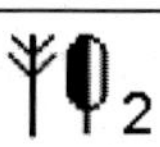	0.594	0.406	0	4.93	927	1.00 M
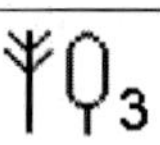	0.405	0.595	0.118	3.36	632	1.00 M
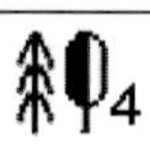	0.497	0.503	0.389	3.98	748	0.91 CP
	0.137	0.863	0.512	2.31	434	0.92 CP
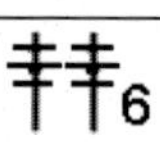	0.747	0.253	0.156	7.94	149	1.00 M
			b) Floristic phratries			
Mm	0.514	0.486	0	4.12	774	0.95 CP
Mk	0.497	0.503	0.171	3.97	746	1.00 M
Mo	0.329	0.671	0.420	2.98	560	1.05 CA
Okh	0.302	0.698	0.439	2.87	539	1.05 CA
MA	0.584	0.416	0.039	4.81	904	0.91 CP
Ang	0.601	0.399	0.039	5.00	940	1.04 CA

Note. Symbols see in the text, Chapter 9.

— « » —

The predictive landscape-ecological models that we have obtained for the Lower-Amur Sub-Pacific experimental plot disclose some particular mechanisms of formation of "multiform and dynamic" (Kolesnikov. 1956) forest cover of this marginal-continental ecoregion under conditions of the warming and cooling waves alternating throughout the Pleistocene (Korotky et al. 2011), providing "the assembly of the flora" (Urusov 1988) as "... a result of its saturation to the maximum of climatic resources" (A.A. Krištofovič; cit. by: Kolesnikov 1969 218).

The response of forests of different formations in the Lower-Amur ecoregion to a particular climatic signal manifests itself in terms of complex inter-species relationships and diverse forest-forming processes composed of age-related, century-long and restorative successions of forest communities. The most adequate for this landscape-ecological prediction are the *age-related successions of forest stands* that proceed (Kolesnikov 1956) by an unclosed helical pattern with no less than a 2–2.5-century period and supposedly caused by superposition (overlap) of external climatic signals and autochthonous change of edaphic conditions of growth sites.

Under conditions of the thermo-arid climatic trend, the buffer Manchurian–Okhotian forests become enriched in broad-leaved species mainly due to their *external (extrazonal)* but not internal transgression. It confirms the known thesis (Krištofovič 1946; Vasilyev 1946) that the *evolutionary climax* represented in this case by the buffer spruce-broad-leaved forests is formed due to *regional migration processes* in the plant cover. The cold arid trend will correspond to the multiplicity of weak phytocoenotic transitions, which suggests the *omnipresent processes of the mixing of floras and complication of forest communities*. It seems that the forests of the Lower Priamurye have been formed precisely due to periodic cold snaps and continue to be formed "... as exceptionally multiform and externally varying" (Kolesnikov 1956 136). Under conditions of the thermo-arid trend, such processes are much less pronounced. Thus, the cold arid ecological scenario particularizes the hypothetical pattern of formation of the boreal Sub-Pacific forest cover proceeding by "... the pathway of wave breaking from the north and northeast, with accumulation of the products ... of floristic import" (Krištofovič 1946 64).

17.5. Sustainability of (Geo)Ecosystems and Its Chronological-Organizing Role in Their Climatogenic Transformation

17.5.1. The Problem of Studying the Sustainability of Natural Ecosystems

The predictive landscape-ecological analysis of a territory cannot go without assessing the sustainability of object under consideration to external effects. The known American ecologist R.E. Ricklefs (1976) summarizes the considerations of this problem as follows: sustainability is the culmination of all ecological dependences; the significance of new data and importance of new ideas will undoubtedly be estimated by their contribution to understanding this ecological synthesis – sustainability of natural systems. The significance of *sustainability as ecological synthesis* is determined by the fact that this property of natural complexes is a master link in the chain of component interconnections, which determine the functional organization of geo(eco)systems. The most part of significant landscape-ecological connections are concentrated on the parameters of sustainability, which determines the high indicative capacity of these parameters (Kolomyts et al. 2000; Kolomyts 2008).

The concept of two alternative types of ecosystem stability (Holling 1973; Odum 1983, Svetlosanov 1990; Thompson et al. 2009) has been established by the present time. These are (1) resilience, which characterizes the capability of an ecosystem to return to its initial status after a disturbance and to regenerate the main structure and functioning; and (2) resistance, which determines the capability of an ecosystem to resist transformation into another system. This property is related to the ecosystem stability, which shows the capability of a system to maintain a dynamic balance for a particular time period by means of absorbing the external effects.

It is also proposed to specify the engineering resilience and ecological resilience (Holling 1973; Peterson et al. 1998; Gunderson 2000). Engineering resilience refers to the existence of only one stable status (Gunderson 2000). Ecological resilience is determined as the capability of a system to absorb effects to a threshold level, at which it is transformed into a qualitatively new system. For example, during a drought, a resilient forest ecosystem corresponding to the status of engineering resilience would recover from this stress without significant changes in the species composition. An ecosystem with the status of ecological resilience may have more than one stable status, and the ecosystem resilience (capability for regeneration) characterizes the capability of forest to withstand a long-term drought. In this period, the forest may undergo succession with transformation into several different stable ecosystems with a new species composition, and the final stage will be represented by a meadow. Forests at many of these successive stages are as valuable for the forestry as the initial ecosystem, and the community may be recognized as a particular forest type. This is the concept of an unbalanced dynamic (Thompson et al. 2009).

The problem of forest stability in our country has been studied for a long time. Early in the 20th century, G.F. Morozov (1949) set an objective for forestry specialists to use forests so rationally that they could preserve their biological stability and productivity for the quick natural regeneration of indigenous forest types, which have, in his opinion, the highest stability. In geoecology, the stability or homeostatic rate (Armand 1983) is still evaluated by qualitative gradations such as slight, medium and high, and rigorous classification criteria are sometimes absent.

Nevertheless, formal characteristics were already used in the evaluation of the stability of zonal-regional geosystems of large areas (Buks 1977). Five stability levels of landscape-ecological complexes were specified with respect to landscape initial biological productivity and the provision in heat. Nevertheless, the stability type and the kinds of external effect are unclear; therefore, the specified stability levels are abstract.

A detailed qualitative analysis of forest stability under the effect of climate changes and economic activity is given by (Rozhkov and Kozak 1989). Never forest ecosystems to an external impact are only generally evaluated on the basis of qualitative parameters of the substrate, and the conclusions are often very subjective.

General ideas on the stability theory of geoecosystems are given in the work by (Armand 1983) and in monographs (Kupriyanova 1989; Glazovsky and Armand 1992). The authors describe the stability mechanisms of various natural components with the use of qualitative and quantitative parameters that characterize (mainly indirectly) the stability of a particular geocomponent system. A principle of levels of biotic stability mechanisms of natural complexes is proposed, and the stability of zonal ecosystems under the impact of a number of natural and anthropogenic factors is evaluated semi-quantitatively by scores (Glazovsky and Armand 1992).

System stability may be also characterized by its dynamic and hierarchic structure on the basis of the oscillation concept (Puzachenko 1986) with consideration of the ratio between its own fluctuation frequency (the periods of relaxation) and the frequencies of the external effects, which cause forced fluctuations of the system.

In the national landscape geography, there are issues devoted to the use of the concepts of reliability theory in the calculation of the partial stability rate of a geosystem (Grodzinskii, 1987). The stability is evaluated by the behavior of particular natural components and is related to the notion of failure. It signifies that the component loses its homeostasis and begins to change within particular limits. The evaluation of these limits is unfortunately difficult.

Hence, determination of the threshold values of the structural and functional parameters of particular geocomponents used for calculation of the failure probability may be subjective.

In foreign works on forest ecology, ecosystem stability is usually qualitatively described with the use of quantitative indicators that characterize stability indirectly. The growth rate, productivity, and compactness of the tree stand, species and functional diversity, plant damage, and other parameters are used for stability evaluation (Holling 1973; Chapin et al. 1997; Diaz and Cabido 2001; Thompson et al. 2009). A number of works of foreign researchers is devoted to the ratio between various types of forest stability: initial resistance and subsequent resilience. For example, research on the response of trees of twelve various species in New Zealand to unfavorable external impact has shown that fast growing species are characterized by low resistance but high resilience (Bee et al. 2007).That is to say, fast growing trees are first damaged first but then quickly restored.

The stability of forest ecosystems is sometimes correlated with topography. For example, in a study of the functional resilience of a subtropical forest in Okinawa, it was revealed that it depended on species diversity, which in turn is maintained by the diversity of features of topography (Kubota et al. 2004).The specification of so-called topographic types of vegetation repeats the classification of geomorphologic complexes of forests of the Far East (Kolesnikov1956). Physical nonuniformity of the environment can be more important for evaluation of plant community resilience compared to species diversity (Zaret 1982).

It is also assumed that there is a basic ratio between the biological diversity, productivity, and stability of forests and that this ratio is of practical importance for the adaptive monitoring of forests upon climate changes (Begon et al. 1986; Thompson et al. 2009). This concept has been already formulated, and it has been proposed to use K. Shannon's measure of negative entropy diversity for estimating ecosystem stability (Margalef1958). There is evidence that forest ecosystems with greater biological diversity are more productive under similar conditions compared to ecosystems with a small number of plant species. The latter are prone to catastrophic degradation, including diseases and invasions (Scherer-Lorenzen et al. 2005). More productive ecosystems are more resilient and hence regenerate more rapidly after disturbance (Stone et al. 1996). Some scientists point to a correlation between the functional diversity of a forest ecosystem and its productivity (Chapin et al. 1997; Diaz and Cabido 2001), and many plant species in forests are unnecessary for total productivity (Scherer-Lorenzen et al. 2005). According to the insurance hypothesis (Naeem 1998; Yachi and Loreau 1999), the presence of excessive plant species is an important feature of most forest ecosystems; it favors their stability and protects the forest from species extinction.

In general, the more complicated the community (the greater the species diversity), the higher its stability (Ricklefs 1976). This theoretical statement was not only adopted in general ecology (Odum1983) but also found its place in landscape ecology (Sochava 1978). Nevertheless, it is often rejected and is still under discussion (Whittaker 1975; Loreau et al. 2002). First, the notions of complicated rate and stability are very indefinite (Svirezhev 1983). Second, the Shannon entropy diversity measure (its analog is represented by the well-known measure of stability by MacArthur (1955)) cannot be applied to the complex natural ecosystems with rather well-developed internal relationships (Svirezhev and Logofet 1978; Svetlosanov 2009).

There are comprehensively elaborated mathematical models of ecosystem stability in general ecology. The models describe a narrow range of relatively simple ecological systems and are usually deterministic. They analyze particular populations or their combination – a biotic community (Smith 1974; Logofet and Svirezhev 1983; Svetlosanov 1990). The stability of these ecosystems is calculated as the ratio between the deviation of the disturbing effect and the response of the chosen parameters to this effect or as the ratio between the value of homeostasis and the parameter of distribution nonuniformity of a dependent variable (Fedorov 1974, 1975). In other works, the stability of ecosystems is evaluated by the balance equations of their equilibrium conditions, which are described by the energy and matter flows in food chains (Svirezhev and Elizarov 1972; Logofet and Svirezhev 1983; Svetlosanov 2009). The stability is calculated by linear differential equations. Great attention is paid to the stability of the hierarchy of food chains (of the predator–prey type in particular).

Mathematical stability, according to A.A. Lyapunov, is also taken into account, though it has been revealed (Svetlosanov, 1990) that it is hardly applicable to evaluation of the stability of natural ecosystems.

Unfortunately, the approaches involving mathematical modeling of the stability of ideal ecosystems are not widely applied in landscape ecology, where investigation objects are considerably more complicated, polycomponent and discrete and are composed of heterogenic elements with different characteristic times. The ecosystems of the landscape level are distinguished by the probabilistic nature of internal and external interactions (Armand et al. 1969) and nonlinear behavior related to a regulated network of positive and negative feedbacks (Zeidis et al. 2001; Svetlosanov 2009). The stability and functioning of these systems is primarily determined by geochemical cycles (annual and long-term turnover of organic matter) but not by the number and dynamics of plant and animal species (Svirezhev 1983).

These natural systems with the so-called poor structure are better described by discrete mathematics on the basis of the probability-statistical approach, which was proved by our landscape-ecological investigations (Kolomyts 1995, 2005, 2008; Sharaya 2007, 2009). Geoecosystem stability should be obviously evaluated with the use of these approaches and on the basis of the most representative parameters of the biological turnover. We have searched for the unified parameters (indices) of potential stability of elementary biochorological units —biogeocoenoses (Timofeeff-Ressovskii and Tyuryukanov 1966) – as integral formations according to the hypothesis by V.N. Sukachev (1972).

17.5.2. Principles of the Analytic Modeling of Forest Ecosystem Sustainability

The developed *evolutionary landscape-ecological concept* concerning the response of terrestrial ecosystems to climatic impacts is closely related to the idea of sustainability. In accordance with the two aspects of natural ecosystem organization (structure and function), it would be logical to consider the two categories of their sustainability – structural and functional.

The stability is usually treated as the constant system structure (or a slight fluctuation, which does not exceed a threshold point) kept by variations in the functional parameters (Smith 1974; Armand 1983; Glazovsky and Armand 1983; Scherer-Lorenzen et al. 1989; Svetlosanov 2009). The concepts of homeostasis and geoecosystem homeostatic reserve, as well as the norms of anthropogenic impact on the system, are related to this approach. Nevertheless, the quantitative evaluation of the structural stability of multicomponent systems with different characteristic times is a difficult problem (Odum 1983). It is much easier to determine the functional stability, which is vital for ecosystems. This raises the efficiency of the energy flow and nutrient turnover, improving the capability of community to withstand different disturbing factors (Ricklefs 1976).

This approach is based on the interpretation of stability as the capacity of a system to change its structural parameters in order to preserve the initial functioning level (mode) (Kerzhentsev 2006, 2010). The ecosystem structure is formed according to its main functions in the biosphere: the maintenance of the closed matter cycle, which provides stability of the ecosystem itself (Gerasimov 1985; Tishkov 2005), and regulation of the biota in the environment according to the principle by Le Chatelier (Gorshkov 1984). A close relationship between soil-phytocenotic components in the indigenous biogeocenoses is formed in the course of long-term evolution and results in optimal substance migration in the biosphere (Dobrovol'skii and Nikitin 2006). Therefore, we will discuss the functional stability(not a structural one) of natural ecosystems as their most fundamental parameter. This approach allows the calculation of ecosystem stability by discrete parameters of biological turnover.

The following three levels of the analysis of dynamic properties of natural complexes combined by the common concept of "sustainability" are outlined. The first level is related to the two initial alternative categories of *structural sustainability*: *areal* (geotopical) and *migrational* (transgressive), which characterize the state of sustained equilibrium of the system and its forced behavior as a response to external effects, respectively (see below). It is a kind of spatial section of the dynamic state of geo(eco)system. The second level concerns its temporal section – the state of *functional sustainability*, which is associated with the characteristic times of different cycles of the biological turnover. Here, the *phytocoenotic (labile) and soil-biotic (inertial) sustainability* can be distinguished. Finally, the *measures of resistant and elastic-plastic (elastic) stability* are calculated and mapped at the third level.

Resistant, or inertial, sustainability is understood as its ability not to vary the parameters of its functioning under external effects, inter alia, not to substantially reduce the level of productivity or, on the contrary, not to exhaust the dead phytomass reserve. Resistant sustainability characterizes the extent of susceptibility of a natural complex to the initial signals of external effects. Transformation of a system begins after the signal has exceeded a particular threshold of its sensitivity.

At the same time, elastic, or dynamic, sustainability characterizes the system ability to completely or partially restore the equilibrium disturbed by an effect, i.e., to restore the above characteristics of metabolism to the initial state. Elastic sustainability manifests itself in the dynamics of ecogenetic (restorative) successions (Sukachev 1972) tending to bring the system to the initial state. The same geo(eco)systems have different degrees of stability of each of the two types depending on the kind and intensity of external effect.

17.5.3. Calculations of Functional Phytocoenotic Sustainability

Biological turnover is a complicated multicycle process that includes numerous elementary soil processes and cycles of organic matter production, decomposition, mineralization, and Humification with different characteristic times (Kononova 1963; Duchaufour 1968; Glazovskaya 1981). In boreal forests, where the soddy-podzolic process and percolation water regime of soils predominate, the lower biological turnover is slightly related to humification processes of the mortmass (Odum 1971). It is represented by a kind of "autonomic turnover of elements between living and dying organisms that continue to get into soil surface in forests" (Ponomareva and Plotnikova 1980, 115). This is a partial phytocenotic cycle, which can be treated as a short-term (and hence accelerated) component of the total biological turnover. This statement is confirmed by the empirical data on coniferous, mixed, and even broad-leaved forests (Kolomyts 2008). We have not revealed any significant partial dependence of humus mass on the annual production of a forest community and on the mass of forest litter or roots. There is no reliable multiple correlation either. A sufficiently close relationship between the humus mass and the mass of roots and forest litter has only been revealed for zonal forest and steppe ecotones, with predominance of the evaporation regime and soddy-meadow processes in soil. Nevertheless, the drop in the humus mass in these systems as a result of global warming will be only related to the drop in the mass of forest litter and roots by 10–25%. The correlation between soil organic matter and annual green mass production is also very slight. The involvement of humus in the short biological cycle is low enough, even under these subboreal zonal-climatic conditions.

Therefore, the phytocenotic (including forest litter) cycle of organic mass turnover for boreal and subboreal forest formations in general may be specified as an autonomous and rather independent component of the total soil-biological turnover. Its most dynamic part is represented by the yearly cycle of regeneration and decomposition of living above-ground phytomass, which is assigned to quick cycles (Duchaufour, 1968), which we term *labile phytocenotic stability*.

The following two normalized indices of phytocoenotic and functional sustainability: resistant (*IRS*) and elastic-plastic (*IES*), – were calculated by the previously developed and sufficiently tested methodological procedure for all of the studied objects of experimental test sites (groups of forest types, floristic phratries and groups of microlandscapes) (Kolomyts 1995, 2008).

Labile stability may be characterized by two complex discrete parameters of metabolism (Bazilevich and Rodin 1969; Krutyko et al. 1982): the coefficient of the annual turnover of the above-ground phytomass ($CR = PV/BL$) and the litter–fall index ($CY = PV/LM$), i.e., the coefficient of annual destruction (Kolomyts 1995, 2008). In the above-mentioned equations, PV is the annual production of green mass, BL is the total living above-ground phytomass, and *ML* is the mass of the forest litter. Both parameters characterize the initial transportation of organic matter and energy in the entire trophic chain and thus characterize the effect of plant mass on the stable functioning of the ecosystem. Spatial variations of a particular function of the status of ecosystems within a statistical sample were calculated to evaluate labile stability (Kolomyts 1995, 2008). Let us denote the indices of resistance and resilience (labile stabilities) as I_{resist} and I_{resil}. The resilience index for a forest biogeocenosis is calculated as a measure of the Euclidian distance from the optimum functional status by the equation:

$$IES_i = 1 - [\sqrt{(\Delta CR_i)^2 + (\Delta CY_i)^2}\,] / \sqrt{2}, \qquad (17.1)$$

where $\Delta CR_i = (CR_{max} - CR_i)/(CR_{max} - CR_{min})$; $\Delta CY_i = (CY_i - CY_{min})/CY_{max} - CY_{min})$; and i is the number of the testing plot. The maximums and minimums of *CR* and *CY* are taken from i.

Therefore, the resilience index is evaluated in dimensionless units. If Cr_i approaches $CRmax$ and CY_i approaches 0, then IRS_i approaches 1. It can be seen that both prognostic parameters are considered to be equivalent and are accepted with the weights equal to 1. The index of potential resistance (IRS_i) of a geosystem was calculated in a similar way. The minimum *CR* and the maximum *CY* were taken as the optimum values.

$$IRS_i = 1 - [\sqrt{(\Delta CR_i)^2 + (\Delta CY_i)^2}\,] / \sqrt{2}, \qquad (17.2)$$

where $(\Delta CR_i) = (CR_i - CR_{min}) / (CR_{max} - CR_{min})$; $(\Delta CY_i = (CY_{max} - CY_i) / (CY_{max} - CY_{min})$.The resistance index will approach one when CR_i approaches CR_{min} and CYi approaches CY_{max}.

The stability indices of an ecosystem characterize its capability to withstand the integral disturbing effect and show its distance from the critical status. The minimum and maximum *CR* and *CY* are taken from a particular statistical sample; hence, it is obvious that each of the indices characterizes a relative and not absolute stability of the ecosystem within a sampling area. The indices show the part of the maximum possible stability, which is typical of a particular natural complex. *IRS* reflects the well-known law of requisite variety (Ashby 1956), which states that the diversity of its states or reactions (adaptations) must be no less than the diversity of external impacts for stable system preservation. The ecosystem stability (invariance) is related to the following two factors: its multiple relations and the numerous tools compensating for external effects (Puzachenko 1983).

The basic peculiarity of the ratio of parameters *IRS* and *IES* is their *alternativeness.* If a geo(eco)system weakly responds to the initial external effect, i.e., its resistant sustainability is enhanced, the potential of its elastic sustainability as the system ability to neutralize the aftereffect is comparatively low. At the minimum *IRS* values, the *IES* parameter reaches its maximum.

When water and temperature conditions become stress factors, the regime of mineral plant nutrition is usually disturbed. On the basis of this regularity, we can determine the period of resistant labile stability of forest biogeocenosis at climate successions (prior to irreversible disturbance of its functioning). It is equal to *CY* expressed in the years needed for complete decomposition of forest litter at a given volume of annual debris. It can be seen that this time period is equal to the first several years after the beginning of external impact, if the intensity of biological turnover is high or moderate (Bazilevich and Rodin 1969; Kovda 1971). In some cases, the length of the period can exceed ten years and even reach 30 years in the stagnation type of biological turnover.

The resilience period of forest biogeocenosis necessary for regeneration of the initial level of the closed short biological cycle can be evaluated by the following approach. The empirical dependence of the utilization coefficient of dead phytomass (*CU*) on *ML* and *BD* (of the total dead above-ground mass) shows that, if mortmass accumulation in a coniferous forest is equal to 40–60 t/ha, no more than 40% of it is utilized at the following trophic levels, while a constant amount of dead mass in the forest at the rate of10–15 t/ha is sufficient to maintain a continuous flow of organic matter in this trophic chain, as well as its utilization up to 98% (Kolomyts et al. 1993). A simple calculation based on these data shows that resilience activity over a 1.5–2-year-long period is sufficient to restore the initial functioning level at a high annual production of green mass (>8–10 t/ha) typical for the most productive coniferous and foliage forests. For the forest communities with relatively low green mass production (below 3–5 t/ha), this period lasts 4–5 years. The functional-regeneration potential of the forest community is obviously more effective compared to the initial forest capability to withstand external impacts, all other factors being equal.

The stability of the annual regeneration of living above-ground phytomass (i.e., labile phytocenotic stability) can be tentatively used as an indicator of the general stability of forest ecosystems, including the inertial stability, which is related to the considerably longer soil-biotic metabolism cycle (Kolomyts et al. 2015).

The analysis of *IRS* and *IES* combinations for each series of objects made it possible to disclose particular *chronological regularities of climatogenic phytocoenological transformations* in island-arc and marginal-continental landscapes, a particular temporal order of landscape-ecological transitions. The exogenous dynamics of functional and structural parameters of forest ecosystems of the region reflects the replacement of their initial adaptive strategy in the changing climate by the strategy of subsequent sustainable development, which tends to bring them to the initial state after "elimination" or even weakening of climatic impacts. In this respect, the ratio of resistant to elastic sustainability of geo(eco)systems characterizes to a particular extent the *chronological organization* of climatogenic transformations of natural-territorial mosaic. Let us consider the peculiarities of this chronological organization on the Pacific Ocean mega ecotone of North Eurasia.

17.5.4. Mendeleev Island-Arc Landscape

The plant communities of island volcanic landscape are characterized by a rather broad range of indices of resistant and elastic sustainability: from 0.15–0.25 to 0.70–0.95 (Figures 17.12 and17.13; Tables 17.9 and 17.10). At the same time, there is a distinct differentiation between these indices not only by floristic phratries and groups of forest types but also by the groups of microlandscapes. The average values of the *IRS* and *IES* parameters for all test plots of the Mendeleev experimental site are 0.316 and 0.456, respectively. It means that the young volcanogenic landscape is still highly sensitive to the initial hydrothermal signals but has particular mechanisms of restorative (ecogenetic) successions representing their elastic-plastic sustainability. In general, the indices of resistant sustainability of forest communities increase and the indices of elastic-plastic sustainability decrease from the top downwards along the landscape profile of the Mendeleev Volcano. Consequently, the communities of the upper and middle strata of low mountains (cedar prostrate shrubs and crooked birch forests) will be pioneers in response to climatic signals; however, the total process of their transformation will be difficult and, to a certain extent, reversible, because they are able to more rapidly restore their previous state after "elimination" of the impact. The piedmont coniferous/broad-leaved forests will exhibit a certain initial inertia but, subsequently, more intensive succession dynamics and, therefore, their climatogenic transformation will probably be maximally irreversible.

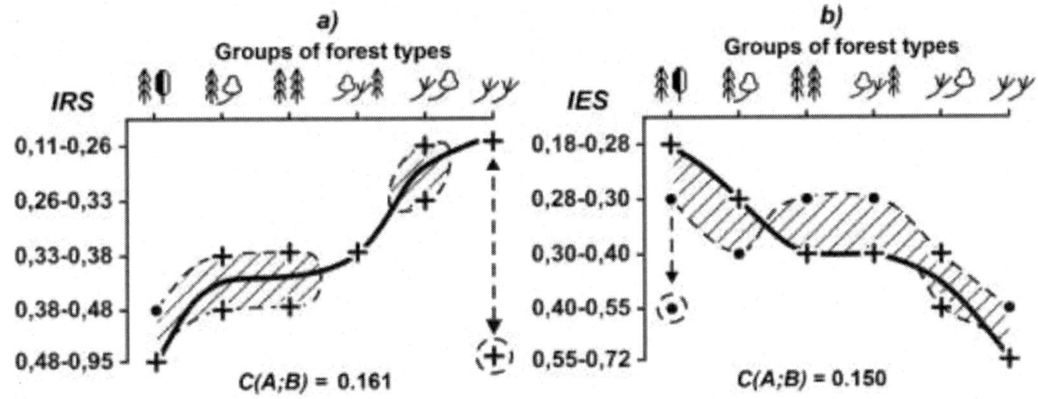

Figure 17.12. The Mendeleev landscape. Ecological niches of the groups of forest types in the space of resistant and elastic-plastic phytocoenotic sustainability.

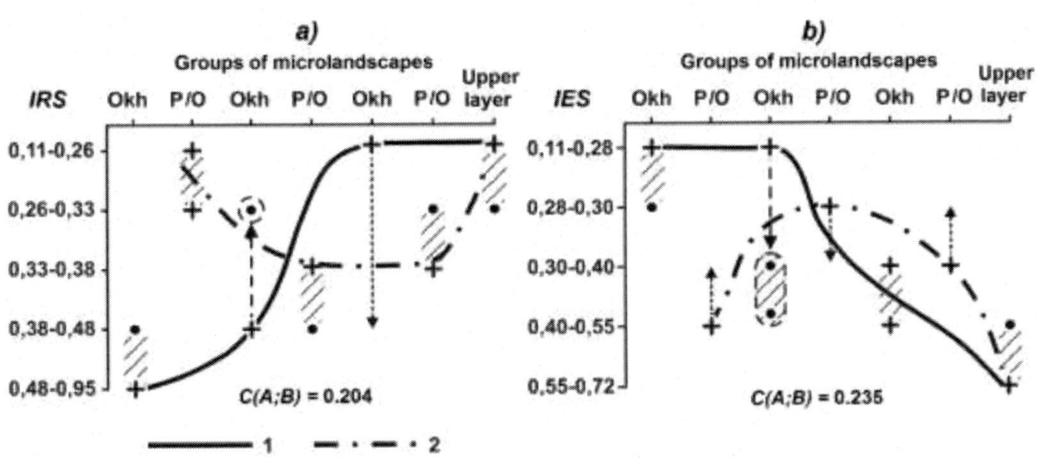

Figure 17.13. The Mendeleev landscape. Ecological niches of the groups of microlandscapes in the space of resistant and elastic-plastic phytocoenotic sustainability in the Sea-of-Okhotsk (1) and Pacific Ocean (2) sectors.

Table 17.9. Distribution of sustainability indices by the floristic phratries of Mendeleev landscape (the matrices of normalized particle coefficients of connection)

a) Resistent stability					
Meaning of index	**Floristic phratries**				
	MO	**Okh**	**BK**	**Ber**	**BKO**
0.11–0.26	0.17	0	**0.41**	**0.44**	0
0.26–0.33	0.17	0.25	**0.41**	0.11	0
0.33–0.38	0.08	**0.37**	0	0	**1.00**
0.38–0.48	0.23	**0.37**	0.18	0	0
0.48–0.95	**0.35**	0	0	0.44	0
b) Elastic stability					
Meaning of index	Floristic phratries				
	MO	Okh	BK	Ber	BKO
0.18–0.28	**0.44**	0.22	0	0	0
0.28–0.30	0.25	**0.33**	0	0	0.47
0.30–0.40	0.05	**0.33**	0.52	0.06	**0.53**
0.40–0.55	0.26	0.13	**0.48**	0.16	0
0.55–0.72	0	0	0	**0.79**	0

Footnote. The domination fields of ecological niches are in bold.

Table 17.10. Distribution of stability indexes by floristic phratries in Komsomolsk Nature Reserve (the matrix of normalized partial coefficients of connection)

a) Resistent stability					*C(A;B)* = 0.095	
Meaning of index	**Floristic phratries**					
	Mm	**Mx**	**MO**	**Okh**	**MA**	**An**
0.08 –0.26	0	0.29	0	0	**0.40**	0.10
0.26 –0.31	**0.41**	0.15	**0.45**	0.09	0.05	0.10
0.31–0.38	0.24	0.05	0.23	**0.43**	0.14	0.10
0.38–0.55	0.35	0.07	0.32	**0.49**	0	0.11
0.55–0.75	0	**0.43**	0	0	0.40	0
0.75–1.00	0	0	0	0	0	**0.59**
b) Elastic stability					*C(A;B)* = 0.083	
Meaning of index	Floristic phratries					
	Mm	Mx	MO	Okh	MA	An
0.04 –0.15	0	0	0	0	0	**0.51**
0.15 –0.33	0.10	0.26	0.39	0.33	0.07	0.08
0.33–0.42	0.28	0.12	**0.48**	0.30	0.07	0.04
0.42–0.50	**0.43**	0.22	0	**0.37**	0.06	0.07
0.50–0.65	0.20	**0.39**	0.13	0	0.22	0.13
0.65–0.88	0	0	0	0	**0.58**	0.17

Footnote. The domination fields of ecological niches are in bold.

There are highly representative parabolic curves of the dependences of sustainability indices on absolute height with the respective predominant areas of forest type groups (see Figure 17.10). First of all, these curves confirm that the bioclimatic altitudinal-zonal factor in a young island-arc landscape is the major one not only in its structural organization but also in the function of its phytobiota (see Chapters 4 and 6). It is also obvious that, with respect to the ratio of the *IRS* and *IES* parameters, the coniferous/broad-leaved forests of piedmont plains of the Mendeleev landscape slightly verge towards birch crooked forests and cedar prostrate shrubs of the middle stratum with not only high sensitivity to climatic signals but also considerable elastic-plastic sustainability. The mountain silver fir/spruce forests occupying the major and very vast range of heights of the volcanic dome are characterized, as it has already been mentioned, by the maximum productivity and the most intensive biological cycle. These functional properties allow them to maintain the relatively high initial (resistant) sustainability. However, their succession-restoration potential is poorly developed (*IES*≈ *IRS* = 0.25–0.35); therefore, their climatogenic transformations may be maximally irreversible. It is just the *mountain dark-coniferous forests that are the weakest biotic element of an island volcanogenic landscape in its confrontation with the century and supercentury climatic fluctuations.*

The chronological organization of climatogenic phytocoenological transformations is revealed by the analysis of *IRS* and *IES* for the entire series of forest types (just as for the series of floristic phratries). The first noticeable response to disturbing climatic signals (of both warming and cooling) should be expected in the upper TE layer of the volcano, in the subgoletz zone of cedar prostrate shrubs, and in the underlying zone of crooked birch forests of the Beringian and Beringian–Kamchatian floras, where resistant sustainability is minimal (*IRS* ≤0.15–0.30). In the presence of a cold-arid climatic signal, the total transgression of the above extrazonal objects will begin towards these communities: heather moors and alpine tundras. The thermoarid signal will give rise to the functional and then structural rearrangement of subgoletz prostrate forests and elfin woodlands from below the silver fir/spruce forests, with a probability of 11–13% (see Figure 17.7), followed by the appearance of mixed and even broad-leaved forests. However, in both cases the development of extrazonal objects in the subgoletz zone and at the upper boundary of dark-coniferous forests will be very difficult and therefore retarded due to rather high succession–restoration potential of the communities of cedar prostrate shrubs and crooked birch forests (*IES* ≥ 0.50–0.70).

The coniferous/broad-leaved forests of the piedmont layering of relief will be in the opposite situation. Under both cooling and warming, they will begin to be transformed last of all, because they have the maximum resistant sustainability (60% dominant of *IES* ≈ 0.40–0.60, with the maximum up to 0.95). However, when the climatic signal exceeds their tolerance limit, there will be quite intensive transformation of these forests due to their weak succession-restoration potential (*IES* ≤ 0.20–0.30).

The situations described concern mostly the more continental Sea-of-Okhotsk sector of Mendeleev landscape (Figure 17.13). In the more oceanic Pacific sector, the coniferous/broad-leaved forests highly sensitive to the primary climatic signals (*IRS* ≤ 0.25–0.35) will begin to transform much earlier but at a lower rate, because they have a considerable reserve of elastic-plastic sustainability. Their changes will be impeded by the continuous recovery of forest- restoration successions.

Under cooling, the piedmont mixed forests of the Manchurian–Okhotian flora, being subreducents, will be occupied by 60% and more with fir/spruce and stone-birch forests with cedar prostrate shrubs coming down the mountains. The transgressive portion of extra sampling objects (heather moor and alpine tundra communities), according to the EGISS model, will be here no more than 25% by 2100 and decrease to zero by 2200. According to the thermoaridization scenario (HadCM3 for the period up to 2100), the coniferous/broad-leaved forests of the piedmont layer after certain inertia will begin to be quite intensively replaced by polydominant nemoral-forest communities with thermophilic species coming from the south (see above). During 50–75 years, they can be transformed by all 100% into these extra sampling objects in their original geotopes. However, having high migrational sustainability, mixed forests per se will become absorbers for local altitudinal-zonal communities of the Mendeleev Volcano and begin to equally actively move to mountain fir/spruce forests, stone birch forests, and even cedar prostrate shrubs (see below Figure 17.7). This process will be facilitated by comparatively low elastic sustainability of these communities of reducers and suberducents (see Figure 17.12 *b*).

The mountain-forest communities taking an intermediate position between the two above-considered “poles” of mesocatena on the volcanic dome: fir/spruce forests, birch/spruce forests and buffer prostrate/birch/spruce communities, – have up to 100% moderate values of both resistant and elastic sustainability (the ranges of values from 0.25–0.30 to 0.40–0.45). Nevertheless, the mountain forests of the Sea-of-Okhotsk sector have a slightly higher restoration potential (see Figure 17.13 *b*) and, therefore, will apparently be transformed at lower rates compared to their analogs in the Pacific Ocean sector. Here, the situation will be opposite to that for the piedmont layer.

17.5.5. The Komsomolsk Nature Reserve

The forest communities and microlandscapes of marginal-continental Sub-Pacific substantially differ from the island-arc Neo-Pacific in the much weaker determinancy of functional sustainability parameters in their structural subdivisions. The coefficients of contingency *C(A;B)* of floristic phratries and groups of forest types with *IES* and *IRS* are low, close to the level of insignificance (Table 17.10; Figures 17.14 and 17.15). Ecological dominants usually do not exceed 35–55% of probability, while ecological niches are "fuzzy" and, which is most important, have numerous and rather considerable 20–30% enclaves. The latter is determined by the spatial differentiation of forest-forming process in this ecoregion by not only the predominant altitudinal-zonal factor but also the complicating effect of circulation-solar exposure, bedrock lithology, ground water depth level, mechanical composition and moisture content of soil-grounds (see Chapters 11–13). As a result, there is substantial “spreading” of the most of phytocoenotic structures over the groups of microlandscapes, with the development of enclaves and areas of their sporadic distribution (Figure 17.15) with functional parameters (including sustainability indices) different from the parameters of the main ecological niche. At the same time, the *IRS* and *IES* enclaves are located in both directions from their ecological dominants, which is indicative of the greater total spread of both types of sustainability.

It seems to be one of the mechanisms of emergence of the known (Kolesnikov 1956; Urusov 1988; Nakamura, Krestov, Glazovsky and Armand 1983; Scherer-Lorenzen 2005) floristic abundance, species richness and "multiformity" of phytocoenological systems of the Priamurye Sub-Pacific. This mechanism has also created a rather broad range of functional sustainability parameters for many structural parts of these systems. The respective diversity of sustainability parameters is also characteristic of low-mountain microlandscapes of the Lower Priamurye, though the coupling coefficients are higher here (see Figure 17.15).

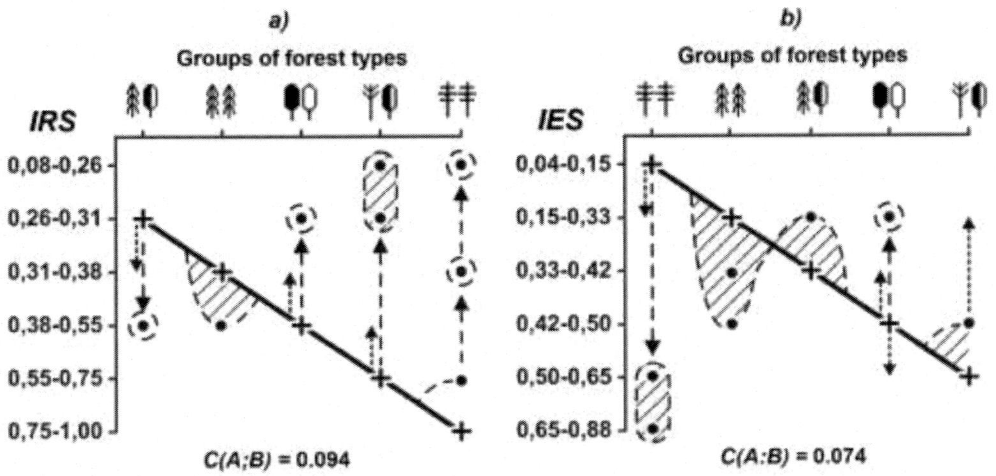

Figure 17.14. The Komsomolsk Nature Reserve. Ecological niches of the groups of forest types in the space of resistant and elastic phytocoenotic sustainability.

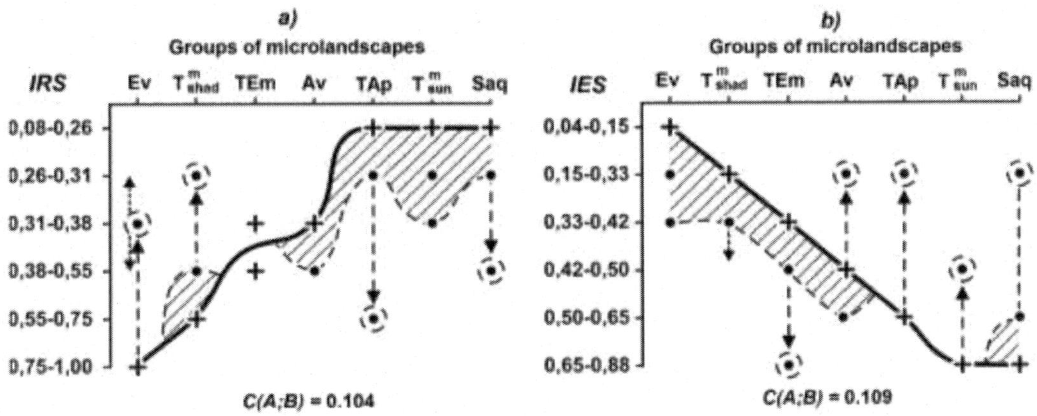

Figure 17.15. The Komsomolsk Nature Reserve. Ecological niches of the groups of microlandscapes in the space of resistant and elastic phytocoenotic sustainability.

Let us give two examples. Cedar/broad-leaved forests, being the dominants in the Ev and TA microlandscapes, are sporadically widespread over all above-lying mountain slopes up to watersheds (Figure 17.16). Accordingly, they are characterized by a very broad range of sustainability parameters (see Figure 17.14): from the maximum dominant values (*IRS* = 0.55–0.75; *IES* = 0.50–0.65) to the minimum values in enclaves and sporadic areas (*IRS* = 0.08–0.26; *IES* = 0.15–0.33). Silver fir/spruce forests with a weakly marked area of dominance (38%) on transeluvial mountain watersheds form their own rather large enclave (with a probability of 31%) on valley flat interfluves and the second, less marked enclave on sunny mountain slopes. Hence, the resistant sustainability of forests has quite a broad range and the elastic sustainability has a very "fuzzy" ecological niche.

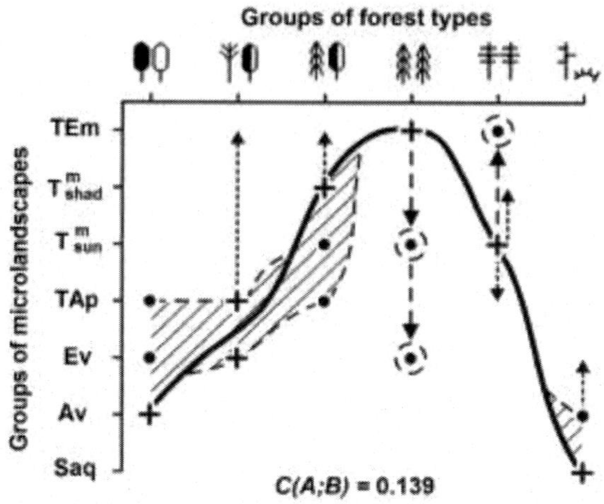

Figure 17.16. The Komsomolsk Nature Reserve. Ecological niches of the groups of forest types in the space of the groups of microlandscapes.

Thus, it can be supposed that the evolution of subaerial landscapes during the transition from insular Neo-Pacific in the marginal-continental Sub-Pacific is accompanied by certain bifurcation of the parameters of their biotic function and the respective "multiplication" of sustainability values. In short, the increase in the biodiversity of phytocoenological structures is repaid by a decreased level of their conformity to the parameters of functional sustainability and transition from rather determined connections to much more probabilistic connections. Simultaneously, there is a general reduction and alignment of taxonomic (average-weighted) norms of both types of sustainability to the range of 0.35–0.45 (Table 17.11). Accordingly, the responses of geo(eco)systems under study to the same climatic signals become more multiform and less definite.

Let us consider the characteristics of sustainability in separate groups of forest types. In the presence of both warming and cooling signals, the priority response should be expected in the Manchurian–Okhotian spruce/broad-leaved forests mostly on the shadowy mountain slopes. Their dominant IRS= 0.26–0.31, with a probability of 51%, while the maximum values of the index are no more than 0.50. The high functional-structural organization of these buffer communities takes a turn for their equally high sensitivity to climatic fluctuations. At the same time, the mechanisms of elastic-plastic sustainability are developed insufficiently (IES = 0.15–0.42, with a probability of 75%); therefore, being disturbed, they very weakly manifest their succession-restoration dynamics and the disturbances may be long-term and persistent.

Compared to spruce/broad-leaved forests, the fir/spruce forests of the Okhotian flora on transeluvial mountain watersheds and valley flat interfluves are slightly less sensitive to the primary climatic signals (IRS = 0.31–0.55, with a probability of 90%). However, their restoration potential is also poorly marked and, therefore, they can be susceptible to irreversible climatogenic transformation. This process can lead to comparatively rapid disappearance of silver fir-spruce forests under both warming and cooling, because both types of the climatic trend will be characterized by a decrease in the atmospheric humidity factor unfavorable for the Okhotian flora (see Table 17.11). As a result, silver fir/spruce forests will become reducers and subreducents (see Figures 17.10 and 17.11) capable of maintaining their self-similarity (P_{ii}) by no more than 30%, while under cooling they will be replaced by one third by extrazonal aliens – subgoletz prostrate woods.

Table 17.11. Taxonomic standards of resistent and elastic stability for floristic phratries and groups of forest types in Komsomolsk Nature Reserve

Phytocoenological complexes			Indexes of sustainability	
			resistent	elastic
Floristic phratries	Manchurian mesophilic	Mm	0.36	0.44
	Manchurian xerophilic	Mx	0.42	0.43
	Manchurian-Okhotian	MO	0.35	0.35
	Okhotian	Okh	0.40	0.36
	Manchurian-Angaridian	MA	0.39	0.64
	Angaridian	An	0.64	0.32
Groups of forest types	Broadleaf		0.38	0.40
	Cedar-broadleaf		0.43	0.43
	Spruce-broadleaf		0.35	0.37
	Fir-spruce		0.39	0.35
	Larch tree	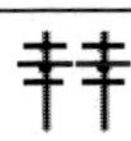	0.61	0.39
	Forest bogs	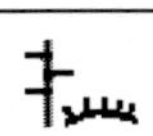	0.29	0.58

The Manchurian broad-leaved forests of valley accumulative and flat-interfluve sites and foot of slopes have the weak and moderate indices of both resistant and elastic sustainability (within a range of 0.15–0.55, with a probability of 70–80%), which must be evidence of particular spatial contrasts in their climatogenic dynamics depending on the type of mesosites.

The larch forests of the Angaridian flora on sunny mountain slopes, having the dominant IRS = 0.75–1.00 with a probability of 45%, have the lowest sensitivity to the primary climatic signals. Their initial sustainability is confirmed by the values of parameter P_{ii} (56% under warming and more than 75% under cooling). On the mountain watersheds, though, larch forests are much more sensitive to the primary hydrothermal signals (a 22% enclave of IRS = 0.08–0.26). The elastic sustainability of larch forests can equally be either maximum or close to the minimum (IES has 47–48% dominants of 0.50–0.88 and 0.04–0.15). It is in agreement with their positions of subabsorbents and mesotrophs in the system of internal landscape-ecological transitions (see Figures 17.10 and 17.11). They are most resistant also to the external phytocoenotic expansion under the cold-arid trend, which corresponds to the ecological-physiological characteristics of species of the Angaridian flora. Under thermoaridization, larch forests are less sustainable and can be by 33% absorbed by extrazonal objects (subboreal broad-leaved forests with thermophilic species).

– « » –

In conclusion of this section, we must touch upon the problem of correlations between intraregional transitions of forest ecosystems and external (extrazonal) transgressions. Transit ecotypes represent the total effect of mutual transitions only within the given statistical sample and do not concern their transformations into extra sampling objects, i.e., into the altitudinal-zonal ecosystems being initially absent in the territory of the experimental site. The given ecotypes of plant communities represent the spatial differentiation and correlation between the two categories of stability of the communities – areal (geotopic) and migrational (transgressive). However, this stability relates only to the elements of local phytocoenological mosaic. Transit geobotanic leaders (absorbents and subabsorbents) absorbing the reduced associations thereby demonstrate the local migrational sustainability. In the meantime, in their basal areas, they still can be to a great extent replaced by external communities – extrazonal aliens. It means that they initially take the marginal place in the hydrothermal fields under consideration and have no prototype objects for them in the studied sample, if the climatic trend brings them beyond the limits of these fields. It is accompanied by the appearance of "residual" transitions of a given object into other prototype objects, which are absent in the given operation system (sample), and these transitions can be rather intensive and suppress the entire system of intrasamling transformations of geo(eco)systems.

For example, in the territory of the Komsomolsk Nature Reserve under thermoaridization conditions, it will be meso-hygrophilic cedar/broad-leaved forests on the super-floodplain terraces of the Amur River and in the valleys of small rivers (see Figure 17.10). These are the valley representatives of northern cedar prostrate shrubs (Kolesnikov 1956) being under the marginal hydrothermal conditions – the highest summer temperatures and excess edaphic moisture content. Being local absorbers and transgressing with a probability up to 10–15% into the areas of cedar meso/xerophytes and silver fir/spruce mesophytes lying above the slopes, they will disappear by more than 65% from their basal geotopes. Here, they will be replaced by the aliens from more southern regions of the Middle Sikhote-Alin: hornbeam liana cedar/broad-leaved forests (Kolesnikov, 1969a) enriched in thermophilic and hydrophilic species of the Manchurian flora.

In the Mendeleev volcanic landscape, according to the EGISS cold-arid climatic scenario, an analogous situation may be observed for the upper-layer subgoletz communities of cedar prostrate shrubs, as well as the buffer prostrate/birch/spruce associations near the upper boundary of the middle layer. This situation is determined by marginal positions of the above ecosystems in the altitudinal-zonal hydrothermal space of Mendeleev volcanic low mountains. Middle-mountain tundras, heather moors and cedar/alder prostrate shrubs may be present here as extrazonal geobotanic objects. At the same time, the reducer and subreducent communities, being yet at a considerable distance from the marginal zone, are absorbed mainly by local absorbents and subabsorbents. The transgression of extrazonal aliens into their areas is substantially limited or not manifested at all. Such associations include the piedmont and low-layer coniferous/broad-leaved forests, as well as middle-layering stone birch/spruce communities.

Under the thermoarid climatic trend, according to the HadCM3 scenario for the period up to 2100, when the more continental and more thermophilic nemoral-forest communities will play the role of extrazonal objects, the situation will be exactly opposite.

Chapter 18

ABSTRACT OF EVOLUTIONARY LANDSCAPE-ECOLOGICAL CONCEPT

18.1. INTRODUCTION

The monograph presents the results of research into the spatial–temporal organization of terrestrial ecotone geo(eco)systems in the Pacific Mobile Belt as a focus of evolutionary ecological and geographical processes and phenomena at the modern stage of development of the biosphere. This trend of complex physical geography has been put forward previously by the author as a conceptual idea (Kolomyts 1987, 1988). The problem of studying the organization of biogeocoenotic and landscape systems of marginal-continental and insular land in the tectonically and climatically active zone of contact between the ocean and the continent is set forth. The proposed statements are factually substantiated on the basis of the known bioclimatic and geological-geomorphological data, as well as the results of large-scale landscape-ecological surveys performed at experimental sites in the marginal-continental and island-arc sectors of the mega ecotone. Hence, it was possible to saturate the concept with particular contents and to develop its fundamental theoretical and methodological aspects on the basis of empirical-statistical modeling and prediction of the structural-functional organization of biogeosystems and landscapes at different stages of geological history of continental biosphere formation in the Pacific Mobile Belt.

In the book we have considered the three spatial-temporal sections of geological history of the formation of continental biosphere in the boreal Eurasian region of the Pacific Ocean mega ecotone: (1) the origin and evolution of island-arc geo(eco)systems of the Neogene/Quaternary period in the oceanic environment, (2) the subsequent ascending development of the continent as a young (Mesozoic) mountain-valley morphostructure, and (3) the final regressive branch of the total trajectory of subaerial landscape genesis – the fragmentation of separate marginal parts of the continent and their sinking below sea level, with the formation of continental islands with a "senile" erosion-denudation relief. Thus, the three main steps of the known Davis–Penck–King geographical cycle have been encompassed.

The scientific and methodological significance of the research can be assessed by the following established regularities of structural-functional organization and dynamics of boreal forest geo(eco)systems in the zone of transition from the continent to the ocean.

18.2. Ecology of Island-Arc Volcanic Landscape

The empirical statistical models of the island-arc stage of continental biosphere development in the North-West Pacific have been created by the example of experimental testing area near the active Mendeleev Volcano on the Kunashir Island (the South Kuril Ridge). It was shown that local geomorphological conditions created the centers of origin of diverse phytocoenological and soil structures of the higher (zonal-regional) level at the initial stages of continental biosphere development.

We consider the plant cover and landscapes of volcanic islands, in contradistinction to the authors who have put forward the basic postulates of island biogeography and island landscape science (E. Hulten; M. Tatewaki; A.G. Voronov; G.M. Ignatyev; K.N. Dyakonov; Yu.G. Puzachenko; A.N. Ivanov; etc.), not in contrast to the analogous formations on continents but in their dynamic unity – as *particular elements of a single evolutionary physical-geographical process* (A.A. Grigoryev) forming the new territories of continental biosphere in the mobile belt of the planet. The main spectrum of altitudinal zonality originated at a rather early stage of development of subaerial landscapes, when the tectonic-volcanic massifs emerging from the oceanic abyss had yet stayed within the confines of low-mountain relief.

18.2.1. Regional Bioclimatic System of the South Kuril Islands

The *main features of bioclimatic system of the boreal island-arc Neo-Pacific* under the conditions of low-mountain volcanic relief (the absolute heights below 900 m) have been described. It is known about the general lowering of altitudinal belts in Kunashir compared to the neighboring marginal-continental Middle Sikhote-Alin Range. It is determined by the substantial difference between these regions in the degree of their oceanicity/continentality, which fits well into the general pattern of these changes for the mountain countries with different levels of atmospheric humidification. The positions of marginal-continental, continental-insular and island-arc landscapes of the south of the Russian Far East (including the region of the Mendeleev Volcano) in the hydrothermal field of the boreal belt of the Pacific Ocean mega ecotone was determined by the sums of active temperatures and annual precipitation.

In the Kuril island ecoregion, the altitudinal-zonal hydrothermal field is more contrasting than the latitudinal-zonal field, which creates more marked phytocoenological difference on the slopes of volcanic mountains and ridges compared to transitions in the latitudinal-zonal bioclimatic system. The calculations of pluviometrical level have shown that the upper borders of the coniferous/broad-leaved and black-coniferous taiga belts in the Mendeleev landscape are lower by 550 and 620 m compared to the Lower Priamurye and Middle Sikhote-Alin, respectively, which is adequate to the increase in annual precipitation in the foothills by 400–650 mm. The mean values of pluviometrical level for the upper border of dark-coniferous forests and for the upper border of the fir/spruce forest belt slightly exceeds the levels calculated for suboceanic mountainous territories in the south of Europe (the Alps and the Western Caucasus), which is quite typical for the monsoon-oceanic region.

Similarly, the thermometric and complex hydrothermal levels were calculated. The thermometric level on the Kunashir Island proved to be more than twofold higher than the altitudinal temperature gradient. That is why the entire system of altitudinal-zonal boundaries is shifted downwards and the spectrum of altitudinal zonality in the Mendeleev volcanic landscape is contracted. The group of weather stations on the continental coast of the Sea of Japan, the South and Middle Sakhalin, and the South Kuril Islands are within a comparatively narrow range of the sums of biologically active temperatures but essentially different in atmospheric humidification, which is the basic climatic factor of sharp biogeographical contrasts between the island-arc Neo-Pacific (V.B. Sochava) and the marginal-continental/continental-insular Sub-Pacific.

Under the climatic conditions of insular boreal Neo-Pacific, the main spectrum of altitudinal zonality proves to be simultaneously lower and contracted if compared to the analogous spectra of zonality of large continental islands and all the more so continental margin. The average altitudes of dark-coniferous and broad-leaved forests on the Kunashir Island are less by 200 and 100 m, respectively, compared to the Lower Priamurye (about 400 m). In contrast, the mountain tundras with elfin woods and light forests occupy the same average altitudes (630–900 m) on the island as cedar elfin wood and larch tree light forest in the Lower Priamurye. On volcanogenic islands, the entire spectrum of altitudinal belts described here is within the low-mountain stratum of macrorelief.

The main spectrum of altitudinal zonality emerges at a rather early stage of development of subaerial landscapes, when the tectonic-volcanic massifs rising from the oceanic abyss yet do not go beyond the low-mountain relief. The well-known concept of distinct and rather complete manifestation of altitudinal bioclimatic zonality on volcanoes of the South Kuril Islands, in spite of their relatively small heights, has been confirmed. Later, during the transition to the marginal-continental regime, there occurs a general raise of bioclimatic borders with the simultaneous extension of altitudinal belts.

Thus, the two most important features of this regional bioclimatic system have been established: (1) the simultaneous reduction and compression (compared to the neighboring continent) of the entire spectrum of altitudinal zonality, which is almost completely (from the forest belts to the subgoltsy belt) falls within the low-mountain layer of macrorelief and (2) the coverage by local structural-functional contrasts of the taxonomic norms of the soil-biotic characteristics of quite a number of superior zonal-regional units.

18.2.2. Tectonic-Volcanogenic Basis of Formation of Island-Arc Landscapes

The processes of the primary subaerial biogenesis occurring at the Pacific Ring of Fire (the mobile belt of the Earth) are the successive stages of modern increment of the continental biosphere. Hence, the "Mendeleev Volcano" experimental test site in Kunashir is undoubtedly representative for perceiving the initial stages of establishment and development of terrestrial geo(eco)systems in the surrounding oceanic environment. The Mendeleev Volcano belongs to an extrusive "bushed" type of isolated stratovolcanoes. This region characterizes one of the initial stages of subaerial landscape genesis. At this stage, *active volcanism* with its intensive eruptive processes is completed and changes into *passive volcanism*. The determining endogenous factor is the energy-consuming deep *fluid systems*, with their conduits (fumaroles) forming solfatare fields. However, even for the *fluid stage*, our model describes a rather stable biogeocoenological structure of this territory aged 39–40000 years. It is just the period when landscape genesis is carried out not only by the morphotectonics and hydro-climatic regime but also by the intensive "work" of soil-biotic components.

The processes of formation of volcanogenic-accumulative morphosculpture created a lithogenic basis of the island-arc stage of establishment and development of the next part of the continental biosphere. All of the newly-formed terrestrial natural complexes in the oceanic mobile belts pass through this stage; therefore, the Mendeleev volcanic landscape is a typical model of one stage of the primary subaerial landscape genesis.

The most pronounced features of this stage are as follows: (a) the direct influence of morphotectonic factor on landscape structure; (b) the high degree of conformity between the age of relief and the age of soil-forming bedrocks; (c) the distinct geological-geomorphological and bioclimatic sectorality of island territory (the Sea-of-Okhotsk and Pacific-Ocean sectors are distinguished over the entire Kuril Ridge); and (d) the youth and dynamicity of the soil-lithogenic unit of geo(eco)systems due to the thick layers of pyroclastic deposits, where soil formation is accelerated.

The rudiments of the regional properties of marginal-continental and intracontinental geographical spaces are formed at the local level in island volcanic geosystems. Island-arc geosystems are the sources of the widely known floristic and phytocoenotic diversity of the Far-East phytobiota, as well as the centers of origin of the buffer biogeocoenological systems: a phenomenal event characteristic of the continent/ocean transition zone.

18.2.3. Structural Organization of Island-Arc Lanscape

The novel methods of in-depth analysis of these relationships using discrete empirical statistical modeling both at the regional level (island physico-geographical provinces and regions) and at the scale of small natural complexes (microlandscapes) have been applied in this work to island-arc volcanic landscapes. Fundamentally new results have been obtained from the information and set-theoretical models of ecological connections between forest vegetation/soils and geomorphological/climatic factors on the Mendeleev volcanic landscape test site. The analysis and synthesis of phytocoenological, soil and microlandscape structures under particular geological-geomorphological and macroclimatic conditions on the basis of these models made it possible to determine the level of structural and functional unity or, on the contrary, disconformity of natural components and complexes formed on the morphostructures of different age and under the conditions of modern endogenous regime of passive volcanism, as well as different degrees of oceanicity of the climate.

On the Mendeleev Volcano experimental test site, the main research objects are plant communities as representatives of five *floristic phratries* and six *forest type groups*. The altitudinal-zonal structure of plant cover has been examined for the two opposite sectors: the northwestern Sea-of-Okhotsk sector, which is considered to be warmer and drier in the vegetation period, and the colder and more humid southeastern Pacific Ocean sector. In addition, the subregional typological natural complexes have been distinguished: seven *groups of microlandscapes*, which are the paragenetic links of mesoscale landscape interfaces – mesocatenas. They were distinguished by the combination of two oro-climatic factors: (1) the vertical landscape microstratification including the particular level of absolute heights and the predominant type of mesolocation and (2) the circulatory macroexposure of volcanic slopes and surrounding plains.

The modeling results provided a much stricter and more streamlined insight into the known concept of "fuzziness/discontinuity" of altitudinal-zonal boundaries on Neo-Pacific islands and the "leveling" of these boundaries by local factors. In addition, the specific role of these factors in the formation of regional natural-territorial structures was revealed. During the formation of young island-arc volcanic landscapes, the topological structural-functional contrasts encompass the taxonomic norms of soil-biotic characteristics of quite a number of zonal-regional units, which are as though *focused on various forms of lowland microrelief.* Thus, at the initial stages of subaerial landscape genesis, the *local geomorphological conditions create the centers of origin of different phytocoenological and soil structures of a higher (zonal-regional) level.* The first cause is the pronounced thermo- and hydroedaphic contrasts formed on local catenas. Such contrasts quite definitely simulate the altitudinal-zonal and expositional differentiation of hydrothermal conditions of forest growth and soil formation.

The inter-component connections in the Mendeleev volcanic landscape are presented by the general information model. These connections are organized into a highly determined system based on the predominant factor of absolute height. It forms low-mountain microlandscapes (by strata), has a direct effect on the floro- and phytocoenotic structures, and determines the lithoedaphic and hydrothermal boundary conditions of their distribution. The altitudinal zonality discloses the basic feature of bioclimatic system at the stage of island-arc biogeocoenosis: *direct subordination of phytobiota to vertical hydrothermal gradients.* In the Priamurye Sub-Pacific and in the Volga River basin, with their more continental climate, the altitudinal zonality in low mountains is weak, being "obscured" by other abiotic factors (expositional, lithogenic, etc.); in the island landscape of Neo-Pacific, it is the leading and strongest abiotic factor of biogeocoenotic organization. This fact indicates that island biogeosystems are exposed to direct influence of the respective background climatic system from the very beginning of their emergence, being yet in the grip of the ocean. The influence of the factor of solar-circulation exposure on floro- and phytogenesis in a volcanogenic-island ecosystem is quite distinct but more than three times weaker compared to the influence of absolute height. The obtained quantitative estimates do not confirm the concepts (which are based only on visual observations) of the exceptionally high disturbance of altitudinal zonality on the Kuril Islands by the exposure factor. It is also obvious that the ecosystems of island-arc Neo-Pacific are highly sensitive to the global climatic changes: the background hydrothermal signal is transmitted to phytobiota directly and only with certain adjustment by the exposure factor and microrelief forms. It seems to be the cause of the known instability of island phytobiota, its climatogenic dynamicity.

The Vysotsky-Ivanov's humidity factor (F_{hum}) is a rather representative index of natural-territorial differentiation in the island volcanic low mountains. In general, the partial effect of solar-circulation exposure is weak, and the basic contribution to moisture content is made by not solar but circulation exposure associated with the sea currents and the monsoon regime. The more sunny NE–SE slopes of the Pacific Ocean sector facing the Kuril-Kamchatka current have the same F_{hum} as the more shadowy NW–NE slopes of the Sea-of-Okhotsk-sector, where the branch of the warm Kuro-Sio current approaches (the compensatory effect). On piedmont plains, as well as in the lower and middle strata, there are no marked sector contrasts in F_{hum}. The most extensive spatial variations of atmospheric humidification are observed in the lower stratum of the Pacific Ocean sector. On the shadowy NW–NE slopes, F_{hum} is higher than on the sunny SW–W slopes. At the same time, the dark-coniferous forests of Mendeleev landscape grow in a sufficiently broad range of solar exposure. *The belt of mountain dark-coniferous forests and the birch elfin wood band* are formed under the comparatively similar conditions of atmospheric humidification, which gives grounds for considering them as a *genetically uniform altitudinal-zonal formation.*

18.3. Functional Organization of Island-Arc Landscape

18.3.1. Primary Productivity and Biological Cycle

The following two phenomena of structural-functional organization of Mendeleev volcanic landscape have been revealed. First, there is an *exceptionally high system-forming role of forest phytobiota in the formation and development of young volcanic landscapes.* The Mendeleev boreal landscape differs from its intracontinental analogs in the higher values of biological productivity and accumulation of living organic matter, which is generally typical of the Pacific Ring of Fire – the mobile belt of the planet. The annual phytiobiota production in South Kunashir boreal forests is even higher, and it is a truly phenomenal event.

The forest communities of insular Neo-Pacific in the south of the boreal belt are characterized by the annual turnover of plant matter falling into the category of intensive metabolism and being typical only of subtropical forests by the zonal standards. The following buffer communities are especially representative in this respect: cedar-elfin-wood/birch/spruce in the middle stratum of low mountains and stone-birch/spruce at the upper border of dark-coniferous forests. The strategy of plant community development with regard to the formation of maximum biomass in forest landscapes of insular Neo-Pacific is altogether much more marked compared to their zonal analogs on the neighboring continent.

The subboreal and boreal forests of the young volcanogenic Neo-Pacific landscape also differ in the *structure of productive process*. The production of ground cover green mass on volcanic islands is especially high: 60–70% of the total above-ground productivity (vs. 35–50% on the continent). The intensive production of green photosynthesizing mass comprising the annual tree waste gives an impetus to considerable acceleration of the biological cycle as a factor of functional stability of forest biogeocoenosis as is confirmed by the complex parameters of metabolism (the coefficient of annual phytomass turnover, the litter-fall index, etc.).

Thus, the extra-zonal (i.e., going beyond the intracontinental zonal norms) values of productivity and biological cycle in the subboreal and boreal forests of young island-arc landscapes demonstrate the highly efficient work of their soil-biotic core. The intensification of metabolism is aimed at providing the functional stability and structural development of forest communities. It allows the living above-ground phytomass to reach the values approaching the maximum of intracontinental ecoregions and even exceeding it. It seems to be one of the most important regularities of the regional functional organization of forest cover at the initial, island-arc stage of subboreal landscape genesis.

Then, the intensive annual turnover of above-ground phytobiota and the high rate of the entire biological cycle in South Kunashir landscapes are the main factors of their stability under the conditions of relatively low temperatures and, together with magmatic geotherms (see below), contribute to the formation of *subboreal "climatically unjustified" forests.* It is the second *phenomenon of organization of forest ecosystems at the initial, island-arc stage of subaerial landscape genesis.* Here, one cannot but see the emergence of progressive evolutionary tendencies in buffer forest communities, so typical of the Pacific Ocean mega ecotone of Northern Eurasia. The known proposition of evolutionary theory of the biosphere (S.S. Schwartz) has been confirmed, stating that *the enhanced intensity of biogeocoenoses and the improvement of biological cycles underlie evolutionary progress and are the main contents of evolution. Such evolutionary tendencies in plant cover emerge already at the initial, volcanogenic stage of formation of the continental biosphere.*

The high level of organic matter production typical of the plains and foothills of the Mendeleev volcanic landscape is determined by the most regional bioclimatic system of the southern Kuril Islands. In the low and middle altitudinal strata of the volcanic dome, the Sea-of-Okhotsk sector differs from the Pacific-Ocean sector not only in the larger reserves of living phytomass in dark-coniferous forests but also in their enhanced total productivity. The negative forest-growth effect of cold sea currents under the conditions of extremely high oceanicity of the climate is quite clearly manifested already at the early, island-arc stage of subaerial landscape genesis, within a comparatively small territory of an island with a low-mountain relief. However, in higher mountains, the productivity contrasts are substantially smoothed over. The maximal production isomorphism is typical of the sectors of the upper stratum of the volcano.

18.3.2. Polysystem Models of Biogeocoenotic Organization

The altitudinal-zonal structure of functional phytocoenotic space of a young volcanic landscape has been analyzed on the basis of "fuzzy" set models. The most common features of this space are its pronounced differentiation, fragmentation, and the relative autonomy of metabolic processes in different groups of forest types. The lateral interactions between separate fragments of phytocoenological mosaics are weekly marked either. There is a highly contrasting distribution of metabolism by the groups of forest types and a marked heterogeneity of mechanisms of the biological cycle (in the structural scheme of dominance, their binary relationships are below the threshold of indiscernibility. At the same time, the coniferous/broad-leaved and dark-coniferous forests are at the opposite "poles" of these contrasts, indicating an obvious divergence between these two major components of forest cover of Mendeleev landscape in metabolic mechanisms. Thus, the initial (island-arc) stage of subaerial landscape genesis differs from the more mature (continental) stages in the general relatively low level of spatial organization of the regional geo(eco)system, with the actual absence of its typological "core," which would provide its qualitative definiteness.

The piedmont coniferous/broad-leaved forests are most age-specific and maximally approach the climax state; however, with their sufficiently high productivity, they have already reached the limit of functional development and belong to the reduced ecotypes of forest communities of this ecoregion, i.e., microbolites. Accordingly, their biological cycle is minimally stable. They also include the most productive piedmont and mountain fir/spruce forests comprising the submicrobolite category. Subalpine cedar elfin wood communities are rather similar to plain subboreal forests in the biological cycle, in spite of their lowest productivity. Hence, they can be also considered to be sufficiently mature pseudoclimax phytocoenoses.

The piedmont-plain coniferous/broad-leaved forests are enormously different from all other forest formations of the Mendeleev landscape as regards the reserves of above-ground phytomass. Their weakly connected position of submacrobolites in the structural schemes of dominance, not only by phytomass but also by total productivity, is also characteristic. It is evidence of certain uniqueness of these subboreal forest formations, their functional isolation from the total altitudinal-zonal bioclimatic space of volcanic landscape. The mountain fir/spruce forests generally hold a leading position of *macrobolites* by total productivity. They represent the altitudinal-zonal space of the Mendeleev landscape to the greatest extent, and their total production process is maximally stable. The exceptionally high functional isomorphism of stone birch elfin woods and cedar elfin woods, as well as their substantial difference in productivity from subboreal and boreal forests, demonstrates the functional unity of these phytocoenotic groups forming the common Okhotsk–Beringian bioclimatic system of the subgoltsy belt.

Their rather high similarity to stone birch/spruce forests is indicative of the buffer state of the latter, which is transitional from dark-coniferous to subalpine belt.

The buffer cedar-elfin-wood/birch/spruce forests appeared near the upper border of the dark-coniferous belt as a result of transgression of spruce and fir into stone birch forests above the slope. These communities acquired particular structural features of dark-coniferous forest but yet retained the characteristics of crooked birch forests as regards biological cycle. Being mesobolites by productivity, they are the dominant macrobolites by the biological turnover rate. In the given spectrum of phytocoenological groups, these more complex buffer formations seem to be the youngest, progressively developing formations with the maximum functional stability. Following the known statement about the functional criteria of evolution, it may be supposed that the buffer cedar-elfin-wood/birch/spruce forests with the most intensive (and, in this sense, the most perfect) and stable biological cycle are at the forefront of evolutionary progress in the forest cover of Mendeleev volcanic landscape.

The solution of the problem of *landscape-ecological standardization* has become a novel aspect of scientific research. Using the *plakor (flat-interfluve) phytocoenoses*, the background zonal-regional standards of metabolic processes, i.e., the framework conditions for implementation of these processes, were revealed in each altitudinal stratum and each sector of Mendeleev landscape. The constructed plakor series of forest phytocoenoses made it possible to consider the background altitudinal-expositional structure of their functioning. The measures of nontransitive relations between the distinguished altitudinal-exposureunits of physical-geographical background were calculated by the three functional clusters: total above-ground productivity, biological cycle, and soil morphology as an indicator of the detritus branch of metabolism. It has been established that the phytocoenological functional space of the Mendeleev landscape is conformal by its organization to the background altitudinal-expositional ecological space, which is also represented by weak lateral connections and an extremely high value of the critical indistinguishability threshold on the orgraphs of the measures of inclusion. Both functional models are altogether adequate by productivity and biological cycle. In addition, the following issues have been confirmed: (i) the conclusion that the buffer stone-birch/spruce and cedar-elfin-wood/birch/spruce communities belong to the upper frontier band of the dark-coniferous belt of low mountains and (ii) the interpretation of genesis of these buffer communities as a result of modern transgression of fir/spruce communities into the subalpine altitudinal belt.

At the same time, the background bioclimatic model emphasizes the differences between the Sea-of-Okhotsk and Pacific Ocean sectors of plains and foothills of Mendeleev landscape both in production processes and in the biological cycle. The steep decline in autotrophic biogenesis with altitude and the weakening of phytobiotic cycle were observed in the colder and more humid Pacific Ocean sector compared to the warmer and drier Sea-of-Okhotsk sector. Then, there is a certain degree of production isomorphism between the zonal representatives of piedmont plains of the Pacific Ocean sector and the lower-stratum slopes of the volcanic dome. Accordingly, dark-coniferous forests are predominant in this sector beginning from the coastal areas. Their submacrobolite state by the biological cycle demonstrates the relative youth of these forests. On piedmont plains of the Sea-of-Okhotsk sector, the zonal bioclimatic background has a very retarded biological cycle, which creates the low values of metabolic index and homeostaticity reserves for subboreal mixed forests that are widespread here.

18.3.3. Structure of Heat Balance of Island-Arc Neo-Pacific

The energy bases of productivity of forest communities of the island-arc Neo-Pacific have been analyzed as a focus of local and regional diversity of terrestrial ecosystems, which is further developed on the continent. The mechanisms of formation at the topological level of new intrazonal types of ecological systems, which are a driving force of evolutionary processes in the biosphere, have been revealed.

The characteristic regularity of insular exctratropical Neo-Pacific is the overall southward shift of the borders of climatic belts and natural zones compared to the neighboring continents, with the simultaneous meridian extension of belts and zones, which is usually accounted for by deficiency of heat-and-energy resources. We have found out the cause of "distortion of natural zonality within the northwestern islands of the Pacific Ocean" in much more complex landscape-geophysical processes. It has been established that the temperatures of vegetation period on the islands decrease much more than the total solar radiation and the radiation balance. With the same balance, the difference in the mean temperatures of the warmest month is about 3.0–3.5°. It is caused by differences in the structure of the expenditure part of the heat balance: mainly in the ratio of heat expenditure for evapotranspiration and turbulent warmth change. The abnormally high productivity of mixed and coniferous forests of the Mendeleev landscape is ensured by much greater heat expenditure for transpiration. This expenditure reaches 71–73% and 52–58% of radiation balance in the vegetation period by total production and green mass, respectively. The heat expenditure on production of ground cover green mass is also high (33–42%), being associated with the phenomenon of gigantism in the herbage of insular Neo-Pacific. At the same time, in the Lower Priamurye, the respective heat expenditure is 26–28%, 17–19% and 2% for mixed forests and as low as 14%, 7% and ≤1% for fir/spruce forests. The similar transpiration expenditures were obtained for mixed and coniferous forests of the boreal ecotone of the Volga river basin.

Such great expenditures of radiant energy for transpiration on the islands of the Neo-Pacific sector are associated with a substantial reduction of the share of turbulent warmth exchange, which causes a decrease in the temperature of vegetation period, with the respective change in zonal hydrothermal parameters and with the southward shift of zonal boundaries mentioned above. However, *the concept of reduction of the latitudinal-zonal norms of heat-and-energy resources in the North-West Neo-Pacific based on temperature values likely needs to be revised*. Otherwise, it is impossible to explain the apparent paradox, i.e., the abnormally high productivity of natural ecosystems of insular Neo-Pacific at a relatively low temperature of vegetation period, which is set up for the indicator of "heat-and-energy resources." Undoubtedly, it is the radiation balance that should be regarded as such indicator.

Thus, *the two basic mega types of the heat balance structure in terrestrial ecosystems have been distinguished: continental and oceanic*. These types significantly differ in the ratios of two main components of the expenditure part of the balance – evapotranspiration and turbulent heat exchange. The predominance of transpiration component of the heat balance in the oceanic type creates a high level of autotrophic biogenesis in forest island communities, providing their stability at the initial stage of subaerial landscape genesis. Forest ecosystems, exerting an interactive influence on the items of heat balance expenditure, are a rather powerful regulator of hydrothermal regime of the atmosphere via their autotrophic biogenesis already at the given stage of continental biosphere development. Later on, this regulatory mechanism of forest cover becomes global.

The original cause is the pronounced thermo- and hydroedaphic contrasts on local catenas. These contrasts simulate the altitudinal-zonal as well as exposure (circulation and solar) differentiation of hydrothermal conditions of forest growth and soil formation.

The energy bases of the original role of island-arc Neo-Pacific as a focus of local and regional diversity of terrestrial ecosystems, which is further developed on the continent, were analyzed. The local contrasts of hydrothermal conditions create the spatiotemporal variability of island ecosystems, their high sensitivity to climatic changes evolutionary transformation ability.

18.4. Effect of Geothermal Energy on Forest Formation in the Island Volcanic Landscape

The special novelty data were obtained as a result of numerical modeling of the effects of volcanic geotherms (termed as hydrotherms in the geological literature) on soil temperature and moisture regimes, water balance and, accordingly, the structure, productivity and biological cycle parameters of forest communities. In the vicinity of the Mendeleev Volcano there are the conduits of magmatic steam-hydrotherms, being close to the day surface, which cause an abnormal increase in soil temperature and thereby have a considerable effect on the structure and functions of forest ecosystems.

The empirically substantiated concept of geothermal energy of passively erupting volcanoes as a factor of accelerated evolution of island phytobiota has been set forth as a generalizing result. Magmatic geotherms (steam-hydrotherms) were favorable for transgression in this region of more southern species and entire communities to this boreal-forest ecoregion over a long period of time, and these species and communities have remained here as intrazonal and relict formations.

The final phenomenon of biogeocoenological organization of volcanogenic landscape has been described. Magmatic geotherms contribute to: (a) the formation of "climatically unjustified" but sufficiently stable subboreal forest ecosystems, with subtropical relicts; (b) the increase in their floro- and phytocoenotic diversity; and (c) the emergence of rudimentary buffer forest communities. The calculations of equivalent air temperatures reflecting the total ecological effect of climatic and geothermal factors show that the forest communities of the Mendeleev volcanic landscape have been formed and function under the same heat and energy conditions as their zonal marginal-continental analogs. Obviously, *the zonal types of geo(eco)systems can be predetermined not only by the atmospheric hydrothermal factor but also by the thermolithoedaphic factor.* Receiving additional heat from the geotherms, the forest communities of the Mendeleev landscape have reached their normal zonal and/or altitudinal-zonal hydrothermal parameters. The active volcano has compensated for the insufficiency of climatic resources and provided the stable function and structural design of these "climatically unjustified" (extrazonal, according to the landscape-geophysical canons) island-arc ecosystems.

At the same time, the vapor-gas flows of heat and chemical solutions change the water balance and soil geochemical processes that reduce the productivity of forest communities and slow down the biological cycle in general – mainly because of rapid decrease in grass cover production.

The geothermal heating of edaphotopes on the slopes and near the foothills of volcanoes of the Kuril Range is not an exceptional but mass phenomenon in the hydrothermal regime of forest island-arc ecosystems. Gas-steam flows existing in the bodies of even ancient (badly destroyed) eruptive cones and being widespread in the Northwest sub-Pacific, have had substantial influence on the thermoedaphic conditions of forest growth throughout the entire Pleistocene history and maintain these conditions at present.

The volcanic island-arc landscapes of Neo-Pacific, which are in the sphere of influence of underground gas-steam-thermal springs and solfataras, become a kind of a cradle for buffer forest communities (on the South Kuril Islands, these are Manchurian–Okhotian and Okhotian–Kamchatian–Beringian). The latter are characterized by further development and more complex spatial differentiation when the territory changes from the insular state of Neo-Pacific to the marginal-continental state of Sub-Pacific, with the generally weakened oceanicity of the climate and the complicated mountain-valley relief.

The spectra of floristic and phytocoenological structures typical of the continent-to-ocean transition zone, which are maximally possible under given climatic conditions, appear in the island-arc landscapes due to passive volcanic energy. The concept of *geothermal energy of passively active volcanoes as a factor of accelerated evolution of island phytobiota* has been empirically substantiated. The magmatic geotherms of the Mendeleev landscape contributed to the transgression of more southern species and entire communities into this boreal forest ecoregion. Under the influence of geotherms, the forest cover was rearranged towards the increase in floro- and phytocoenotic diversity. *It is precisely due to the passive volcanogenic energy that the spectra of floristic and phytocoenological structures typical of the continent-to-ocean transition zone, which are maximally possible under these climatic conditions, emerge in island-arc landscapes.*

18.5. Evolutionary Ecology of Boreal Forests in Marginal-Continental Landscapes

18.5.1. Structural Organization of Forest Ecosystems

The multidimensional empirical-statistical modeling of landscape connections in the marginal-continental sector of the Pacific Ocean mega ecotone was performed on the basis of experimental test site "Komsomolysky Nature Reserve" in the Lower Priamurye ecoregion. This ecoregion belongs to Sub-Pacific (Sochava 1980) and is the eastern branch of the transcontinental boreal bioclimatic ecotone of Eurasia (Basilevich et al. 1986; Kolomyts 1995). The ecology of a boreal-forest landscape of the continental margin characterizes the mature stage of evolution of the continental biosphere. The information and fuzzy set theory-based models were used to determine the main features of mono- and polysystemic spatial organization of sub-Pacific floristic and phytocoenotic formations under the conditions of low-mountain relief and boreal bioclimatic ecotone, with the aggravated monsoon-type interaction between the continent and the ocean.

The empirical-statistical modeling of the structure and functions of boreal forest ecosystems of the marginal-continental sub-Pacific in the South of the Russian Far East has been performed for the first time. The quantitative indices of mono- and polysystem organization of forest cover made it possible *to more strictly substantiate and considerably deepen the known concepts of the evolutionary pattern of forest-formation of this place, which is based on the floristic richness of forest phytocoenoses and involves a variety of their successions* (A.N. Krištofovič; V.N. Vasilyev; B.P. Kolesnikov). *The regularities and causal mechanisms of formation of buffer forest communities so characteristic of ecotone systems have been represented more integrally.*

The general information model of monosystemic organization of plain-lowland subregional and local geo(eco)systems of the Lower Amur sub-Pacific has been used to characterize the most common regularities of formation of phytocoenological and landscape structures under the conditions of epigeosynclinal mountainous country of the Mesozoic age and the boreal bioclimatic ecotone.

The altitudinal bioclimatic zonality in the marginal-continental Sub-Pacific is manifested very selectively and is altogether substantially disturbed by the solar-circulation exposure contrasts. The characteristic ecological consequence of dislocation (altitudinal-exposure) zonality is selective penetration of representatives of different floras into some forest types by altitudinal belts and solar-circulation exposures of the slopes of different steepness, as well as in association with bedrock lithology. The combinations of these oro-climatic conditions have created certain "ecological corridors" for the migration flows of phytobiota and for the selection of novel communities – the main processes of evolutionary transformations of forest cover in the South of the Far East, according to the Krištofovič–Vasilyev–Sochava–Kolesnikov concepts.

The typical ecological consequence of dislocation geographical zonality in sub-Pacific is the mutual penetration of representatives of different floras into different forest communities, selectively by altitudinal bands and slope exposures. The correlation between the groups of forest types and the floristic phratries they are formed of is exclusively high, demonstrating the oroclimatically predetermined stability of florogenesis that creates different types of forest.

It has also been shown that the polysystem space of the Priamurye boreal ecotone has a high degree of intercomplex relationships. It was represented by mutual penetration of floras and phytocoenotic structures and resulted in the formation of buffer forest formations – a phenomenal event in the temperature belt of the Asian sub-Pacific marginal continental land.

The coupled analysis of polysystem models of microlandscapes showed certain disconformity between an integrated ("unipolar") spatially ordered system of structural phytocoenotic connections between microlandscapes, on the one hand, and a disintegrated "multipolar" functional (metabolic) space of phytobiota, on the other hand. In the territorial mosaic of this disconformity, the foci of modern forest-forming processes and the main directions of forthcoming functional and structural changes in the forest cover of Priamurye Sub-Pacific have been outlined.

The causal mechanisms of appearance of the known floristic richness and phytocoenotic diversity of forest cover typical for marginal-continental lands of the southern Far East have been established. The more detailed mechanisms of these biospheric processes have been revealed using discrete partial models of the binary ordination of geocomponent features. These models emphasize the *evolutionary nature* of the *forest-forming process* based on the floristic richness of forest phytocoenoses and including the diversity of their successions.

18.5.2. Functional Organization of Boreal Forests

The forest-formation process was differentiated by the combination of absolute heights, bedrock lithology and circulation-solar exposure of slopes, with formation of the respective phytocoenological structures. Here we can see one of the mechanisms of origin of the floristic diversity and species richness of the forests of the Piamurye sub-Pacific. It has been established that *the forests of the Priamurye boreal ecotone have comparatively similar values of organic matter production* – the basic characteristic of their functioning – under the conditions of their rather marked floristic and phytocoenological contrasts. Here we can see *one of the phenomenal ecological consequences of unique forest-forming processes* described by B.P. Kolesnikov for the forests of the Primorye–Priamurye Sub-Pacific.

The functioning of forest ecosystems is characterized by the discrete parameters of autotrophic biogenesis, according: the reserves of different live phytomasses and the productivity. The cedar and fir/broad-leaved and nemoral fir/spruce forests of sunny and neutral mountain slopes and flat crests of ridges have the maximum reserves of live phytomass. The reserves of phytomass decrease 1.5 times due to insufficiency of summer warmth on the shadowy slopes of lowlands occupied mainly by fir/spruce, spruce/broad-leaved and larch forests.

The *forests of eluvial plain sites (plakors)* of the Priamurye boreal ecotone are characterized by both the highest total annual production and the maximal values of total live phytomass. Consequently, they *possess the most developed mechanisms of adaptive functioning*, which allow them to maintain the maximum possible total production of organic matter under the given hydrothermal and edaphic conditions.

The xerophilic broad-leaved forests (with oak) were formed through a violent evolutionary strategy – the tendency to gaining the maximum levels of productivity and biomass. Such strategy of biotic community could be ensured only by the stable favorable conditions of the abiotic environment. At the same time, mesophilic broad-leaved forests were formed under less favorable conditions. Hence, their development was guided by the patience strategy, with keen interspecies competition reducing the productivity but resulting in the formation of more diverse phytocoenotic structures.

18.5.3. Buffer Boreal Forests as a Phenomenon of Evolutionary Forest Formation Processes

The regularities and causal mechanisms of formation of *buffer forest communities* typical of ecotone systems have been represented more integrally. On the basis of empirical/statistical models, we have revealed the mechanisms of emergence of such communities with the most developed structure and the high level of functioning. It has been established that *the buffer dark-coniferous/broad-leaved forests of the Priamurye sub-Pacific are a phenomenon of evolutionary landscape genesis.* The Manchurian–Okhotian spruce/broad-leaved and nemoral fir/spruce forests differ from the forests of not only their basal but also other phratries, first of all, in the more intensive development of tree layer and the highest quality class, with maintaining the maximum floristic diversity of forest stand, undergrowth and ground cover of Manchurian forests.

These buffer forest formations are also characterized by the highest indices of accumulation of live organic matter – the maximal skeletal, green and total phytomass (both above-ground and cumulative). Forest litter reaches maximum thickness due to slower mortmass decomposition; however, the weakening of mineralization processes is accompanied by intensification of the synthesis of humic substances; therefore, the humus layer of the soil also becomes very thick. Finally, the processes of illuviation, lessivage and intrasoil weathering are most intensive here and, hence, the soils of these forests are closest to the evolutionary climax.

Thus, there was not merely the transzonal mixing of the Manchurian mesophilic and Okhotian floras. The buffer forest ecosystems formed had more developed structure and more diverse functions compared to ecosystems of the basal floras. The forests of the Manchurian–Okhotian phratrie proved to be the most similar to the *evolutionary climax* (Krištofovič 1946), as the last, the highest stage of vegetation development under the given zonal-regional climatic conditions. The climax is formed by *migration processes* in the plant cover (Vasilyev, 1946), i.e., the driving forces for its evolution that create buffer forest communities.

The emergence of another buffer phratrie, Manchurian–Angaridian, had no such structural/functional effect. Its communities take an intermediate position between the forests of the basal phratries by all of the above parameters and are similar to Manchurian xerophilic forests only in the measures of floristic diversity of forest stand and undergrowth.

It follows from the above that the latitudinal-zonal (north ↔ south) transgression of tree and shrub species and phytocoenoses in general in the marginal-continental sector of the Pacific Ocean mega ecotone of Eurasia was much more marked and had an incomparably greater structural and functional effects in the forest cover of the region than the longitudinal-sectoral (east ↔ west) transgression. Such an obvious prevalence of the forest-forming process of the first group over the process of the second group could only be a result of long-term sustainable preservation of oceanic features in the regional bioclimatic system of the Primorye/Priamurye sub-Pacific.

It can also be assumed that the main biosphere function of the Manchurian–Okhotian mixed forests on the Pacific Ocean mega ecotone is the high indices of biogenic accumulation of substances, which prevents the leaching of chemical components. According to A.I. Perelman (1975), it is nothing else but the most developed *mechanism of negative reverse bio-inert connection* under the given zonal-regional conditions, which contributes to stabilization of the soil-biotic "core" of natural complexes. The phenomenon of Manchurian–Okhotian phratrie – a novel, more highly organized system of forest formation – is considered as one of the results of evolutionary phytocoenological processes that occurred in the recent geological past on the boreal ecotone of the Primorye/Priamurye sub-Pacific.

There is convincing evidence of the previously proposed concepts (Yurtsev, 1974; Kolomyts, 1987; Panfilov, 2005) about the Pacific Ocean mega ecotone of Northern Eurasia as a focus of evolutionary processes in the continental biosphere.

18.6. Continental-Island Landscape of Fading Evolution

18.6.1. Common Thesis

The structural-functional analysis of continental-island landscape of fading evolution was performed by the example of Putyatin Island (Peter the Great Bay, South Primorye). Such landscape characterizes the mega state of this part of continental biosphere on the regressive branch of its geological development. The island geosystem is at the stage of fading evolution: the "senile" erosion-denudation relief, the uniform anthropogenically modified forest cover (secondary prariefied oak forests), and the active manifestation of exogenous morpholithogenesis. According to palynological data, the modern plant cover of Putyatin Island underwent cardinal transformations in the Late Holocene: first climatogenic (from polydominant broad-leaved forests to coniferous/broad-leaved forests) and then anthropogenic (to the absolute dominance of secondary oak forests with very monotonous phytocoenotic structure). The oak forests of Putyatin Island, as a subarid phytocoenological structure, do not correspond to the modern zonal-climatic conditions of southern coastal-island regions of Primorsky Krai. Here, such *zonal representative is the soil cover* which has yet preserved many morphological features since the time of dominance of coniferous/broad-leaved forests. Under the conditions of anthropogenically altered and, to a considerable extent, degraded forest vegetation of the continental-island geosystem, *the main object of landscape structural-morphological analysis is the soil and the related processes of exogenous morpholithogenesis.*

18.6.2. Structural and Functional Organization of Continental-Island Landscape

The general information-statistical model of monosystem spatial organization of the Putyatin landscape has demonstrated: (1) the highly developed system of chain reactions traced from the original abiotic factors-edificators (geomorphological and hydrothermal) to the structural soil-biotic features, then to the parameters of productivity of forest communities and, finally, to the discrete characteristics of the biological cycle; (2) the distinct echelonment of intercomponent relationships, creating the hierarchic ordering of monosystem organization of forest biogeocoenotic cover of a continental-island land; and (3) the relative homogeneity of intercomponent relationships, with the dominance of the higher contingency coefficients, which provide a certain stability of the achieved monosystem structure. The presented model seems to be at the initial stage of its "damped" evolution under the conditions of disintegration of continental margin.

The role of slope steepness as the most powerful abiotic factor determining the entire system of local intercomponent relationships has been revealed. In essence, it is a phenomenon of morpholitho- and biogenesis in the regressive continental-island subboreal landscape, and it shows that the main factor of edaphic humidification under the conditions of "senile" relief is the degree of drainage of the territory but not its radiation-thermal regime. The hydrothermal structure of the geotopological space of continental-insular sub-Pacific was analyzed in comparison with other regions. Under the conditions of monsoon-continental climate of marginal-continental sub-Pacific, the cross-correlation between soil temperature and moisture is very weak, being evidence of their certain mutual independence, autonomy and, consequently, separate contributions of each of them to the system of intercomponent relationships. The local hydrothermal diversity creates a specific monsoon-continental type of catenary spectra of forest biogeocoenoses, which is clearly represented in the landscape of Putyatin Island.

The analysis of soil morphology made it possible to elucidate the main features of soil formation under the conditions of monsoon-continental climate of subboreal South Primorye. The hydromorphic factor is more powerful here than the thermal factor. With respect to phytomasses, productivity and biological cycle, Putyatin secondary oak forests are at a much lower level of development compared to the much more northern primary and quasi-primary broad-leaved and mixed forests of the Lower Priamurye and Kunashir (see Kolomyts, 2015, 2016). By these parameters, the forest communities of the Putyatin island landscape form a single, "unipolar" spatially ordered functional system, in the center of which there is a transit group of biogeocoenoses. The flat-interfluve (eluvial) topogeosystems in this case have the lower bioproductivity and much lower reserves of homeostaticity. The mechanisms that close the biological cycle in oak forests of Putyatin Island are inadequate to the processes of production and accumulation of live organic matter. These are substantially different characteristics of the functioning of forest communities in the continental-island landscape.

18.6.3. Exogenous Morpholithogenesis

The Putyatin continental-island landscape is a scene of extensive development of the processes of *exogenous morpholithogenesis,* which perform the total planation of the relief to a residual-denudation, "senile" state, with the development of pediments and pediplains. Slope as an element of erosion-accumulative morphosculpture is interpreted (Florensov 1978) as a *morpholithodynamic system* (MLDS). In view of the known soil-geomorphological interrelationships (B.B. Polynov; M.A. Glazovskaya; S.S. Voskresensky; A.J. Gerrarg; Yu.G. Simonov), we consider MLDS *as a morphosculptural analog of the catenary system of the types of sites and groups of forest biogeocoenoses, which we have distinguished on Putyatin Island.* Exogenous morpholithogenesis is characterized by the four types of structures of the morpholithodynamic system, which differently involve eluvial and defluction/solifluction processes, with the respective development of the microterrace structure of slopes. Terracettes are an impressive indicator of defluction/solifluction processes in the upper soil layers. There is a relationship between the distribution of terracettes and the pasture degradation of ground cover. It has been shown that *pasture degradation is a secondary factor of microterrace structure of slopes. The primary factors are the natural defluction/solifluction processes caused by the cyclic (in the periods of monsoon rains) viscous-fluid states of sandy-loamy soils under the conditions of significantly steep slopes.*

The complex model of a combined meadow/forest continental-island ecosystem was constructed (by calculating, first, the measures of similarity and inclusion relations between objects and, second, the information-statistical indices of diversity of these relations). Based on this model, *the coupled mono-polysystem analysis of spatial organization of meadow/forest ecosystem* at the experimental site was carried out. It was established that *the stability of the "slope" system is provided by the dynamicity of slope morpholithogenesis per se – inasmuch as the deluvial and the related eluvial–deluvial structures are predominant in the latter, with the appropriate thickness of the deluvial layer.* Meadow soils drastically differ from forest soils in more intensive humification, and this difference increases down the catena with the increase in edaphic moistening. Accordingly, the entire humus layer is more developed. There is no significant relationship between the thickness of soil layer B and the types of sites in the forest-meadow ecosystem. It means that *all topological couplings of the Putyatin island landscape have maintained an approximately integrated level of development of illuvial-metamorphic processes achieved in the previous historical periods of Holocene.*

18.7. The Climatogenic Mechanisms of Evolutionary Landscape-Ecological Processes

These mechanisms have been described in different sectors of the Pacific Ocean mega ecotone using the *simulation of these processes by landscape-ecological predictions.* The basis was the regional climatic scenarios of the global prediction models EGISS and HadCM3 (version A2) of island-arc Neo-Pacific and marginal-continental sub-Pacific for the forthcoming 100–200 years. These models give two predictive climatic trends: cold-arid and thermoarid. The instrument of simulation was the previously developed technique of regional and local landscape-ecological prediction (Kolomyts 2003, 2005, 2008). The mechanisms of evolutionary processes were revealed by the trajectories of functional-structural transformations of biogeosystems and landscapes in light of the given hydrothermal trends for a certain period of climatic prediction, with particular temperature and precipitation deviations from the basal period. The probabilistic nature of landscape-ecological prediction per se made it possible to reveal the local and regional diversity of forest geo(eco)system responses to the same background climatic signal, their stability, and the trajectories of their mutual and extrazonal transitions, which represented the multiple-valued pattern of climatogenic mechanisms of evolutionary processes. The predictive ecological scenarios that we have developed for the boreal forest ecosystems of the Pacific Ocean mega ecotone so far have no such analogs in domestic or foreign literature.

The Mendeleev island-arc landscape is characterized by: (1) the extremely low geotypical stability of floristic, phytocoenotic and landscape units and (2) the exceptionally high rate of functional absorption of the areals of almost all basal plant communities by extrazonal objects characterizing already the regional scales of climatogenic transformation of the biogeocoenological structure of island-arc landscape. The cold-arid trend anticipates the large-scale transgression of cedar and alder elfin woods, heather tundras, heathlands and subalpine meadows, while thermal aridization anticipates the transgression of mixed and broad-leaved forests growing nowadays in Hokkaido and Northeastern China.

The modern analogs of subboreal forests anticipated under thermal aridization in the lower and even middle layers of Mendeleev landscape can be found in the vicinity of the neighboring Golovnin Caldera at the southwestern tip of Kunashir. Here, the extrusive domes are located, with intensive fumarolic activity in their periphery. The intensive geothermal heating of soils determined the emergence of two *areas-refugiums of extrazonal broad-leaved forest associations with subtropical elements* in this region, against the background of fir/spruce forests predominant in the neighborhood.

In the low-mountain Lower Priamurye, against the background of analogous but much weaker transformations, the cooling will cause expansion of the areas of larch forests, while the warming will cause transgression of polydominant broad-leaved forests with the involvement of thermophilic species from South Primorye. The calculations and assessments of the parameters of stability of forest communities made it possible to reveal certain chronological regularities of climatogenic phytocoenological transformations in island-arc and marginal-continental landscapes. It has been shown that, in case of the thermoarid climatic trend, the buffer Manchurian–Okhotsk forests become enriched with broad-leaved species mainly due to their *external (extrazonal)* but not internal *transgression*. It confirms the known thesis (Krištofovič, 1946; Vasilyev, 1946) that the *evolutionary climax* represented in this case by the buffer spruce/broad-leaved forests is formed due to *regional migration processes* in the plant cover.

The cold-arid predictive ecological scenario specifies another aspect of the hypothetical pattern of forest cover formation in the boreal sub-Pacific. This formation seemed to occur "…by way of breaking waves from the north and northeast, with deposition of the products … of floristic input" (Krištofovič1946, p. 64). The mixing of floras in the cold-arid trend is favored by the lower (compared to thermoaridization) general pattern of phytocoenotic and microlandscape transformations, the more quiet climatogenic dynamics of geo(eco)systems.

18.8. Sustainability of (Geo-)Ecosystems and Its Chronological-Organizing Role in Their Climatogenic Transformation

The problem of ecosystem stability in forest ecology is briefly reviewed, and the applicability of well-known stability mathematic models for landscape-ecological systems is evaluated. The ideology and methodology of the quantitative assessment of stability of forest ecosystems are given. The methods of calculating of the two main types (resilience and resistance) of potential forest ecosystem stability were elaborated. They permit the evaluation of the stability of biogeocenoses as integral elementary chorological units rather than by some structural or functional characteristics. The elaborated approaches may be inherently assigned to landscape-ecological methods. They are effective because they are based on discrete parameters of biological turnover. These parameters are responsible for the main mechanisms of natural ecosystem stability and provide a great amount of empirical data for stability calculation. The parameters of resistant and elastic-plastic sustainability have been calculated for the forest geo(eco)-systems of Mendeleev island-arc landscape and the Lower Priamurye marginal-continental region. The chronological-organizing role of sustainability of forest communities in their climatogenic transformation has been revealed.

CONCLUSION

The problems of the study of geo(eco)system organization of insular and marginal-continental land in different morphotectonic and macroclimatic sectors of the Pacific ocean mega-ecotone of Northern Eurasia are set forth. The ecological and phytocoenotic effects of oro-climatic interactions along the mega-ecotone are described. The issues of the theory of evolutionary landscape study are considered as a novel trend of complex physical geography; its subject must be the processes and events of landscape formation in the tectonically and climatically active ocean-continent interface. A comprehensive evolutionary landscape-evolutionary concept of organization of terrestrial geo(eco-)systems in the Pacific Ocean Mobile Belt – the tectonically and climatically active ocean-continent interface as a focus of evolutionary ecological and geographical processes and phenomena at the modern stage of development of the biosphere has been set forth. The empirical statistical modeling of forest natural complexes has been performed using the materials of large-scale landscape-ecological surveys performed at experimental sites on the Pacific Ocean mega ecotone of Northern Eurasia. The theses put forward on the basis of evolutionary landscape studies have been substantiated.

The evolutionary model of continental biosphere presented in this book is substantially different from the theses about the "law of geoecosystem evolution of the biosphere" proposed in the works (Shilo 1967, 2000; Shilo and Shumilov 2004). This concept, applying for the development of V.I. Vernadsky's theory of biosphere, involves consideration of the development of the latter as "... a close spatial-temporal contingency of tectonic-geomorphological evolution of the Earth's surface with the formation of landscape complexes and the related biotic ... associations" (Shilo, Shumilov 2004 5). The two ecological potentials (ecosystemic and biogeochemical) are proposed to be the integral indicators of the following paragenetic triad: "tectonics – type of continental lithogenesis – biotic associations." This model corresponds to the "tectonic" hypothesis of "evolutionary–biological effects" on the Kuril Islands (Urusov, Lauve 1980; Urusov 1988, 1993; Urusov and Chipizubova 2008), which also states the existence of a direct and moreover synchronous relationship between tectonic and phytocoenological processes.

Let us say right away that the logical construction of the authors per se is static and dimensionless (and, in this sense, not geographical). It is hardly able to describe evolution as a multilevel process, because it includes geocomponents with highly diverse, merely incomparable characteristic times. Therefore, there is no point in speaking about the one-to-one (isomorphic) spatial-temporal contingency of the members of this triad, and the authors perceive the basic content of evolution in this very contingency. The representations of lithogenesis in morphotectonics and both of them in soil-biotic components are multiple-valued (homomorphic), and these representations should be considered only as the framework conditions for creating the variety of contacts between different natural environments but not as the biosphere evolution mechanisms per se. The evolutionary mechanisms are concentrated primarily in the transformations of the most high-frequency geocomponents, as it has already been mentioned by V.P. Semenov-Tyanshansky (1928).

The morphotectonic predetermination of soil-biotic components is too rough and generalized, adequate only for the largest subdivisions of the biosphere, which are ranked as physical-geographical countries and geographical belts, with their characteristic times of tens and hundreds of millions of years. This determination prevents from disclosing the entire spatial-temporal diversity of the states and dynamics of biota, which occurs within the same morphostructure and conceals the mechanisms and driving forces of evolutionary process per se.

The most important peculiarities of functioning of the "lithogenic geome – pedone – phytobiota" triad are the incomparability of the temporal frequencies of oscillations, or times of relaxation, of its components (Armand and Targulyan 1974), as well as the absence of any reliable correlations between them with a more than 3–4-fold difference between their relaxation periods (Puzachenko 1986), including the age of their modern state. *Such are the real fundamental laws of formation of the biosphere.*

Strictly speaking, our empirical model is also based on such a triad. However, as opposed to the mentioned authors who confine to the scales of large zonal-regional units of Northern Eurasia along the meridian transects, the "center of gravity" of our evolutionary analysis is shifted towards the local (topological) level of natural complexes. It is a basically novel model disclosing rather deep mechanisms of evolutionary processes in the biosphere. As is known, the sphere of ecosystems at the topological level is the most complex and active part of natural environment, its functional "core" (Sochava 1974, 1978). The "matter and energy turnovers of biogeocoenoses are interrelated and form a giant cycle of biosphere of the Earth" (Timofeeff-Ressovsky 1970 90). The structure and function of topogeo(eco)systems are the sources of global biosphere processes and driving forces of the biosphere evolution (Shvarts 1973, 1975), and here one should search for the "laws of geoecosystem evolution." These sources are the main object of evolutionary landscape-ecological research that we have carried out.

REFERENCES

Abaturov, B. D., and Lopatin, V. N. 2002. "Svyasy zonalynoy structury rastitelynogo pokrova s osobennostyami uglerodnogo obmena i energetiki derevyev, trav i kustarnikov (Connection of zonal structure of vegetation cover with the pecularity of carbon change of trees, grass and bushes)." *Povolzhsky ekologichesky zhurnal* 1: 3–18 (in Russian).

Abdurakhmanov, A. I., Razzhigaeva, and N. G., Rybkin, A. V. 2004. "Vulkan Mendeleeva – istoriya razvitiya i sovremennoe sostoyanie (o. Kunashir, Kurilyskie ostrova)." In *Vzaimosvyasy mezhdu tektonikoy, seysmichnostyu, magmoobrazovaniem i izverzheniem vulkanov*, edited by Evgeny I. Gordeev, 45–47. Petropavlovsk-Kamchatsky: Kamcht. nauch. tsentr RAN (in Russian).

Abramov, A. A. 2009. *Zakonomernosti formirovaniya geokriologicheskih usloviy rayonov aknivnogo vulkanizma (na primere vulkanizma Klyuchevskoy gruppy vulkanov).* Avtoreferat …, kand. geol.-miner. nauk. Moskva: Geologich. fak-t MGU (in Russian).

Abramov, A. A., and Gilichinsky, D. A. 2008. "Geokriologicheskie usloviya rayona Klyuchevskoy gruppy vulkanov (Kamchatka)." *Kriosphera zemli* XII (1): 29–40 (in Russian).

Akhmetyeva, N. P. 1977. *Paleogeographiya Nizhnego Priamurya (Paleogeography of Low Priamurye).* Moskva: Nauka (in Russian).

Albritton, D. L., Barker, T., and Bashmakov, I. 2003. *Izmenenie klimata, 2001: Obobshchenny doklad MGIK*, by edited Robert T. Uwotson. Geneva: World Meteorogical Opganization (in Russian).

Alisov, B. P. 1956. *Klimat SSSR*. Moskva: Izd-vo MGU (in Russian).

Alisov, B. P., and Poltoraus, B. V. 1974. *Klimatologiya*. Moskva: Izd-vo MGU. (in Russian).

Allen, D. M. 1974. "The relationship between variable selection and data augmentation and a method for prediction." *Technometrics* 16: 125–127.

Andreev, V. L. 1980. *Klassifikatsionnye postroeniya v ekologii i sistematike (Classification Constructions in Ecology and Systematics)*. Moskva: Nauka (in Russian).

Aprodov, V. A. 1982. *Vulkany*. Moskva: Mysly (in Russian).

Armand, A. D. 1975. *Informatsionnye modely prirodnyh kompleksov*. Moskva: Nauka (in Russian).

Arkhipov, S. A., and Nikolaev, V. A., eds. 1972. *Istoriya razvitiya relyefa Sibiri i Dalynego Vostoka. Yug Dalynego Vostoka.* Moskva: Nauka (in Russian).

Armand, A. D. 1983. "Resistance (homeostatics) of geographic systems to different external influences." In *Ustoichivosty geosistem (Stability of Geosystems)*, edited by Aleksey D. Armand and Ivan. Yu. Dolgushin, 14–32. Moskva: Nauka (in Russian).

Armand, A. D. 1988. *Samoorganizatsiya i samoregulirovaniye geograficheskih system (Self-organization and self-regulation of Geographical Systems).* Moskva: Nauka (in Russian).

Armand, A. D. 1989. "General concepts." In *Ekosystemy v kriticheskih sostoyaniyah (Ecosystems in Critical States)*, edited by Yury G. Puzachenko, 10–41. Moskva: Nauka (in Russian).

Armand, A. D. 1992. "Sharp and gradual timberlines as a result of species interaction." In *Landscape Boundaries. Consequences for biotic diversity and Ecological Flows*, edited by A. J. Hansen, F. and di Castri, 360–378. New York Berlin Heidelberg, Springer–Verlag.

Armand, A. D., and Targulyan, V. O. 1974. *Nekotorye printsipialynye ogranicheniya experimenta i modelirovaniya v geographii.* Izvestiya AN SSSR. Ser. Geogr.4: 129–138 (in Russian).

Armand, A. D., and Vedyushkin, M. A. 1989. *Triggernye geosistemy.* Moskva: Institut geografii AN SSSR (in Russian).

Armand, D. L. 1967. "Nekotorye zadachi i metody phyziki landshafta." In *Geophysika landshafta*, edited by David L. Armand, 7–24. Moskva: Nauka (in Russian).

Armand, D. L. 1975. Nauka o landshafte. Moskva: Mysly (in Russian).

Armand, D. L., Preobrazhensky, V. S., and Armand, A. D. 1969. "Prirodnye kompleksy i sovremennye metody ih izucheniya." *Izvestiya AN SSSR. Ser. geographich* 5: 5–16 (in Russian).

Ashby, W. R. 1956. *An Introduction to Cybernetics*. London: Oxford University Press.

Avdeyko, G. P., Popruzhenko, S. V., and Polueva, A. A. 2001. "*Sovremennaya tektonicheskaya struktora Kurilo-Kamchatskogo regiona i usloviya magmoobrazov aniya.*" [*The modern tectonic structure of the Kuril-Kamchatka region and the conditions of magma formation.*] In *Geodinamika i vulkanizm Kurilo-Kamchatskoy ostrovoduzhnoy sistemy* [*Geodynamics and volcanism of the Kuril-Kamchatka island arc system*], edited by Boris V. Ivanov, 9–33. Petropavlovsk-Kamchatsky: Institut vulkanich. geologii i geokhimii RAN (in Russian).

Bailey, R. G. 1998. *Ecoregions. The Ecosystem Geography of the Oceans and Continents.* New-York: Springer-Verlag.

Barkalov, V. Yu. 2009. *Flora Kurilyskih ostrovov* [*Flora of the Kuril Islands*]. Vladivostok: Dalynauka.

Barsh, G. 1980. "Rezulytaty i problem tipizatsii i klassifikatsii khoricheskih system." [*Results and problems of typing and classification of choric systems*] In *Struktura, dinamika i razvitie landshavtov*, edited by Vladimir S. Preobrazhensky, 101–109. Moskva: In-t geograifii AN SSSR (in Russian).

Baskov, E. A., and Surikov, S. N. 1975. *Gidrotermy Tikhookeanskogo segmenta Zemli* [Hydrothermal Pacific Segment]. Moskva: Nedra.

Bazilevich, N. I. 1981. "Produktivnosty, energetika i biogeokhimiya nazemnyh ekosistem Ti hookeanskogo kolytsa (Productivity, power engineering and biogeochemistry of terres trial ecosystens of Pacific Ocean ring)." In *Voprosy geografii. Sbornik. 117*, edited by Kirill N. Dyakonov and Roman I. Zlotin, 146–208. Moskva: Mysly. (in Russian).

Bazilevich, N. I. 1993. *Biologicheskaya produktivnosty ecosystem Severnoy Evrasii [Biological Productivity of Northern Eurasian Ecosystems*]. Moskva: Nauka.

Bazilevich, N. I., Grebenshchikov, O. S., and Tishkov, A. A. 1986. *Geographicheskie zakonomernosti struktury i funktsionirovaniya ekosistem (Geographical conformities to natural laws of ecosystem structure and functioning).* Moskva: Nauka (in Russian).

Bazilevich, N. I., and Rodin L. E. 1969. "Geographicheskie zakonomernosti produktivnosti i krugovorota khimicheskih elementov v osnovnyh tipah rastitelynosti Zemli." [Geographical patterns of productivity and the cycle of chemical elements in the main types of vegetation of the Earth.] In *Obshchie teoreticheskie problemy biologicheskoy produknivnosti*, edited by Oleg V. Zelensky and Leonid. E. Rodin, 24–33. Leningrad: Nauka (in Russian).

Bazilevich, N. I., and Rodin, L. E. 1971. "Produktivnosty i krugovorot elementov v estestvennyh i kulyturnyh fitotsenozah (po materialam SSSR)." In *Biologicheskaya produktivnosty i krugovorot khimicheskih elementov v rastitelynyh soobshchestvah* [*Biological productivity and the cycle of chemical elements in plant communities*], edited by Natalya I.

Bazilevich and Leonid E. Rodin, 5–32. *Moskva: Nauka.*

Bazilevich, N. I., and Titlyanova, A. A. 2008. *Biologichesky krugovorot na pyati kontinentah: azot i zolynye elementy v prirodnyh nazemnyh ekosistemah (Biological cycle at five continents).* Novosibirsk: Izd-vo SO RAN (in Russan).

Bee, J. N., Kunstler, G., and Coomes, D. A. 2007. 'Resistance and resilience of New Zealand tree species to browsing." *Journ. Ecol.* 95: 1014–1026.

Begon, M., Harper, J. L., and Towensend, C. R. 1989. *Ecology: Individuals, Populations and Communities*. Vol. 2. Oxford–London–Edinburg: Blackwell Sci. Publications.

Belousov, V. V. 1978. *Endogennye rezhimy materikov [Endogenous regimes of the continents].* Moskva: Nedra (in Russian).

Berg, L. S., 1958. *Izbrannye trudy. V. 2. Phizicheskaya geodraphiya.* M.: Izd-vo AN SSSR (in Russian).

Bezrukov, P. L., Zenkevich, N. L., Kanaev, V. Ph., and Elbytsev, G. B. 1958. "Podvodnye gory i vulkany Kurilyskoy ostrovnoy gryady." In *Trudy Laojratorii Vulkanologii,* Vyp. 13, edited by Boris P. Piyp, 71–88. Moskva: Izd-vo AN SSSR (in Russian).

Blüthgen J. 1966. *Allgemeine klimatogeographie*. Vol. 2. Berlin: Leherbuch der allgemeinen Geographie.

Bobrov, E. G.1980. "Nekotorue cherty noveyshey istorii flory i rastitelynosti Yuzhnoy chasti Dalynego Vostoka. [*Some features of the recent history of flora and vegetation of the southern part of the Far East*]" *Bot. zhurnal* 65 (2): 172–183.

Borzov, A. A. 1954. *Geographicheskie raboty [Geographical work].* Moskva: Geografgiz.

Boul, S. W., Hole, F. D., and McCracken, R. J. 1973. *Soil genesis and classification.* North Carolina – Visconsin: The Lowa State Univer. Press.

Braytseva, O. A., Melekestsev, I. V., and Erlikh, E. N. 1974. "Noveyshaya vulkanicheskaya deyatelynosty, Kamchatka, Kurilyskie i Komandorskie ostrova." In *Istorya razvitiya re lyefa Sibiri i Dalynego Vostoka*, edited by Ivan V. Luchitsky, 369–394. Moskva: Nauka (in Russian).

Brillouin L. 1956. *Science and information theory*. New York: Acad. Press INC.

Brigham-Grette, J. 2000. "New perspectives on Beringian Quatenary paleogeography, stratigraphy, and glacial history." *Quatenary Sci. Rev.* 20: 15–24.

Budyko, M. I. 1971. *Klimat i zhizny [Climate and live]*. Leningrad: Gidrometeoizdat.

Budyko, M. I. 1977. *Globalynaya ekologiya [Global ecology]*. Moskva: Mysly (in Russian).

Budyko, M. I. 1984. *Evolyutsiya biosphery* [*Biosphere evolution*]. Leningrad: Gidrometeoizdat.

Budyko, M. I. 2002. "Globalynoe poteplenie [Global warming]" In *Izmeneniya klimata i ego posledstviya* [*Climate change and its effects*], edited by Mikhail I. Budyko, 7–12. Sunkt-Peterburg: Nauka.

Buks, I. I. 1977. "Landscape-ecological characteristics of the Baikal-Amur Mainline zone and resistance of environment." In *Voprosy geografii. Sbornik. 105. Baikalo-Amurskaya magistraly [Problems of Geography, Vol. 105: Baikal-Amur Mainline]*, edited by Fedor V. Dyakonov and Oleg A. Kibalychich, 81–97. Moscow: Mysly.

Bykov, B. A. 1973. *Geobotanichesky slovar [Geobotanic Dictionary]*. Alma-Ata: Akad. nauk KazSSR (in Russian).

Chapin F. S., Walker B. H., and Hobbs R. J. 1997. "Biotic control over the functioning of eco-systems." *Science* 277: 500–504.

Chudaev, O. V. and Chudaeva, V. A. 2009. "Sostav i genesis gidrotermalynyh system oblastey sovremennogo vulkanizma.*" Vestnik DVO RAN* 4: 24–29 (in Russian).

Connor, C. B., Lichtner, P. C., and Conway, F. M. 1977. "Cooling of an igneous dike 20 yr after intrusion." *Geology* 25: 711–714.

Diaz, S., and Cabido, M. 2001. "Vive la difference: plant functional diversity matters to ecosystem processes." *Trends Ecol. Evol.* 16: 646–655.

Dimo, V. N. 1967. "Raschetny metod opredeleniya temperatury pochvy [*Calculation method for determining soil temperature*]." *Byulleteny Pochvennogo instituta im. V. V. Dokuchaeva* 1: 88–99.

Dobrovolysky, G. V., and Nikitin, E. D. 1990. *Funktsii pochv v biosfere i ekosistemah [Soil functions at biosphere and ecosystems]*. Moskva: Nauka.

Dobrovol'skii, G. V., and Nikitin, E. D. 2006. *Ekologiya pochv [Ecology of Soils]*, Moskva: Moscow. State. University.

Dokuchaev V. V. 1949. *Izbrannye sochineniya. Vol. III. Kartografiya, genesis i klassifikatsiya pochv*. Moskva: Gos. Izd-vo selyskokhoz. literatury.

Dokuchaev, V. V. 1953. *Nashi stepi prezhge i tepery [Our steppes present and now]*. Moskva: Selykhozgiz.

Droznin, V. A., Dubrovskaya, I. K., and Kiryukhin, A. V. 2004. "Diagnostika geotermalynogo rezervuara Avachinskogo vulkana po fumarolynoy aktivnosti i seysmologicheskim dannym." In *Vzaimosvyasi mezhdu tektonikoy, seysmichnostyu, magmoobrazovaniem i izverzheniyami vulkanov v vulkanicheskih dugah [Interrelations between tectonics, seismicity, magma formation and volcanic eruptions in volcanic arcs]*, by edited Evgeny I. Gordeev, 37–39. Petropavlovsk-Kamchatsky: Kamcht. nauch. tsentr RAN (in Russian).

Diaz S., and Cabido M. 2001. "Vive la difference: plant functional diversity matters to ecosystem processes." *Trends Ecol. Evol.* l. 16: 646–655.

Duchaufour Ph. 1968. *Precis de Pedologie. L'evolution des Soils. Essai sur la dynamique des profils* [*Precis of Pedology. The evolution of Soils. Dynamic profile test*]. Deuxieme edition. Paris.

Dyakonov, K. N. 1990. "Faktory fiziko-geograficheskoy differentsiatsii ekvatorialyno-tropicheskih ostrovov zapadnoy chasti Indiyskogo okeana. [*Factors of physico-*

geographical differentiation of equatorial-tropical islands of the western Indian Ocean]" *Zemlevedenie. Novaya seriya* XVII: 71–80.

Dyakonov, K. N. 2002. "Vzaimodetstvie strukturnogo, evolyutsionnogo i funktsionalynogo napravleny v landshaftnyh issledovaniyah. [*The interdependence of structural, evolutionary and functional directions in landscape research*]" *Vestnik MGU. Ser. R. Geografiya* 1: 13–21.

Dyakonov, K. N., and Puzachenko, Yu. G. 2005. "Teoreticheskie voprosy ostrovnogo landshaftovedeniya." In *Gorizonty geographii. K 100-letiyu K. K. Markova*, edited by Tatyana A. Yanina, 14–17. Moskva: Geogr. fak-t MGU (in Russian).

Dylis, N. V. 1978. *Osnovy biogeotsenologii (Fundamentals of biogeocoenology).* Moskva: Izd-vo MGU.

Erlikh, E. N., Melekestsev, I. V., and Shantser, A. E. 1974. "Noveyshaya tektonika." In *Kamchatka, Kurilyslie i Komandorskie ostrova. Istoriya razvitiya relyefa Sibiri i Dalynego Vostoka* [*Kamchatka, the Kuril Islands and the Commander Islands. The history of the development of the relief of Siberia and the Far East*], edited by Ivan V. Luchitsky, 345–368. Moskva: Nauka.

Fedina, A. E. 1971. *Landshaftnaya struktura Prielybrusya [Landscape structure of Prielibrus].* Moskva: Iazd-vo MGU.

Fedorov, V. D. 1974. "Stability of ecological systems and its measurement." *Izvestiya. Akad. Nauk SSSR, Ser. Biol.* 3: 402–415 (in Russian).

Fedorov, V. D. 1975. "Concept of stability of biological systems." In *Trudy Sov.-Amer. Simpoziuma. Vsestoronniy analiz okruzhayushchyi sredy [Trans. Soviet.-American Symp. Global Environmental Analysis]*, edited by Yury A. Izraely, 207–2017. Leningrad: Gidrometeoizdat.

Fedotov, S. A. 1991. "O mekhanizme vylkanicheskoy deyatelynosti na Kamchatke, Kurilo-Kamchatskoy duge i v skhodnyh strukturah [About the mechanism of volcanic activity in Kamchatka, the Kuril-Kamchatka arc and similar structures]." In *Deystvuyushchie vukkany Kamchatki [Active Wiccans of Kamchatka],* Edited by Sergey A. Fedotov and Yury P. Masurenkov, 1: 18–29. Moskva: Nauka.

Florensov, N. A. 1978. *Ocherki strukturnoy geomorfologii [Essays on Structural Geomorphology].* Moskva: Nauka.

Fridland, V. M. 1970. "Vliyunie stepeni vyvetrelosti pochvoobrazuyushchih porod na processy formirovaniya pochv v razlichnyh bioklimaticheskih zonah." *Pochvovedenie [Soil science]* 12: 5–15.

Galerkin, L. I., Barash, M. B., and Sapozhnikov, V. V. 1982. *Tikhiy ocean [Pacific Ocean].* Moskva: Mysly.

Ganzey, K. S. 2009. "Landshaftnaya tipologiya Kurilyskih ostrovov [Landscape typology of the Kuril Islands]." *Vestnik DVO RAN* 4 (146): 153–159 (in Russian).

Ganzey, K. S. 2010. *Landshaphty i fiziko-geographicheskoe rayonirovanie Kurilyskih ostrovov* [*Landscapes and physical-geographical zoning of the Kuril Islands*]. Vladivostok: Dalynauka.

Ganzey, K. S. 2014. "*Razvitie ostrovnyh geosistem pod vliyaniem vulkanizma* (*na primere ostrovov-vulkanov Kurilyskoy dugi*) [*The development of island geosystems under the influence of volcanism (for example, the volcanic islands of the Kurilian arc*)]." In *Voprosy geogrephii, Sbornik. 138. Voprosy landshaftovedeniya,* edited by Vladimir M. Kotlyakov, 295–309. Moskva: Kodex".

Gartsman, I. N. 1971. "Problemy geographicheskoy zonalynosti i diskretnosty gidrome teorologicheskih poley v gornyh usloviyah mussonnogo klimata [Problems of the ge ograhical zonality and discrete properities of hydro-meteorological fields in mountain conditions of moonson climate]." *Trudy DVNIGMI,* Leningrad, Gidrometeoizdat 35: 3–31.

Gvozdetsky, N. A. 1972. "Landshaftnaya karta i skhema fiziko-geograficheskogo rayonirovaniya Zakavkazya." In *Landshaftnoe kartografirovanie i fiziko-geograficheskoe rayonirovanie gornyh oblastey [Landscape mapping and physical-geographical zoning of mountain regions],* edited by Nikolay A. Gvozdetsky, 97–118.Moskva: Izd-vo MGU.

Gerasimov, I. P. 1976. *Geneticheskie, geograficheskie i istoricheskieproblemy sovremennogo pochvovedeniya [Genetic, Geographical, and Historical Problems of Modern Soil Science].* Moskva: Nauka.

Gerasimov, I. P. 1985. *Ecologicheskie problemy v proshloy, nastoyashchey i budushchey geographii Mira [Ecological Problems of the Past, Present, and Future World Geography].* Moskva: Nauka.

Gerasimov, I. P. 1986. *Problemy globalynoy geomorfologii (Problems of global geomorphogy).* Moskva: Nauka (in Russian).

Gerasimov, I. P., Bravar, I, and Dresh, Zh., eds. 1980. *Alypy–Kavkaz. Sovremennye problemy konstruktivnoy geographii gornyh stran [Modern problems of constructive geography of mountain countries].* Moskva: Nauka.

Gerasimov, I. P., and Grin, A. M. 1976. "Experimentalyny poligon dlya izucheniya prirodnyh i antropogennyh geosistem tsentralynoy chasti lesostepi Russkoy ravniny [An experimental training ground for the study of natural and anthropogenic geosystems in the central part of the forest-steppe of the Russian Plain](Kharakteristika, programma, pervye rezulytaty." *Izvestiya. AN SSSR. Ser. geogr.* 1: 18–26.

Gerasimov, I. P, and Velichko, A. A., eds. 1982. *Paleogeografiya Evropy za poslednie sto tysyach let (Atlas-mionografiya).* Moskva: Nauka.

Gerenchuk, K. I., Miller, G. P., and Trokhimchuk, S. V. 1963. "O morfologicheskoy structure gornyh landshaftov (na primere Ukrainskih Karpat)." In *Voprosy landshaftovedeniya [Issues of landscape science]*, edited by Nikolay A. Gvozdetsky, 33–40. Alma-Ata: Otdel geografii AN KazSSR (in Russian).

Gerrard, A. J. 1981. *Soil and landform. An Integratio of Geomorphology and Pedology.* London, Boston, Sydney: George Allen & Unwin.

Gettner, A. 1930. *Geographiya: ee istoriya, sushchnosty i metody. Transl. from German [Geography: its history, essence and methods].* Leningrad –Moskva // Gosizdat.

Gladkova, G. A., and Butovets, G. K. 1988. "Lesnye vulkanicheskie pochvy ostrova Kuna shir." *Pochvovedenie [Soil science]* 1: 54–67.

Glazovskii, N. F. and Armand, A. D., eds. 1992. *Mekhanizmy ustoychivosti geosistem [Mechanisms of Geosystem's Resistance].* Moscow: Nauka.

Glazovskaya, M. A. 1953. "Osobennosti vyvetrivaniya i pochvoobrazovaniya vo vnutrennem Tyany-Shane." In *Geographicheskie issledovaniya v Tsentralynom Tyany-Shane [Geographical exploration in the Central Tien Shan],* edited by Innokenty P. Gerasimov, 32–41. Moskva: Izd-vo AN SSSR.

Glazovskaya, M. A. 1964. *Geokhimicheskie osnovy tipologii i metodiki issledovaniy prirodnykh landshaftov [Geochemical Bases of Typology and Methodology in the Study of Natural Landscapes].* Moskva: Izd. MGU (Moscow State University).

Glazovskaya, M. A. 1973. *Pochvy Mira. Geographiya pochv* [*Soil geography.*] Moskva: Izd-vo MGU.

Glazovskaya, M. A. 1981. *Obshchee pochvovesdenie i geografiya pochv [Fundamental Soil Sciences and Soil Geography].* Moscow: Vysshaya Shkola.

Glazovskaya, M. A. 1988. *Geokhimiya prirodnyh i antropogennyh landshaftov SSSR [Geochemistry of nature and anthropogenic landscape of USSR].* Moskva: Vysshaya shkola.

Glazovskaya, M. A. 1992. "Biogeokhimicheskaya organizovannosty ekologicheskogo prostranstva v prirodnyh i antropogennyh landshaftah kak kriteriy ih ustoychivosti [Biogeochemical organization of the ecological space in natural and man-made landscapes as a criterion for their stability]." *Izvestiya RAN. Ser. Geogr.* 5: 5–12.

Golubchikov, Yu. N., and Mandych, A. F. 1979. "Landshaftno-geofizicheskie osobennosti lesnyh i bezlesnyh biogeotsenozov na yuge Khabarovskogo kraya. [*Landscape-geophysical features of forest and treeless biogeocenoses in the south of the Khabarovsk Territory.]"* In *XIV Tikhookeansky naych. Congress. Komitet CD – geografiya. Tezisy dokladov,* edited by Vladimir O. Targulyan and Yury P. Badenkov, 50–51. Khabarovsk: Tikhookean. nauch. assotsiatsiya.

Gordon C., Cooper C., and Senior C. A. 2000. "The simulation of SST, sea ice extents and ocean heat transport in a version of the Hadley Centre coupled model without flux adjustments." *Climate Dynamics* 16: 147–168.

Gorshkov, G. S. 1958. "Deystvuyushchie vulkany Kurilyskoy dugi (Volcanic activity of Kuril arc)." In *Trudy Laboratorii vulkanologii,* edited by Boris P. Piyp, 13: 5–70. Moskva: Izd-vo AN SSSR.

Gorshkov, G. S. 1967. *Vulkanizm Kurylyskoy ostrovnoy dugi (Volcanism of Kuril insular arc).* Moskva: Nauka.

Gorshkov, V. G. 1994. *Physical and Biological Bases of Life Stability. Man, Biota, Environment.* New York, Springer Verlag.

Grabkov, V. K., Berzan, A. P., Alekseeva, L. M., and Kapustyan, A. L. 1984. "O rastitelynlosti ostrova Simushir." In *Itogi issledovany po voprosam ratsionalynogo ispolyzivaniya resursov Sakhaslina i Kurilyskih ostrovov*, edited by Vladimir M. Urusov, 49–50. Yuzhno-Sakhalinsk.

Grabkov, V. K., and Khudyakov, G. I. 1993. "Vulkanicheskie lanshafty i fiziko-geogra ficheskoe rayoniriovanie Kurilyskih ostrovov." In *Issledovanie i konstruirovanie landshaftov Dalynego vostoka i Sibiri (Investigation and constructing of Landscapes of Russian Far East and Siberia)*, edited by Vladimir M. Urusov, 28–47. Vladivostok: Tikhokean. institut geografii DVNTS RAN.

Grigoryev, A. A. 1966. *Zakonomernosti stroeniya i razvitiya geographicheskoy sredy [Conformities to natural lows of struvture and development of geographical environment].* Moskva: Mysly.

Grishin, S. Yu. 2003. "Krupneyshie izverzheniya XX veka na Kamchatke i Kurilyskih ostrovah i ikh vliyanie na rastitelynosty. [*The largest eruptions of the twentieth century in Kamchatka and the Kuril Islands and their impact on vegetation*]" *Izvestiya Rus. Geogr. Ob-va* 135 (3):19–28.

Grodzinskii, M. D. 1987. "Resistance of geosystems: theoretical approach to quantitative analysis." *Izvestiya. Ross. Akad. Nauk, Ser. Geogr [News. Ross Acad. Science].* 6: 5–15.

Gvozdetsky, N. A. 1963. "O vysotnoy zonalynosti kak osnovnoy zakonomernosti landshaftnoy differentsiatsii gornyh oblastey." In *Voprosy landshaftovedeniya [Landscaping Issues],* edited by Nikolay A. Gvozdetsky, 14–22. Alma-Ata: Otdel geografii AN Kaz. SSR.

Gunderson, L. 2000. "Ecological resilience: in theory and application." *Ann. Rev. Ecol. Syst.* 31: 425–439.

Haaze, G. 1980. "Izuchenie topicheskih i khoricheskih struktur, ih dinamiki i razvitiya v landshaftnyh sistemah." In *Struktura, dinamika i razvitie landshavtov*, by edited Vladimir. S. Preobrazhensky, 57–81. Moslva: Institut geographii AN SSSR.

Haggett, P. 1975. *Geography: A Modern Synthesis. Second Edition*. New York, London: Harper Internat. Edition.

Hansen A. J., and di Castry Fr., eds. 1992. *Landscape Boundaries: Consequences for Biotic Diversity and Ecological Flows*. New York: Springer-Verlag.

Harbaugh, J. W., Bonham-Carter, G. 1970. *Computer simulation in geology.* New York – London – Sydnye – Toronto: Wiley-Interscience.

Holling C. S. 1973. "Resilience and stability of ecosystems." *Ann. Rev. Ecol.* 4: 1–23.

Holten, J. I., Paulsen, G., and Oechel, W. C., eds. 1993. *Impacts of climatic change on natural ecosystems (with emphasis on boreal and arctic/alpine areas).* Trondheim: NINA and DN.

Houghton, L. G., Meira Filho, L. G., Callander, B. A., et al., eds. 1996. *Climate Change 1995. The Science of Climatic Change.* Cambridge, UK: Cambridge Univ. Press.

Hopkins D. M. 1972. "The paleogeography and climatic history of Beringia during late Cenozoic Time." *Inter-Nord.* 12: 48 –56.

Hulten E. 1960. *Flora of the Aleutian Islands and westernmost Alaska Peninsula with notes on the flora of the Commander Islands*. 2end ed. Weinheimp.

Hulten E. 1968. *Flora of Alaska and neighboring territories*. Stanford.

Hutchinson, M. F. 2004. *Anusplin, Version 4.3. Centre for Resource and Environmental Studies.* Canberra, Australia: The Australian National University.

Igarashi, Y. 1993. "History of environmental change in Hokkaido from the viewpoint of Palynological research." In *Biodiversity and Ecology in the Northernmost Japan*, 1–19. Sapporo: Hokkaido University Press.

Ignatyev, G. M. 1979. *Tropicheskie ostrova Tikhogo okeana*. Moskva: Mysly.

Isachenko, A. G. 1963. "Osnovnye problemy landshaftovedeniya gornyh stran." In *Voprosy landshaftovedeniya*, edited by Nikolay A. Gvozdetsky, 6–13. Alma-Ata: Otdel geografii AN Kaz. SSR.

Isachenko, A. G. 1965. *Osnovy landshaftovedeniya i fiziko-geographicheskoe rayonirovanie*. [*Fundamentals of landscape science and physical-geographical zoning*] Moskva: Vysshaya shkola.

Isachenko, A. G. 1985. *Landshafty SSSR [Landscapes of the USSR]*. Leningrad: Izd-vo LGU.

Isachenko, A. G. 1988. "Sistema landshaftov i soderzhaniye landshaftnoy karty Mira." *Izvestiya Vsesoyuz. Geogr. Ob-va* 120 (6): 489–50.

Isakov, Yu. A., Kazanskaya, N. S., and Tishkov, A. A. 1986. *Zonalynye zakononernosti dinamiki geosystem [Zonaliny patterns of the dynamics of geosystems]*. Moskva: Nauka.

Ivanov, A. N. 2005. "Landshaftnye osobennosti ostrovov Severo-Zapadnoy Patsifiki *[Landscape features of the Northwest Islands]*." *Izvestiya Russ. Geogr. Ob-va* 141 (4): 4–11.

Ivanov, A. N. 2014. "Problemy ostrovnogo landshaftovedeniya (Problems of island landscape study)." In *Voprosy geographii, Vyp. 138, Gorizonty landshaftovedeniya*, edited by Vladimir M. Kotlyakov and Kirill N. Dyakonov, 138–158. Moskva: "Kodex" (in Russian).

Ivanov, G. I. 1976. *Pochvoobrazovaniye na Yuge Dalynego Vostoka* [*Soil Formation in the South of the Far East*]. Moskva: Nauka.

Ivanov, N. I. 1958. "Atmosfernoe uvlazhnenie tropiceskih i sopredelynyh stran Zemnogo shara [*Atmospheric humidification of tropical and adjacent countries of the globe*]." In *Zapiski Geogr. ob-va Soyuza SSR. Novaya seriya*, 18: 1–311. Moskva–Leningrad: Izd-vo AN SSSR.

Ivashkevich, B. A. 1933. *Dalynevostochnye lesa i ih promyshlennoe budushchee.* Moskva–Khabarovsk: Dalygiz.

Ivlev, A. M. 1977. *Osobennosti genezisa i biogeojhimii pochv Sakhalina.* Moskva: Nauka (in Russian).

Ivlev, A. M., Ignatenko, I. V., and Targulyan, V. O. (1987). "Osobennosti pochvoobrazovaniya v zone perekhoda ot kontinenta k okeanu [Features of soil formation in the transition zone from continent to ocean.]" In *Pochvenny pokrov Dalynego Vostoka,* edited byVladimir O. Targulyan, 4–10. Vladivostok: Nauch. Sovet po problemam pochvovedeniya (in Russian).

Ivashutina, L. I., and Nikolaev, V. A. 1972. "Izuchenie kontrasnosti landshaftnyh sopryazheny v tselyah obosnovaniya fiziko-geographicheskogo rayonirovaniya." In *Landshaftove-deniye*, by edited Alida A. Vidina and Irina I. Mamay, 59–73. Moskva: Izd-vo MGU.

Izrael, Yu. A., *e*d. 1992. *Otsenki ekologicheskih i sotsialno-ekonomicheskih posledstviy iz meneniy klimata. Doklad Rabochey gruppy II MGEIK.* Sankt-Peterburg: Gidrometeoizdat.

Jeffers J. N. R. 1978. *An introduction to systems analysis: with ecological applications.* Cumbria, UK: Inst. of Terrestrial Ecology. Edward Arnold.

Kabata-Pendias, A, and Pendias, H. 1986. *Trace elements in soil and plants.* Boca Raton, Florida, USA: CRC Press, Inc. Forth Printing.

Karpukhin, S. S., and Karyagin, P. M. 1984. "Atmosphernye osadki i raspredelenie vlagi v chekhle sklonovyh otlozheny." In *Prognozno-geographichesky analiz territorii adminis trativnogo rayona*, edited by Andrey P. Kapitsa and Yury G. Simonov, 66–70. Moskva: Nauka.

Karpukhin, S. S., Karyagin, P. M., and Rototaeva, T. P. 1984." Sklonovye protsessy." In *Prognozno-geographichesky analiz territorii administrativnogo rayona*, edited by An drey P. Kapitsa and Yury G. Simonov, 63–66. Moskva: Nauka.

Kato R. 1952. "Hokkaido Enshuurin no shinrin shokusei [Vegetation of Hokkaido Experimental Forest]." *Res. Rep. Tokyo Univ. Exp. Forest.* 43: 1–18.

Kerzhentsev, A. S. 1999. "Mechanism of spatial-temporal variability of soils and ecosystems." In *Ekologiya i pochvy [Ecology and Soilos].* edited by Anatoly S. Kerzhentsev, 3: 31–58.

Kerzhentsev, A. S. 2001. "Fluktuatsiya, metamorfozy, evolutsiya i ekologiya pochv." In *Ecologiya i pochvy. Izbrannye lektsii X Vserossiyskoy shkoly,* by edited Anatoly S. Kerzhentsev, IV: 18–24.. Pushchino: Pushchinsky nauch tsentr RAN.

Kerzhentsev, A. S. 2006. *Funktsionalynaya ekologiya*. Moskva: Nauka.

Kerzhentsev, A. S. 2010. "Soil functionality and ecosystem sustainability." *Herald Russ. Acad. Sci.* 80 (4): 360–364.

Khain, B. E., 1973. *Obshchaya geotektonika.* Moskva: Nedra.

Khilmi, G. F. 1966. *Osnovy fyziki biosphery*. Leningrad: Gidrometeoizdat.

Khotinsky, N. A. 1977. *Golotsen Severnoy Evrasii*. Moskva: Nauka.

Khromov, S. P., Mamontova, L. I. 1974. *Meteorologichesky slovar*. Leningrad: Gidrometeoizdat.

Khudyakov, G. I. 1972. "Istoriya razvitiya relyefa." In *Istoriya razvitiya relyefa Sibiri i Dalynego Vostoka. Yug Dalynego Vostoka*, by edited Sergey A. Arkhipov and Vladimir A. Nikolaev, 64–234. Moskva: Nauka.

Khudyakov, G. I. 1977. *Geomorphotectonika Yuga Dalynego Vostoka [Geomorphotectonics of South of Far East]*. Moscow: Nauka.

Klijn, F., & Udo de Haes, H. 1994. "A hierarchical approach to ecosystems and its implication for ecological land classification." *Landscape Ecology* 9 (2): 89–104.

Kojima S. 1991. "Classification and ecological characterization of coniferous forest phyto geocoenoses of Hokkaido, Japan." *Vegetation* 96: 25–42.

Kolesnikov, B. P. 1956. *Kedrovye lesa Dalynego Vostoka (Cedar forests of Far East).* Moskva-Leningrad: Izd-vo AN SSSR.

Kolesnikov, B. P. 1961. "Rastitelynosty." In *Dalyny Vostok, Fiziko-geographicheskaya kharakteristika [Physico-geographical characteristic]*, dited by Gavriil D. Rikhter, 183–245. Moskva: Izd-vo AN SSSR.

Kolesnikov, B. P. 1969 a. "Rastitelynosty." In *Yuzhnaya chasty Dalynego Vostoka [Southern part of the Far East]*, by edited Vera. V. Nikolyskaya and Aleksandr S. Khomentovsky, 206–250. Moskva: Nauka.

Kolesnikov, B. P. 1969 b. *Vysokogornaya rastitelynosty Srednego Sikhote-Alinya.* Vladivostok: Dalynevost. knizhnoe Izd-vo.

Kolomyts, E. G. 1977. "Gidrotermicheskaya struktura landchaftov Khingano-Bureinskogo nizkogorya." In *Landchaftnye kontrasty v geosistemah*, edited by Andrey P. Kapitsa, 19–76. Vladivostok: Tikhookean. institt geografii DVNTS AN SSSR.

Kolomyts, E. G. 1979. "Ekologo-geograficheskie aspekty izucheniya gornoy taygi Srednego Urala." In *Temnokhvoynye lesa Srednego Urala*, by edited Boris P. Kolesnikov. Trudy In-ta ekologii rasteny i zhivotnyh, 128: 51–83.

Kolomyts, E. G. 1983. "Teoretiko-mnozhestvennoe modelirovanie vysokogornyh ecosystem." *Trugy Vysokogornogo Geophys. Instituta* 52: 65–83.

Kolomyts, E. G. 1987. *Landshaftnye issledovaniya v perekhodnyh zonah (Landscape research in transitional zones)*. Moskva: Nauka.

Kolomyts, E. G. 1988. "Ecoton kak obyekt fiziko-geographicheskogo issledovaniya." *Izvestiya AN SSSR. Ser. Geogr.* 5: 24–36.

Kolomyts, E. G. 1995. "Organizatsiya i ustoychivosty khvoynolesnyh ecosystem na borealy nom ecotone Russkoy ravniny (Organization and resistance of coniferous ecosystems in boreal ecotone of the Russian Plain)." *Izvestiya RAN. Ser. geogr.* 3:37–51.

Kolomyts, E. G. 1998. *Polimorfizm landshaftno-zonalynyh system*. Pushchino: ONTI Pushchinskogo nauch. tsentra RAN.

Kolomyts E. G. 2003. *Regionalnaya model globalnyh izmeneniy prirodnoy sredy (Regional model of environment global changes).* Moskva: Nauka (in Russian. Abstract by English, 306–318).

Kolomyts, E. G. 2005. *Borealyny ecoton i geographicheskaya zonalynosty. Atlasmonografiya (Boreal ecotone and geographical zonality. Atlas-monograph).* Moskva: Nauka (in Russian).

Kolomyts, E. G. 2008. *Lokalnye mekhanizmy globalnykh izmeneniy prirodnykh ekosistem (Local mechanisms of global changes in natural ecosystems).* Moskva: Nauka (in Russian. Abstract in English374–380).

Kolomyts, E. G. 2011. "Polymorphism of landscape-zonal systems (Empirical-statistical modeling)."*Internat. Journ. Ecol. Econ. Stat.* 21 (11): 58–81.

Kolomyts, E. G. 2012. *Forest ecosystems and global climate changes (Experience of predicttive modeling).* Saarbrücken, Germany: LAP LAMBERT Acad. Publishing GmbH & Co. Kg.

Kolomyts, E. G. 2015 a. "Buffer boreal forests as an evolutionary phenomena on the Pacific Ocean mega ecotone of northern Eurasia." *Journ. of Global Ecology and Environment* 2 [Issue 2]: 90–115.

Kolomyts, E. G. 2015 b. *Pacific ocean mega ecotone of northern Eurasia, Part I, Evolutionary ecology of Boreal Forests.* Raleigh, North Carolina, USA: Lulu Press, Inc. (in Russian Abstract in English 222–225).

Kolomyts E. G. 2015 c. "Polysystem ecological space and stability of alpine meadows (Central Caucasus)." *Journ. of Global Ecology and Environment* 2 [Issue 1]: 12–33.

Kolomyts, E. G. 2016 a."Buffer boreal forests as an evolutionary phenomenon on the Pacific Ocean mega ecotone of Northern Eurasia." *Environmental Research Journal* 10 [Issue 1]: 1–53.

Kolomyts, E. G. 2016 b. "Effect of geothermal energy on forest formation in insular volcanic landscapes." *American science journal* 2 (1): 25–44.

Kolomyts E. G. 2016 c. "Geotherms and forest formations in insular volcanic landscape." *Journ. of Global Ecology and Environment* 4 (4): 195–211.

Kolomyts, E. G., Kerzhentsev, A. S., and Sharaya, L. S. 2015. "Analytic and cartographic models of the functional stability of Forest Ecosystems." *Biology Bulletin Reviews* 5 (4): 311–330.

Kolomyts, E. G., Rozenberg, G. S., Glebova, O. V., Surova, N. A., Sidorenko, M. V., and Yunina, V. P. 2000. *Prirodny complex bolyshogo goroda. Landshaftno-ekologichesky analiz (Natyre complex of large city. Landscape-ecological analysis).* Moskva: Nauka (in Russian).

Kolomyts, E. G., Rozenberg, G. S., Saksonov, S. V., and Sharaya, L. S. 2012. *Forests of Volga river basin under global warming (Landscape-ecological analysis and prognosis).* New York: Published by Nova Science Publishers, Inc.

Kolomyts, E. G., Rozenberg, G. S., and Sharaya, L. S. 2009. "Methods of landscape ecology in prognostic estimation of the biotic regulation of the carbon cycle under conditions of global climate warming." *Russ. Journ. Ecology* 40 (6): 379–386.

Kolomyts, E. G., Sharaya, L. S., Surova, N. A. 2016. "Effect of geothermal energy on forest formation in insular volcanic landscapes." *Geografy and natural resources* 4: 29–40.

Kolomyts, E. G., and Surova, N. A. 2014. "Modeli ekologicheskogo prostranstva i ustoychivosti vysokogornyh lugov (na primere Tsentralynogo Kavkaza)." *Geografiya i prirodnye resursy* [*Geography and Natural Resources*] 4: 120–131.

Kolomys, E. G., and Sharaya, L. S. 2015. "Kolichestvennaya otsenka funktsionalynoy ustoychivosti lesnyh ecosistyem [Quantitative evaluation of functional sustainability of forest ecosystems]." *Russ. Journ. Ecology* 2: 83–94 (in Russian).

Kolomyts, E. G. and Surova, N. A. 2018 a. "Island volcanic ecosystems in Pacific mobile belt as sources of continental biosphere formation." *Biology Bulletin Reviews* 8: 328–349.

Kolomyts, Erland G., and Surova, Natalya A. 2018 b. "Regional bioclimatic system of the South-Kuril island ridge and its evolutionary role on the insular-arc stage of continental biosphere formation." *Journ. of Global Ecology and Environment* 2: 57–85.

Kolomyts, E. G. and Surova, N. A. 2018 c. "Island volcanic ecosystems in the Pacific mobile belt as sources of continental biosphere formation." *Biology Bulletin Reviews* 8 (4): 328–349.

Kolomyts, E. G., and Tsepkova, N. L. 2009. "Informatsionnaya modely monosistemnoy organ izatsii vysokogornyh lugov (na primere Prielybrusya." *Izvestiya Samar. nauch. tsentra RAN* 12 (1): 9–20.

Kolomyts, E. G., Yunina, V. P., Sidorenko, M. V., and Vorotnikov, V. P. 1993. *Ekosistemy khvoinogo lesa na zonalynoi granites. Organizatsiya, ustoichivosty, antropogennaya dinamika (Ecosystems of coniferous forest on the zonal edge: Organization, resistance, and anthropogenic dynamics).* Nizhnii Novgorod: Institute ekologii Volzh. basseyna RAN.

Komarov, A. S., Chertov, O. G, and Zudin, S. L. 2003. "EFIMOD 2 – a model of growth and elements cycling of boreal forest ecosystems." *Ecol. Modell.* 170: 373–392.

Komarov, V. L. 1940. *Uchenie o vide rasteny (ctranitsa iz istorii biologii).* Moskva–Leningrad: Izd-vo AN SSSR.

Komarov, V. L. 1947. *Vvedenie k floram Kitaya i Mohgolii. Izbr. Sochineniya. T. 2.* Moskva–Leningrad: Izd-vo AN SSSR.

Komedchikov, N. N, Asoyan, D. S, and Loginova, L. V. 2012. "Atlas Kurilyskih ostrovov – novy tip regionalynogo atlasa: osobennosti soderzhniya i ekologo-geograficheskoy kharakteristiki ostrovnoy geosistemy.[Atlas of the Kuril Islands - a new type of regional atlas: features of the content and ecological and geographical characteristics of the island geosystem]" *Isvestiya RAN. Ser. Geogr.* 3: 109–117.

Kononova, M. M. 1963. *Organicheskoe veshchestvo pochvy: ego priroda, svoistva i metody izucheniya (Soil Organic Matter: Nature, Properties, and Study Methods).* Moscow: Akad. Nauk SSSR.

Korotky A. M. 1972. "Chetvertichny period." In *Istoriya rasvitiya relyefa Sibiri i Dalynego Vostoka. Yug Dalynego Vostoka*, edited by Sergey A. Arkhipov and Vladimir A. Nikolaev, 264 – 304. Moskva: Nauka.

Korotky A. M. 2008. "Kurilyskie ostrova." In *Geosistemy Dalynego vostoka na rubezhe XX–XXI vekov. Vol. I. Prirodnye geosistemy i ih komponenty [Geosystems of the Far East at the turn of the twentieth and twelfth century. Volume. I. Natural geosystems and their components]*, edited by Sergey S. Ganzey, 110–118. Vladivostok: Dalynauka.

Korotky A. M., Korobov, V. V., and Skrylynik, G. P. 2011. *Anomalynye prirodnye protsessy i ih vliyanie na sostoyanie geosistem Yuga Rossiyskogo Dalynego Vostoka [Anomaly natural processes and their influence on the state of geosystems in the South of the Russian Far East].* Vladivostok: Dalynauka.

Korotky, A. M., and Skrylynik, G. P. 2008. "Opasnye prirodnye protsessy i yavleniya Rossiyskogo Dalynego Vostoka." In *Geosistemy Dalynego vostoka na rubezhe XX–XXI vekov. Vol. I. Prirodnye geosistemy i ih komponenty*, edited by Sergey S. Ganzey, 348–383. Vladivostok: Dalynauka.

Korotky, A. M., Volkov, V. G., Grebennikova, T. A., Razzhigaeva, N. G., Pushkary, V. S., Ganzey, L. A., and Mokhova, L. M. 1999. "Dalyny Vostok (Far East)." In *Izmenenie klimata i landshaftov za poslednie 65 millionov let [Climate and landscape change over the past 65 million years]*, edited by Andrey A. Velichko, 146–164. Moskva: GEOS.

Korsunskaya, G. V. 1946. "Vulkany yuzhnoy gruppy Kurilyskih ostrovov." *Izvestiya Vsesoyuz* [*News of the All-Union*]. *Geogr. Ob-va* LXXX (4): 433–442 (in Russian).

Korsunskaya, G. V. 1958. *Kurilyskaya ostrovnaya duga (fiziko-geographichesky ocherk).* Mockva: Geographgiz (in Russian).

Korzhuev, S. S. 1990. "Morfostrukturnye aspekty goroobrazovaniya i tektoniki plit." *Zemlevedenie. Novaya seriya* XVII (LVII): 114–126 (in Russian).

Kotlyakov, V. M., and Komarova A. I. 2007). *Geographiya. Ponyatiya i terminy [Geography. Concepts and terms].* Moskva: Nauka.

Kotlyakov, V. M., Baklanov, P. Ya., and Komedcikov, N. N., eds. 2009. *Atlas Kurilyskih ostrovov (Atlas for Kuril islands).* Institut geographii i Tikhookean. institute geographii RAN. Moskva–Vladivostok: IPTS "DIK".

Kovda, V. A. 1971. "Biogeochemical concept of biosphere." In *Resursy biosfery na territorii SSSR [Resources of Biosphere in the Soviet Union]*, edited by Innokenty P. Gerasimov, 16–26. Moskva: Nauka.

Kovda, V. A. 1973. *Osnovy ucheniya o pochvah. Obshchaya teoriya pochvoobrazovatelyskogo protsessa. Kniga 1.* Moskva: Nauka (in Russian).

Kozlov, D. N. 2013. *Osobennosti morfologii kraternyh ozer Kurilyskih ostrovov.* Avtoreferat … kand. geogr. nauk. Sankt-Peterburg: Ross. Gos. Ped. Univers. im A. I. Gertsena (in Russian).

Kozlov, D. N., and Zharkov, R. V. 2012. "Teplovizionnaya syomka aktivnyh vulkanov Kurilyskih ostrovov v 2009–2011 gg." *Vestnik KRAUNTS, Nauki o Zemle [Herald KRAUNZ, Earth Sciences]* 1 (19): 1–7.

Krambein, W. C., and Graybill, Fr. A. 1965. *An introduction to statistical models in geology.* New-York: McGram-Hill Book Company.

Krauklis, A. A. 1974. "Osobennosti geograficheskih gradatsiy topicheskogo poryadka." In *Topologicheskie aspekty ucheniya o geosistemah [Topological aspects of the study of geosystems]*, edited by Viktor B. Sochava, 87–137. Novosibirsk: Nauka.

Krauklis, A. A. 1979. *Problemy experimentalynogo landshaftovedeniya [Problems of experimental landscape science].* Novosibirsk: Nauka.

Krestov, P. V. 1993. "Ekologo-fitotsenologicheskaya kharakteristika lesov shirokolistvenno-khvoynogo poyasa v basseyne reki Bolyshaya Ussurka, Primorskiy kray." *Bot. zhurnal [journal]* 78 (4): 116–122.

Krestov P. V. 2004. "Rastitelyny pokrov Komandorskih ostrovov." *Bot. zhurnal* 89 (11): 1740–1962.

Krestov, P. V. 2006. *Rastitelyny pokrov i phitogeographicheskie linii Severnoy Patsifiki [Vegetation cover and phytogeographical lines in Northern Pacific].* Vladivostok: Biologo-pochvenny institut DVO RAN.

Krishtofovich, A. N. 1946. "Evolutsiya rastitelynogo pokrova v geologicheskom proshlom i ego osnovnye factory [Evolution of plant cover in the geological past and its main factors].” In*Materialy po istorii flory i rastitelynosty SSSR. Vyp. II,* by edited Vladimir L. Komarov, 21–87. Moskva–Leningrad: Izd-vo AN SSSR.

Krishtofovich, A. N. 1958. “Proiskhozhdenie flory Angarskoy sushi.” In *Materialy po istorii flory i rastitelynosty SSSR. Vyp. III,* edited by Vladimir N. Vasilyev, 7–63. Moskva–Leningrad: Izd-vo AN SSSR.

Krutyko, V. N., Pegov, S. A., Khomyakov, D. M., and Khomyakov, P. M. 1982. *The formalization of quality assessment of the environmental components.* Preprint of the Russian Inst. of Systemic Studies, Russ. Acad. Sci.

Kosugi, Y., Tanaka, H., and Takanashi, S. 2005. “Three years of carbon and energy fluxes from Japanese evergreen broad-leaved forest.” *Agricultural and Forest Meteorology* 132: 329–343.

Kubota, Y., Murata, H., and Kikuzawa, K. 2004. “Effects of topographic heterogeneity on tree species richness and stand dynamics in a subtropical forest in Okinawa Island, southern Japan. *”Journ. of Ecology* 92: 230–240.

Kulytiasov, I. M. 1982. *Ecologiya rasteniy*. Moskva: Izd-vo MGU (in Russian).

Kupriyanova, T. P., ed. 1989. *Faktory i mekhanizmy ustoychivosti geosistem [Factors and Mechanisms of Resistance of Geosystems]*. Moskva: Inst. Geogr., Akad. Nauk SSSR.

Kurentsova, G. E. 1973. *Estestvennye i antropogennye smeny rastitelynosty Primorya i Yuzhnogo Priamurya (Nature and human vegetation changes on Primorye and South Priamurye).* Novosibirsk: Nauka.

Kustler, G. 1957. ”ABC of information theory.” In *Information Theory in Biology*, edited by Hubert P. Yockey, 5–48. London–New York–Los Angeles: Pergamon Press.

Lappo, S. S., Lebedev, V. L. 2005. “Geograficheskaya zonalynosty v vodah Mirovogo okeana.” In *Voprosy geographii, Vyp. 138, Gorizonty landshaftovedeniya*, edited by Vladimir M. Kotlyakov and Kirill N. Dyakonov, 230–237. Moskva: Geogr. fakulytet MGU.

Larcher, W. 1976. *Okologie der Pelanzen.* Stuttgart, Germany: Verlag Eugen Ulmer.

Lashkov, A. N. 1948. “K morphologii pochv Yuzhnyh Kurilyskih ostrovov.” *Izvestiya Vsesoyuz [News of the All-Union]. Geogr. Ob-va* 80 (1): 61–68.

Letnikov, Ph. A. 2000. “Phlyuidny rezhim endogennyh protsessov v kontinentalynoy lithosphere i problemy metallogenii” In *Problemy globalynoy geodinamiki [Problems of global geodynamics],* edited by Dmitry. V. Rundkvist, 204–224. Moskva: GEOS.

Leonov, V. L. 1991. “O nekotoryh zakonomernostyah razvitiya gidrotermalynoy i vulkanicherskoy deyatelynosti na Kamchatke.” *Vulkanologiya i seysmologiya* 2: 28–40.

Liverovsky, Yu. F. 1969. “Pochvy.” In *Yuzhnaya chasty Dalynego Vostoka*, by edited Vera. V. Nikolyskaya and Aleksandr S. Khomentovsky, 159–204. Moskva: Nauka.

Liverovsky, Yu. A. 1974. *Pochby SSSR. Geograficheskaya kharakteristika*. Moskva: Mysly.

Logofet, D. O., and Svirzhev, Yu. M. 1983. “Concept of stability of biological systems.” In *Problemy ekologicheskogo monitoringa i modelirovaniya ekosistem [Problems of Environmental Monitoring and Modeling of Ecosystems]*, edited by Yury A. Izraely, 6:159–171.Leningrad: Gidrometeoizdat.

Loreau M., Downing A., and Emmerson M. 2002. “A new look at the relationship between diwersity and stability.” In *Biodiversity and ecosystem functioning*, by edited Mitchel Loreau, Shahin Naeem and Pablo Inchausti, 79–91. Oxford University Pres. Oxford. UK.

Luchitsky, I. V., ed. 1974. *Kamchatka, Kurilyskie i Komandorskie ostrova. Istoriya razvitiya relyefa Sibiri i Dalynego Vostoka. Yug Dalynego Vostoka [Kamchatka, the Kuril and*

Commander Islands. The history of the development of the relief of Siberia and the Far East. South Far East]. Moskva: Nauka.

Lyashchevskaya, M. S., Kiseleva, A. G., Rodnikova, I. M., and Pshenichnikova, N. F. 2010. "Razvitie pochvenno-rastitelynogo pokrova ostrovnyh ekosistem v pozdnem golotsene." In *Evolyutsiya zhizni na Zemle*, edited by Vera. M. Podobina, 159– 171. Tomsk: TML-Press.

Lyashchevskaya, M. S., Kiseleva, A. G., and Rodnikova, I. M. 2014. "Razvitie pochvenno-rastitelynogo pokrova ostrova Puptyatina v pozdnem golotsene (Yaponskoe more)." *Geografia i prirodnye resursy* 1: 124–133.

MacArtur R. U.1955. "Fluctuations of animal populations and a measure of community stability." *Ecology* 36: 533–536.

Maksimova, V. F. "Phitoindikatsiya v sisteme okean – materik." In *Teoretucheskie i prikladnye aspekty biogeographii*, edited by Anatoly G. Voronov, 149 – 162. Moskva: Nauka.

Mamay, I. I. 1994. *Dinamika landshaftov*. Moskva: Izd-vo MGU. 167 p.

Manyko, Yu. I. 1974. "Vliyanie sovremennogo vulkanizma na rastitelynosty Kamchatki i Kurilyskih ostrovbov." *Komarovskie chteniya* XXII: 5–31.

Manyko, Yu. I., and Rozenberg, V. A. 1970. "Vysotnaya poyasnosty rastitelynosti na vulkane Mendeleeva." In *Biologicheskie Resursy o-va Sakhalin i Kurilyskih ostrovov*, edited by Vyacheslav Yu. Barkalov, 65–71. Vladivostok: Biologo-pochvenny institut DVNTS AN SSSR. P. 65–71.

Margalef R.1958. "Information theory in ecology." *Gen. Syst.* 3: 36–71.

Markhinin, E. K. 1956. "Vulkan Mendeleeva i yego sostoyanie letom 1954 g." *Byulleteny Vulkanologich. Stantsii* 24: 33–38 Moskva: Izd-vo AN SSSR.

Markhinin, E. K. 1980. *Vulkany i zhizny [Problemy biovulkanologii]* Moskva: Mysly.

Markhinin, E. K., and Stratula, D. S. 1977. *Gidrotermy Kurilyskih ostrovov*. Moskva: Nauka.

Markov, K. K., ed. 1980. *Fizicheskaya geographiya Mirovogo Okeana (Physical geography of World Ocean)*. Moskva: Nauka.

Markov, K. K., Lazukov, G. I., and Nikolaev, V. A. 1965. *Chetvertichny period. Vol. 2.* Moskva: Izd-vo MGU.

Markov, Yu. D., and Utkin, I. V. 2009. "The Peter the Great Bay (Sea of Japan) shalf evolution in Late Pleistocene–Holocene." In *Geology and geoecology of Eurasian continental margins. Viol. 1*, edited by Yury A. Lavrushin, 113 – 123. Moscow: GEOS.

Martynov, Yu. A., Rybin, A. V., Drily, S. I., and Martynov, A. Yu. 2009. "Zony anomalynogo vulkanizma Kurilyskih ostrovov, ostrov Paramushir." *Vestnik DVO RAN* (4): 17– 23.

Masurenkov, Yu. P. 1991. "Zakonomernosti gipsometricheskogo i lateralynogo raspredeleni ya geystvuyushchih vulkanov." In *Deystvuyushchie vulkany Kamchatki, Vol. 1*, edited by Sergey A. Phedotov and Yury P. Masurenkov, 46–57. Moskva: Nauka.

Mazing, V. V. 1970. "K voprosu evolyutsii biotsenoticheskih system." In *Teoreticheskie problemy phitotsenologii i biogeotsenologii*, by edited Evgeny M. Lavrenko and Tikhon A. Rabotnov, 95–107. Moskva: Nauka. P. 95–107.

Melekestsev, I. V. 1974. "Fluvialynaya deyatelynosty." In *Kamchatka, Kurilyskie i Komandorskie ostrova. Istoriya razvitiya relyefa Sibiri i Dalynego Vostoka*, edited by Ivan V. Luchitsky, 394–402. Moskva: Nauka.

Melekestsev, I. V. 1980. *Vulkanizm i relyefoobrazovanie*. Moskva: Nauka.

Melekestsev, I. V., Braytseva, O. A., and Ponomareva, V. V. 2001. "Novy podkhod k opredeleniuy ponyatiya "deystvuyushchy vulkan." In *Geodinamika i vulkanizm Kurilo-Kamchatskoy ostrovoduzhnoy sistemy [Geodynamics and volcanism of the Kurile-Kamchatka island-ars system],* edited by Boris V. Ivanov, 191–203. Petropavlovsk-Kamchatsky: Institut vylkaubcheskoy geologii i geokhimii RAN.

Melekestsev, I. V., Braytseva, O. A., Erlikh, E. N., and Kozhemyaka, N. N. 1974. "Vulkanich-eskie gory i ravniny." In *Kamchatka, Kurilyskie i Komandorskie ostrova. Istoriya razviti-ya relyefa Sibiri i Dalynego Vostoka [Kamchatka, the Kuril and Commander Islands. The history of the development of the relief of Siberia and the Far East]*, edited by Ivan V. Luchitsky, 162–233. Moskva: Nauka.

Melekestsev, I. V., Volynets, O. N., and Ermakov, V. A. 1991. "Vulkan Shiveluch." In *Deystvuyushchie vulkany Kamchatki, Vol 1*, edited by Sergey A. Phedotov and Yury P. Masurenkov, 84–97. Moskva: Nauka.

Meshcheryakov, Yu. A. 1981. *Izbrannye trudy. Relyef i sovremennaya geodinamika [Selected works. Relief and contemporary geodynamics]*. Moskva: Nauka.

Milkov, F. N. 1981. *Fizicheskaya geografiya: sovremennoe sostoyanie, zakonomernosti, problemy [Physical geography: recent state, conformity to natural laws, problems]*. Voronezh: Izd-vo Voronezh. Univ.

Miller, G. P. 1974. *Landshaftnye issledovaniya gornyh i predgornyh territory [Landscape studies of mountain and foothill territories]*. Lyvov: Izd-vo Lyvov. Un-ta.

Mirkin, B. M., and Rozenberg G. A. 1983. *Tolkovy slovar sovremennoy fitotsenologii*. Moskva: Nauka. 134 p.

Mitropolysky, A. K. 1971. *Tekhnika statisticheskih vychisleny [Statistical Computing Technique]*. Moskva: Nauka.

Moiseev, N. N., Aleksandrov, V. V., and Tarko, A. M. 1985. *Chelovek i biosphere. Opyt systemnogo analiza i eksperimenty s modelyami [System analysis experience and model experiments]*. Moskva: Nauka.

Molchanov, A. A. 1961. *Les i klimat [Forest and climate]*. Moskva: Izd-vo AN SSSR.

Molchanov, A. A., ed. 1975. *Dubravy lesostepi v biogeotsenoticheskom otnoshenii*. Moskva: Nauka.

Morozov, G. F.1949. *Uchenie o lese [Theory about Forest]. Izd-e 7-e.* Moskva–Leningrad: Goslesbumizdat.

Mukha, D. E. 2007. "Vliyanie vulkanizma na prirodu Kamchatki." In *Geographicheskie i geoekologicheskie issledovaniya na Dalynem Vostoke [Geographical and geoecological studies in the Far East.]. Vyp.3,* edited by Petr Ya. Baklanov and Svetlana I. Kozhenkova,86–88. Vladivostok: Dalynauka.

Naeem S. 1998. "Species redundancy and ecosystem reliability." *Conserv. Biol.* (12): 39–45.

Nakamura, Y., and Krestov, P. V. 2005. "Coniferous forests of the temperate zone of Asia. Coniferous forests." *Ser. Ecosystemsof the World* 6: 163–220.

Nef E. 1968. "O nekotoryh voprosah sravnitelynoy ekologii landshafta.[On some issues of comparative landscape ecology]" In *Doklady In-ta geographii Sibiri i Dalyn. Vostoka*, 19: 44–53.

Neyshtadt, M. I. 1952. "O koreyskom kedre na Dalynem Vostoke kak "relikte" tretichnogo vremeni [On the Korean cedar in the Far East as a "relic" of the Tertiary time]." *Doklady AN SSSR* 86 (2): 425–428.

Neyshtadt, M. I. 1957. *Istoriya lesov i paleogeografiya SSSR v golotsene [The history of forests and the paleogeography of the USSR in the Holocene]*. Moskva: Izd-vo AN SSSR. 404 p.

Nikolaev, V. A. 1978. *Klassifikatsiya i melkomasshtabnoe kartografirovanie landshaftov [Classification and small-scale mapping of landscapes]*. Moskva: Izd-vo MGU.

Nikolyskaya, V. V. 1969. "Paleogeografiya geomorfologicheskogo etapa i sovremennye relyefoobrazuyushchie protsessy." In *Yuzhnaya chasty Dalynego Vostoka [Southern part of the Far East]*, edited by Vera. V. Nikolyskaya and Aleksandr S. Khomentovsky, 49–64. Moskva: Nauka.

Nikolyskaya V. V., and Khomentovsky A. S., eds. 1969. *Yuzhnaya chasty Dalynego Vostoka (Southern Part of Far East)*. Moskva: Izd-vo AN SSSR (in Russian).

Nikolyskaya, V. V., and Timofeev, D. A. 1966. "Sravnenie sovremennoy fiziko-geograficheskoy zonalynosty s zonalynostyu epokhi maksimalynogo oledeneniya."*DokladyIn-ta ge ografii Sibiri i Dalynego Vostoka*, 12: 35–46 (in Russian).

Nilolyskaya, V. V., Timofeev, D. A., and Chichagov, V. P. 1969. "Prirodnoe rayonirovanie." In *Yuzhnaya chasty Dalynego Vostoka*, edited by Vera. V. Nikolyskaya and Aleksandr S. Khomentovsky, 301–344. M.: Nauka.

Nikonov, V. I. 1975. "Prirodnye landshafty Nizhnego Priamurya." In *Sibirsky geographich. Sbornik. Vyp. 10*, by edited Viktor B. Sochava, 128–175. Novosibirsk: Nauka.

Odum, Eu. P. 1971. *Fundamentals of Ecology.* Third edition. Philadelphia–London–Toromto: W. B. Saunders Company.

Odum, Eu. P. 1983. *Basic Ecology. Vol. 2.* Philadelphia: CBS College Publishing, USA.

Osipov, S. V., Urusov, V. M., Chipizubova, M. N., and Seledets, V. P. 2008. "Rastitelyny pokrov." In *Geosistemy Dalynego Vostoka na rubezhe XX–XXI vekov. Vol. I. Prirodnye geosistemy i ih komponenty*, edited by Sergey S. Ganzey, 203–235. Vladivostok: Dalynauka.

Osipov, V. V., ed. 2004. *Ecosystemy tellermanovskogo lesa*. Moskva: Nauka (in Russian).

Paffengolts, K. N., ed. 1973. *Geologichesky slovar (Geological Dictionary).* Moskva: Nedra.

Panfilov, V. D. 2005. *Tsentry evolyutsii i istoricheskie migratsii bioty Zemnogo Shara. Nauchny dnevnik (Evolutionary centers and history migrations of biota of Earth Ball).* Moskva: Institutegeography RAN.

Perelman, A. I. 1975. *Geokhimiya landshaphtov (Landscape geochemistry).* Moskva: Vysshaya shkola.

Peterson, G., Allen, C. R., and Holling, C. S. 1998. "Ecological resilience, biodiversity and scale." *Ecosystems*, 1: 6–18.

Pianka, E. R. 1978. *Evolutionary ecology.* Second Edition. New York: Harper and Row, Publishers.

Pietsch1, T. W., Bogatov, V. V., and Amaoka, K. 2003. "Biodiversity and biogeography of the islands of the Kuril Archipelago." *Journ. of Biogeogr*, 30: 1297–1310.

Pogrebnyak, P. S. 1955. *Osnovy lesnoy tipologii.* Kiev, Izd-vo Akademii Nauk USSR (in Russian).

Pojar J., Klinka K., and Meidiger D. V. 1987. "Bioclimatic ecosystem classification in British Columbia." *For. Ecol. Manage.*, 22: 119–154.

Polivanov, V. S. 1984. "Pochvenny pokrov kak slozhnaya geograficheskaya sistema." In *Prognozno-geografichesky analiz territorii administrativnogo rayona*, edited by Andrey P. Kapitsa and Yury G. Simonov, 82–97. Moskva: Nauka.

Polynov, B. B. 1934. *Kora vyvetrivaniya. Chasty pervaya*. Leningrad: Izd-vo AN SSSR.

Polynov, B. B. 1956. Izbrannye trudy (Selected works). Moskva: Izd-vo AN SSSR.

Ponomarenko, A. G. 1998. '"Paleontologiya i balansy biogeokhimicheskih krugovorotov." In *Ekosistemnye perestroyki i evolyutsiya biosphery,* edited by Aleksey Yu. Rozanov, 3: 9–4.

Ponomareva, V. V. and Plotnikova, T. A. 1980. *Gumus i pochvoobrazovanie: metody i rezulytaty izucheniya (Humus and Soil Formation Processes: Methods and Results of Study).* Leningrad: Nauka.

Pope V. D., Gallani M. L., Rowntree P. R., and Stratton R. A. 2000. "The impact of new physical parametrizations in Hadley Centre climate model – HadCM3." *Climate Dynamics*, 16: 123–146.

Popov, N. A.1963. *Lesa Yuzhnyh Kurilyskih ostrovov i perspektivy ih ispolyzovaniya.* Krasnoyarsk: In-t lesa i drevesiny SO AN SSSR (in Russian).

Preobrazhensky, V. S. 1969. "O sisteme metodov obshchey fizicheskoy geografii." In *Metody landshaftnyh issledovany.* by edited Vladimir S. Preobrazhensky, 7–34. Moskva: Nauka.

Protopopov, V. V. 1975. *Sredoobrazuyushchaya roly temnokhvoinogo lesa*. Novosibirsk: Nauka.

Puzachenko, Yu. G. 1983. "Invariance of geosystems and their components." In *Ustoichivosty geosistem (Stability of Geosystems)*, edited by Aleksey D. Armand and Ivan Yu. Dolgushin, 32–41. Moskva: Nauka (in Russian).

Puzachenko, Yu. G. 1986. "Prostranstvenno-vremennaya ierarhiya geosystem s pozitsii teorii kolebany (Spatio-temporal hierarchy of geosystems in terms of theory of fluctuations)." In *Voprosy geographii. Vol. 127. Modelirovanie geosystem*, edited by Georgy M. Lappo, 96 –111. Moskva: Mysly.

Puzachenko, Yu. G. 1996. *Osnovy obshchey ekologii*. Moskva: Izd-vo MGU (in Russian).

Puzachenko, Yu. G. 1998. *Metodologicheskie osnovy geographicheskogo prognoza i okhrany sredy*. M.: Izd-vo URAO.

Puzachenko, Yu. G., and Moshkin, A. V. 1969. "Informatsionno-logichesky analis v medico-geographicheskih issledovaniyah." In *Itogi nauki*, edited by Kim. S. Losev, 3. Med-. itsinskaya geographiya. M.: VINITI.

Puzachenko, Yu. G., and Skulkin, V. S. 1981. *Struktyra rastitelnosti lesnoy zony SSSR. Sistemny analiz (Structure of forest zone vegetation for USSR)*. Moskva: Nauka.

Rabotnov, T. A. 1978. *Fitotsenologiya*. Moskva: Izd-vo MGU.

Rakovskaya, E. V., Lyuboshkin, S. V, and Pashkang, K. V. 1984. "Landshaftnaya karta Nechernozemnoy zony RSFSR i vozmozhnosti ee ispolyzovaniya dlya tseley selyskogo khozyaystva." In *Voprosy geographii. Sbornik 124. Prirodnye kompleksy i selyskoe khozyaystvo*, edited by Vladimir M. Chupakhin, 107–120. Moskva: Mysly (in Russiasn).

Ramensky, L. G. 1971. *Izbrannye raboty (Selected works). Problemy i metody izucheniya rastitelnogo pokrova (Selected Works. Problems and Methods of Studying Vegetation Cover)*. Leningrad: Nauka.

Rauner, Yu., L. 1972. *Teplovoy balans rastitelynogo pokrova (Warm balance for vegetation cover).* Leningrad: Gidrometeoizdat.

Razzhigaeva, N. G., Ganzey, L. A., Belyanina, N. I., et al. 2013. "Roly klimaticheskogo i vulkangennogo faktorov v formirovanii i razvitii landshaftov o. Simushir (Tsentralynye Kurily), v srednem-pozdnem golotsene." *Tikhookeanskaya geologiya* 32 (3): 55–76.

Razzhugaeva, N. G., Ganzey, K. A., Grebennikova, T. A., et al. 2009. "Evolyutsiya i rubezhi perestroyki landshaftov Kurilyskih ostrovov v golotsene." In *Geographiya i geoekologiya na sovremennom etape vzaimogeystviya prirody i obshchestva*, edited by K. V. Chistyakov, 727–733. Cankt-Peterburg: S-PbGU.

Razzhugaeva, N. G., Ganzey, K. A., Mokhova, L. M., and Pshenichnikova, N. Ph. 2011. "Lugovye landshafty Yuzhnyh Kuril: proiskhozhdenie, vozrast i razvitie." *Geographiya i prirodnye resursy* 3: 96–104.

Reymers, N. F. 1990. *Prirodopolyzovanie. Slovar-spravochnik*. Moskva: Mysly (in Russian).

Reutt, A. T. 1970. "Rastitelynosty" In *Sever Dalynego Vostoka*, edited by Nikolay A. Shilo, 257–299. Moskva: Izd-vo AN SSSR.

Rice, R. L. 1978. *Fundamentals of Geomorphology*. London.

Ricklefs, R. E. 1976. *A Textbook in Basic Ecology*. University Pennsylvani, Portland, Oregona: Chiron Press Inc.

Rode, A. A., ed. 1975. *Tolkovy slovar po pochvovedeniyu*. Moskva: Nauka.

Rodnikova, I. M., Lyashchevskaya, M. S., Kiseleva, A. G., and Pshenichnikova, N. F. 2012. "Sostoyanie i dinamika pochvenno-rastitelynogo pokrova maslyh ostrovov zaliva Petra Veliogo (Yaponskoe more)." In *Geografiya i prirodnye resursy*, 1: 96–103.

Rototaeva, T. P., and Simonov, Yu. G. 1984. "Elementarnaya morphosistema "sklon." In *Prognozno-geographichesky analiz territorii administratuvnogo rayona*, by edited An drey P. Kapitsa and Yury G. Simonov, 57–59. Moskva: Nauka.

Rozen, R.1969.*Printsip optimalnosti v biologii*. Moskva: Mir.

Rozenberg, G. S. 1984. *Modeli v phytocenologii*. Moskva: Nauka.

Rozhkov, A. A. and Kozak, V. T. 1989. *Ustoichivosty lesov (Resistance of Forests)*. Moscow: Agropromizdat.

Rudnev, N. I. *Radiatsionny i teplovoy balans fitotsenozov*. 1984. Moskva: Nauka.

Samner, G. 1978. *Matematics for physical geographers*. London: Edward Arnold.

Safonov, Yu. G., Znamensky V. S., Prokhvisneva E. A., et al. 1998. "Rasprostranennosty, sostav i usloviya sovremennyh term Baykalo-Charskoy i Kurilo-Kamchatskoy ostrovoduzhnyh oblastey." In *Globalynye izmeneniya prirodnoy sredy,* edited by Nikolay L. Dobretsov and Viktor I. Kovalenko, 177–179. Novosibirsk: Izd-vo SO RAN NITS OIGGM.

Santer, B. 1985. "The use general circulation models in climate impact analysis – a preliminary stay of the impacts of a CO_2 – indicated climatic change on west european agriculture." *Climatic Ghanges* 7 (1): 71–93.

Scherer-Lorenzen, M., Korner, C., and Schulze, E. D., eds. 2005. *Forest Diversity and Function: Temperate and Boreal Systems*. Berlin: Springer-Verlag.

Semenov-Tyan-Shansky, V. P. 1927. *Rayon i strana*. Moskva–Leningrad: Gos. izd-vo.

Sharaya, L. S. 2007. "Kolichestvenny analiz prostranstvennoy izmenchivozti parametrov sostouaniya ecosystemy Zhyguli." In *Samarskaya Luka. Bulleten*, by edited Sergey V. Saksonov, 16 (4): 639–659.

Sharaya, L. S. 2009. Predskazatelynoe kartographirivanie lesnyh ekosistem v geoekologii." *Povolzhsky ekologichesky zhurnal* 3: 249–257.

Shary, P. A., Sharaya, L. S., and Mitusov, A. V. 2002. "Fundamental quantitative methods of land surface analysis." *Geoderma* 107 (1–2): 1–32.

Shchukin, I. S. 1980. *Chetyreh'yazychny entsiklopedichesky slovar terminov po fizicheskoy geografii*. Moskva: Sov. Entsiclopediya.

Shestakov, A. S., and Drozdov, A. V. 1992. "Dinamika landshaftov i degradatsiya okruzhayushchey sredy." In *Globalynye izmeneniya i regionalynye vzaimosvyazi*, by Nikita Ph. Glazovsky, 98-121. Moskva: In-t geografii AN SSSR (in Russian).

Shikhlinsky E. M. 1969. *Teplovoy balans Azerbaydzhanskoy SSR*. Baku; Izd-vo "ELM".

Shilo, N. A. 1967. *Tektono-geomorfologicheskaya evolyutsiya poverkhnosti Severo-Vostoka i rossypeobrazovaniye*. Magadan: SVKNII SO AN SSSR (in Russian).

Shilo, N. A. 2000. *Uchenie o rossypyah*. Moskva: Izd-vo Akad. gornyh nauk..

Shilo, N. A., and Shumilov, Yu, V. 2001. "O zakone geosistemnoy evolyutsii biosphery." In *Prostranstvennaya i vremennaya izmenchivosty prirodnoy sredy Severo-Vostochnoy Azii v chetvertichny period*, edited by Konstantin V. Simakov, 4–11.Magadan: Severo-Vostoch. nauch. tsentr DVO RAN.

Shmalygauzen, I. I. 1968. *Kiberneticheskie voprosy biologii*. Novosibirsk: Nauka.

Shvidenko, A. Z., Shchepachenko, D. G., Nilsson, S., and Boluy, Yu. I. 2008. *Tablitsy i modeli khoda rosta i produktivnosti nasazhdeny osnovnyh lesoobrazuyushchih porod Severnoy Evrazii (normativno-spravochnye materialy)*. Moskva: Mezhdunar. In-t prikladnogo sistem. analiza.

Shwarts, S. S. (1970). "Evolyutsiya i biosphera." In *Teoreticheskie problemy fitotsenologii i biogeotsenologii,* by edited Evgeny M. Lavrenko and Tiknon A. Rabotnov, 213–228. Moskva: Nauka.

Shwarts, S. S. 1975. "Ekologicheskie osnovy ojhrany biosphery." In *Metodologichesie aspekty issledovaniya biosphery,* by edited Vladimir E. Sokolov, 100–112. Moskva: Nauka (in Russian).

Shwarts, S. S. (1980). *Ecologicheskie zakonomernosty evolyutsii*. Moskva: Nauka (in Russian).

Simonov, Yu. G. 1972. *Regionalyny geomprfologichesky analiz*. Moskva: Izd-vo MGU.

Simonov, Yu. G. 1982. "Osnovnye svoystva obyektov geograficheskogo prognozirovaniya i sposoby ih formalizovannogo opisaniya." In *Problemy regionalynogo geografich eskogo prognoza*, edited by Andrey P. Kapitsa and Yury G. Simonov, 112–193. Moskva: Nauka..

Smith, J. M. 1974. *Models in Ecology*. Cambridge: At the University of Sussex Press.

Smith, T. M., Leemance, R., and Shugart, H. H. 1992. "Sensitivity of Terrestrial Carbon Stor age to CO_2-Induced Climate Change: Comparison of Four Scenarios Based on General Circulation Models." *Climatic Change* 21: 367–384.

Smith T. M. & Shugart H. H. 1993. "The transient response of terrestrial carbon to perturbed climate." *Nature,* 361 (6412): 523–526.

Sochava, V. B. 1944 a. "O genezisei fitotsenologii ayanskogo temnokhvoynogo lesa." *Bot. zhurn*al 29 (5): 205–217 (in Russian).

Sochava, V. B. 1944 b. "O proiskhozhdenii prirody severnyh polyarnyh stran." *Priroda* 4: 45–55 (in Russian).

Sochava, V. B. 1946. "Voprosy filogeneza i fitotsenologii Manchzhurskogo smeshannogo lesa (Questions of phylogenesis and phytocoenology for Manchurian mixed forest)." In *Materialy po istorii flory i rastitelynosty SSSR. Vyp. II*, edited byVladimir L. Komarov, 283 – 302. Moskva– Leningrad: Izd-vo AN SSSR (in Russian).

Sochava, V. B. 1950. "Noveyshie vertikalynye dvizheniya zemnoy kory i rastitelyny pokrov." *Zemlevedenie. Novaya seriya,* III (XLIII): 32–45 (in Russian).

Sochava, V. B. 1962. "Iskhodnye polozheniya tipizatsii tayezhnyh zemely na landshaftno-geograficheskoy osnove." *Doklady In-ta geographii Sibiri i Dalyn. Vostoka* 2: 14–23 (in Russian).

Sochava, V. B. 1963. "Opredelenie nekotoryh ponyaty i terminov fizicheskoy geografii." *Doklady In-ta geographii Sibiri i Dalyn. Vostoka* 3: 50–59 (in Russian).

Sochava, V. B. 1970. *Geographiya i ecologiya.* Leningrad: Materialy V c'ezda Geogr. ob-va Soyuza SSR (in Russian).

Sochava, V. B. 1974. "Geotopologiya kak razdel ucheniya o geosistemah." In *Topologicheskie aspekty ucheniya o geosistemah*, edited by Viktor B. Sochava, 3–86. Novosibirsk: Nauka (in Russian).

Sochava, V. B., ed. 1977. *Ecologo-fitotsenoticheskie kompleksy Aziatskoy Rossii (Ecological-phytocoenotical complexes of Asian Russia).* Irkutsk: Institut Geographii Sibiri i Dalynego Vostoka (in Russian).

Sochava, V. B. 1978. *Vvedenie v uchenie o geosistemakh (Introduction into the Theory of Geosystems).* Novosibirsk: Nauka (in Russian).

Sochava, V. B. 1979. *Rastitelny pokrov na tematicheskih kartah (Vegetation cover on the tematic maps)*, Novosibirsk, Nauka (in Russian).

Sochava, V. B. 1980. *Geographicheskie aspekty sibirskoy taygi (Geographical aspects of Siberian taiga).* Novosibirsk: Nauka (in Russian).

Sochava, V. B. 2005. *Izbrannye Trudy. Teoreticheskaya i prikladnaya geografiya.* Novosibirsk: Nauka. 288 p. (in Russian).

Sochava, V. B., and Bayborodin, V. N., eds. 1977. *Korrelyatsionnaya ekologo-fitotseno-ticheskaya karta Aziatskoy Rossii.* M-b 1:7500000. Irkutsk: Intstitut geographii Sibiri i Dalynego Vostoka SO AN SSSR (in Russian).

Sokolov, I. A. 1973. *Vulkanizm i pochvoobrazovanie (Volcanism and soil formation).* Moskva: Mysly (in Russian).

Solntsev, V. N. 1981 a. *Sistemnaya organizatsiya landshaftov.* Moskva: Mysly (in Russian).

Solntsev, V. N. 1981 b. "Khronoorganizatsiya geograficheskih yavleny." In *Voprosy geografii. Sbornic 117*, *Geofizika landshafta*, edited by Kirill N. Dyakonov and Roman. I Zlotin. Moskva: Mysly. P. 40–68 (in Russian).

Solntsev V. N. 1997. *Strukturnoe landshaftovedenie. Osnovy kontseptsii, nekotorye argumenty.* Moskva: Izd-vo MGU (in Russian).

Solntsev, N. A. 2001. *Uchenie o landshafte. Izbrannye trudy.* Moskva: Izd-vo MGU (in Russian).

Sprygin, I .I. 1986. "Materialy k poznaniyu rastitelynosty Srednego Povolzhya." *Nauchnoe nasledstvo* 11: 22–512, by edited Anatoly G. Voronov. Moskva: Nauka. (in Russian).

Stepanova, K. D., Belaya, G. A., Kachura, N. N., Morozov, V. L., Sokirka, A. I. 1981. *Biologicheskaya produktivnosty lugovyh soobshchestv Dalynego Voctoka (priokeanicheskie rayony).* Moskva: Nauka (in Russian).

Stinson G., Kurz W. A., Smyth C. E., et al. 2011. "An inventory-based analysis of Canada's managed forest carbon dynamics, 1990 to 2008." *Global Change Biology*, 17: 2227–2244.

Stone L., Gabric A., and Berman T. 1996. "Ecosystem resilience, stability, and productivity: Seeking a Relationship." *Amer. Natur.* 148: 892–903.

Sugrobov, V. M., and Yanovsky, Ph. A. 1991. "Geotermicheskoe pole Kamchatki, vynos tepla vulkanami i gidrotermami." In *Deystvuyushchie vulkany Kamchatki, Vol. 1*, by edited Sergey A. Phedotov and Yury P. Masurenkov, 58– 66. Moskva: Nauka (in Russian).

Sukachev V. N. 1960. "Correlation between the concept "forest ecosystem' and "forest biogeocoenose" and their importance for the classification of forests." *Silva Fenn.* 105: 15–27.

Sukachev, V. N. 1972. *Izbrannye trudy (Selected Scientific Works). Vol.1. Osnovy lesnoy tipologii i biogeotsenologii.* Leningrad: Nauka (in Russian).

Suzuki S. N., Ishihara M. I., and Hidaka A. 2015. "Regional-scale directional changes in abundance of tree species along a temperature gradient in Japan: *Global Change Biology*, 21: 3436–3444.

Svetlosanov, V. A. 1990. "Resistance and stability of natural ecosytems: modeling." *Itogi Nauki i Tekhniki, Ser.: Teor. Obshchie Voprosy Geografii. Vol. 8*, editied by Kim S. Losev. Moskva: VINITI (in Russian).

Svetlosanov, V. A. 2009. *Ustoichivosty prirodnyh sistem k prirodnym i antropogennym vozde-ystviyam (uchebnoe posobie) (Stability of Natural Systems to Natural and Anthropogenic Influences: Manual)*, Moscow: Mosk. Gos. Univer. (in Russian).

Svirezhev, Yu. M. 1977. "Matematicheskoe modelirovanie biologicheskih soobshchestv." In *Chislo i mysly*, 1: 63–117. Moskva: Znanie (in Russian).

Svirezhev, Yu. M. 1982. "Matematicheskie modeli v ekologii." In *Chislo i myly*, 5: 16–55. Moskva: Znanie (in Russian).

Svirezhev, Yu. M. 1983. "Stability and complexity in mathematical ecology." In *Ustoichivosty geosistem* (*Stability of Geosystems)*, edited by Aleksey D. Armand, and IvanYu. Dolgushin, 41–50. Moscow: Nauka (in Russian).

Svirezhev, Yu. M. and Elizarov, E. Ya. 1972. "Mathematical modeling of biological systems (Problems of Space Biology)." In *Problemy kosmicheskoi biologii,* XX: 1–159. Moscow: Nauka (in Russian).

Svirzhev, Yu. M. and Logofet, D. O. 1978. *Ustoichivosty biologicheskih soobshchestv (Stability of Biological Communities).* Moscow: Nauka (in Russian).

Syomkin, B. I. (1973). "Descriptivnye mnozhestva i ih prilozheniya." In *Issledovaniya system. I. Analiz slozhnykh system*, by editedBoris I. Syomkin, 83–94. Vladivostok: Tikhookeansky institut geografii AN SSSR (in Russian).

Syomkin, B. I., Komarova, T. A. 1977. "Analiz fitotsenoticheskih opisany s ispolyzovaniem mer vklyucheniya (na primere rastitelynyh soobshchestv doliny r. Amguemy na Chukotke)."*Bot. jurnal* 62 (7): 54–63 (in Russian).

Sysuev, V. V. 2002. "Strukturoobrazuyushchie geosistemnye protsessy: kharakternye masshtaby i modelirovanie." *Vestnik MGU. Ser. 5. Geographiya* 1: 22–28 (in Russian).

Tarankov, V. I. 1974. *Mikroklimat lesov Yuzhnogo Primorya*. Novosibirsk: Nauka (in Russian).

Targulyan, V. O. 1982. "Osnovnye zakonomernosti genezisa i geografii pochv gumidnyh oblastey Tikhookeanskogo regiona." In *Pochvy ostrovov i priokeanicheskih regionov Tikhogo Okeana,* edited byAnatoly M. Ivlev, 5–18. Vladovostok: DVNTS AN SSSR (in Russian).

Targulyan, V. O. 1984. "O sootnoshenii ponyaty "struktura" i "funktsionirovanie" v issledovanii po geografii ecosystem.: In *Sovremennye problemy geografii ecosystem*, by edited RomanI Zlotin, and Yury A. Isakov, 10–15. Moskva: In-t geographii AN SSSR. (in Russian).

Targulyan, V. O., and Sokolov, I. A. 1978. "Strukturny i funktsionalyny podkhod k pochve: pochva–pamyaty i pochva–momemt." In *Matematicheskoe modelirivanie v ekologii*, bu edited Aleksandr M. Molchanov, 17–33. Moskva: Nauka (in Russian).

Tarko, A. M. 2005. *Antropogennye izmeneniya globalnyh biosfernyh protsessov. Matematicheskoe modelirovanie (Anthropogenic changes of global biosphere processes. Mathematic modeling)*. Moskva: Phizmatlit (in Russian).

Tatewaki M. 1933. "The phytogeography of the Middle Kuriles." *Journ. Fac. Agr. Hokkaido Univ*. 28 (5): 191–363. Sapporo.

Tatewaki M. 1957. "Geobatanic studies on the Kurile island." *Acta Horti Gotoburgensis* 21, (2): 43–123.

Tatewaki M. 1963. "Hultenia." *Journ. Fac. Agricult. Hokkasido Univ*. 53 (2): 133–199. Sapporo.

Theurillat, J. P., Felber, F., Geissler, P., et al. 1998. "Sensitivity of plant and soil ecosystems of the Alps to climate changes." In *Views from the Alpe regional perspectives on climate change*, edited by Peter Cebon, 225–308. Cambridge, Massachusetts, USA: MIT Press.

Thompson I., Mackey B., McNulty S., and Mosseler A. 2009. *Forest Resilience, Biodiversity, and Climate Change. A synthesis of the biodiversity/resilience/stability relationship in forest ecosystems*. Technical Series, 43: 1–67. Montreal: Secretariat of the Convention on Biological Diversity.

Tikhomirov, B. A. 1946. "K proiskhozhdeniyu assotsiatsii kedrovogo stlanika (*PINUS PU MILA* RGL.)." In*Materialy po istorii flory i rastitelynosty SSSR. Vyp. II,* by edited Vladimir L. Komarov), 491–511. Moskva-Leningrad, Izd-vo AN SSSR (in Russian).

Timopheev-Resovsky, N. V. and Tyuryukanov, A. N. 1966. "On elementary bichorologic subdivisions of the biosphere." *Byull. MOIP (Moscow Society of Nature Investigators). Biological section* LXXI (1): 123–132 (in Russian).

Timofeev-Resovsky, N. V., Vorontsov, N. N., and Yablokov, A. V. 1969. *Kratky ocherk teorii evolyutsii (Short sketch of evolutionary theory)*. Moskva: Nauka (in Russian).

Tishkov, A. A. 1988. "Podkhody k issledovaniyam dinamiki bioty kak obyekta geograficheskogo prognozirovaniya." In *Geograficheskoe prognozirovanie i prirodookhrannye problemy,* edited by Vladimir S. Preobrazhensky, 49–60. Moskva: Inctitut geografii AN SSSR (in Russian).

Tishkov, A. A. 1991. "Roly suktsssiy vo vremennoy organizatsii geosistem." In *Geosistemy vo vremeny*, by edited Aleksandr M. Grin, 118–132. Moskva: Institutgeografii AN SSSR (in Russian).

Tishkov, A. A. 2005. *Biosphernye funktsii prirodnyh ekosistem Rossii (Biospheric Finctions of Natural Ecosystems of Russia)*. Moskva: Nauka (in Russian).

Titov, I. A. 1952. *Vzaimodeystvie rastitelynyh soobshchestv i usloviya sredy*. Moskva: Sov. nauka (in Russian).

Tolmachev, A. I. 1954. *K istorii formirovaniya i razvitiya temnokhvoynoy taygi*. Moskva–Leningrad: Izd-vo AN SSSR (inRussian).

Tooming, Kh. G. 1984. *Ecologicheskie printsipy maksimalynoy produktivnosti posevov*. Leningrag: Gidrometeoizdat (in Russian).

Tunegolovets, V. M., Gartsman, B. I., and Krokhin, V. V. 2008. "Klimat i gidrologiya." In *Geosistemy Dalynego Vostoka na rubezhe XX–XXI vekov. Vol. I. Prirodnye geosistemy i ih komponenty*, edited by Sergey S. Ganzey, 119– 143. Vladivostok: Dalynauka (in Russian).

Tsyganenko, A. F. 1972. *Geographiya pochv*. Leningrag: Izd-vo LGU (in Russian).

Tsymek, A. A., Solovyev, K. P., and Chumin, V. T. 1969. "Lesa Khabarovskogo kraya" In *Lesa SSSR. Vol. 4. Lesa Urala, Sibiri i Dalynego Vostoka,* edited by Aleksey B. Zhukov, 553–620. Moskva: Nauka (in Russian).

Turner, M. G., and Gardner, R. H., eds. 1990. *Quantitative Methods in Landscape Ecology. The analysis and Interpritation of Lanscape Heterogeneity*. New-York – Berlin – Heidelberg: Springer-Verlag.

Udra, I. F. 1975. "K istorii rasprostraneniya koreyskogo kedra na Dalynem Vostoke." *Izvestiya AN SSSR. Ser. Geogr.* 5: 35–42 (in Russian)

Urusov, V. M. 1988. *Genezis rastitelynosti i ratsionalynoe prirodopolyzovanie na Dalynem Vostoke*. Vladivostok: TIG DVNTS AN SSSR (in Russian).

Urusov, V. M. 1993. *Struktura raznoobraziya i proiskhozhdenie phlory i rastitelynosti Yuga Dalynego Vostoka*. Vladivostok: TIG DVO RAN (in Russian).

Urusov, V. M., and Chipizubova, M. N. 2008. "Obshchie zakonomernosti geograficheskogo raspredeleniya formatsy i tipov rastitelynosti. Genezis rastitelynosti" In *Geosistemy Dalynego Vostoka na rubezhe XX–XXI vekov. Vol. I. Prirodnye geosistemy i ih komponenty*, edited by Sergey S. Ganzey, 203–220. Vladivostok: Dalynauka (in Russian).

Urusov, V. M., and Lauve, L. S. 1980. "O vysotnyh poyasah rastitelynosty i formatsionnyh reliktah v Primorye v svyazi s izmeneniyami klimata." *Bot. zhurnal* 65 (2): 185–197 (in Russian).

Ushakov, S. A. and Yasamanov, N. A. 1984. *Dreyf materikov i klimaty Zemli*. Moskva: Mysly (in Russian).

Utechin, V. D. 1977. *Pervichnaya produktivnosty lesostepnyh ekosistem (Primary Productivity of Forest-Steppe Ecosystems)*. Moscow: Nauka (in Russian).

Valter H. 1968. *Die vegetation der Erde. In öko-physiologischer Betrachtung. Bd II. Die gemäßigten und arktischen Zonen*. Jena: VEB Gustav Fischer Verlag.

Vasilevich, V. I. 1983. *Ocherki teoreticheskoy fitotsenologii*. Leningrad: Nauka (in Russian).

Vasilyev, N. G., Efremov, D. F., Borisova, V. A., et al. 1976. "Kratky ocherk lesnoy rastitelynosti basseyna reki Yay (Severny Sikhote-Aliny)." *Komarovskie chteniya* XXIV: 3–29 (in Russian).

Vasilyev, N. G., and Rozenberg, V. A. 1977. "Vysotnye predely rasprostraneniya drevesnoy rastitelynosti na Kurilyskih ostrovah." *Problemy botaniki* XIII: 69–77. Baku: "ELM" (in Russian).

Vasilyev, V. N. 1944 a. "O vzaimootnosheniyah "manychzhurskoy" i "okhotskoy" rastitelynosti i flory." *Bot. zhurnal* 29 (5): 161–170 (in Russian).

Vasilyev, V. N. 1944 b. "Rastitelynosty severnoy chasti vulkanicheskogo kolytsa Tikhogo okeana." *Izvestiya Vsesoyuz. Geogr. Ob-va* 76 (5): 223–240 (in Russian).

Vasilyev, V. N. 1946. "Zakonomernosti protsessa smen rastitelynosti (Conformity to natural laws of process of plain changes)." In*Materialy po istorii flory i rastitelynosty SSSR, Vyp. II*, edited byVladimir L. Komarov, 365–403. Moskva-Leningrad: Izd-vo AN SSSR (in Russian).

Vasilyev, V. N. 1958. "Proiskhozhdenie flory i rastitelynosti Dalynego Vostoka i Vostochnoy Sibiri." In *Materialy po istorii flory i rastitelynosty SSSR, Vyp. III,* edited by Vladimir N. Vasilyev, 361–457. Moskva-Leningrad: Izd-vo AN SSSR (in Russian).

Vedyushkin, M. A. 1989. "Gisterezis v konkurentnyh sistemah." In *Factory i mekhanizmy ustoichivosty geosystem,* edited by Nikita Ph. Glazovsky, 215–225. M., Institut geographii AN SSSR (in Russian).

Vedyushkin, M. A. 1992. "Modelirovanie prostranstvennyh perekhodov mezhdu fitotsenozami." In *Matematicheskoe mogelirovanie populyatsy rasteny i fitotsenozov, by* edited Elena V. Zubkova, 24–30. Moskva: Nauka (in Russian).

Velichko, A. A., ed. 1999. *Izmenenie klimata i landshaftov za poslednie 65 millionov let.* Moskva: GEOS (in Russian).

Velichko A. A., Frenzel, B., and Pesci, M., eds. 1991. *Atlas of Paleoclimates and Paleoenvironments of the Northern Hemisphere. Late Pleistocene – Holocene.* Budapest: Geogr. Res. Inst. Hungar. Academy of Sci.

Viktorov, A. S. 1986. *Risunok landchafta.* Moskva: Mysly (in Russian).

Viktorov, A. S. 2006. *Osnovnye problemy matematicheskoy morphologii landshafta.* Moskva: Nauka (in Russian).

Vitvitsky, G. N. 1969. "Klimat." In *Yuzhnaya chasty Dalynego Vostoka,* by edited Vera V. Nikolyskaya and Aleksandr S. Khomentovsky, 70–96. Moskva: Nauka (in Russian).

Volobuyv, V. R. 1973. *Sistema pochv Mira.* Baku: Izd-vo "ELM" (in Russian).

Vorobyev, D. P. 1963. *Rastitelynosty Kurilyskih ostrovov (Vegetation of Kuril islands).* Moskva-Leningrad: Izd-vo AN SSSR (in Russian).

Voronov, A. G. 1972. "Biogeographicheskie nablyudeniya na ostrovah Tikhogo okeana." *Vestnik MGU. Ser. 5. Geographiya,* 2: 11–21 (in Russian).

Voskresensky S. S. 1971. *Dinamicheskaya geomorfologiya. Formirovanie sklonov.* Moskva: Izd-vo MGU (in Russian).

Vyatkina, M. P., Gimelbrat, D. E., Golovneva, L. B., et al. 2014. *Rastitelny pokrov vulkanicheskih plato Tsentralynoy Kamchatki (Klyuchevskaya gruppa vulkanov).* Moskva: KMK (in Russian).

Vygodskaya, N. N. 1981. *Radiatsionny rezhim i structura gornyh lesov.* Leningrad: Gidrome teoizdat (in Russian).

Vysotsky, G. N. 1909. "O fitotipologicheskih kartah, sposobah ih sostavleniya i ih praktich-eskom znachenii." *Pochvovegenie* 2: 98–119 (in Russian).

Vysotsky, G. N. 1960. *Izbrannye trudy (Selected works).* Moskva: Selykhozgiz (in Russian).

Whittaker, R. H. 1975. *Communities and ecosystems.* New-York: Macmillan Publi-shing Co., Inc.

Wood J., Hengl T., and Reuter H. I., eds. 2009. "Overview of software packages used in geomorphometry // Geomorphometry: Concepts, Software, Applications." *Developments in Soil Science*, 33: 257–267. Amsterdam, etc.: Elsevier.

Yablokov, A. V., and Yusufov, A. G. 1976. *Evolyutsionnoe uchenie (Evolutionary doctrine).* Moskva: Vysshaya shkola (in Russian).

Yachi, S. and Loreau, M. 1999. "Biodiversity and ecosystem productivity in a fluctuating environment: the insurance hypothesis." *Proc. Natl. Acad. Sci. U.S.A.* 96: 1463–1468.

Yurtsev, B. A. 1974. *Problemy botanicheskoy geographii Severo-Vostochnoy Azii (Problems of botanic geography of North-East Asia).* Leningrad: Nauka (in Russian).

Yurtsev, B. A. 1978. "Botaniko-geograficheskaya kharakteristika Yuzhnoy Chukotki." *Komarovskie chteniya* XXVI: 3–6 (in Russian).

Zaletaev, V. S. 1997. "Aktualynye problemy izucheniya ekotonov." In *Ekotony v biosphere*, by edited by Vladimir S. Zaletaev, 5–9. Moskva: Institut vodnyh problem RAN (in Russian).

Zalihanov, M. Ch., Kolomyts, E. G., L. S. Sharaya, et al. 2010. *Vysokogornaya geoecologiya v modelyah (High Mountain Geoecology in Models)*. Moskva: Nauka.

Zaret, T. M. 1982. "The stability/diversity controvers: a test of hypotheses." *Ecology* 63: 721–731.

Zeidis, I. M., Kruzhalin, V. I., Simonov, Yu. G., et al. 2001. "General properties of geosystems dynamics." *Vestnik MGU. Ser. 5. Geografiya* 4: 3–8 (in Russian).

Zhuchkova, V. K., Zonov, Yu. B., and Goryachenkov, V. A. 1973. "Metodicheskie priyomy landchaftnyh issledovany vulkanicheskih rayonov Kamchatki" In *Landshaftny sbornik,* by edited Nikolay A. Solntsev, 117–137. Moskva: Izd-vo MGU (in Russian).

Zlobin, T. K., Abdurakhmanov, L. I., and Zlobina L. M. 1997. "Glubinnye seysmicheskie issledovaniya vulkana Mendeleeva na Yuzhnyh Kurilah." *Tikhookeanskaya geologiya* 16 (4): 79–87 (in Russian).

Zlotin, R. I. 1987. "Ekologicheskie problemy bioty i ustoychivosty geosystem." *Izvestiya AN SSSR. Ser. Geographich* 6: 45–51 (in Russian).

Zonn, S. V. 1969. "O nekotoryh itogah i perspectivah izucheniya proizvoditelynosti lesnyh biogeotsenozov." In *Obshchie teoreticheskie problemy biologicheskoy produktivnosti,* by edited Natalya I. Bazilevich, 39–45. Leningrad: Nauka (in Russian).

Zonov, Yu. B. 1977. *Formirovanie pervichnyh landshaftov payonov aktivnogo vulkasnizma Vostochnoy Kamchatki.* Diss. na soisk. uch. stepeni kand. geogr. nauk. Vladivostok: DVGU (in Russian).

Zubenok, L .I. 1976. *Isparenie na kontinentah.* Leningrad: Gidrometeoizdat (in Russian).

About the Author

Kolomyts Erland Georgievich
Institute of Ecology of the Volga River Basin of Russian Academy of Sciences,
Head of Laboratory of Landscape Ecology, Moscow Region Russia
Email: egk2000@mail.ru

Kolomyts Erland Georgievich (birch 1936), the graduate of Geographical Faculty of Moscow State University (1960), Doctor of Geographical Sciences (1972); Professor of Physical Geography and Ecology (1986); Meritorious Scientific Worker of the Russian Federation (2003); the Grigoryev Prize Laureate of the Russian Academy of Science (2015); Honorary Doctorate of Letters, the International Biographical Centre Cambridge, England (2016); Honorary Doctorate of Philosophy, the International Biographical Centre Cambridge, England (2018). Since 1989, the chief scientist and head of the Laboratory of Landscape Ecology at the Institute of Ecology of the Volga River Basin (Russian Academy of Sciences). The sphere of scientific interests: complex physical geography, geophysics and ecology of landscape, structural and evolutionary snow studies, ecological-geographical prognosis, and evolutionary biosphere processes.

INDEX

A

B

C

D

E

F

G

Geotopic areal stability – 325
Geotope (location types) and its phytocoenotic significance – xxii, 23–24, 101

H

I

L

M

N

O

P

R

S

T

V

Z

Related Nova Publications

THE ANDES: GEOGRAPHY, DIVERSITY, AND SOCIOCULTURAL IMPACTS

EDITOR: Casey D. Allen

SERIES: Earth Sciences in the 21st Century

BOOK DESCRIPTION: Written by regional experts and drawing on field-based endeavors of several researchers, each chapter encompasses a timely topic, from climate-related studies and historical occupations to indigenous manufacturing and recent archaeological evidence. If you want to know what types of research are currently being conducted in the Andes, this book will surely satisfy.

HARDCOVER ISBN: 978-1-53611-094-4
RETAIL PRICE: $160

RED SEA: HISTORICAL SIGNIFICANCE, PROPERTIES AND ECONOMIC IMPORTANCE

EDITOR: Delbert Robbins

SERIES: Earth Sciences in the 21st Century

BOOK DESCRIPTION: *Red Sea: Historical Significance, Properties and Economic Importance* opens with a discussion on oceanic spreading beginning in the Pliocene (~5 Ma) (after advanced continental drifting) and continuing to the present day, as demonstrated by the intense volcanic and seismic activity in the area during the last 10,000 years.

SOFTCOVER ISBN: 978-1-53613-200-7
RETAIL PRICE: $82

To see a complete list of Nova publications, please visit our website at www.novapublishers.com

Related Nova Publications

Tribology in Geology and Archaeology

Author: Robert G. Bednarik

Series: Earth Sciences in the 21st Century

Book Description: This volume explores the geological applications of tribology in some detail, before introducing the entirely new subdisciplines of archaeotribology and the tribology of rock art.

Hardcover ISBN: 978-1-53614-909-8
Retail Price: $230

Glaciology for Glacial Geologists

Author: Terence J. Hughes

Series: Earth Sciences in the 21st Century

Book Description: We live in the Quaternary Ice Age, the last million years when large ice sheets covered much of North America and Eurasia, with successive glaciations lasting about 90,000 years interspersed with interglaciations lasting about 10,000 years, such as our preset Holocene interglaciation.

Hardcover ISBN: 978-1-53612-793-5
Retail Price: $195

To see a complete list of Nova publications, please visit our website at www.novapublishers.com